MILITARY AIRCRAFT, TANKS & WARSHIPS
VISUAL ENCYCLOPEDIA
MORE THAN 850 ILLUSTRATIONS

JIM WINCHESTER, ROBERT JACKSON & DAVID ROSS

This edition first published in 2023

Reprinted in 2025

Copyright © 2017 Amber Books Ltd

Published by
Amber Books Ltd
United House
North Road
London
N7 9DP
United Kingdom
www.amberbooks.co.uk
Instagram: amberbooksltd
Facebook: amberbooks
Twitter: @amberbooks
Pinterest: amberbooksltd

All rights reserved. With the exception of quoting brief passages for the purpose of review no part of this publication may be reproduced without prior written permission from the publisher. The information in this book is true and complete to the best of our knowledge. All recommendations are made without any guarantee on the part of the author or publisher, who also disclaim any liability incurred in connection with the use of this data or specific details.

ISBN: 978-1-83886-409-5

Project Editors: Sarah Uttridge and Michael Spilling
Picture Research: Terry Forshaw
Design: Andrew Easton

Printed in China

Picture Credits
All © Art-Tech/Aerospace with additional artworks from DeAgostini, Alcaniz Fresno's SA, Tony Gibbons, Vincent Bourguignon, Ray Hutchins, Oliver Missing, Military Visualizations, Inc. and US Department of Defense.

CONTENTS

Introduction	6
Cold War Military Aircraft	8
Military Aircraft in the Modern Era	132
Cold War Tanks and AFVs	214
Modern Era Tanks and AFVs	328
Warships 1945–Present	350
Aircraft Index	440
Tank Index	443
Warship Index	444
General Index	446

INTRODUCTION

In the years since the end of World War II, we have witnessed the Korean War, the Vietnam War, the Indo-Pakistan Wars, Arab-Israeli conflicts, the Falklands War, the Yugoslav Wars, and the recent conflicts in Iraq and Afghanistan, among others. And then during the Cold War there were also the conflicts that were not fought: the NATO forces in western Europe and the Warsaw Pact troops in eastern Europe did not fire on each other, but their aircraft, armoured fighting vehicles and warships were steadily updated.

Military Aircraft

Jet aircraft were only just emerging in the final stages of World War II and it wasn't until the Korean War that the first generation jet fighters really proved their mettle. In the 1960s, supersonic fighters featuring guided missiles were introduced. By the late 1970s, digital electronics were the new feature, while from the 1990s stealth technology was the great leap forward.

The journey of military aircraft development over the past 70 years also includes experimental X-Planes, the last of the flying boats and the innovation of inflight refueling.

But novelty is not everything: although the Boeing B-52 bomber was introduced as long ago as 1955, it is still in service with the United States Air Force.

Tanks and Armoured Fighting Vehicles

The Cold War saw Soviet tanks such as the World War II classic T-34 exported to China, North Korea and the Middle East, while in the West the American Sherman and British cruiser tanks were replaced by heavier designs such as the Patton, Centurion and AMX-13. These and later tanks possessed the traditional advantages of the tank –firepower, armour protection and mobility – while also serving as platforms for the latest military technology.

In addition to the speed and firepower of main battle tanks suited to the open battlefield, smaller, lighter and more manoeuvrable armoured personnel carriers and infantry fighting vehicles have become a feature of modern war zones, especially in urban areas.

Warships

During the early years of the Cold War, it became apparent that the aircraft carrier was becoming increasingly vulnerable to attack from the air and under the sea, and measures were begun to protect it. The Americans and Russians both began developing nuclear-powered ballistic missile submarines, with the first one, the American Nautilus, launched in 1954. Able to remain submerged – and often undetected – for lengthy periods, these new subs would be a great deterrent against attacks on their nation's carriers.

Meanwhile, developments in aircraft-carrier design continued, with nuclear propulsion and nuclear weapons making the US 'super carriers' the most powerful military vehicles ever created. But with the proliferation of Soviet nuclear submarines threatening US aircraft carriers, new escort vessels, mainly frigates, were developed. These were fast enough to pursue high-speed nuclear-powered submarines.

In the 21st century, both aircraft carriers and amphibious assault ships, able to transport troops and vehicles, continue to be built.

Above: USS *Carl Vinson* is one of ten US Navy Nimitz-class supercarriers.

COLD WAR MILITARY AIRCRAFT

A new conflict sprang up within a few years of the end of World War II as former allies the USSR and the Western powers competed for influence in Europe, Asia and elsewhere.

Atomic weapons threatened the end of civilization, but thankfully remained unused. Actual fighting in the Cold War years was largely constrained to proxy wars in countries like Vietnam and Afghanistan, although aerial spying reached new heights. Aircraft development was rapid, with new materials such as titanium needed to cope with the heat caused by supersonic speeds and afterburning jet engines.

Left: The F-4 Phantom II first entered service in 1960 and continued to form a major part of US military air power throughout the 1970s and 1980s.

Early Jets

The first generation of turbine-powered aircraft that entered service immediately after the war were essentially piston-engined aircraft with jets in them. This was literally true of the J 21R and Fireball, but in general early jets were quite conservative designs. British and German engine designs dominated this first generation. Although the Vampire's centrifugal flow design was almost universal initially, the German axial-flow engine design was to become dominant over time.

De Havilland Spider Crab (Vampire)

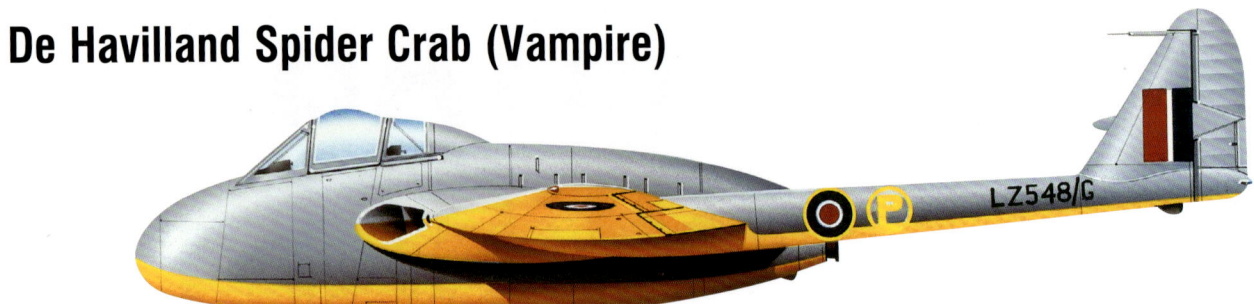

Originally named Spider Crab, but fortunately changed to Vampire in April 1944, the de Havilland DH.100 was designed around the Halford H.1 centrifugal-flow turbojet and had a relatively tubby fuselage. The Vampire entered squadron service until March 1946 and became a great export success.

SPECIFICATIONS	
Country of origin:	United Kingdom
Type:	single-engined jet fighter
Powerplant:	one 12kN (2700lb) thrust Halford H.1 turbojet engine
Performance:	maximum speed 915km/h (570mph); range unknown; service ceiling unknown
Weights:	(approx) empty 3290kg (7520lb); maximum 5620kg (12,390lb)
Dimensions:	span 11.6m (38ft); length 9.37m (30ft 9in); height 2.69m (8ft 10in); wing area 24.34 sq m (262 sq ft)
Armament:	none

Lockheed P-80 Shooting Star

The USAF's first operational jet fighter first flew as the XP-80 in January 1944. Crashes of early test and production aircraft killed several experienced pilots. Several P-80s were rushed to Europe in 1945 and two examples saw limited service in Italy before the war's end.

SPECIFICATIONS	
Country of origin:	United States
Type:	single-seat fighter bomber
Powerplant:	one 17.1kN (3850lb) Allison J33-GE-11 turbojet
Performance:	maximum speed at sea level 966km/h (594mph); service ceiling 14,265m (46,800ft); range 1328km (825 miles)
Weights:	empty 3819kg (8420lb); maximum take-off weight 7646kg (16,856lb)
Dimensions:	wingspan 11.81m (38ft 9in); length 10.49m (34ft 5in); height 3.43m (11ft 3in); wing area 22.07 sq m (237.6sq ft)
Armament:	six .5in machine guns, plus two 454kg (1000lb) bombs and eight rockets

EARLY JETS TIMELINE

1943  1944

EARLY JETS

Ryan FR-1 Fireball

The US Navy took a more cautious approach for its first jet-propelled aircraft, commissioning the mixed-powerplant Fireball in 1942. The jet engine was regarded as an adjunct to the radial to give the dash speed for catching Kamikazes, but the Fireball was too late for operational war service.

SPECIFICATIONS	
Country of origin:	United States
Type:	mixed-powerplant carrier-based fighter
Powerplant:	one 7.1kN (1600lb) thrust General Electric J31-GE-3 turbojet engine and one 1060kW (1350hp) Wright R-1820-72W Cyclone radial piston engine
Performance:	maximum speed 686km/h (426mph); range 2100km (1300 miles); service ceiling 13,137m (43,100ft)
Weights:	empty 3590kg (7915mph); maximum 4806kg (10,595lb)
Dimensions:	span 12.19m (40ft); length 12.19m (32ft 4in); height 4.15m (13ft 8in); wing area 25.6 sq m (275 sq ft)
Armament:	four .5in (12.7mm) Browning M2 machine guns; up to 908kg (2000lb) of bombs or eight 127mm (5in) rockets

Mikoyan-Gurevich MiG-9 'Fargo'

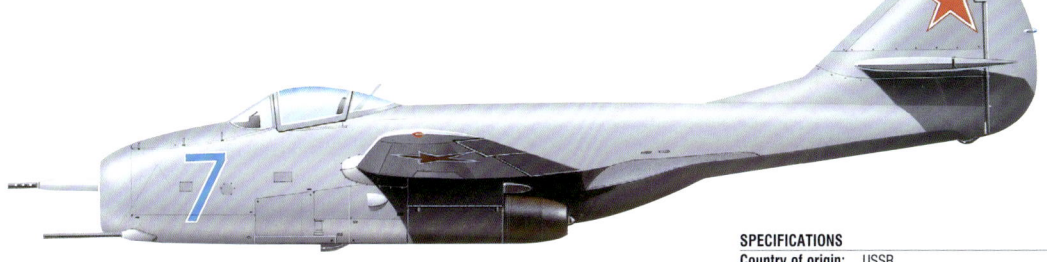

The MiG-9 was the first Soviet jet to fly, powered by Russian copies of the German BMW 003 turbojet. It was built in relatively large numbers (610 examples) and received the reporting name 'Fargo', but was never regarded as a satisfactory fighter despite its heavy cannon armament.

SPECIFICATIONS	
Country of origin:	USSR
Type:	twin-engined jet fighter
Powerplant:	two 7.8kN (1533lb) thrust Kolesov RD-20 afterburning turbojet engines
Performance:	maximum speed 909km/h (565mph); range 800km (495 miles); service ceiling 13,000m (42,650ft)
Weights:	empty 3420kg (7540lb); maximum 5500kg (12,125lb)
Dimensions:	span 10m (32ft 10in); length: 9.83m (32ft 3in); height 3.22m (10ft 7in); wing area 18.2 sq m (196 sq ft)
Armament:	one 37mm NL-37 37mm cannon, two NS-23 23mm cannon

Saab J 21R

Saab's J 21A fighter was a pusher-engined twin-boom fighter that entered service in 1945. Then, in 1947 and as a stepping-stone to new jet designs, Saab created the J 21R (R standing for *Reaktion* or jet) with a Goblin engine. The 64 built were used mainly as ground-attack aircraft.

SPECIFICATIONS	
Country of origin:	Sweden
Type:	single-seat jet fighter-bomber
Powerplant:	one 13.8kN (3100lb) thrust De Havilland Goblin 2 turbojet engine
Performance:	maximum speed 800km/h (497mph); range 720km (450 miles); service ceiling 12,000m (39,400ft)
Weights:	empty 3200kg (7055lb); maximum 5000kg (11,023lb)
Dimensions:	span 11.37m (37ft 4in); length 10.45m (34ft 3in); height 2.9m (9ft 8in); wing area 22.3 sq m (240 sq ft)
Armament:	one 20mm Bofors cannon, four 13.2mm M/39A machine guns; 10 100mm or five 180mm rockets

1946  1947

US Carrier Air Power

In August 1945, the US Navy had several advanced piston-engined designs ready for introduction to the fleet, but none made it to the front before the war ended. Hellcats, Avengers and especially Helldivers were soon retired from frontline service and scrapped, converted to target drones or relegated to reserve units. The new aircraft replaced them in those squadrons that were not immediately disbanded.

Grumman F8F-1 Bearcat

Lightweight and powerful, the F8F Bearcat was one of the fastest single-seat piston-engined fighters ever built. Although in service by May 1945, it missed out on wartime combat. Highly regarded by pilots, the Bearcat had a short frontline career before its replacement by jets.

SPECIFICATIONS	
Country of origin:	United States
Type:	single-engined carrier-based fighter
Powerplant:	one 156kW (2100hp) Pratt & Whitney R-2800-34W Double Wasp radial piston engine
Performance:	maximum speed 678km/h (421mph); range 1778km (1105 miles); service ceiling 11,796m (38,700ft)
Weights:	empty 3207kg (7070lb); maximum 5873kg (12,947lb)
Dimensions:	span 10.92m (35ft 10in); length 8.61m (28ft 3in); height 4.21m (13ft 9in);
Armament:	four .5in (12.7mm) M2 machine guns; up to 454kg (1000lb) of bombs or four 5in (127mm) rockets

Grumman F7F-2N Tigercat

Designed for operation off the new 'Midway'-class carriers, the Tigercat was Grumman's first twin-engined combat aircraft. Marine units received some before the war's end, but it missed out on combat until Korea. The F7F-2N was a two-seat night-fighter variant, used mainly from land bases.

SPECIFICATIONS	
Country of origin:	United States
Type:	twin-engined night fighter
Powerplant:	two 1790kW (2400hp) Pratt & Whitney R-2800-22W radial piston engines
Performance:	maximum speed 582km/h (362mph); range 1545km (960 miles); service ceiling 12,131m (39,800ft)
Weights:	empty 7380kg (16,270lb); loaded 11,880kg (26,190lb)
Dimensions:	span 15.7m (51ft 6in); length 13.8m (45ft 4in); height 4.6m (15ft 2in); wing area 42.3 sq m (455 sq ft)
Armament:	four 20mm cannon and four .5in (12.7mm) Browning machine guns

US CARRIER AIR POWER

North American FJ-1 Fury

Although it had a relatively undistinguished career, the Fury was the first aircraft to complete an operational tour at sea, and paved the way for the more aesthetically pleasing F-86 Sabre. For a brief period, it could also claim to be the fastest aircraft in US Navy service.

SPECIFICATIONS

Country of origin:	United States
Type:	single-seat carrier-borne fighter
Powerplant:	one 17.8kN (4000lb) Allison J35-A-2 turbojet
Performance:	maximum speed 880km/h (547mph); service ceiling 9754m (32,000ft); range 2414km (1500 miles)
Weights:	empty 4011kg (8843lb); maximum loaded 7076kg (15,600lb)
Dimensions:	wingspan 9.8m (38ft 2in); length 10.5m (34ft 5in); height 4.5m (14ft 10in); wing area 20.5 sq m (221sq ft)
Armament:	six .5in machine guns

McDonnell F2H-2 Banshee

The success of the FH-1 Phantom meant that it was almost inevitable that McDonnell would be asked to submit a design to succeed the Phantom in service. The new aircraft was larger, incorporating folding wings and a lengthened fuselage to accommodate more fuel and more powerful engines. The F2H-1 was delivered in 1948. The F2H-2 had wingtip fuel tanks.

SPECIFICATIONS

Country of origin:	United States
Type:	carrier-based all-weather fighter
Powerplant:	one 14.5kN (3250llb) Westinghouse turbojet
Performance:	maximum speed 933km/h (580mph); range 1883km (1170 miles); service ceiling 14,205m (46,600ft)
Weights:	empty 5980kg (13,183lb); maximum take-off weight 11,437kg (25,214kg)
Dimensions:	span 12.73m (41ft 9in); length 14.68m (48ft 2in); height 4.42m (14ft 6in); wing area 27.31 sq m (294sq ft)
Armament:	four 20mm cannon; underwing racks with provision for two 227kg (500lb) or four 113kg (250lb) bombs

Martin AM-1 Mauler

Able to carry a heavy load of bombs or torpedoes, the Mauler was intended to replace several types, including the Avenger and Helldiver. It proved difficult to operate from carriers and was passed to reserve units in favour of the smaller but more reliable Douglas AD Skyraider.

SPECIFICATIONS

Country of origin:	United States
Type:	single-engined carrier-based multi-purpose bomber
Powerplant:	one 2237kW (3000hp) Pratt & Whitney R-4360-4 Wasp Major radial piston engine
Performance:	maximum speed 591km/h (376mph); range 2885km (1800 miles); service ceiling 9296m (30,500ft)
Weights:	empty 6557kg (14,500lb); maximum 10,608kg (23,386lb)
Dimensions:	span 15.24m (50ft); length 12.55m (41ft 2in); height 5.13m (16ft 10in); wing area 46.1 sq m (496 sq ft)
Armament:	four 20mm cannon; up to 4488kg (10,698lb) of bombs

English Electric Canberra

First flown in 1949, the Canberra became one of the great post-war successes of the British aircraft industry, with 782 built for the RAF and RN, and another 120 for export customers. Canberras saw action in numerous wars in Africa, Southeast Asia and South America. Photoreconnaissance Canberras flew missions around (and sometimes inside) the borders of the Soviet Union and Warsaw Pact countries, and in conflicts with British involvement up to Afghanistan and Iraq in the 2000s.

Canberra B.2

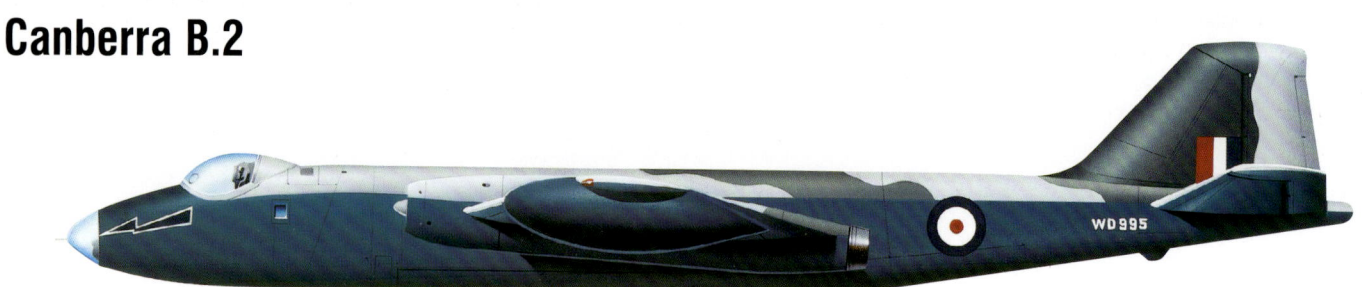

To meet specification B.3/45, W.E.W. Petter planned a straightforward unswept jet bomber with a broad wing for good behaviour at high altitude. Like the Mosquito, the A.1 bomber was to be fast enough to escape interception, whilst carrying a 2727kg (6000lb) bomb load over a radius of 750 nautical miles. It was to have a radar sight for blind attacks in all conditions. The first B.2s, tactical day bombers, were delivered in May 1951.

SPECIFICATIONS	
Country of origin:	United Kingdom
Type:	two-seat interdictor aircraft
Powerplant:	two 28.9kN (6500lb) Rolls-Royce Avon Mk 101 turbojets
Performance:	maximum speed 917km/h (570mph); service ceiling 14,630m (48,000ft); range 4274km (2656 miles)
Weights:	empty not published approx. 11,790kg (26,000lb); maximum take-off weight 24,925kg (54,950lb)
Dimensions:	wingspan 29.49m (63ft 11in); length 19.96m (65ft 6in); height 4.78m (15ft 8in); wing area 97.08 sq m (1045 sq ft)
Armament:	internal bomb bay with provision for up to 2727kg (6000lb) of bombs, plus an additional 909kg (2000lb) of underwing pylons

Canberra PR.3

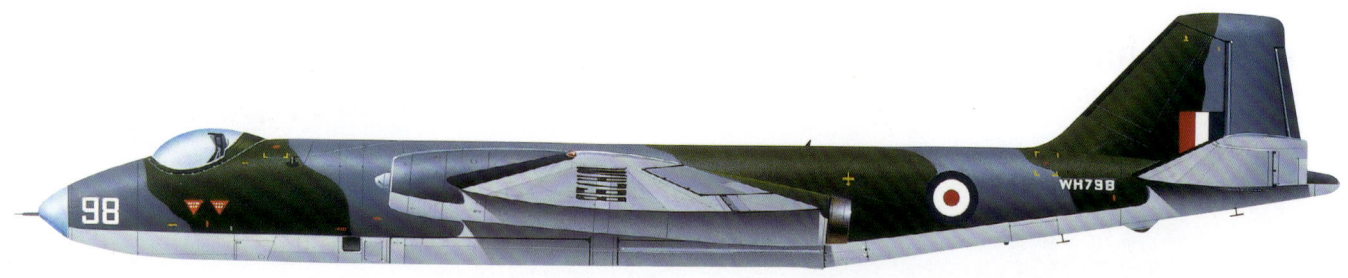

The PR.3 was a photographic reconnaissance version of the B.2 and entered service in late 1952. Instead of weapons, it had mounts for one vertical camera and six oblique cameras. The bomb bay itself was replaced by an extra fuel tank and flares for night photography.

SPECIFICATIONS	
Country of origin:	United Kingdom
Type:	twin-engined jet photoreconnaissance aircraft
Powerplant:	two 28.9kN (6500lb) Rolls-Royce Avon Mk 101 turbojets
Performance:	maximum speed 917km/h (570mph); service ceiling 14,630m (48,000ft); range 4274km (2656 miles)
Weights:	empty not published approx 11,790kg (26,000lb); maximum take-off weight 24,925kg (54,950lb)
Dimensions:	wingspan 29.49m (63ft 11in); length 20.3m (66ft 8in); height 4.78m (15ft 8in); wing area 97.08 sq m (1045 sq ft)
Armament:	none

ENGLISH ELECTRIC CANBERRA

Canberra B(I).8

The B(I).8 was an interdictor/strike variant and the first to have the offset fighter-style canopy. The bomb aimer was replaced by a low-altitude bombing system computer. For strike missions it could carry nuclear weapons, or a cannon pack in the interdictor role.

SPECIFICATIONS	
Country of origin:	United Kingdom
Type:	twin-engined jet bomber
Powerplant:	two 33.23kN (7490lb) thrust turbojet engine Avon R.A.7 turbojet engines
Performance:	maximum speed 933km/h (580mph); range 5440km (3380 miles); service ceiling 15,000m (48,000ft)
Weights:	9820kg (21,650lb); maximum 24,948kg (55,000lb)
Dimensions:	span 19.51m (65ft 6in); length 19.96m (65ft 6in); height 4.77m (15ft 3in); wing area 88.19 sq m (960 sq ft)
Armament:	four 20mm cannon

Canberra TT.18

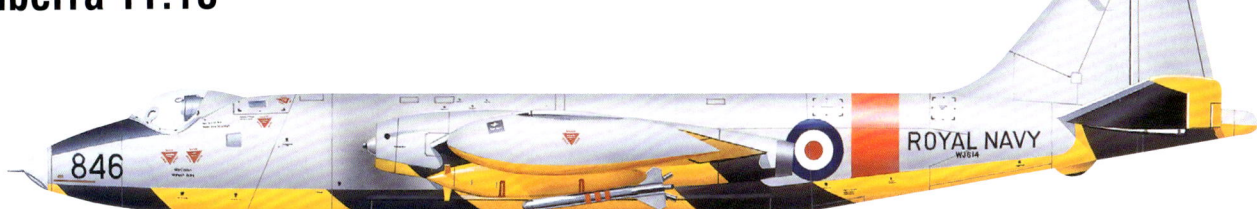

The Royal Navy operated a number of Canberra variants for roles such as target towing. The TT.18 was converted from the B.2, with underwing winches allowing the streaming of target drogues for aerial or naval gunnery. It could also operate as a target itself to train radar operators.

SPECIFICATIONS	
Country of origin:	United Kingdom
Type:	twin-engined jet target tug
Powerplant:	two 28.9kN (6500lb) Rolls-Royce Avon Mk 101 turbojets
Performance:	maximum speed 917km/h (570mph); service ceiling 14,630m (48,000ft); range 4274km (2656 miles)
Weights:	empty not published approx. 11,790kg (26,000lb); maximum take-off weight 24,925kg (54,950lb)
Dimensions:	wingspan 29.49m (63ft 11in); length 19.96m (65ft 6in); height 4.78m (15ft 8in); wing area 97.08 sq m (1045 sq ft)
Armament:	none

Canberra PR.9

Longest-lived of the UK-built Canberras, the PR.9s were delivered in 1959 and were finally retired in 2006. During their career, they received many camera and sensor fits, including the Hycon B or System III camera used in the U-2 and an infrared line scan (IRLS) unit.

SPECIFICATIONS	
Country of origin:	United Kingdom
Type:	photoreconnaissance aircraft
Powerplant:	two 46.7kN (10,500lb) Rolls-Royce Avon Mk 206 turbojets
Performance:	maximum speed about 650mph (1050km/h) at medium altitude; service ceiling 14,630m (48,000ft); range 5842km (3630 miles)
Weights:	empty not published approx. 11,790kg (26,000lb); maximum take-off weight 24,925kg (54,950lb)
Dimensions:	wingspan 20.68m (67ft 10in); length 20.32m (66ft 8in); height 4.78m (15ft 8in); wing area 97.08 sq m (1045 sq ft)
Armament:	none

COLD WAR MILITARY AIRCRAFT

Israeli War of Independence

The state of Israel was formally created from the former British Mandate of Palestine in 1948. On the day of independence, Israel was attacked by Egypt, Iraq, Syria and Lebanon. The Israeli Air Force, Hel HaAvir, was quickly established in the face of an international arms embargo. Israel acquired aircraft by various means, and its pilots – many of them with much combat experience – also came from far and wide.

Boeing B-17G Fortress

Israel obtained four B-17Gs from a broker in Florida in early 1948 and flew them to Czechoslovakia for modification and refurbishment, although one was impounded en route. Targets in Egypt were bombed on the delivery flight to Israel in July 1949. They flew about 200 sorties in the following year.

SPECIFICATIONS	
Country of origin:	United States
Type:	(B-17G) 10-seat heavy bomber
Powerplant:	four 895kW (1200hp) Wright R-1820-97 nine-cylinder radial engines
Performance:	maximum speed 486km/h (302mph); service ceiling 10,850m (35,600ft); range 2897km (1800 miles)
Weights:	empty 20,212kg (44,560lb); maximum take-off weight 32,659kg (72,000lb)
Dimensions:	span 31.63m (103ft 9.4in); length 22.78m (74ft 9 in); height 5.82m (19ft 1in)
Armament:	12 .5in machine guns (in chin turret, cheek positions, in dorsal turret, in roof position, in ventral turret, in waist positions, in tail), plus a bomb load of 7983kg (17,600lb)

De Havilland Mosquito FB.VI

The French government sold Israel 67 surplus and rather worn Mosquitoes in 1951, among them 40 FB.VI fighter bombers as illustrated. More Mosquitoes later arrived from Britain. They flew mainly reconnaissance missions over Israel's neighbours before retirement in 1957.

SPECIFICATIONS	
Country of origin:	United Kingdom
Type:	(B.IV Series 2) twin-engined bomber
Powerplant:	Two 954kW (1280hp) Rolls-Royce Merlin 23 V-12 piston engines
Performance:	maximum speed 612km/h (380mph); service ceiling 9144m (30,000ft); range 3384km (2040 miles)
Weights:	empty 6396kg (14,100lb); loaded 10,151kg (22,380lb)
Dimensions:	span 16.5m (54ft 2in); length 12.35m (40ft 10in); height 4.66m (15ft 4in); wing area 42.18 sq m (454 sq ft)
Armament:	up to 1814kg (4000lb) of bombs

ISRAELI WAR OF INDEPENDENCE

Supermarine Spitfire

Newly communist Czechoslovakia supplied Israel with 76 of its ex-RAF Spitfires in 1948–49. Serving alongside Avia S.199s in 101 Squadron, Israeli Spitfires claimed a number of Egyptian MC.205s and also three RAF Spitfires and a Tempest in incidents of mistaken identity.

SPECIFICATIONS

Country of origin:	United Kingdom
Type:	(Spitfire F.Mk IX) single-seat fighter and fighter-bomber
Powerplant:	one 1167kW (1565hp) Rolls-Royce Merlin 61 or 1230kW (1650hp) Merlin 63 12-cylinder Vee engine
Performance:	maximum speed 655km/h (408mph); service ceiling 12,105m (43,000ft); range 1576km (980 miles)
Weights:	empty 2545kg (5610lb); maximum take-off weight 4309kg (9500lb)
Dimensions:	span 11.23m (36ft 10in); length 9.46m (31ft); height 3.85m (12ft 7.75in)
Armament:	two 20mm fixed forward-firing cannon and four .303in fixed forward-firing machine guns in the leading edges of the wing, plus an external bomb load of 454kg (1000lb)

Macchi MC.205 Veltro

Egypt bought 24 Macchi fighters from Italy in 1948. Some were MC.205 Veltros and others were older MC.202s rebuilt to MC.205V standard. This example served with 2 Squadron, Royal Egyptian Air Force, at Al Arish. Egyptian Macchis claimed three kills in 1948–49.

SPECIFICATIONS

Country of origin:	Italy
Type:	(MC.205V) single-seat fighter and fighter-bomber
Powerplant:	one 1100kW (1475hp) Fiat RA.1050 RC.58 Tifone 12-cylinder inverted-Vee engine
Performance:	maximum speed 642km/h (399mph); service ceiling 11,000m (36,090ft); range 1040km (646 miles)
Weights:	empty 2581kg (5691lb); maximum take-off weight 3408kg (7514lb)
Dimensions:	span 10.58m (34ft 8.5in); length 8.85m (29ft 0.5in); height 3.04m (9ft 11.5in)
Armament:	two 12.7mm fixed forward-firing machine guns in the upper part of the forward fuselage, and two 20mm forward-firing cannon in the leading edges of the wing, plus bomb load of 320kg (705lb)

Supermarine Spitfire

Egypt's Spitfire Mk Vs and IXs (illustrated) fought Israel's own Spitfires on several occasions. Although they are not believed to have won any engagements, they did destroy several Avias and other IAF aircraft. At least one crash-landed REAF Spitfire was captured and put into Israeli service.

SPECIFICATIONS

Country of origin:	United Kingdom
Type:	(Spitfire F.Mk IX) single-seat fighter and fighter-bomber
Powerplant:	one 1167kW (1565hp) Rolls-Royce Merlin 61 or 1230kW (1650hp) Merlin 63 12-cylinder Vee engine
Performance:	maximum speed 655km/h (408mph); service ceiling 12,105m (43,000ft); range 1576km (980 miles)
Weights:	empty 2545kg (5610lb); maximum take-off weight 4309kg (9500lb)
Dimensions:	span 11.23m (36ft 10in); length 9.46m (31ft); height 3.85m (12ft 7.75in)
Armament:	two 20mm cannon and four .303in machine guns in the wing, plus an external bomb load of 454kg (1000lb)

Korean War

In June 1950, North Korea launched a surprise attack on its southern neighbour, with massive ground forces supported by air power. The North Korean Air Force had about 180 aircraft, the majority of them Yak fighters and Il-10 ground attackers. The nearest and strongest United Nations forces were those of the USAF in Japan, who had one wing of F-80 jets and various piston-engined aircraft. B-29s soon flattened most of North Korea's industry, and by November UN troops were nearing victory.

North American F-51D Mustang

Flown by Arnold 'Moon' Mullins, this F-51D Mustang of the 67th Fighter Bomber Squadron shot down at least one North Korean Yak-9 in February 1951. Mustangs were used mainly for ground-attack missions and proved more vulnerable to ground fire than radial-engined aircraft.

SPECIFICATIONS	
Country of origin:	United States
Type:	(F-51D) single-engined fighter-bomber
Powerplant:	one 1264kW (1695hp) Packard V-1650-7 12-cylinder Vee engine
Performance:	maximum speed 703km/h (437mph); service ceiling 12,770m (41,900ft); range 3703km (2301 miles)
Weights:	empty 3103kg (6840lb);maximum take-off weight 5493kg (12,100lb)
Dimensions:	span 11.28m (37ft 0.25in); length 9.84m (32ft 3.25in); height 4.16m (13ft 8in) with the tail down
Armament:	six .5in fixed forward-firing machine guns in the leading edges of the wing, plus an external bomb and rocket load of 907kg (2000lb)

Boeing B-29B Superfortress

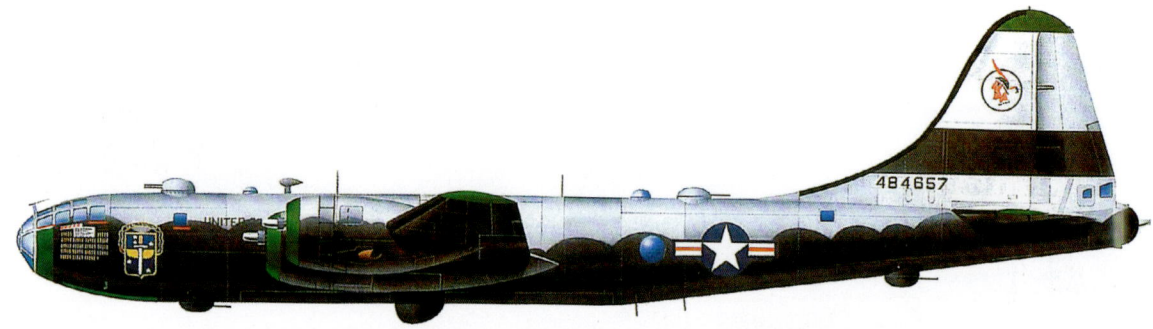

'Command Decision' was a B-29B Superfortress of the 19th Bomb Group, 28th Bomb Squadron, based on Okinawa. Its claim to fame is that its gunners shot down five North Korean fighters. In total, B-29 gunners were credited with 27 victories over North Korean aircraft.

SPECIFICATIONS	
Country of origin:	United States
Type:	(B-29) nine-seat long-range heavy bomber
Powerplant:	four 1640kW (2200hp) Wright 18-cylinder two-row radial engines
Performance:	maximum speed 576km/h (358mph); service ceiling 9710m (31,850ft); range 9382km (5830 miles)
Weights:	empty 31,816kg (70,140lb); maximum take-off 56,246kg (124,000lb)
Dimensions:	span 43.05m (141ft 2.75in); length 30.18m (99ft); height 9.02m (29ft 7in)
Armament:	one 20mm cannon and six .5in MGs (in tail, and dorsal and ventral barbettes), plus an internal bomb load of 9072kg (20,000lb)

KOREAN WAR

Lockheed F-80C

SPECIFICATIONS

Country of origin:	United States
Type:	single-engined fighter-bomber
Powerplant:	one 24.0kN (5400lb) thrust Allison J33-A-35 turbojet engine
Performance:	maximum speed 965 km/h (600 mph); service ceiling 14,000m (46,000ft); range 1930km (1200 miles)
Weights:	empty 3819kg (8420lb); maximum 7646kg (16,856lb)
Dimensions:	wingspan 11.81m (38ft 9in); length 10.49m (34ft 5in); height 3.43m (11ft 3in); wing area 22.07 sq m (237.6 sq ft)
Armament:	six .5in (12.7mm) machine guns; two 454kg (1000lb) bombs or eight 2.75in rockets

Jet squadrons were sent to South Korea as soon as North Korean troops were driven out. 'Eagle Eyed Fleagle/Miss Barbara Ann' was an F-80C of the 36th FBS based at Suwon. F-80s of this squadron claimed several Yak-9s in air combat in July 1950.

Republic F-84E Thunderjet

SPECIFICATIONS

Country of origin:	United States
Type:	single-engined jet fighter-bomber
Powerplant:	one 21.8kN (4900lb) thrust Allison J35-A-17 turbojet engine
Performance:	maximum speed 986km/h (613mph); range 2390km (1485 miles); service ceiling 13,180m (43,240ft)
Weight:	loaded 10,185kg (22,455lb)
Dimensions:	span 11.1m (36ft 5in); length 11.41m (36ft 5in); height 3.91m (12ft 10in); wing area 214.8 sq m (260 sq ft)
Armament:	six .5in (12.7mm) machine guns; two 454kg (1000lb) bombs or eight 2.75in rockets

The F-84E was one of the stalwarts of the tactical ground-attack effort in Korea, attacking trains, convoys, artillery positions and other targets. It was not noted as a dogfighter, however, and this 9th FBS Thunderjet was shot down by North Korean MiGs in September 1952.

Gloster Meteor F.8

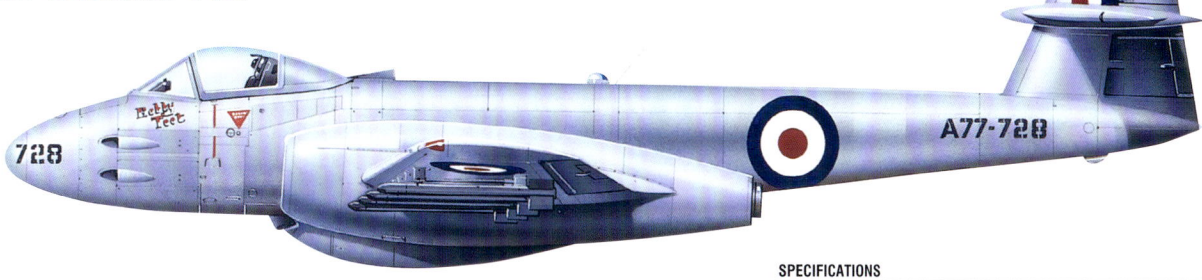

SPECIFICATIONS

Country of origin:	United Kingdom
Type:	single-seat fighter
Powerplant:	two 16.0kN (3600lb) Rolls-Royce Derwent 8 turbojets
Performance:	maximum speed (33,000ft) 962km/h (598mph); service ceiling 13,106m (43,000ft); range 1580km (980 miles)
Weights:	empty 4820kg (10,626lb); loaded 8664kg (19,100lb)
Dimensions:	wingspan 11.32m (37ft 2in); length 13.58m (44ft 7in); height 3.96m (13ft)
Armament:	four 20mm Hispano cannon, foreign F.8s often modified to carry two iron bombs or eight rockets

Australia first committed Mustangs and then Meteor F.8s to Korea in July 1951. No 77 Squadron found its straight-wing jets inferior to the MiG-15 and lost more Meteors in combat than it claimed MiGs. Reassigned to ground-attack duties, the RAAF Meteors were much more successful.

Korean War: Later Types

Carrier-based air power played an important part in the Korean conflict from the start. Both Royal Navy prop-driven fighters and USN jets began strikes on the North in early July 1950. The situation changed in November 1950 with two events – the Chinese intervention in the land battle and the arrival of swept-wing MiG-15s supplied by the Soviet Union and largely flown by Soviet pilots.

Hawker Sea Fury FB.11

Despite its pioneering work introducing jets to carriers, the Fleet Air Arm deployed only piston-engined aircraft in Korea. Flown by Peter Carmichael of No 805 Squadron on HMS *Ocean*, this Sea Fury FB.11 destroyed a MiG-15 in air combat in August 1952.

SPECIFICATIONS	
Country of origin:	United Kingdom
Type:	single-engined carrier-based fighter-bomber
Powerplant:	one 1850kW (2480hp) Bristol Centaurus XVIIC radial piston engine
Performance:	maximum speed 740 km/h (460 mph); range 1127km (700 miles); service ceiling 10,900m (35,800ft)
Weights:	empty 4190kg (9240lb); maximum 5670kg (12,500lb)
Dimensions:	span 11.7m (38ft 4in); length 10.6m (34ft 8in); height 4.9m (16ft 1in); wing area 26 sq m (280 sq ft)
Armament:	four 20mm Hispano Mk V cannon; up to 908kg (2000 lb) of bombs or 12 3in rockets

Vought F4U-4B Corsair

The wartime-vintage Corsair was brought back into production for Korea, where its ruggedness and heavy warload were valuable. Marine units like VMA-312 used cannon-armed F4U-4Bs from land bases in their traditional close air-support role.

SPECIFICATIONS	
Country of origin:	United States
Type:	(F4U-4) single-seat carrierborne and land-based fighter and fighter-bomber
Powerplant:	one 1678kW (2250hp) Pratt & Whitney R-2800-18W Double Wasp 18-cylinder two-row radial engine
Performance:	maximum speed 718km/h (446mph); service ceiling 12,650m (41,500ft); range 2511km (1560 miles)
Weights:	empty 4175kg (9205lb); maximum take-off weight 6149kg (13,555lb) as fighter or 8845kg (19,500lb) as fighter-bomber
Dimensions:	span 12.49m (40ft 11.75in); length 10.27m (33ft 8.25in); height 4.50m (14ft 9in)
Armament:	six .5in MGs, external bomb and rocket load of 907kg (2000lb)

KOREAN WAR: LATER TYPES

Grumman F9F-2 Panther

Grumman's Panther was the US Navy's primary strike jet of the war. Panthers had relatively few encounters with North Korean fighters. This F9F-2 of VF-781 on the USS *Bon Homme Richard*, however, shot down a MiG-15 flown by a Russian pilot in November 1952.

SPECIFICATIONS

Country of origin:	United Kingdom
Type:	single-engined carrier-based jet fighter-bomber
Powerplant:	one 26.5kN (5950lb) thrust Pratt & Whitney J42-P-6/P-8 turbojet engine
Performance:	maximum speed 925 km/h (575 mph); range 2100km (1300 miles); service ceiling 13,600m (44,600ft)
Weights:	empty 4220kg (9303lb); maximum 7462kg (16,450lb)
Dimensions:	span 11.6m (38ft); length 11.3m (37ft 5in); height 3.8m (11ft 4in); wing area 23 sq m (250 sq ft)
Armament:	four 20mm M2 cannon; up to 910kg (2000lb) bombs, six 5in (127mm) rockets

Mikoyan-Gurevich MiG-15 'Fagot'

The MiG-15 was very much the equivalent of the F-86 Sabre with some advantages, such as heavy armament, and weaknesses, such as a poorer turn radius. This MiG-15bis was flown by a North Korean defector to Kimpo airfield near Seoul in September 1953 and later evaluated in the United States.

SPECIFICATIONS

Country of origin:	USSR
Type:	single-seat fighter
Powerplant:	one 26.3kN (5952lb) Klimov VK-1 turbojet
Performance:	maximum speed 1100km/h (684 mph); service ceiling 15,545m (51,000ft); range at height with slipper tanks 1424km (885 miles)
Weights:	empty 4000kg (8820lb); maximum loaded 5700kg (12,566lb)
Dimensions:	wingspan 10.08m (33ft 0.75in); length 11.05m (36ft 3.75in); height 3.4m (11ft 1.75in); wing area 20.60 sq m (221.74 sq ft)
Armament:	one 37mm N-37 cannon and two 23mm NS-23 cannon, plus up to 500kg (1102lb) of mixed stores on underwing pylons

Ilyushin Il-28 'Beagle'

One of the most important early jet bombers, the Il-28 'Beagle' was supplied to North Korea late in the war. The threat of such bombers worried the UN commanders, but they flew no known bombing missions against the South. North Korea still uses the Il-28 in the form of the Chinese-built Harbin H-5.

SPECIFICATIONS

Country of origin:	USSR
Type:	three seat bomber and ground attack/torpedo carrier
Powerplant:	two 26.3kN (5952lb) Klimov VK-1 turbojets
Performance:	max speed 902 km/h (560mph); service ceiling 12,300m (40,355ft); range 2180km (1355 miles); with bomb load 1100km (684 miles)
Weights:	empty 12890kg (28,418lb); maximum take-off 21,200kg (46,738lb)
Dimensions:	wingspan 21.45sq m (70ft 4.5in); length 17.65m (57ft 10.75in); height 6.70m (21ft 11.8in); wing area 60.80 sq m (654.47 sq ft)
Armament:	four 23mm cannon; internal bomb capacity 1000kg (2205lb), max bomb capacity 3000kg (6614lb); torpedo version: provision for two 400mm light torpedoes

COLD WAR MILITARY AIRCRAFT

Korean Sabres

Emblematic of the Korean War, the North American F-86 Sabre escorted bombers and flew ground-attack missions, but is best remembered for its duels with North Korean MiGs. The introduction of the MiGs saw a wing of F-86As shipped from the United States to Korea in November 1950, and the first victory against a MiG was recorded in mid-December. The Sabres were restricted by their modest endurance and rules that prevented them pursuing the enemy too close to Chinese airspace.

F-86A-5

Joseph E. Fields, 1st Lieutenant of the 336th Fighter Inceptor Squadron flew this F-86A. Fields scored a MIG-15 kill outright on 21 September 1952. In total, Sabre pilots claimed 792 MiGs destroyed for about 80 losses in air combat, but modern research suggests the kill-loss ratio was less favourable.

SPECIFICATIONS	
Country of origin:	United States
Type:	single-engined jet fighter
Powerplant:	one 23.7kN (5340lb) thrust General Electric J47-GE-7 turbojet engine
Performance:	maximum speed 965km/h (600mph); range 530km (329 miles); service ceiling 14,600m (48,000 ft)
Weights:	empty 4700kg (10,500lb); maximum 6300kg (13,791lb)
Dimensions:	span 11.3m (37ft 1in); length 11.4m (37ft 6in); height 4.4m (14ft 8in); wing area 26.76 sq m (288 sq ft)
Armament:	six .5in (12.7mm) machine guns

F-86A-5

The majority of Korean Sabres flew in an unpainted natural metal finish, but several aircraft flew in an experimental camouflage scheme, as seen on this F-86A of the 335th FIS. Another project included the fitting of 20mm cannon to Sabres. The cannon-armed Sabres claimed six MiGs before the war's end.

SPECIFICATIONS	
Country of origin:	United States
Type:	single-engined jet fighter
Powerplant:	one 23.7kN (5340lb) thrust General Electric J47-GE-7 turbojet engine
Performance:	maximum speed 965km/h (600mph); range 530km (329 miles); service ceiling 14,600m (40,000 ft)
Weights:	empty 4700kg (10,500lb); maximum 6300kg (13,791lb)
Dimensions:	span 11.3m (37ft 1in); length 11.4m (37ft 6in); height 4.4m (14ft 8in); wing area 26.76 sq m (288 sq ft)
Armament:	six .5in (12.7mm) machine guns

KOREAN SABRES

F-86E-10

SPECIFICATIONS	
Country of origin:	United States
Type:	single-engined jet fighter
Powerplant:	one 23.1kN (5200lb) thrust General Electric J47-GE-13 turbojet engine
Performance:	maximum speed 965km/h (600mph); range 530km (329 miles); service ceiling 14,600 m (48,000 ft)
Weights:	empty 4700kg (10,500lb); maximum 6300kg (13,791lb)
Dimensions:	span 11.3m (37ft 1in); length 11.4m (37ft 6in); height 4.4m (14ft 8in); wing area 26.76 sq m (288 sq ft)
Armament:	six .5in (12.7mm) machine guns

The F-86E Sabre introduced an all-moving tailplane that improved handling as the Sabre neared the speed of sound. The pilot of F-86E 'Four Kings and a Queen', Cecil Foster, was 1st Lieutenant of the 16th Fighter Interceptor Squadron, he was the top-scoring 16th FIS pilot, with nine victories.

F-86E-10

SPECIFICATIONS	
Country of origin:	United States
Type:	single-engined jet fighter
Powerplant:	one 23.1kN (5200lb) thrust General Electric J47-GE-13 turbojet engine
Performance:	maximum speed 965km/h (600mph); range 530km (329 miles); service ceiling 14,600 m (48,000 ft)
Weights:	empty 4700kg (10,500lb); maximum 6300kg (13,791lb)
Dimensions:	span 11.3m (37ft 1in); length 11.4m (37ft 6in); height 4.4m (14ft 8in); wing area 26.76 sq m (288 sq ft)
Armament:	six .5in (12.7mm) machine guns

Walker 'Bud' Mahurin was a World War II ace and commander of the 4th Fighter Interceptor Group. His final score was 24.5, including 3.5 MiGs. 'Half' scores were credited for kills shared between pilots, each of whom damaged the target.

F-86F-30

SPECIFICATIONS	
Country of origin:	United States
Type:	single-engined jet fighter-bomber
Powerplant:	one 26.3kN (5910lb) thrust General Electric J47-GE-27 turbojet engine
Performance:	maximum speed at sea level 1091km/h (678mph); service ceiling 15,240m (50,000ft); range 914km (568 miles)
Weights:	empty 5045kg (11,125lb); maximum loaded 9350kg (20,611lb)
Dimensions:	wingspan 11.3m (37ft 1in); length 11.43m (37ft 6in); height 4.47m (14ft 8.75in); wing area 28.15 sq m (302.3 sq ft)
Armament:	six .5in (12.7mm) machine guns; 908kg (2000lb) of bombs

The first fighter-bomber variant of the Sabre was the F-86F. The F-86F-30 had a wider chord wing root and an extra pylon for bombs or drop tanks. This machine was flown by US Marine exchange pilot John Glenn, who shot down three MiGs and went on to later fame as an astronaut and politician.

Nightfighters Over Korea

As on the Eastern Front, the Po-2 biplane was used for night-harassment raids. Destroying these 'Bedcheck Charlies' consumed a lot of resources, including deployment of several types of night fighters. Some were soon more gainfully employed in the night-interdiction role, attacking North Korean truck convoys, trains and other nocturnal activities. Jet night fighters were also used as escorts for night B-29 raids, and there were a small number of night victories over MiGs.

F-82G Twin Mustang

One of the most unusual aircraft of the era, the F-82 was essentially two P-51 fuselages joined with a new wing and powered by Allison engines with counter-rotating propellers. F-82Gs of the 68th Fighter (All-Weather) Squadron scored the first three aerial victories of the war, all by daylight.

SPECIFICATIONS	
Country of origin:	United States
Type:	twin-engined night fighter
Powerplant:	two 1781kW (1380hp) Allison V-1710-143/145 V-12 piston engines
Performance:	maximum speed 740km/h (460mph); range 3605km (2520 miles); service ceiling 11,855m (38,900ft)
Weights:	empty 7271kg (15,997lb); maximum 11,632kg (25,591lb)
Dimensions:	span 15.62m (51ft 3in); length 12.93m (42ft 9in); height 4.22m (13ft 10in); wing area 37.9 sq m (408 sq ft)
Armament:	six .5in (12.7mm) Browning M2 machine guns; 25 127mm (5in) rockets or up to 1800kg (4000lb) of bombs

Lockheed F-94B Starfire

The F-94B, a much-changed derivative of the F-80, arrived in Korea in mid-1951 as a replacement for the F-82G. It struggled against the slower Po-2s and did not destroy one until January 1953, when this 319th FIS jet claimed one. Another was destroyed in a collision, which was fatal to all involved.

SPECIFICATIONS	
Country of origin:	United States
Type:	tandem-seat all-weather interceptor
Powerplant:	one 26.7kN (6000lb) Allison J33-A-33 turbojet
Performance:	maximum speed at 30,000ft, 933km/h (580mph); service ceiling 14,630m (48,000ft); range 1850km/h (1150 miles)
Weights:	empty 5030kg (11,090lb); maximum take-off weight 7125kg (15,710lb)
Dimensions:	wingspan not including tip tanks 11.85m (38ft 10.5in); length 12.2m (40ft 1in); height 3.89m (12ft 8in); wing area 22.13 sq m (238sq ft)
Armament:	four .5in machine guns

Douglas F3D-2 Skyknight

The Douglas Skyknight replaced the Tigercat in Korea. Between November 1952 and the end of the war, Marine F3Ds shot down five MiG-15s and a Po-2. The first kill was scored by an aircraft of VMF-(N)-513, whose victim was credited as a Yak-15, although this type was not in the North's inventory.

SPECIFICATIONS

Country of origin:	United States
Type:	twin-engined jet night fighter
Powerplant:	two 15.1kN (3400lb) thrust Westinghouse J34-WE-36 turbojet engines
Performance:	maximum speed 852km/h (460mph); range 2212km (1374 miles); service ceiling 11,200m (36,700ft)
Weights:	empty 6813kg (14,989lb); maximum 12,151kg (26,731lb)
Dimensions:	span 15.24m (50ft); length 13.85m (45ft 5in); height 4.9m (16ft 1in); wing area 37.2 sq m (400 sq ft)
Armament:	four 20mm Hispano-Suiza M2 cannon

Grumman F7F-3N Tigercat

Tigercats scored only two kills against Po-2s, partly because of their higher speed, which meant they quickly overtook the target once identified. With bombs, rockets and napalm, however, they proved effective at interdiction and night-attack missions.

SPECIFICATIONS

Country of origin:	United States
Type:	twin-engined night fighter
Powerplant:	two 1566kW (21000hp) Pratt & Whitney R-2800-34W radial piston engines
Performance:	maximum speed 700km/h (435mph); range 1545km (960 miles); service ceiling 12,405m (40,700ft)
Weights:	empty 7380kg (16,270lb); loaded 11,880kg (26,190lb)
Dimensions:	span 15.7m (51ft 6in); length 13.8 (45ft 4in); height 4.6m (15ft 2in); wing area 42.3 sq m (455 sq ft)
Armament:	four 20mm cannon and four .5in (12.7mm) Browning machine guns; up to 1814kg (4000lb) of bombs

Vought F4U-5N Corsair

Marine Corps and Navy squadrons flew F4U-5N Corsairs from austere land bases in Korea. The most successful pilot was Guy Bordelon of the Navy's VC-3, who scored five kills. The AN/APS-6 radar of the F4U-5N had a range of about 8km (5 miles).

SPECIFICATIONS

Country of origin:	United States
Type:	single-engined night fighter
Powerplant:	one 1827kW (2450hp) Pratt & Whitney radial piston engine
Performance:	maximum speed 718km/h (446mph); range 2425km (1507 miles); service ceiling 11,308m (41,500ft)
Weights:	empty 4175kg (8873lb); maximum 6654kg (13,846lb)
Dimensions:	10.16m (41ft); length 10.27m (33ft 8in); height 4.5m (14ft 9in); wing area 29.17 sq m (314 sq ft)
Armament:	four 20mm M3 cannon; 10 127mm (5in) rockets or up to 2390kg (5200lb) of bombs

COLD WAR MILITARY AIRCRAFT

British V-Bombers

Three manufacturers responded to a 1946 requirement for a new British heavy jet bomber, with three designs – the Valiant, Victor and Vulcan – entering service in that order from 1954 to 1956. The Valiant was the most conservative but was put into production partly in case the others failed. The crescent-winged Victor was futuristic in appearance, but still basically conventional, unlike the tailless delta Vulcan. All three saw action as conventional bombers and later use as aerial tankers.

Vickers Valiant B(K).1

The Valiant was the first to enter service, the first to see action (in the Suez crisis in 1956) and the first to be retired. The introduction of Soviet surface-to-air-missiles (SAMs) saw a switch to low-level operations, which caused excessive fatigue, forcing its complete retirement by 1965.

SPECIFICATIONS	
Country of origin:	United Kingdom
Type:	strategic bomber
Powerplant:	four 44.7kN (10,050lb) Rolls-Royce Avon 204 turbojets
Performance:	maximum speed at high altitude 912km/h (567mph); service ceiling 16,460m (54,000ft); maximum range 7242km (4,500 miles)
Weights:	empty 34,4191kg (75,881lb); maximum loaded with drop tanks 79,378kg (175,000lb)
Dimensions:	wingspan 34.85m (114ft 4in); length 33m (108ft 3in); height 9.8m (32ft 2in); wing area 219.43 sq m (2,362 sq ft)
Armament:	internal weapons bay with provision for up to 9525kg (21,000lb) of conventional or nuclear weapons

Handley Page Victor B.2

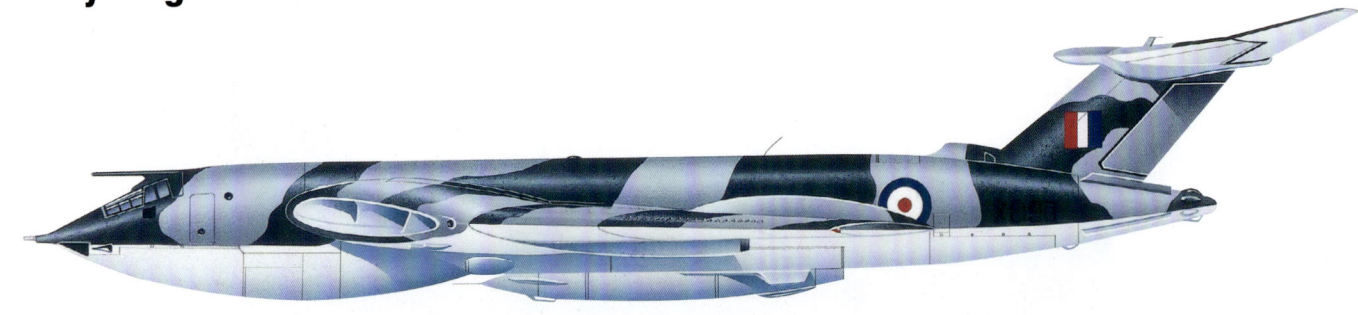

The initial Victor B.1 was replaced by the B.2 with revised wings, more powerful engines and a wider range of weapons options. Conventional bombs were dropped by Victors over Borneo on a single mission during the Confrontation with Indonesia in the early 1960s.

SPECIFICATIONS	
Country of origin:	United Kingdom
Type:	four-seat air-refuelling tanker
Powerplant:	four 91.6kN (20,600lb) Rolls-Royce Conway Mk 201 turbofans
Performance:	max speed 1030km/h (640mph); maximum cruising height 18,290m (60,000ft); range with internal fuel 7400km (4600 miles)
Weights:	empty 41,277kg (91,000lb); maximum take-off 105,687kg (233,000lb)
Dimensions:	wingspan 36.58m (120ft); length 35.05m (114ft 11in); height 9.2m (30ft 1.5in); wing area 223.52 sq m (2,406 sq ft)
Armament:	up to 35 1000lb (450kg) bombs

BRITISH V-BOMBERS

Handley Page Victor K.2

When the Victor B.1 was retired from the bombing role, a number were refitted as tankers. The same happened with the B.2, which became the K.2 tanker, with three refuelling points. Victor K.2s became the last V-bombers in service, retiring from RAF service in 1993.

SPECIFICATIONS	
Country of origin:	United Kingdom
Type:	four-seat air-refuelling tanker
Powerplant:	four 91.6kN (20,600lb) Rolls-Royce Conway Mk 201 turbofans
Performance:	max speed 1030km/h (640mph); maximum cruising height 18,290m (60,000ft); range with internal fuel 7,400km (4,600 miles)
Weights:	empty 41,277kg (91,000lb); maximum take-off 105,687kg (233,000lb)
Dimensions:	wingspan 36.58m (120ft); length 35.05m (114ft 11in); height 9.2m (30ft 1.5in); wing area 223.52 sq m (2406 sq ft)
Armament:	none

Avro Vulcan B.1

The Vulcan was the most distinctive of the V-bombers. The B.1 version entered service in 1957, and was used for a number of record-breaking long-range flights. Able to carry a range of British and American weapons, most B.1s were painted white to reflect the flash of a nuclear explosion.

SPECIFICATIONS	
Country of origin:	United Kingdom
Type:	Four-engined strategic bomber
Powerplant:	four 48.93kN (11,000lb) Bristol Siddeley Olympus 101 thrust turbojet engines
Performance:	maximum speed 1017km/h (632mph); range 6293km (3910 miles); service ceiling 17,000m (55,000ft)
Weights:	empty 37,909kg (83,573lb); maximum 86,000kg (190,00 lb)
Armament:	21,000lb (9500kg) of conventional bombs or single nuclear weapon
Dimensions:	span 30.3m (99ft 5in); length 30.5 m (97ft 1in); height 8.1m (26ft 6in); wing area 330.2 sq m (3554 sq ft)

Avro Vulcan B.2

The smoother leading edge identified the B.2 version of the Vulcan, which also had more powerful engines and strengthened structure. The Vulcan's only combat missions were flown in the Falklands War, including three by Vulcan B.2A XM607, illustrated here.

SPECIFICATIONS	
Country of origin:	United Kingdom
Type:	low-level strategic bomber
Powerplant:	four 88.9kN (20,000lb) Olympus Mk.301 turbojets
Performance:	maximum speed 1038km/h (645mph) at high altitude; service ceiling 19,810m (65,000ft); range with normal bomb load about 7403km/h (4600 miles)
Weights:	maximum take-off weight 113,398kg (250,000lb)
Dimensions:	wingspan 33.83m (111ft); length 30.45m (99ft 11in); height 8.28m (27ft 2in); wing area 368.26 sq m (3964 sq ft)
Armament:	internal weapon bay for up to 21,454kg (47,198lb) bombs

NORAD

North American Air Defense Command (NORAD) was established to counter the threat of Soviet bombers and missiles attacking over the North Pole. Its assets included Arctic radar stations, interceptors and missiles. Fighters were armed with barrages of unguided rockets or even nuclear-tipped rockets for destroying bomber formations. Canada's contribution included the CF-100. America convinced the Canadian government to abandon the much more sophisticated CF-105 Arrow in favour of Bomarc SAMs, which was then cancelled, and Canada was sold Voodoo fighters instead.

North American F-86D Sabre Dog

SPECIFICATIONS	
Country of origin:	United States
Type:	single-seat night and all-weather interceptor
Powerplant:	one 33.3kN (7500lb) General Electric J47-GE-17B turbojet
Performance:	maximum speed 1138km/h (707mph); range 1344km (835 miles); service ceiling 16,640m (54,600ft)
Weight:	loaded 7756kg (17,100lb)
Dimensions:	span 11.3m (37ft 1in); length 12.29m (40ft 4in); height 4.57m (15ft)
Armament:	24 7mm (2.75in) 'Mighty Mouse' air-to-air unguided rocket projectiles

The most famous of the early generation jet fighters, the F-86 Sabre flew for the first time in October 1947, with the first operational aircraft delivered two years later. During the Korean War, Sabres claimed the destruction of 810 enemy aircraft, 792 of them MiG-15s. Numerous variants were built, with most made by Canadair destined for NATO airforces. It was also built under licence in Australia.

Avro Canada CF-100 Mk 4b

The CF-100 Canuck, usually called the 'Clunk', was Canada's main interceptor for most of the 1950s. Just under 700 were built in five marks. Some had an eight-gun pack under the fuselage, but the primary weapons were rocket pods, which were fired when the radar judged the target was in range.

SPECIFICATIONS	
Country of origin:	Canada
Type:	twin-engined all-weather fighter
Powerplant:	two 32.5kN (7300lb) thrust Avro Canada Orenda 11 turbojet engines
Performance:	maximum speed 888km/h (552mph) ;range 3200km (1988 miles); service ceiling 13,700m (45,000ft)
Weight:	loaded 15,170kg (33,450lb)
Dimensions:	span 17.4m (57ft 2 in); length 16.5m (54ft 2in); height 4.4m (14ft 6in); wing area 54.9 sq m (591 sq ft)
Armament:	58 70mm (2.75in) rockets

McDonnell F-101B Voodoo

The F-101B was a two-seat, all-weather, long-range interceptor version of the Voodoo, accommodating a pilot and radar operator to work the MG-13 fire-control system and more powerful engines. By fitting a tandem cockpit, internal fuel capacity was sacrificed. A total of 407 were built, with final delivery taking place in March 1961.

SPECIFICATIONS	
Country of origin:	United States
Type:	two-seat all-weather long-range interceptor
Powerplant:	two 75.1kN (16,900lb) Pratt & Whitney J57-P-55 turbojets
Performance:	maximum speed at 12190m (40,000ft) 1965km/h (1221mph); service ceiling 16,705m (54,800ft); range 2494km (1550 miles)
Weights:	empty 13,141kg (28,970lb); maximum take-off 23,768kg (52,400lb)
Dimensions:	wingspan 12.09m (39ft 8in); length 20.54m (67ft 4.75in); height 5.49m (18ft); wing area 34.19 sq m (368 sq ft)
Armament:	two Mb-1 Genie missiles with nuclear warheads and four AIM-4C,-4D, or 4G Falcon missiles, or six Falcon missiles

COLD WAR MILITARY AIRCRAFT

Air Sea Rescue

In an era before long-range helicopters, the only way to find and aid survivors of foundering vessels, downed military pilots and passengers of ditched airliners was to drop supplies or lifeboats, or land a flying boat nearby. The USAF Air Rescue Service built on its wartime experience by developing rescue versions of the B-29 and B-17. Flying boats and amphibians remained in service as rescue craft long after their replacement in other roles.

Boeing B-17H Fortress (later SB-17G)

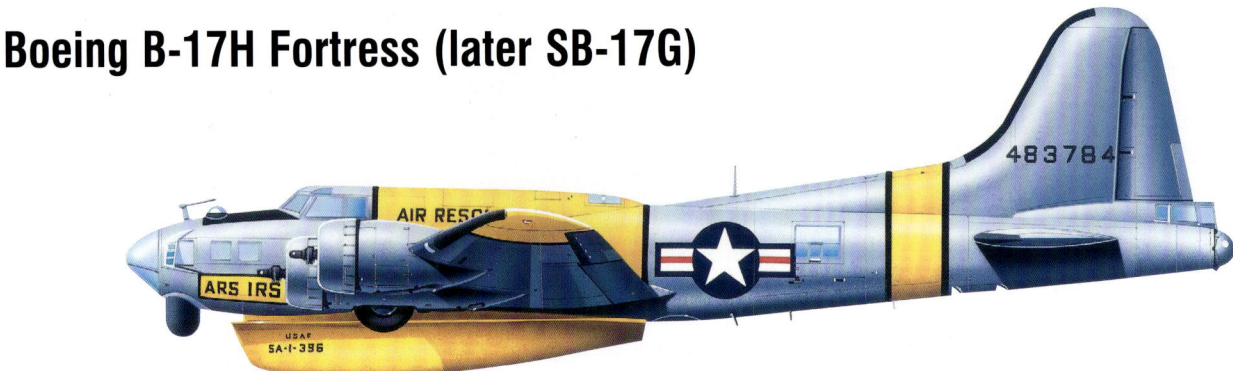

First known as the SB-17G and later redesignated B-17H, surplus Flying Fortresses were converted to the ASR role with the addition of search radar and droppable A-1 lifeboat. This laminated mahogany vessel had twin engines and berths, and supplies for 20 survivors.

SPECIFICATIONS	
Country of origin:	United States
Type:	10-seat heavy bomber
Powerplant:	four 895kW (1200hp) Wright R-1820-97 nine-cylinder radial engines
Performance:	maximum speed 486km/h (302mph); service ceiling 10,850m (35,600ft); range 2897km (1800 miles)
Weights:	empty 20,212kg (44,560lb); maximum take-off 32,659kg (72,000lb)
Dimensions:	span 31.63m (103ft 9.4in); length 22.78m (74ft 9 in); height 5.82m (19ft 1in)
Armament:	none

Martin PBM-5 Mariner

The PBM Mariner was built as a patrol and bomber flying boat, but variants were created for the Coast Guard and surplus aircraft sold to several nations. The Uruguayan Navy obtained three PBM-5Es in the mid-1950s and used them for search-and-rescue duties.

SPECIFICATIONS	
Country of origin:	United States
Type:	nine-seat maritime patrol and anti-submarine flying-boat
Powerplant:	two 1566kW (2100hp) Wright Double Wasp radial piston engines
Performance:	maximum speed 340km/h (211mph); range 4345km (2700 miles); service ceiling 6160m (20,200ft)
Weights:	empty 15,422kg (34,000lb); maximum take-off 27,216kg (60,000lb)
Dimensions:	span 35.97m (118ft); length 24.33m (79ft 10in); height 8.38m (27ft 6in); wing area 130.8 sq m (1408 sq ft)
Armament:	nacelle bays and underwing racks with provision for up to 3628kg (8000lb) of bombs or depth charges

AIR SEA RESCUE

Grumman HU-16 Albatross

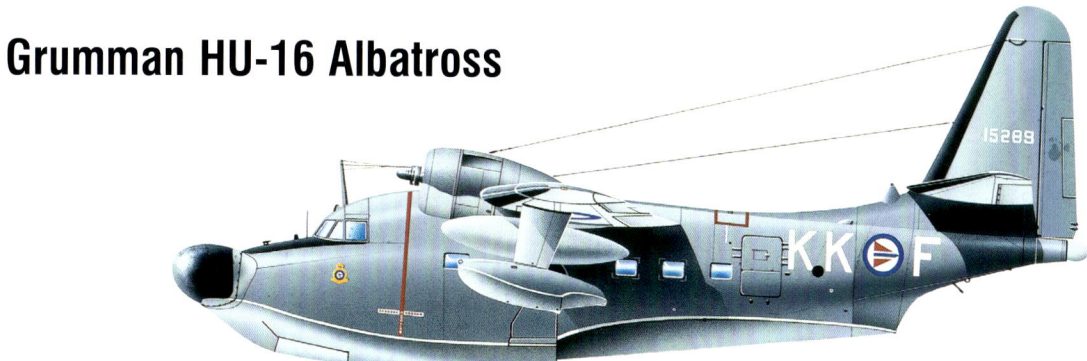

Norway was one of over 20 countries to use the Albatross. Two units, including 333 Skvadron at Andøya (illustrated), used the aircraft for SAR and other duties, including medical evacuation, postal delivery and polar-bear tracking. The Norwegian SA-16Bs were passed on to Greece in 1968.

SPECIFICATIONS

Country of origin:	United States
Type:	twin-engined amphibian
Powerplant:	two 1063kW (1425hp) Wright R-1820-76 Cyclone radial piston engines
Performance:	maximum speed 380km/h (236mph); range 4587km (2850 miles); service ceiling 6553m (21,500ft)
Weights:	empty 9100kg (20,000lb); maximum 14,969lb (33,000kg)
Dimensions:	span 24.4m (80ft); length 19.16m (62ft 10in); height 7.8m (25ft 10in); wing area 82 sq m (883 sq ft)
Armament:	none

Dornier Do 24T-2

In autumn 1937, licensed production of 48 Do 24K-2 aircraft was started by Aviolanda in the Netherlands. Only 25 had been delivered before the German occupation of the Netherlands in May 1940. Manufacturing was re-established under German control, with 159 examples built and additional aircraft supplied by French production throughout and after the war.

SPECIFICATIONS

Country of origin:	Germany
Type:	air-sea rescue and transport flying boat
Powerplant:	three 746kW (1000hp) BMW-Bramo 323R-2 Fafnir radial piston engines
Performance:	maximum speed 340km/h (211mph); range 2900km (1802 miles); service ceiling 5900m (19,355ft)
Weights:	empty 9200kg (20,286lb); maximim take-off weight 18,400kg (40,565lb)
Dimensions:	span 27m (88ft 7in); length 22.05m (72ft 4in); height 5.75m (18ft 10in); wing area 108 sq m (1163 sq ft)
Armament:	one 20mm MG 151 cannon in dorsal turret, two 7.92mm MG 15 machine guns (in bow turret and tail turret)

Douglas SC-54D Skymaster

Convair converted 38 Douglas C-54s for the USAF Air Rescue Service as the SC-54G. Most of the changes were internal, but a large transparent blister was added to the starboard rear fuselage to aid visual searches. Their main role was dropping supplies and rescuers to people lost in Arctic regions.

SPECIFICATIONS

Country of origin:	United States
Type:	four-engined military search and rescue aircraft
Powerplant:	four 1014kW (1360hp) Pratt & Whitney R-2000-11 radial piston engines
Performance:	maximum speed 442km/h (275mph); range 6400km (4000 miles); service ceiling 6800m (22,300ft)
Dimensions:	span 35.8m (117ft 6in); length 28.6m (93ft 10in); height 8.38m (27ft 6in); wing area 136 sq m (1460 sq ft)
Weights:	empty 17,660kg (38,930lb); maximum 33,000kg (73,000lb)
Armament:	none

Hawker Hunter

The Hunter was one of the most successful British jet fighters, with nearly 2000 delivered to the RAF, FAA and 20 foreign users. Hunters were built under licence in Belgium and the Netherlands as well as by Hawker in the United Kingdom, who also refurbished and resold export aircraft to new customers. The Hunter saw action in numerous conflicts, in particular with India against Pakistan and with the RAF at Suez, in Aden and Borneo.

Hunter F.1

Without question the most successful post-war British fighter aircraft, the Hunter has a grace that complements its effectiveness as a warplane. Entering service in July 1954, the aircraft was produced in numerous guises and remained in service for 40 years. The F.1 was easily supersonic in a shallow dive.

SPECIFICATIONS	
Country of origin:	United Kingdom
Type:	single-seat fighter
Powerplant:	one 28.9kN (6500lb) Rolls-Royce Avon 100 turbojet
Performance:	maximum speed at sea level 1144km/h (710mph); service ceiling 15,240m (50,000ft); range on internal fuel 689km (490 miles)
Weights:	empty 5501kg (12,128lb); loaded 7347kg (16,200lb)
Dimensions:	wingspan 10.26m (33ft 8in); length 13.98m (45ft 10.5in); height 4.02m (13ft 2in); wing area 32.42 sq m (349 sq ft)
Armament:	four 30mm Aden cannon; underwing pylons with provision for two 1000lb bombs and 24 3in rockets

Hunter F.5

Unlike other Hunters, the F.5 was built with Armstrong-Siddeley Sapphire engines rather than Rolls-Royce Avons. It entered service in 1954, and only 45 were built. Like most subsequent models, they featured collectors for empty cannon shell cases under the fuselage, known as 'Sabrinas'.

SPECIFICATIONS	
Country of origin:	United Kingdom
Type:	single-seat fighter
Powerplant:	one 35.59kN (8000lb) Armstrong-Siddeley Sapphire turbojet engine
Performance:	maximum speed at sea level 1144km/h (710mph); service ceiling 15,240m (50,000ft); range on internal fuel 689km (490 miles)
Weights:	empty 5501kg (12,128lb); loaded 8501kg (18,742lb)
Dimensions:	wingspan 10.26m (33ft 8in); length 13.98m (45ft 10.5in); height 4.02m (13ft 2in); wing area 32.42 sq m (349 sq ft)
Armament:	four 30mm Aden cannon; up to 2722kg (6000lb) of bombs or rockets

Hunter F.58

The Swiss Air Force was the last European user of the Hunter, retiring the type in 1994 after over 30 years of service. Switzerland's F.58s, which were a version of the RAF Hunter F.6, could be equipped with air-to-air and air-to-ground missiles. They also equipped the Patrouille Swiss aerobatic team.

SPECIFICATIONS	
Country of origin:	United Kingdom
Type:	single-seat fighter
Powerplant:	one 45.13kN (10,145lb) thrust Rolls-Royce Avon 207 turbojet engine
Performance:	maximum speed at sea level 1144km/h (710mph); service ceiling 15,240m (50,000ft); range on internal fuel 689km (490 miles)
Weights:	empty 6405kg (14,122lb) maximum 17,750kg (24,600lb)
Dimensions:	wingspan 10.26m (33ft 8in); length 13.98m (45ft 10.5in); height 4.02m (13ft 2in); wing area 32.42 sq m (349 sq ft)
Armament:	four 30mm Aden cannon; up to 2722kg (6000lb) of bombs or rockets; AIM-9 Sidewinder AAMs or AGM-65 ASMs

Hunter T.8M

Production models of the Hunter T.7 trainer began entering service in 1958, with the T.8 derived for naval use. Common to all versions were the side-by-side seating and dual controls of the enlarged cockpit. Two-seat trainers were supplied to a number of countries.

SPECIFICATIONS	
Country of origin:	United Kingdom
Type:	dual-seat advanced trainer
Powerplant:	one 35.6kN (8000lb) Rolls-Royce Avon 122 turbojet
Performance:	maximum speed at sea level 1117km/h (694mph); service ceiling 14,325m (47,000ft); range on internal fuel 689km (429 miles)
Weights:	empty 6406kg (14,122lb); loaded 7802kg (17,200lb)
Dimensions:	wingspan 10.26m (33ft 8in); length 14.89m (48ft 10in); height 4.02m (13ft 2in); wing area 32.42 sq m (349 sq ft)
Armament:	two 30mm Aden cannon with 150 rounds

Hunter T.75A

Singapore ordered Hunters in 1968 and operated them into the early 1990s in the air defence role. Singapore's eight two-seat T.75s were converted from single-seat F.4 models before delivery, whereas its 38 single-seaters had mostly been FGA.9 models.

SPECIFICATIONS	
Country of origin:	United Kingdom
Type:	dual-seat advanced trainer
Powerplant:	one 45.13kN (10,145lb) thrust Rolls-Royce Avon 207 turbojet
Performance:	maximum speed at sea level 1117km/h (694mph); service ceiling 14,325m (47,000ft); range on internal fuel 689km (429 miles)
Weights:	empty 6406kg (14,122lb); loaded 7802kg (17,200lb)
Dimensions:	wingspan 10.26m (33ft 8in); length 14.89m (48ft 10in); height 4.02m (13ft 2in); wing area 32.42 sq m (349 sq ft)
Armament:	four 30mm Aden cannon

COLD WAR MILITARY AIRCRAFT

X-Planes

X stands for the unknown in mathematics and physics, and the US aircraft designated in the X series from the 1940s until the present day pushed the boundaries of flight to extremes of speed and altitude that had never before been achieved. Early successes of the programme, mainly conducted from Edwards Air Force Base in the Mojave Desert, included the first scientifically measured flights past the speed of sound and then double that speed, or Mach 2.

Bell X-1A

This development of the classic X-1 duplicated its predecessor's record-breaking achievements, but throughout its life lived under the shadow of accidents. The X-1A had a longer fuselage than the X-1, improved cockpit visibility and turbo-driven fuel pumps. In 1953, veteran high-speed test pilot Major Charles 'Chuck' Yeager flew the rocket plane to 2560km/h (1650mph) at a height of 21,350m (70,000ft), smashing the previous world record.

SPECIFICATIONS	
Country of origin:	United States
Type:	high-altitude high-speed research aircraft
Powerplant:	one 26.7kN (6000lb) thrust (at sea level) four-chamber Reaction Motors XLR11-RM-5 rocket engine
Performance:	maximum speed Mach 2.44 or 2655.4km/h (1646.35mph); endurance approximately 4 min 40 sec; service ceiling over 27432m (90,000ft)
Weights:	empty 3296kg (7251lb); loaded 7478kg (16,452lb)
Dimensions:	span 8.53m (28 ft); length 10.9m (35ft 8in); height 3.3m (10ft 10in); wing area 39.6 sq m (426 sq ft)
Armament:	none

Bell X-2

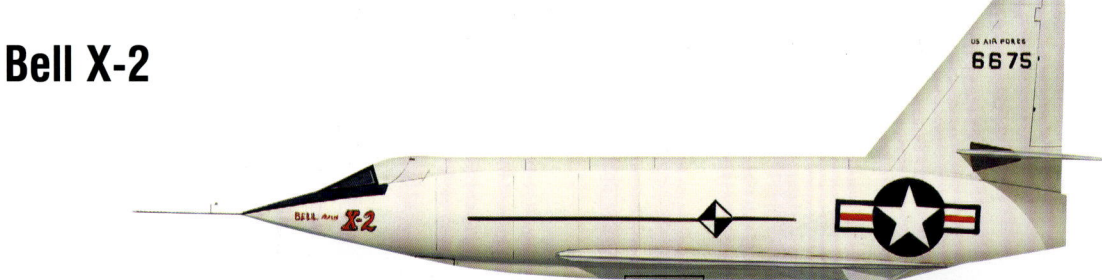

Bell began working on its X-2 in October 1945. The aircraft were rocket-powered, and designed for the analysis of structural and heating effects at speeds up to Mach 3.5 and altitudes up to 38,100m (125,000ft). Captive trials began in July 1951. The first aircraft exploded, while the second completed 12 very successful flights, before crashing on its thirteenth. Nonetheless, the type paved the way for future aircraft programmes.

SPECIFICATIONS	
Country of origin:	United States
Type:	single-seat supersonic research aircraft
Powerplant:	one 66.7kN (15,000lb) thrust Curtiss-Wright XLR25-CW-1 rocket engine
Performance:	maximum speed 3058km/h (1896mph); service ceiling 38405m (126,000ft); endurance 10 min 55 sec of powered flight; fuel capacity 2960 litres (782 gal) liquid oxygen; 3376 litres (892 gal) ethyl alcohol and water
Weights:	empty 5314kg (11,690lb); maximum take-off 11299kg (24,858lb)
Dimensions:	span 9.75m (32 ft); length 13.41m (44ft); height 4.11m (13ft 6in); wing area 24.19 sq m (260 sq ft)
Armament:	none

X-PLANES TIMELINE

 1948  1951  1952

X-PLANES

Douglas X-3A Stiletto

First flying in 1952, the X-3 looked strange, with the pilot positioned in a pressurized cabin on a downwards-firing ejection seat. It was hard to handle when taxiing, tricky on take-off and very difficult to fly. Design of the X-3 was of unprecedented complexity because of the use of titanium and other advanced materials. Unfortunately, it was underpowered and offered little to researchers.

SPECIFICATIONS	
Country of origin:	United States
Type:	single-seat research aircraft
Powerplant:	two 21.6kN (4860lb thrust) Westinghouse J34-WE-17 turbojet engines
Performance:	maximum speed 1136km/h (704mph); take-off speed 418 km/h (260mph); endurance one hour; range 805km (500 miles); service ceiling 11580m (38,000ft)
Weights:	empty 7312kg (16,086lb); maximum take-off 10813kg (23,788lb)
Dimensions:	span 6.91m (22ft 8in) length 20.35m (66ft 9in) height 3.81m (12ft 6in) wing area 15.47 sq m (166 sq ft)
Armament:	none

Bell X-4

The diminutive X-4 tested the tailless configuration at speeds of Mach 0.85 and higher. Drawing on experience gleaned from Northrop flying wings of the 1940s – and sharing some characteristics with today's B-2 stealth bomber – the two X-4s accumulated a wealth of data for the US Air Force and the National Advisory Committee for Aeronautics (NACA, the forerunner to NASA).

SPECIFICATIONS	
Country of origin:	United States
Type:	single-seat experimental research aircraft
Powerplant:	two 7.12kN (1600lb thrust) Westinghouse XJ30-WE-7 turbojets; later two Westinghouse J30-W-9 turbojets, using standard JP-4 jet engine fuel
Performance:	Mach 0.92 or 1123km/h (630mph) at 10,000m (33,000 ft) under extreme test conditions; range 676km (420 miles); service ceiling 13906m (42,300 ft)
Weights:	empty 2294kg (5507lb); maximum loaded 3547kg (7803lb)
Dimensions:	span 8.18m (27ft) length 7.19m (23ft) height 4.58m (15ft) wing area 18.58 sq m (300 sq. ft)
Armament:	none

Bell X-5

US troops occupying the town of Oberammergau, Germany, in 1945, discovered an experimental facility with the almost complete Messerschmitt P.1101 prototype. The P.1101 was brought to the US and Bell won a contract to build two test machines based on the German design. The two X-5s were very similar in layout to the P.1101, but considerably more complex.

SPECIFICATIONS	
Country of origin:	United States
Type:	single-seat experimental research aircraft
Powerplant:	one 21.8kN (4890lb-thrust) Allison J35-A-17 turbojet engine
Performance:	maximum speed approx. 1046km/h (650mph); range 1207km (750 miles); service ceiling 13000m (42,650ft)
Weights:	empty 2880kg (6336lb); maximum take-off weight 4536kg (9980lb)
Dimensions:	span unswept 9.39m (31ft) span swept 5.66m (19ft); length 10.16m (33ft); height 3.66m (12ft); wing area 16.26 sq m (175 sq ft)
Armament:	none

1953

X-Planes Part Two

The US Navy contributed to research flying with the equally experimental Skyrocket series and other record-breakers. The National Advisory Committee on Aeronautics (NACA) became NASA (National Air and Space Administration), to better reflect its work at the edges of the atmosphere and beyond. While one branch of NASA worked towards the Moon landings with ballistic rockets, Edwards AFB saw the limits of wingborne flight expanded to the threshold of space by the X-15.

Douglas D-558-1 Skystreak

Douglas designed the Skystreak and then the Skyrocket in response to a requirement for high-speed research aircraft. The first Skystreak flew in 1947, and broke the world speed record in August of that year. The supersonic D-558-II Skyrocket had a swept wing and fuselage that contained the pilot's jettisonable compartment, turbojet, rocket engine, landing gear and fuel.

SPECIFICATIONS	
Country of origin:	United States
Type:	swept-wing research aircraft
Powerplant:	one 13.61kN (3,059lb) thrust Westinghouse J34-W-22 plus one 27.2 kN (6117lb) thrust Reaction Motors XLR-8 rocket motor
Performance:	maximum speed (turbojet only) 941km/h (583mph); (mixed power) 1159km/h (718mph); (rocket power only) 2012km/h (1247mph)
Weights:	Maximum take-off 6925kg (15,267lb) (turbojet) (mixed powerplant) 7171kg (15,800lb)
Dimensions:	span 7.62m (25ft); length 13.79m (45ft); height: 3.51m (11ft) wing area 16.26 sq m (125 sq ft)
Armament:	none

Ryan X-14B

An experimental aircraft, the X-14 constantly evolved, being powered by three different engines throughout its long life. As a VTOL research aircraft capable of hovering, weight and balance was of immense importance. The fuel tanks were external and underwing, keeping them close to the centre of gravity, and the ejector seat was excluded for weight considerations.

SPECIFICATIONS	
Country of origin:	United States
Type:	VTOL research aircraft
Powerplant:	two 13.4kN (3015lb) thrust General Electric J85-GE-19 non-afterburning turbojets, fitted late in the X-14 programme
Performance:	maximum speed 277km/h (172mph); range 480km (300 miles) service ceiling 5500m (18000ft)
Weights:	empty 1437kg (3160lb); maximum take-off weight 1934kg (4255lb)
Dimensions:	span 10.3m (33ft. 9in); length 7.92m (26ft); height 2.68m (8ft 9in); wing area 16.68 sq m (179 sq ft)
Armament:	none

X-PLANES TIMELINE   1947  1957  1959

X-PLANES PART TWO

North American X-15

After Chuck Yeager piloted the Bell X-1 beyond Mach 1 in 1947, a line of rocket-powered record-breakers followed, climaxing in the North American X-15. A sleek black rocket with tiny wings, the X-15 flew higher and faster than anything before or since. Most of the 199 powered missions between 1960 and 1968 probed the limits of possibility, smashing all previous records in the process.

SPECIFICATIONS	
Country of origin:	United States
Type:	single-seat hyper-velocity rocket-powered research aircraft
Powerplant:	one 313kN (70,4000lb) thrust Thiokol (Reaction Motors) XLR99-RM-2 single-chamber throttleable liquid-propellant rocket engine
Performance:	maximum speed 7297km/h (4534mph); maximum altitude 10,7960m (354,000ft); time to height 140 sec from launch at 15,000m (49,212ft) to 10,0000m (320,000ft); range 450km (280 miles)
Weights:	loaded 25,460kg (56,000lb)
Dimensions:	span 6.81m (22ft 4in); length 15.47m (50ft 9in); height 3.96m (12ft 11in); wing area 18.58 sq m (200 sq ft)
Armament:	none

Northrop X-24

To withstand the frictional heating that occurs during re-entry to the Earth's atmosphere, a blunt-nosed, wingless, lifting body was designed. A forerunner of today's Space Shuttle, the X-24 was taken to altitude under the wing of an NB-52B, making its first gliding flight in April 1969. Twenty-eight powered flights were completed before the aircraft was modified to become the X-24B.

SPECIFICATIONS	
Country of origin:	United States
Type:	lifting body aircraft for research into reusable spacecraft approach patterns
Powerplant:	one 43.64kN (9820lb) thrust Thiokol XLR-RM-13 four-chamber regeneratively cooled rocket engine, plus two 2.22kN (500lb) thrust hydrogen-peroxide rocket engines
Performance:	maximum speed 1873km/h (1161mph); service ceiling 22,595m (74,100ft)
Weights:	maximum 6260kg (13,772lb)
Dimensions:	span 5.8m (19ft); length 11.43m (37ft 6in); height 3.15m (10ft 4in); lifting surface area 30.66 sq m (330 sq ft)
Armament:	none

Grumman X-29

Grumman won a 1981 contract to build X-29 research planes, and managed to slash costs by using parts from many aircraft: F-5 fuselage and nosewheel, F/A-18 Hornet engine and F-16 main undercarriage. The result was more than the sum of its parts. In several years of experiments, the X-29 demonstrated an ability to manoeuvre at angles of attack as high as 67°. The key feature was that the wings were swept forwards rather than back.

SPECIFICATIONS	
Country of origin:	United States
Type:	single-seat forward-swept-wing high-agility research aircraft
Powerplant:	one 71.17kN (15,965lb) thrust General Electric F404-GE-400 turbofan
Performance:	maximum speed Mach 1.87 or 1900km/h at 10,000m (1178mph at 33,000ft); range 560km (347 miles); service ceiling 15,300m (50,000ft)
Weights:	empty 6260kg (13,772lb); maximum 8074kg (17,763lb)
Dimensions:	span 8.29m (27ft) length 16.44m (54ft) height 4.26m (14ft) wing area 188.8 sq m (2031 sq ft)
Armament:	none

1973

1984

COLD WAR MILITARY AIRCRAFT

Shooting Stars

Lockheed's P-80 was the first successful American jet fighter. It was designed in fewer than 180 days around the British Halford H.1 (Goblin) engine, by a Lockheed team led by Clarence 'Kelly' Johnson. Some 1718 were made in many different variants, from reconnaissance platforms to trainers to supersonic interceptors. The prefix was later changed from 'P' (Pursuit) to 'F' (Fighter). The F-80 C-5 was the final production model, with more powerful engines.

F-80C

The Lockheed P-80 first flew in January 1944, and P-80s were flying under combat conditions in Italy a year later. By the time of the Korean War, the aircraft were considered somewhat obsolete, but nonetheless flew 15,000 sorties in the first four months and shot down the first MiG-15 in November 1950, in what is thought to have been the first jet combat.

SPECIFICATIONS	
Country of origin:	United States
Type:	single-engined fighter-bomber
Powerplant:	one 24kN (5400lb) thrust Allison J33-A-35 turbojet engine
Performance:	maximum speed 965km/h (600mph); service ceiling 14,000m (46,800ft); range 1930km (1200 miles)
Weights:	empty 3819kg (8420lb); maximum 7646kg (16,856lb)
Dimensions:	wingspan 11.81m (38ft 9in); length 10.49m (34ft 5in); height 3.43m (11ft 3in); wing area 22.07 sq m (237.6 sq ft)
Armament:	six .5in (12.7mm) machine guns; two 454kg (1000lb) bombs or eight 2.75in rockets

QF-80 Shooting Star

Many surplus F-80s were converted into pilotless drones at the end of their service lives. The Sperry Gyroscope Company undertook many QF-80 conversions, creating an aircraft that could be shot at by air, land and sea weapons systems or used for various test missions.

SPECIFICATIONS	
Country of origin:	United States
Type:	single-seat fighter bomber
Powerplant:	one 24kN (5400lb) Allison J33-A-35 turbojet
Performance:	maximum speed at sea level 966km/h (594mph); service ceiling 14,265m (46,800ft); range 1328km (825 miles)
Weights:	empty 3819kg (8420lb); maximum take-off weight 7646kg (16,856lb)
Dimensions:	wingspan 11.81m (38ft 9in); length 10.49m (34ft 5in); height 3.43m (11ft 3in); wing area 22.07 sq m (237.6 sq ft)
Armament:	none

SHOOTING STARS

RF-80A Shooting Star

Originally designated FP-80A, the RF-80A was a modification of Shooting Star fighters into reconnaissance aircraft by removing armament and installing a new camera nose. The 66 converted aircraft were widely used in Korea.

SPECIFICATIONS

Country of origin:	United States
Type:	single-seat photo reconnaissance
Powerplant:	one 24kN (5400lb) Allison J33-A-35 turbojet
Performance:	maximum speed at sea level 966km/h (594mph); service ceiling 14,265m (46,800ft); range 1328km (825 miles)
Weights:	empty 3819kg (8420lb); maximum take-off weight 7646kg (16,856lb)
Dimensions:	wingspan 11.81m (38ft 9in); length 10.49m (34ft 5in); height 3.43m (11ft 3in); wing area 22.07 sq m (237.6 sq ft)
Armament:	none

T-33A

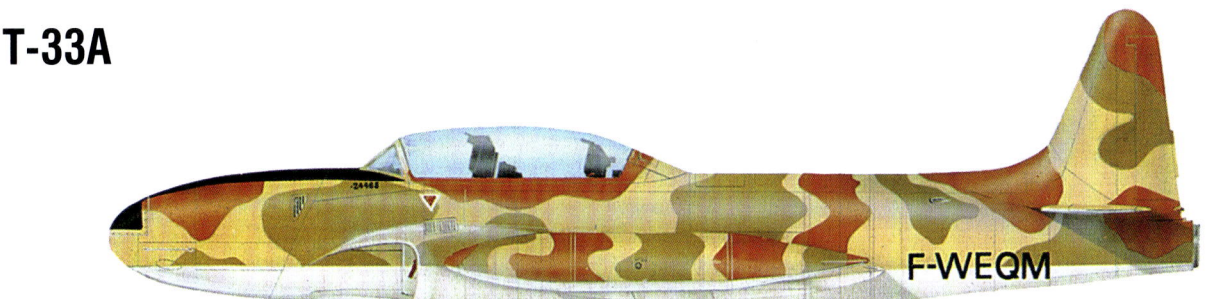

The Lockheed T-33, also named Shooting Star, was the most numerous Western jet trainer of the post-war era, used by nearly 40 nations. This civil-marked ex-Canadian T-33 was refurbished with modern avionics for Bolivia in the 1990s, and still serves in the ground-attack role.

SPECIFICATIONS

Country of origin:	United States
Type:	two-seat jet trainer
Powerplant:	one 24kN (5400lb) Allison J33-A-35 turbojet
Performance:	maximum speed 879km/h (546mph); service ceiling 14,630m (48,000ft); endurance 3 hours 7 minutes
Weights:	empty 3667kg (8084lb); maximum take-off weight 6551kg (14,442lb)
Dimensions:	wingspan 11.85m (38ft 10.5in); length 11.51m (37ft 10in); height 3.56m (11ft 8in); wing area 21.81 sq m (234.8 sq ft)
Armament:	two .5mm machine guns; wide variety of ordnance in COIN role

F-94 Starfire

Retaining many of the features of the F-80 and T-33 aircraft from which it was developed, the tandem-seat Starfire was conceived in 1949 as a radar-equipped, all-weather interceptor. The first deliveries began in June 1950. Two improved variants were produced – the F-94B with a blind landing system, and the F-94C, with 24 Mighty Mouse unguided air-to-air rockets in the nose.

SPECIFICATIONS

Country of origin:	United Kingdom
Type:	tandem-seat all-weather interceptor
Powerplant:	one 26.7kN (6000lb) Allison J33-A-33 turbojet
Performance:	maximum speed at 30,000ft 933km/h (580mph); service ceiling 14,630m (48,000ft); range 1850km/h (1150 miles)
Weights:	empty 5030kg (11,090lb); maximum take-off weight 7125kg (15,710lb)
Dimensions:	wingspan not including tip tanks 11.85m (38ft 10.5in); length 12.2m (40ft 1in); height 3.89m (12ft 8in); wing area 22.13 sq m (238 sq ft)
Armament:	four .5in machine guns

Navy Jets

Britain's Royal Navy made innovations in naval aviation, including jet operations from carriers, but it was the United States that first put operational jet squadrons on carriers. The greater speeds and slower engine response of jets contributed to terrible accident rates during carrier operations. Further British developments included angled carrier decks, which allowed aircraft to miss arrestor wires without threatening others parked forward, and the advanced mirror landing system.

McDonnell F4H-1 Phantom II

In 1942, the Bureau of Aeronautics entrusted McDonnell with the task of designing and building the two prototypes of what would become the US Navy's first carrier-based, turbojet-powered, single-seat fighter. The resulting prototypes were low-wing monoplanes with retractable landing gear, with power provided by two turbojets buried in the wing roots. After first flying in January 1945, the aircraft became the first US jet to be launched and recovered from a carrier.

SPECIFICATIONS

Country of origin:	United States
Type:	carrier-based fighter
Powerplant:	two 7.1kN (1600lb) Westinghouse J30-WE-20 turbojets
Performance:	maximum cruising speed 771km/h (479mph); service ceiling 12,525m (41,100ft); combat range 1118km (695 miles)
Weights:	empty 3031kg (6683lb); maximum take-off weight 5459kg (12,035lb)
Dimensions:	wingspan 12.42m (40ft 9in); length 11.35m (37ft 3in); height 4.32m (14ft 2in); wing area 24.64 sq m (276 sq ft)
Armament:	four .5in machine guns

Grumman F9F-2 Panther

The Grumman Panther was the first truly successful American carrier-based jet. Nearly 1400 were built for the US Navy, Marines and Argentina. Panthers were the main naval strike aircraft in Korea, where the F9F-2 illustrated flew with VF-781 aboard USS *Oriskany* in 1952.

SPECIFICATIONS

Country of origin:	United States
Type:	single-engined carrier-based jet fighter-bomber
Powerplant:	one 26.5kN (5950lb) thrust Pratt & Whitney J42-P-6/P-8 turbojet engine
Performance:	maximum speed 925 km/h (575 mph); range 2100km (1300 miles); service ceiling 13,600m (44,600ft)
Weights:	empty 4220kg (9303lb); maximum 7462kg (16,450lb)
Dimensions:	span 11.6m (38ft); length 11.3m (37ft 5in); height 3.8m (11ft 4in); wing area 23 sq m (250 sq ft)
Armament:	four 20mm M2 cannon; up to 910kg (2000lb) bombs, six 5in (127mm) rockets

McDonnell F2H-2 Banshee

The success of the FH-1 Phantom in US Navy and Marine Corps service meant that it was almost inevitable that McDonnell would be asked to submit a design to succeed the Phantom in service. The first F2H-1 aircraft was delivered to the Navy in August 1948, and was followed by seven sub-variants. The F2H-2 was the second production version, with wingtip fuel tanks. Production total was 56.

SPECIFICATIONS	
Country of origin:	United States
Type:	carrier-based all-weather fighter
Powerplant:	one 14.5kN (3,250lb) Westinghouse J34-WE-34 turbojet
Performance:	maximum cruising speed 933km/h (580mph); service ceiling 14,205m (46,600ft); combat range 1883km (1170 miles)
Weights:	empty 5980kg (13,183lb); maximum take-off weight 11,437kg (25,214lb)
Dimensions:	wingspan 12.73m (41ft 9in); length 14.68m (48ft 2in); height 4.42m (14ft 6in); wing area 27.31 sq m (294 sq ft)
Armament:	four 20mm cannon; underwing racks with provision for two 227kg (500lb) or four 113kg (250lb) bombs

North American AJ-2 Savage

The AJ Savage was designed to carry a single large atomic bomb from carriers. Its two piston engines were supplemented by a turbojet in the fuselage to give a higher dash speed over the target. It was not a great success as a bomber, and most Savages were converted into tankers.

SPECIFICATIONS	
Country of origin:	United States
Type:	mixed-powerplant carrier-based bomber
Powerplant:	two 1864kW (2500hp) Pratt & Whitney R-2800-48 piston engines and one 20.46kN (4600lb) thrust Allison J33-A-19 turbo-jet engine
Performance:	maximum speed 758km/h (471mph); range 2623km (1630 miles); service ceiling 12,192m (40,000ft)
Weights:	empty 12,500kg (27,558lb); maximum 23,973kg (52,852lb)
Dimensions:	span 22.91m (75ft 2in); length 19.22m (63ft 1in); height 6.52m (21ft 5in); wing area 78 sq m (836 sq ft)
Armament:	one 3856kg (8500lb) Mk VI nuclear bomb

North American FJ-3M Fury

Both Army and Navy contracts were awarded to North American in 1944 for a jet fighter. The FJ-1 Fury was unremarkable and the FJ-2 was a navalized version of the land-based F-86E Sabre, with folding wings, catapult hitches and arrestor gear. They were superseded by the FJ-3, which had a more powerful engine, a deeper fuselage, a new canopy and increased weapon load.

SPECIFICATIONS	
Country of origin:	United States
Type:	single-seat fighter-bomber
Powerplant:	one 34.7kN (7800lb) Wright J65-W-2 turbojet
Performance:	maximum speed at sea level 1091km/h (678mph); service ceiling 16,640m (54,600ft); range 1344km (835 miles)
Weights:	empty 5051kg (11,125lb); maximum loaded 9350kg (20,611lb)
Dimensions:	wingspan 11.3m (37ft 1in); length 11.43m (37ft 6in); height 4.47m (14ft 8.75in); wing area 27.76 sq m (288 sq ft)
Armament:	six .5 Colt-Browning M-3 with 267 rpg, underwing hardpoints for two tanks or two stores of 454kg (1000lb), plus eight rockets

COLD WAR MILITARY AIRCRAFT

The Immortal Dakota

First flown in December 1935, the Douglas Sleeper Transport, or DC-3, went on to become the most produced transport aircraft in history. The vast majority were built for military use as C-47s, C-49s, C-53s and other designations, usually known by their British name, Dakota. So many 'Daks' were available as surplus in 1945 that they hampered post-war development of new airliners and light military transports. Turboprop conversions are still used by air forces in Africa and South America.

Dakota III

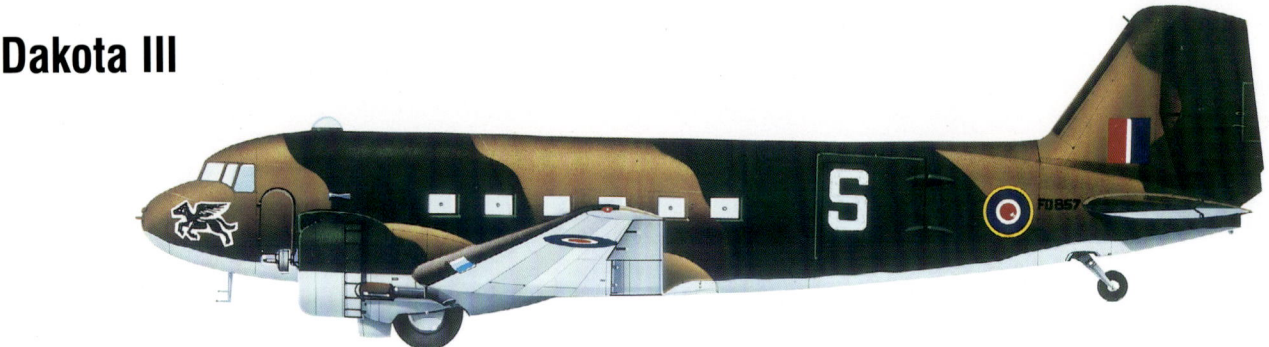

Among the many roles of the C-47 was to drop supplies to partisans and resistance groups. This Dakota III, equivalent to the C-47A, was based at Araxos, Greece, and used by No 267 Squadron RAF to drop supplies over Romania, Albania and Yugoslavia in 1944.

SPECIFICATIONS	
Country of origin:	United States
Type:	(C-47) two/three-seat transport with accommodation for 28 troops, or 14 litters plus three attendants or 10,000lb (4536kg) of freight
Powerplant:	two 895kW (1200hp) Pratt & Whitney R-1830-92 14-cylinder two-row radial engines
Performance:	maximum speed 370km/h (230mph); service ceiling 7315m (24,000ft); range 2575km (1600 miles)
Weights:	empty 8103kg (17,865lb); maximum take-off weight 14,061kg (31,000lb)
Dimensions:	span 28.9m (95ft); length 19.63m (64ft 5.5in); height 5.2m (16ft 11in)
Armament:	none

Lisunov Li-2

The Li-2 was a licence-built Russian version of the DC-3 with more than 1200 changes incorporated, notably relocation of the passenger door to the starboard side, a new freight door and Shvetsov engines. More than 6000 Li-2s were built from 1938 to 1952, some being used as bombers in wartime.

SPECIFICATIONS	
Country of origin:	United States
Type:	twin-engined military transportt
Powerplant:	two 736kW (1000hp) Shvetsov ASh-62IR radial piston engines
Performance:	maximum speed 300km/h (186mph); service ceiling 7315m (24,000ft); range 2500km (1550 miles)
Weights:	empty 7750kg (17,185lb); loaded 10,700kg (23,580lb)
Dimensions:	span 28.9m (95ft); length 19.63m (64ft 5.5in); height 5.2m (16ft 11in)
Armament:	(some versions): three 7.62mm ShKAS and one 12.7mm UBK machine guns; up to 2000kg (908kg) of bombs

THE IMMORTAL DAKOTA

Dakota IV

Belgium operated various C-47s and Dakotas from 1944 until 1976. This unusual example was one of two BAF C-47Bs fitted with the radar of the F-104G to train pilots in navigating the Starfighter. They were nicknamed 'Pinocchio' and used for a short time in the late 1960s.

SPECIFICATIONS

Country of origin:	United States
Type:	twin-engined navigation trainer
Powerplant:	two 895kW (1200hp) Pratt & Whitney R-1830-90C radial engines
Performance:	maximum speed 370km/h (230mph); service ceiling 7315m (24,000ft); range 2575km (1600 miles)
Weights:	empty 8103kg (17,865lb); maximum take-off 14,061kg (31,000lb)
Dimensions:	span 28.9m (95ft); length unknown; height 5.2m (16ft 11in)
Armament:	none

Dakota

The Dakota Mk IV was the RAF equivalent of the C-47B, which varied in having supercharged engines for better high-altitude performance. South Africa has used Dakotas since the 1940s. All the SAAF's remaining Dakotas have been given PT-67 turboprop engines and serve with No 35 Squadron.

SPECIFICATIONS

Country of origin:	United States
Type:	(C-47) two/three-seat transport with accommodation for 28 troops, or 14 litters plus three attendants or 10,000lb (4536kg) of freight
Powerplant:	two 895kW (1200hp) Pratt & Whitney R-1830-90C radial engines
Performance:	maximum speed 370km/h (230mph); service ceiling 7315m (24,000ft); range 2575km (1600 miles)
Weights:	empty 8103kg (17,865lb); maximum take-off 14,061kg (31,000lb)
Dimensions:	span 28.9m (95ft); length 19.63m (64ft 5.5in); height 5.2m (16ft 11in)
Armament:	none

LC-47 Skiplane

Several nations modified C-47s with skis for use in polar exploration. The US Navy's Operation Deep Freeze made the first landings at the South Pole with LC-47s in 1956. Argentina maintained C-47 skiplanes for supporting its Antarctic bases until 1983.

SPECIFICATIONS

Country of origin:	United States
Type:	(C-47) two/three-seat transport with accommodation for 28 troops, or 14 litters plus three attendants or 10,000lb (4536kg) of freight
Powerplant:	two 895kW (1200hp) Pratt & Whitney R-1830-92 14-cylinder two-row radial engines
Performance:	maximum speed 370km/h (230mph); service ceiling 7315m (24,000ft); range 2575km (1600 miles)
Weights:	empty 8103kg (17,865lb); maximum take-off 14,061kg (31,000lb)
Dimensions:	span 28.9m (95ft); length unknown; height 5.2m (16ft 11in)
Armament:	none

Suez Crisis

In 1956, Egypt's President Nasser nationalized the Suez Canal against the wishes of Britain and France, who undertook an operation to reclaim the Canal Zone that October. The action involved one of the first mass helicopter assaults in history and a large parachute landing. Egyptian airfields were bombed by carrier- and land-based aircraft, neutralizing air opposition to the landings. Although a military victory, the Suez Crisis was regarded as a political failure for Britain.

De Havilland Venom FB.4

Aircraft operating in what the British called Operation Musketeer were painted with distinctive black and yellow 'Suez stripes' to distinguish them from Egyptian aircraft, as seen on this Venom FB.4 of one of the RAF squadrons on Cyprus during Musketeer.

SPECIFICATIONS	
Country of origin:	United Kingdom
Type:	single-seat fighter bomber
Powerplant:	one 22.9kN (5150lb) de Havilland Ghost 105 turbojet
Performance:	maximum speed 1030km/h (640mph); service ceiling 14,630m (48,000ft); range with drop tanks 1730km (1075 miles)
Weights:	empty 4174kg (9202lb); maximum loaded 6945kg (15,310lb)
Dimensions:	wingspan (over tip tanks) 12.7m (41ft 8in); length 9.71m (31ft 10in); height 1.88m (6ft 2in); wing area 25.99 sq m (279.75 sq ft)
Armament:	four 20mm Hispano cannon with 150 rounds, two wing pylons capable of carrying either two 454kg (1000lb) bombs or two drop tanks; or eight 27.2kg (60lb) rocket projectiles carried on centre-section launchers

English Electric Canberra B.2

RAF Canberras were based on Cyprus and Malta, and began strikes on Egyptian airfields on the night of 31 October 1956. Within five days of attacks, the Egyptian Air Force was effectively destroyed. As well as B.2 bombers, PR.3 reconnaissance variants took pre- and post-strike photos.

SPECIFICATIONS	
Country of origin:	United Kingdom
Type:	two-seat interdictor aircraft
Powerplant:	two 28.9kN (6500lb) Rolls Royce Avon Mk 101 turbojets
Performance:	maximum speed 917km/h (570mph); service ceiling 14,630m (48,000ft); range 4274km (2656 miles)
Weights:	empty approx. 11,790kg (20,000lb); max. take-off 24,925kg (54,950lb)
Dimensions:	wingspan 29.49m (63ft 11in); length 19.96m (65ft 6in); height 4.78m (15ft 8in); wing area 97.08 sq m (1045 sq ft)
Armament:	internal bomb bay with provision for up to 2727kg (6000lb) of bombs, plus an additional 909kg (2000lb) of underwing pylons

SUEZ CRISIS

Republic F-84F Thunderstreak

SPECIFICATIONS	
Country of origin:	United States
Type:	single-seat fighter bomber
Powerplant:	one 32kN (7220lb) Wright J65-W-3 turbojet
Performance:	maximum speed 1118km/h (695mph); service ceiling 14,020kg (46,000ft); combat radius with drop tanks 1304km (810 miles)
Weights:	empty 6273kg (13,830lb); maximum take-off 12,701kg (28,000lb)
Dimensions:	wingspan 10.24m (33ft 7.25in); length 13.23m (43ft 4.75in); height 4.39m (14ft 4.75in); wing area 30.19 sq m (325 sq ft)
Armament:	six .5in Browning M3 machine-guns, external hardpoints with provision for up to 2722kg (6000lb) of stores

France contributed carrier-based Corsairs, and Dassault Mystères and F-84F Thunderstreaks on Cyprus. F-84s were also based in Israel to prevent an Egyptian counterattack. A large number of Egyptian tanks were destroyed by the French jets, as well as Il-28 bombers on the ground.

Gloster Meteor NF.11

SPECIFICATIONS	
Country of origin:	United Kingdom
Type:	two-seat night fighter
Powerplant:	Two 16kN (3600lb) Rolls-Royce Derwent 8 turbojets
Performance:	maximum speed at 10,000m (33,000ft) 931m/h (579mph); service ceiling 12,192m (40,000ft); range 1580km (980 miles)
Weights:	empty 5400kg (11,900lb); loaded 9979kg (22,000lb)
Dimensions:	wingspan 13.1m (43ft); length 14.78m (48ft 6in); height 4.22m (13ft 10in)
Armament:	four 20mm Hispano cannon

The Royal Egyptian Air Force had mostly Soviet-designed equipment supplied by Czechoslovakia, but also six Meteor night fighters bought from Britain in 1955. Although one pursued an RAF Canberra, which evaded, they claimed no victories during the crisis.

Hawker Sea Hawk FB.Mk.3

SPECIFICATIONS	
Country of origin:	United Kingdom
Type:	single-seat carrier-based fighter-bomber
Powerplant:	one 24kN (5000lb) Rolls-Royce Nene turbojet
Performance:	maximum speed at sea level 958km/h (599mph); or 939km/h (587mph) at height; service ceiling 13,560m (44,500ft); standard range 1191km (740 miles)
Weights:	empty 9720lb; maximum take-off 7355kg (16,200lb)
Dimensions:	wingspan 11.89m (39ft); length 12.09m (39ft 8in); height 2.64m (8ft 8in); wing area 25.83sq m (278sq ft)
Armament:	four 20mm Hispano cannon in nose, underwing hardpoints for two 227kg (500lb) bombs

Six squadrons of Sea Hawks, flying from the carriers HMS *Eagle*, *Albion* and *Bulwark*, attacked Egyptian airfields by daylight and provided close air support for the helicopter and seaborne landings with their cannon and rockets.

COLD WAR MILITARY AIRCRAFT

Fleet Air Arm in the 1950s

The British Royal Navy's Fleet Air Arm was still a large organization in the 1950s, operating over 30 squadrons ashore and afloat. Most of these used piston-engined aircraft, but jets and helicopters were arriving by 1954. FAA squadrons saw action in Korea and at Suez, and acquitted themselves well, but the RN sometimes struggled to get its share of the defence budget. By 1959, however, new fleet carriers such as HMS *Hermes* and supersonic fighters and strike aircraft were entering service.

Skyraider AEW.1

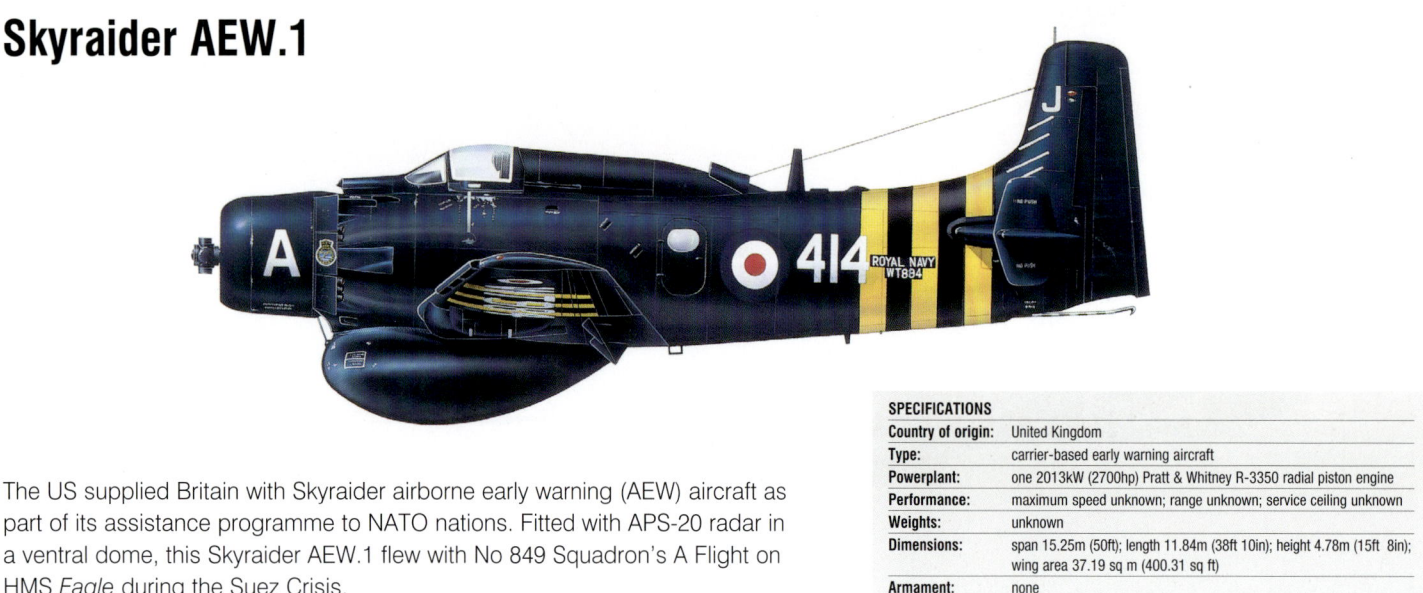

The US supplied Britain with Skyraider airborne early warning (AEW) aircraft as part of its assistance programme to NATO nations. Fitted with APS-20 radar in a ventral dome, this Skyraider AEW.1 flew with No 849 Squadron's A Flight on HMS *Eagle* during the Suez Crisis.

SPECIFICATIONS	
Country of origin:	United Kingdom
Type:	carrier-based early warning aircraft
Powerplant:	one 2013kW (2700hp) Pratt & Whitney R-3350 radial piston engine
Performance:	maximum speed unknown; range unknown; service ceiling unknown
Weights:	unknown
Dimensions:	span 15.25m (50ft); length 11.84m (38ft 10in); height 4.78m (15ft 8in); wing area 37.19 sq m (400.31 sq ft)
Armament:	none

Fairey Firefly AS.5

The Fairey Firefly was in production from 1941 to 1955, and was produced in over 25 sub-types. The AS.5 was intended as an anti-submarine strike aircraft, but also saw duties in the ground-attack and fighter roles. This AS.5 was with No 810 Squadron on HMS *Theseus* off Korea in 1950.

SPECIFICATIONS	
Country of origin:	United Kingdom
Type:	carrier-based fighter/anti-submarine aircraft
Powerplant:	one 1678kW (2250hp) Rolls-Royce Griffon 74 V-12 piston engine
Performance:	maximum speed 618km/h (386mph); range 2092km (1300 miles); service ceiling : 9450m (31,000ft)
Weights:	empty 4388kg (9674lb), maximum 7301kg (16,096lb)
Dimensions:	span 12.55m (41ft 2n); length 8.51m (37ft 11in); height 4.37m (14ft 4in); wing area 30.66 sq m (330 sq ft)
Armament:	four 20mm Hispano cannon; up to 908kg (2000lb) of bombs or 16 27kg (60lb) rockets

FLEET AIR ARM IN THE 1950S

Blackburn Firebrand IV

Conceived in 1939, the Firebrand heavy torpedo-carrying fighter was ready for service just too late for the war. Due to its many teething and operational problems, including lack of speed (particularly with a torpedo), only two frontline squadrons were equipped with Firebrands.

SPECIFICATIONS

Country of origin:	United Kingdom
Type:	carrier-based fighter/torpedo bomber
Powerplant:	one 1865kW (2500hp) Bristol Centaurus IX radial piston engine
Performance:	maximum speed: 560km/h (350 mph); range 200 km (1250 miles); service ceiling 10,363m (34,000ft)
Weights:	empty 5150kg (11,357lb); maximum 7360kg (16,227lb)
Dimensions:	span 15.62m (51ft 4in); length 12m (39ft 1in); height 4.08m (13ft 3in); wing area 35.44 sq m (381.5 sq ft)
Armament:	four 20mm Hispano Mk II cannon; one 840kg (1850lb) torpedo or 908kg (2000lb) of bombs

Hawker Sea Fury FB.2

Derived from the Tempest Mk II, the Sea Fury was one of the fastest and best post-war piston-engined fighters. Canada, Australia and the Netherlands also flew carrier-based Sea Furies, and land-based versions without folding wings or arrestor hooks were exported to a number of nations.

SPECIFICATIONS

Country of origin:	United Kingdom
Type:	single-engined carrier-based fighter-bomber
Powerplant:	one 1850kW (2480hp) Bristol Centaurus XVIIC radial piston engine
Performance:	maximum speed 740 km/h (460 mph); range 1127km (700 miles); service ceiling 10,900m (35,800ft)
Weights:	empty 4190kg (9240lb); maximum 5670kg (12,500lb)
Dimensions:	span 11.7m (38ft 4in); length 10.6m (34ft 8in); height 4.9m (16ft 1in); wing area 26 sq m (280 sq ft)
Armament:	four 20mm Hispano Mk V cannon; up to 908kg (2000lb) of bombs or 12 3in rockets

Hawker Sea Hawk FGA.4

Although the design of the bifurcated jet pipe caused some concern among defence staff when the prototype P.1040 was first unveiled, the Sea Hawk has a well-earned reputation as a reliable, good-handling fighter. It remained in service with the FAA until 1960. In 1959, the Indian Navy ordered 24 aircraft similar to the Mk 6. Some were new-build and the rest were refurbished ex-RN Mk 6s.

SPECIFICATIONS

Country of origin:	United Kingdom
Type:	single-seat carrier based fighter-bomber
Powerplant:	one 24kN (5400lb) Rolls-Royce Nene 103 turbojet
Performance:	max speed 969km/h (602mph); service ceiling 13,565m (44,500ft); combat radius (clean) 370km (230 miles)
Weights:	empty 4409kg (9720lb); maximum take-off weight 7348kg (16,200lb)
Dimensions:	wingspan 11.89m (39ft); length 12.09m (39ft 8in); height 2.64m (8ft 8in); wing area 25.83 sq m (278 sq ft)
Armament:	four 20mm Hispano cannon; plus provision for four 227kg (500lb) bombs, or two 227kg (500lb) bombs and 20 3in or 16 5in rockets

COLD WAR MILITARY AIRCRAFT

Vampires and Venoms

De Havilland's compact twin-boomed Vampire was not the most sophisticated or highest-performing jet fighter of the late 1940s, and became comparatively less so over time. Nevertheless, it was enjoyable to fly, simple to maintain, and cheap to operate, and was widely exported. The Venom was a conservative development of the Vampire and as such it was not competitive as a fighter, but it filled a useful niche as a ground-attack platform.

Vampire FB.5

No 112 Squadron was famous for the exploits of its sharksmouth-decorated P-40s in North Africa. A similar marking was sported by its Vampire FB.5s, which operated in the day-fighter role as part of RAF Germany from 1951 to 1956, before replacement by Sabres.

SPECIFICATIONS	
Country of origin:	United Kingdom/Switzerland
Type:	single-seat fighter-bomber
Powerplant:	one 13.8kN (3100lb) thrust de Havilland Goblin 2 turbojet engine
Performance:	maximum speed 883km/h (548mph); service ceiling 13,410m (44,000ft); range with drop tanks 2253km (1400 miles)
Weights:	empty 3266kg (7200lb); loaded with drop tanks 5600kg (12,290lb)
Dimensions:	wingspan 11.6m (38ft); length 9.37m (30ft 9in); height 2.69m (8ft 10in); wing area 24.32 sq m (262 sq ft)
Armament:	four 20mm Hispano cannon with 150 rounds, wing pylons capable of carrying either two 227kg (500lb) bombs or 60lb rocket projectiles

Vampire FB.9

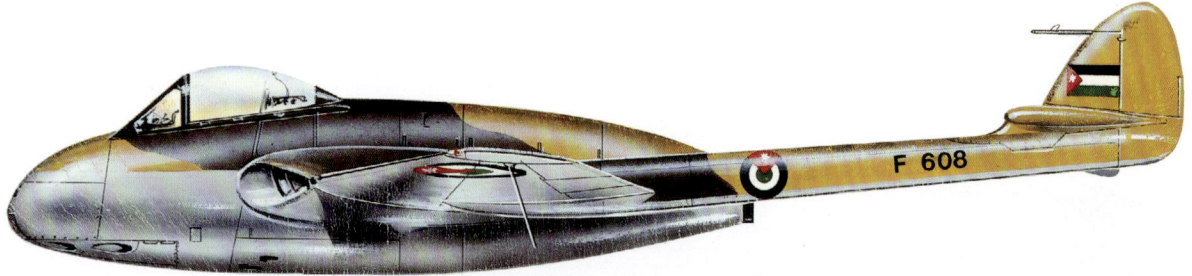

King Abdullah of Jordan was very impressed with RAF Vampires demonstrated in his country, and selected them in 1951 as the first equipment of the new Royal Jordanian Air Force. A mix of FB.9s, FB.52s and T.11s served with the RJAF. The single-seaters were retired in 1967.

SPECIFICATIONS	
Country of origin:	United Kingdom/Switzerland
Type:	single-seat fighter-bomber
Powerplant:	one 14.9kN (3350lb) thrust De Havilland Goblin 3 turbojet engine
Performance:	maximum speed 883km/h (548mph); service ceiling 13,410m (44,000ft); range with drop tanks 2253km (1400 miles)
Weights:	empty 3266kg (7200lb); loaded with drop tanks 5600kg (12,290lb)
Dimensions:	wingspan 11.6m (38ft); length 9.37m (30ft 9in); height 2.69m (8ft 10in); wing area 24.32 sq m (262 sq ft)
Armament:	four 20mm Hispano cannon with 150 rounds, wing pylons capable of carrying either two 227kg (500lb) bombs or 60lb rocket projectiles

Vampire T.11

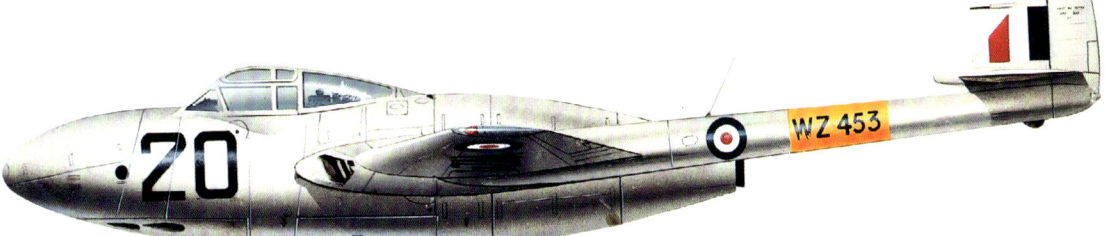

The success of the two-seat night-fighter version of the Vampire led to the development of a trainer. The nose radar was removed and full dual flight controls were added to the pressurized cockpit. In 1956, the T.11 became the standard jet trainer of the Royal Air Force. The production run totalled 731, with export deliveries to 19 countries.

SPECIFICATIONS

Country of origin:	United Kingdom
Type:	two-seat basic trainer
Powerplant:	one 15.6kN (3500lb) de Havilland Goblin 35 turbojet
Performance:	maximum speed 885km/h (549mph); service ceiling 12,200m (40,000ft); range on internal fuel 1370km (853 miles)
Weights:	empty 3347kg (7380lb); loaded (clean) 5060kg (11,150lb)
Dimensions:	wingspan 11.6m (38ft); length 10.55m (34ft 7in); height 1.86m (6ft 2in); wing area 24.32 sq m (262 sq ft)
Armament:	two 20mm Hispano cannon

Venom FB.1

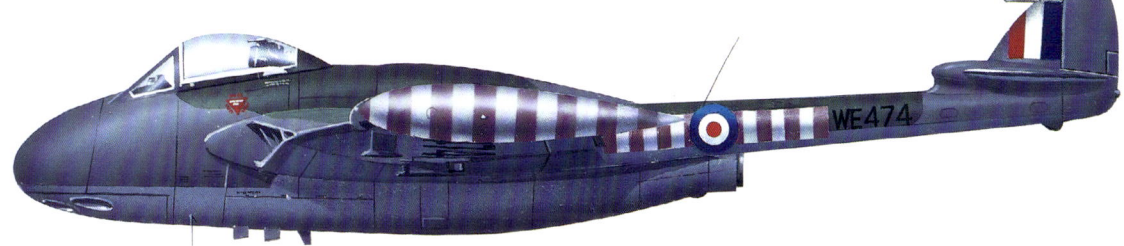

Outwardly similar to the Vampire except for its thinner wing with swept leading edges and wingtip tanks, the Venom also introduced the more powerful Ghost engine and had considerably better performance. The FB.1 entered service with RAF Germany in 1954.

SPECIFICATIONS

Country of origin:	United Kingdom
Type:	single-seat fighter bomber
Powerplant:	one 21.6kN (4850lb) thrust de Havilland Ghost 103 turbojet engine
Performance:	maximum speed 1030km/h (640mph); service ceiling 14,630m (48,000ft); range with drop tanks 1730km (1075 miles)
Weights:	empty 4174kg (9202lb); maximum loaded 6945kg (15,310lb)
Dimensions:	wingspan (over tip tanks) 12.7m (41ft 8in); length 9.71m (31ft 10in); height 1.88m (6ft 2in); wing area 25.99 sq m (279.75 sq ft)
Armament:	four 20mm Hispano cannon with 150 rounds, two wing pylons capable of carrying either two 454kg (1000lb) bombs or two drop tanks; or eight 27.2kg (60lb) rocket projectiles

Venom FB.50

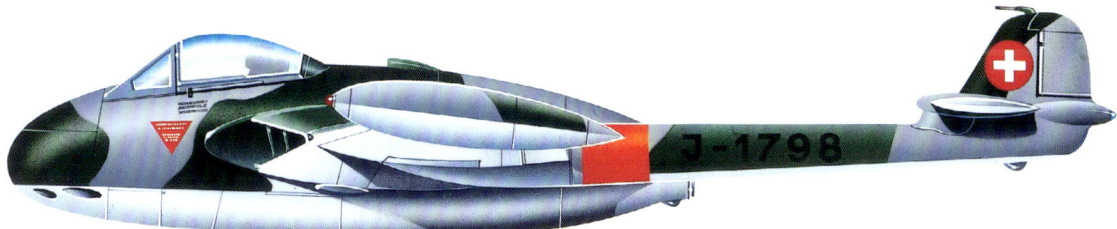

Switzerland has been one of the most enthusiastic of de Havilland's export customers, where they were built under licence. The FB.1, of which 100 were completed, was adopted in 1952 as a successor to the Vampire. These aircraft had long service careers and the final aircraft were not retired until 1983, albeit with substantially altered systems and structures.

SPECIFICATIONS

Country of origin:	United Kingdom/Switzerland
Type:	single-seat tactical reconnaissance with secondary attack capability
Powerplant:	one 21.6kN (4850lb) de Havilland Ghost 103 turbojet
Performance:	maximum speed 1030km/h (640mph); service ceiling 13,720m (45,000ft); range with drop tanks 1730km (1075 miles)
Weights:	empty 3674kg (8100lb); maximum loaded 6945kg (15,310lb)
Dimensions:	wingspan (over tip tanks) 12.7m (41ft 8in); length 9.71m (31ft 10in); height 1.88m (6ft 2in); wing area 25.99 sq m (279.75 sq ft)
Armament:	four 20mm Hispano cannon with 150 rounds, either two 454kg (1000lb) bombs, or two drop tanks, or eight 27.2kg (60lb) rocket projectiles

COLD WAR MILITARY AIRCRAFT

French Jets of the 1950s

The French aircraft industry began its post-war recovery by completing unfinished German aircraft left in its factories, then by buying and licence-producing British equipment. Soon reorganized into regional groupings, French manufacturers competed to produce ever-more exotic jet, rocket and ramjet-powered aircraft. The Dassault concern emerged as dominant, creating a family of fighters and attack jets that formed the backbone of L'Armée de l'Air and Aéronavale from the mid-1950s.

De Havilland Vampire (SNCASE Mistral)

France acquired British-built Vampire 5s before the Société Nationale de Constructions Aéronautiques de Sud-Est, or SNCASE, began production. SNCASE produced several versions, including the SE.535 Mistral, with a Rolls-Royce Nene engine that substantially improved performance.

SPECIFICATIONS	
Country of origin:	United Kingdom
Type:	single-seat fighter-bomber
Powerplant:	one 13.8kN (3100lb) thrust De Havilland Goblin 2 turbojet engine
Performance:	maximum speed 883km/h (548mph); service ceiling 13,410m (44,000ft); range with drop tanks 2253km (1400 miles)
Weights:	empty 3266kg (7200lb); loaded with drop tanks 5600kg (12,290lb)
Dimensions:	wingspan 11.6m (38ft); length 9.37m (30ft 9in); height 2.69m (8ft 10in); wing area 24.32 sq m (262 sq ft)
Armament:	four 20mm Hispano cannon with 150 rounds, wing pylons capable of carrying either two 227kg (500lb) bombs or 60lb rocket projectiles

Sud-Ouest Aquilon 203

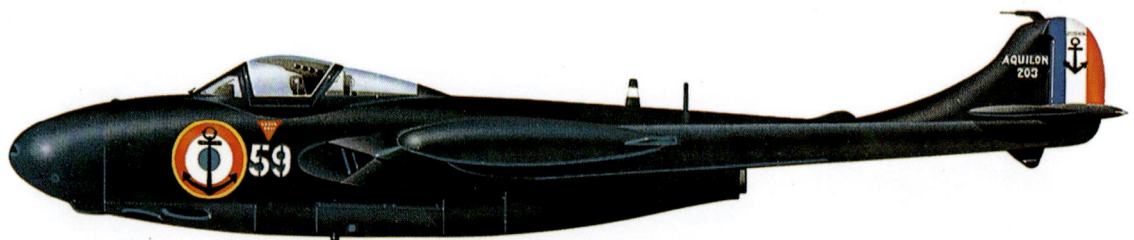

The Aéronavale adopted the De Havilland Sea Venom as a carrier-based night fighter, most of which were built in France as the SNCASE Aquilon. The Aquilon 203 version had an American radar that partly occupied the cockpit, forcing removal of the second seat.

SPECIFICATIONS	
Country of origin:	France/United Kingdom
Type:	single-seat carrier-based fighter
Powerplant:	one 23.4kN (5150lb) de Havilland Ghost 48 turbojet
Performance:	maximum speed 1030km/h (640mph); service ceiling 14,630m (48,000ft); range with drop tanks 1730km (1075 miles)
Weights:	empty 4174kg (9202lb), maximum loaded 6945kg (15,310lb)
Dimensions:	wingspan (over tip tanks) 12.7m (41ft 8in); length 10.38m (32ft 4in); height 1.88m (6ft 2in); wing area 25.99 sq m (279.75 sq ft)
Armament:	four 20mm Hispano 404 cannon with 150 rpg, two wing pylons for Nord 5103 (AA.20) air-to-air missiles

FRENCH JETS OF THE 1950S

MD.450 Ouragon

SPECIFICATIONS	
Country of origin:	France
Type:	single-seat fighter/ground attack aircraft
Powerplant:	one 22.5kN (5070lb) Hispano-Suiza Nene 104B turbojet
Performance:	maximum speed 940km/h (584mph); service ceiling 15,000m (49,210ft); range 1000km (620 miles)
Weights:	empty 4150kg (9150lb); maximum take-off 7600kg (17,416lb)
Dimensions:	wingspan over tip tanks 13.2m (43ft 2in); length 10.74m (35ft 3in); height 4.15m (13ft 7in); wing area 23.8sq m (256.18sq ft)
Armament:	four 20mm Hispano 404 cannon; underwing hardpoints for two 434kg (1000lb) bombs, or 16 105mm rockets, or eight rockets and two 458 litre (101 Imp gal) napalm tanks

First of the jet fighters produced by Marcel Dassault, the MD.450 Ouragan (Hurricane) entered service with L'Armée de l'Air in 1952. This Ouragan wears the colours of the Patrouille de France aerobatic team, which used the type as its first French-made equipment in 1954–57.

Dassault Mystère IVA

SPECIFICATIONS	
Country of origin:	France
Type:	single-seat fighter bomber
Powerplant:	one 27.9kN (6280lb) Hispano Suiza Tay 250A turbojet; or 3500kg (7716lb) Hispano Suiza Verdon 350 turbojet
Performance:	maximum speed 1120km/h (696mph); service ceiling 13,750m (45,000ft); range 1320km (820 miles)
Weights:	empty 5875kg (11,514lb); loaded 9500kg (20,950lb
Dimensions:	wingspan 11.1m (36ft 5.75in); length 12.9m (42ft 2in); height 4.4m (14ft 5in)
Armament:	two 30mm DEFA 551 cannon with 150 rounds, four underwing hardpoints with provision for up to 907kg (2000lb) of stores, including tanks, rockets, or bombs

The US Air Force tested the prototype in September 1952, and placed an off shore contract for 225 of the production aircraft in April 1953. Export orders were won from Israel and India, in addition to the aircraft supplied to the Armée de l'Air. The French aircraft saw action during the Suez conflict, while there were also many sales of the aircraft to the United States, Israel and India.

SO.4050 Vautour IIB

SPECIFICATIONS	
Country of origin:	France
Type:	two-seat medium bomber
Powerplant:	two 24.3kN (7716lb) SNECMA Atar 101E-3 turbojets
Performance:	maximum speed 1105km/h (687mph); service ceiling more than 15000m (49,210ft)
Weights:	empty 10,000kg; maximum take-off weight 20,000kg (44,092lb)
Dimensions:	wingspan 15.09m (49ft 6in); length 15.57m (51ft 1in); height 4.5m (14ft 9in)
Armament:	internal bomb bay with provision for up to 10 bombs, underwing pylons for two bombs up to 450kg (992lb), or two drop tanks

In mid-March 1951, Sud-Ouest flew the prototype of an advanced high-performance twin-jet bomber, designated the SO.4000. From this developed the SO.4050, with swept-wing surfaces and engines mounted in nacelles beneath the wing. Reworked as a two-seat bomber, with a glazed bomb-aiming position in the nose, the aircraft was designated SO.4050-3, first flying in 1954.

Reconnaissance Bombers

With the falling of the 'Iron Curtain' around the Soviet Union and its client states soon after World War II, the activities of the USSR became invisible to the West. The need arose for long-range reconnaissance aircraft, which regularly forayed into Eastern Bloc airspace until the USSR completed its air defences. Converted bombers gave way to specialized aircraft able to fly above fighters and missiles, or look deep inside denied territory using powerful oblique cameras and radars.

Boeing RB-29A Superfortress

Designated F-13s and F-13As before 1948, 117 B-29s and B-29As were modified to carry a suite of six large cameras for photo work over Japan and the Far East. Several RB-29s were shot down by Soviet fighters in 1949–52.

SPECIFICATIONS	
Country of origin:	United States
Type:	four-engined reconnaissance aircraft
Powerplant:	four 1641kW (2200hp) R-3350-57 radial piston engines
Performance:	maximum speed 576km/h (358mph); climb to 6095m (20,000ft) in 38 minutes; service ceiling 9710m (31,850ft); range 9382km (5830 miles)
Weights:	empty 31,816kg (70,140lb); maximum take-off 56,246kg (124,000lb)
Dimensions:	span 43.05m (141ft 2.75in); length 30.18m (99ft); height 9.02m (29ft 7in)
Armament:	one 20mm cannon and six .5in MGs (in tail, each of two dorsal and two ventral barbettes), internal bomb load of 9072kg (20000lb)

Martin P4M-1 Mercator

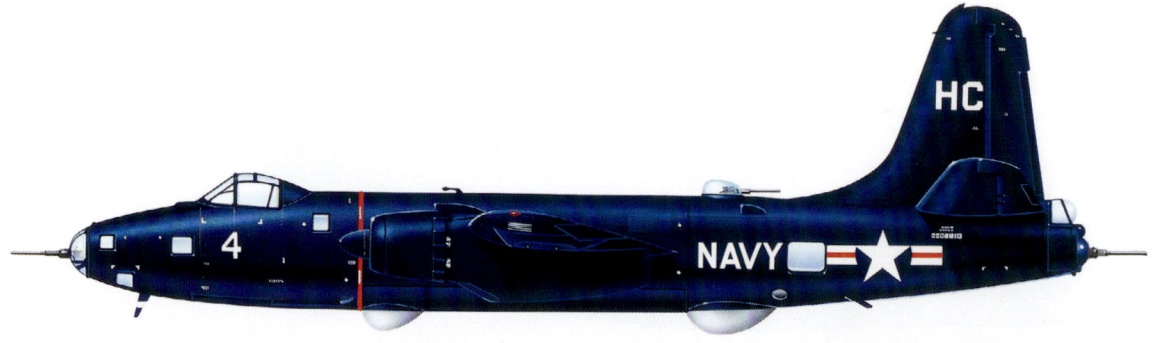

US Naval Aviation's contribution to surveillance of the Soviet Union in the early 1950s included the P4M-1 Mercator. These modified patrol planes were based in the Philippines, Japan and Morocco for electronic surveillance, or 'ferret' missions, of Soviet radar and radio signals.

SPECIFICATIONS	
Country of origin:	United States
Type:	land-based signals intelligence aircraft
Powerplant:	two 2420kW (3250hp) Pratt & Whitney R-4360 radial piston engines and two 20kN (4600lb) thrust Allison J33-A-23 turbojet engines
Performance:	max speed 660km/h (410 mph); service ceiling 10,500m (34,600 ft); range 4570km (2,840 miles)
Weights:	empty 22,016kg (48,536lb); loaded 40,088kg (88,378lb)
Dimensions:	span 34.7m (114ft); length 26m (85ft 2in); height 8m (26ft 1in); wing area 122 sq m (1311 sq ft)
Armament:	four 20mm cannon, two .5in (12.7mm) MGs; 5400kg (12,000lb) bombs

RECONNAISSANCE BOMBERS

North American B-45A Tornado

The USAF's first jet bomber, the B-45 Tornado had a fairly short career in its original role, but was developed into the RB-45C with capacity for up to 12 cameras. As well as USAF operations, one RAF squadron flew secret reconnaissance missions over Russia using borrowed RB-45s.

SPECIFICATIONS

Country of origin:	United States
Type:	four-engined jet bomber
Powerplant:	four 25kN (5200lb) thrust General Electric J47-GE-13 turbojet engines
Performance:	maximum speed 920km/h (570mph); range 1600Km (1000 miles); service ceiling 14,100m (46,400ft)
Weights:	empty 20,726kg (45,694lb); maximum 50,000kg (110,000lb)
Dimensions:	span 27.1m (89ft); length 22.9m (75ft 4in); height 7.7m (25ft 2in); wing area 105 sq m (1125 sq ft)
Armament:	two .5in (12.7mm) machine guns; up to 9997kg (22,000lb) bombs

Boeing RB-47H Stratojet

First bought by the USAAF in October 1945, the jet bomber Model 450 was at its peak in the mid 1950s, with hundreds converted for specialist roles. Thirty-two RB-47H models were made for electronic reconnaissance missions, with the bomb bay converted to accommodate equipment and three EW officers.

SPECIFICATIONS

Country of origin:	United States
Type:	strategic reconnaissance aircraft
Powerplant:	six 32kN (7200lb) General Electric J47-GE-25 turbojets
Performance:	maximum speed at 4970m (16,300ft) 975km/h (606mph); service ceiling 12,345m (40,500ft); range 6437km (4,000 miles)
Weights:	empty 36,630kg (80,756lb); maximum take-off 89,893kg (198,180lb)
Dimensions:	wingspan 35.36m (116ft); length 33.48m (109ft 10in); height 8.51m (27ft 11in); wing area 132.66 sq m (1428 sq ft)
Armament:	two remotely controlled 20mm cannon in tail

North American RA-5C Vigilante

The RA-5C flew in prototype in June 1962. Fifty-five new aircraft were built and all but four of the A-5A Vigilantes were converted to this reconnaissance version. Integrated into the aircraft were all the improvements in range and aerodynamic design that had been developed for the abandoned A-5B project.

SPECIFICATIONS

Country of origin:	United States
Type:	carrier-based long-range reconnaissance aircraft
Powerplant:	two 79.4kN (17,860lb) General Electric J79-GE-10 turbojets
Performance:	maximum speed at altitude 2230km/h (1385mph); service ceiling 20,400m (67,000ft); range with drop tanks 5150km (3200 miles)
Weights:	empty 17,009kg (37,498lb); maximum loaded 36,285kg (80,000lb)
Dimensions:	wingspan 16.15m (53ft); length 23.11m (75ft 10in); height 5.92m (19ft 5in); wing area 70.05 sq m (754 sq ft)

Soviet Interceptors

The Soviet Union put great store on interceptor aircraft for defence of the Motherland. These normally sacrificed manoeuvrability and pilot view for speed and climb performance. The Soviet approach to interceptors emphasized control by ground stations. Positive ground control at all times up to missile launch took the place of flying skill or initiative. The USSR exported this doctrine to most of the nations it supplied fighters to during the Cold War years.

Mikoyan-Gurevich MiG-19S 'Farmer'

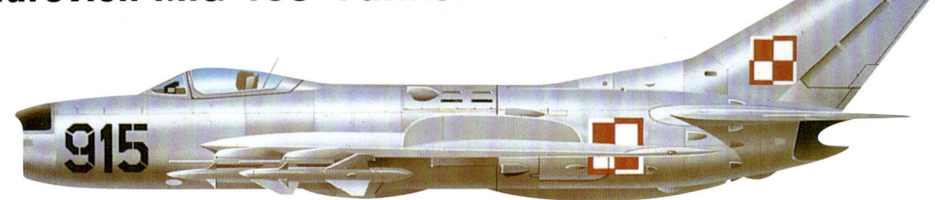

With the unveiling of the MiG-19, the Mikoyan-Gurevich bureau established itself at the forefront of the world's fighter design teams. The first flew in September 1953. With afterburning engines, the MiG-19 became the first supersonic engines in Soviet service. Steadily improved versions culminated in the MiG-19PM, with pylons for four beam-rider air-to-air missiles.

SPECIFICATIONS	
Country of origin:	USSR
Type:	single-seat all-weather interceptor
Powerplant:	two 31.9kN (7165lb) Klimov RD-9B turbojets
Performance:	maximum speed at 9080m (20,000ft) 1480km/h (920mph); service ceiling 17,900m (58,725ft); maximum range at high altitude with two drop tanks 2200km (1367 miles)
Weights:	empty 5760kg (12,698lb); maximum take-off weight 9500kg (20,944lb)
Dimensions:	wingspan 9m (29ft 6.5in); length 13.58m (44ft 7in); height 4.02m (13ft 2.25in); wing area 25 sq m (269.11 sq ft)
Armament:	underwing pylons for four AA-1 Alkali air-to-air-missiles, or AA-2 Atoll

Yakovlev Yak-28P 'Firebar'

The Yak-28P had a similar configuration to the earlier Yak-25/26 family, but had a high shoulder-set wing with the leading edge extended further forward, a taller fin and rudder, revised powerplant in different underwing nacelles and different nosecone. Designed in the late 1950s, it was produced in tactical attack, reconnaissance, electronic countermeasures and trainer versions.

SPECIFICATIONS	
Country of origin:	USSR
Type:	two-seat all-weather interceptor
Powerplant:	two 66.8kN (13,669lb) Tumanskii R-11 turbojets
Performance:	maximum speed 1180km/h (733mph); service ceiling 16,000m (52,495ft); maximum combat radius 925km (575 miles)
Weights:	maximum take-off 19,000kg (41,890lb)
Dimensions:	wingspan 12.95m (42ft 6in); length (long-nose late production) 23m (75ft 7.5in); height 3.95m (12ft 11.5in); wing area 37.6 sq m (404.74 sq ft)
Armament:	four underwing pylons for two AA-2 'Atoll', AA-2-2 ('Advanced Atoll') or AA-3 ('Anab') air-to-air missiles

INTERCEPTORS TIMELINE	1953	1962	1967

SOVIET INTERCEPTORS

Sukhoi Su-15 'Flagon'

The Su-15 was developed to a requirement for a successor to the Sukhoi Su-11 and strongly resembles that aircraft in the wings and tail. The 'Flagon A' entered IA-PVO Strany service in 1967, and 1500 Sukhoi 15s in all versions are estimated to have been built. All these aircraft served with Soviet air arms, since the aircraft was never made available for export.

SPECIFICATIONS

Country of origin:	USSR
Type:	single-seat all-weather interceptor
Powerplant:	two 60.8kN (13,668lb) Tumanskii R-11F2S-300 turbojets
Performance:	maximum speed above 11,000m (36,090ft) approximately 2230km/h (1386mph); service ceiling 20,000m (65,615ft); combat radius 725km (450 miles)
Weights:	empty (estimated) 11,000kg (24,250lb); maximum take-off 18,000kg (39,680lb)
Dimensions:	wingspan 8.61m (28ft 3in); length 21.33m (70ft); height 5.1m (16ft 8.5in); wing area 36 sq m (387.5 sq ft)
Armament:	four external pylons for two R8M medium-range air-to-air missiles ouboard and two AA-8 'Aphid' short-range AAMs inboard, plus two under-fuselage pylons for 23mm UPK-23 cannon pods or drop tanks

Mikoyan-Gurevich MiG-25P 'Foxbat-A'

Reports of the development of a long-range, high-speed strategic bomber in the US in the late 1950s (the B-70 Valkyrie) prompted the Soviet authorities to prioritize the development of an interceptor to match it. The prototypes blazed a trail of world records in 1965–67, and when the MiG-25P entered service in 1970 it far outclassed any Western aircraft in terms of speed and height.

SPECIFICATIONS

Country of origin:	USSR
Type:	single-seat interceptor
Powerplant:	two 100kN (22,487lb) Tumanskii R-15B-300 turbojets
Performance:	maximum speed at altitude about 2974km/h (1848mph); service ceiling over 24,385m (80,000ft); combat radius 1130km (702 miles)
Weights:	empty 20,000kg (44,092lb); maximum take-off 37,425kg (82,508lb)
Dimensions:	wingspan 14.02m (45ft 11.75in); length 23.82m (78ft 1.75in); height 6.10m (20ft 0.5in); wing area 61.40 sq m (660.9 sq ft)
Armament:	external pylons for four air-to-air missiles in the form of either two each of the IR- and radar-homing AA-6 'Acrid', or two AA-7 'Apex' and two AA-8 'Aphid' weapons

Mikoyan-Gurevich MiG-31 'Foxhound-A'

The MiG-31 was developed during the 1970s from the impressive MiG-25 'Foxbat' to counter the threat from low-flying cruise missiles and bombers. In fact, the new aircraft was a vast improvement over its older stablemate, with tandem seat cockpit, IR search and tracking sensor, and the Zaslon 'Flash Dance' pulse-Doppler radar providing fire-and-forget engagement capability.

SPECIFICATIONS

Country of origin:	USSR
Type:	two-seat all weather interceptor and ECM aircraft
Powerplant:	two 151.9kN (34,171lb) Soloviev D-30F6 turbofans
Performance:	maximum speed 3000km/h (1865mph); service ceiling 20,600m (67,600ft); combat radius 1400km (840 miles)
Weights:	empty 21,825kg (48,415lb); max take-off 46,200kg (101,850lb)
Dimensions:	wingspan 13.46m (44ft 2in); length 22.68m (74ft 5.25in); height 6.15m (20ft 2.25in); wing area 61.6 sq m (663 sq ft)
Armament:	one 23mm cannon, provision for missiles, ECM pods or drop tanks

1975

MiG-21 'Fishbed'

Eclipsed only by the MiG-15, the Mikoyan MiG-21 'Fishbed' is the most numerous jet fighter ever produced, with an estimated 11,000 built in Russia, Czechoslovakia, India and China, many of the latter unlicensed copies made after the Sino-Soviet split. The first, unsophisticated, MiG-21F models with simple ranging radars entered service in 1959. Continuous development saw armament options increased and various spines and bulges added to house fuel, avionics and electronic equipment.

MiG-21 'Fishbed'

Czech aircraft maker Aero Vodochody began licence production of the initial MiG-21F-13 as the S-106 in 1960. They completed 194 of them, but the aircraft shown is a later Russian-built MiG-21MF. The Czech Republic replaced its last MiG-21s in 2006 with Saab Gripens.

SPECIFICATIONS	
Country of origin:	USSR
Type:	single-seat all-weather multi role fighter
Powerplant:	one 60.8kN (13,668lb) thrust Tumanskii R-13-300 afterburning turbojet
Performance:	maximum speed above 11,000m (36,090ft) 2229km/h (1385mph); service ceiling 17,500m (57,400ft); range on internal fuel 1160km (721 miles)
Weights:	empty 5200kg (11,464lb); maximum take-off 10,400kg (22,925lb)
Dimensions:	wingspan 7.15m (23ft 5.5in); length (including probe) 15.76m (51ft 8.5in); height 4.1m (13ft 5.5in); wing area 23 sq m (247.58 sq ft)
Armament:	one 23mm cannon, provision for about 1500kg (3307kg) of stores, including air-to-air missiles, rocket pods, napalm tanks, or drop tanks

MiG-21PF 'Fishbed-D'

This Indian Air Force MiG-21PFL is depicted wearing an improvised 'tiger' camouflage scheme at around the time of the 1971 Indo-Pakistan War. It carries AA-2 'Atol' air-to-air missiles and a centreline GSh-9 30mm cannon pod.

SPECIFICATIONS	
Country of origin:	USSR
Type:	single-seat all-weather multi role fighter
Powerplant:	one 60.8kN (13,668lb) thrust Tumanskii afterburning turbojet
Performance:	max speed 2050km/h (1300mph); service ceiling 17,000m (57,750ft); range 1800km (1118 miles)
Weights:	max loaded 9400kg (20,723lb)
Dimensions:	wingspan 7.15m (23ft 5.5in); length (including probe) 15.76m (51ft 8.5in); height 4.1m (13ft 5.5in); wing area 23 sq m (247.58 sq ft)
Armament:	one 23mm cannon, provision for about 1500kg (3307kg) of stores, including air-to-air missiles, rocket pods, napalm tanks, or drop tanks

MiG-21PFMA 'Fishbed'

The MiG-21 was supplied to approximately 55 international operators past and present, not counting Chinese variants. Poland had several hundred 'Fishbeds' in service, including the MiG-21PF.

SPECIFICATIONS

Country of origin:	USSR
Type:	single-seat all-weather multi role fighter
Powerplant:	one 60.8kN (13,668lb) thrust Tumanskii R-11F2S-300 afterburning turbojet
Performance:	maximum speed above 11,000m (36,090ft) 2229km/h (1385mph); service ceiling 17,500m (57,400ft); range on internal fuel 1160km (721 miles)
Weights:	empty 5200kg (11,464lb); maximum take-off 10,400kg (22,925lb)
Dimensions:	wingspan 7.15m (23ft 5.5in); length (including probe) 15.76m (51ft 8.5in); height 4.1m (13ft 5.5in); wing area 23 sq m (247.58 sq ft)
Armament:	one 23mm cannon, provision for about 1500kg (3307kg) of stores, including air-to-air missiles, rocket pods, napalm tanks or drop tanks

MiG-21U 'Mongol'

Aside from the airframe modifications necessary to accommodate the instructor, the 21U is similar in configuration to the initial major production version, the 21F. The first prototype is reported to have flown in 1960. Variations from the single-seater include a one-piece forward airbrake, repositioning of the pilot boom, and adoption of larger mainwheels first introduced on the MiG-21PF.

SPECIFICATIONS

Country of origin:	USSR
Type:	two-seat trainer
Powerplant:	one 5950kg (13,118lb) Tumanskii R-11F2S-300 turbojet
Performance:	maximum speed 2145km/h (1333mph); service ceiling 17,500m (57,400ft); range 1160km (721 miles)
Weights:	not released
Dimensions:	wingspan 7.15m (23ft 5.5in); length (including probe) 15.76m (51ft 8.5in); height 4.1m (13ft 5.5in); wing area 23 sq m (247.58 sq ft)
Armament:	six 7.92mm (.3in) MGs; 1000kg (2205lb) bomb load

Chengdu F-7

China and Russia's initial cooperation on MiG-21 development ended when relations soured in 1962. Chinese manufacturers Shenyang, then Chengdu carried on with production of unlicensed models as the Jiang-7 (J-7) into the 1990s. China supplied J-7s (under the export designation F-7) to a dozen nations, mainly in Africa.

SPECIFICATIONS

Country of origin:	USSR
Type:	single-seat all-weather multi role fighter
Powerplant:	one 66.7kN (14,815lb) thrust Liyang Wopen-13F afterburning turbojet
Performance:	max speed 2229km/h (1385mph); service ceiling 17,500m (57,400ft); range 1160km (721 miles)
Weights:	empty 5200kg (11,464lb); maximum take-off 10,400kg (22,925lb)
Dimensions:	wingspan 7.15m (23ft 5.5in); length (including probe) 15.76m (51ft 8.5in); height 4.1m (13ft 5.5in); wing area 23 sq m (247.58 sq ft)
Armament:	one 23mm cannon, provision for about 1500kg (3307kg) of stores, including air-to-air missiles, rocket pods, napalm tanks, or drop tanks

Flying Boxcars

From the late 1940s, Western air forces started to introduce some of the first purpose-designed military transport aircraft. The Fairchild C-82 Packet began a fashion for twin-boomed transports with capacious fuselages that could be loaded via an integral ramp at the rear. A Boxcar could lift a light tank, armoured car or artillery gun and roll it straight off at a forward airstrip, or drop large numbers of paratroopers.

Fairchild R4Q-1 Boxcar

The C-82 Packet was inadequate for some military needs and the design was soon revised into the C-119 Flying Boxcar, of which nearly 1200 were built for the US and export. The US Marine Corps operated 140 Boxcars, including 41 R4Q-1s like the one illustrated here in the markings of VMR-253.

SPECIFICATIONS	
Country of origin:	United States
Type:	twin-engined military airlifter
Powerplant:	two 2535kW (3400hp) Wright R-3350-36W piston engines
Performance:	maximum speed 402km/h (250mph); range 3219 km (2000 miles); service ceiling 7300m (23,950ft)
Weights:	empty 18,136kg (39,983lb); maximum 33,747kg (74,700lb)
Dimensions:	span 33.3m (109ft 3in); length 26.37m (86ft 6in); height 8m (26ft 3in); wing area 134.43 sq m (1447 sq ft)
Armament:	none

Fairchild C-119G Packet

The C-119G was used by India from the early 1960s until the mid-1980s. Indian C-119s were modified with an auxiliary jet engine on top for boost in 'hot and high' operations. This was a Bristol Orpheus, rather than the Westinghouse J34 jet pack used on US versions.

SPECIFICATIONS	
Country of origin:	United States
Type:	twin-engined military airlifter (C-119G Indian)
Powerplant:	two 2610kW (3500hp) Wright R-3350-85 piston engines and one 22kN (4850lb) thrust Bristol Orpheus turbojet
Performance:	maximum speed 470km/h (292mph); range 3669km (2280 miles); service ceiling 7300m (23,950ft)
Weights:	empty 18,136kg (39,983lb); maximum 33,747kg (74,700lb)
Dimensions:	span 33.3m (109ft 3in); length 26.37m (86ft 6in); height 8m (26ft 3in); wing area 134.43 sq m (1447 sq ft)
Armament:	none

FLYING BOXCARS TIMELINE

1944    1947    1949

FLYING BOXCARS

Blackburn Beverley C.1

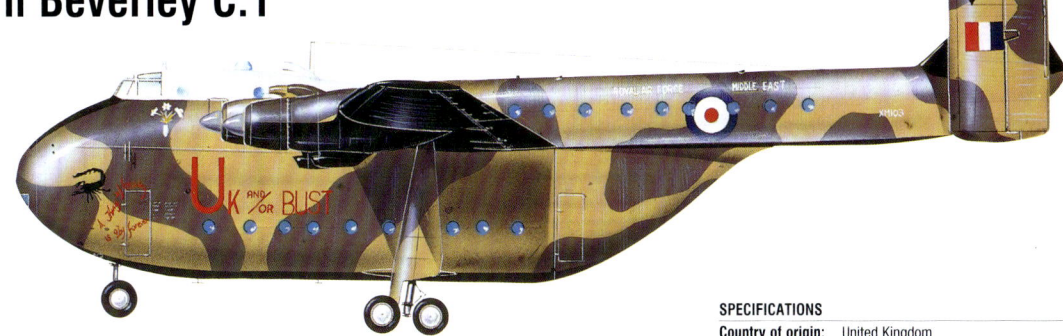

The lumbering Beverley was used mainly to service British garrisons in the Middle East and Far East. It could carry 94 troops (some of them seated in the tail booms) or various vehicles. It could drop 70 paratroopers or up to 11,340kg (25,000lb) of supplies and operate from rough airstrips.

SPECIFICATIONS

Country of origin:	United Kingdom
Type:	twin-engined military airlifter
Powerplant:	four 2125kW (2850hp) Bristol Centaurus 173 radial piston engines
Performance:	maximum speed 383km/h (238mph); range 5938km (3690 miles); service ceiling 4875m (16,000ft)
Weights:	35,940kg (79,234lb); maximum 64,864kg (143,000lb)
Dimensions:	span 49.38m (162ft); length 30.3m (99ft 5in); height 11.81m (38ft 9in); wing area 271 sq m (2916 sq ft)
Armament:	none

Nord 2501D Noratlas

First flown in 1949, the Nord Noratlas was slow to get into service but became an important European tactical transport built in both France and Germany. This Noratlas was one of many supplied from Luftwaffe stocks to African countries, and served with the Niger National Flight.

SPECIFICATIONS

Country of origin:	Germany
Type:	twin-engined military transport
Powerplant:	two 1520kW (2040lb) Bristol Hercules 738/739 radial engines
Performance:	maximum speed 582km/h (362mph); range 2500km (1550 miles); service ceiling 7100m (23,300ft)
Weights:	13,075kg (28,825lb); maximum 22,000kg (48,500lb)
Dimensions:	span 32.5m (106ft 8in); length 21.96m (72ft 1in); height 6m (19ft 8in); wing area 101.2 sq m (1089 sq ft)
Armament:	none

Armstrong Whitworth Argosy C.1

The RAF bought 59 Argosy airlifters from 1962. Known as the 'Whistling Wheelbarrow' for its high-pitched Dart turboprop engines, the Argosy could carry 69 troops or light vehicles. The civilian version had a clamshell door in the nose, but the Argosy C.1 had this sealed and a weather radar in the nose.

SPECIFICATIONS

Country of origin:	United Kingdom
Type:	four-engined military transport aircraft
Powerplant:	four 1820kW (2440hp) Rolls-Royce Dart RDa.8 Mk 10 turboprop engines
Performance:	maximum speed 433km/h (269mph); range 5230km (3250 miles); service ceiling 5500m (18,000ft)
Weights:	empty 4360kg (10,200lb); maximum 46,700kg (103,000lb)
Dimensions:	span 35.05m (115ft); length 27.18m (45ft 3in); height 8.96m (29ft 3in); wing area 135.5 sq m (1458 sq ft)
Armament:	none

1950  1959

COLD WAR MILITARY AIRCRAFT

Grumman Carrier Twins

Grumman has had a presence on almost every US carrier since the 1930s. No less important than its fighter/attack aircraft are its twin-engined support aircraft, which used common components to build a variety of specialized models for different roles. The S2F Tracker ASW aircraft spawned the Tracer AEW aircraft and the Trader transport. The E-2 Hawkeye AEW aircraft and the C-2 Greyhound transport have both served for 45 years.

S-2F Tracker

The S2F Tracker, or 'Stoof', was the US Navy's primary anti-submarine warfare (ASW) platform from the 1950s to the 1970s. It was also sold to several nations for carrier and land-based use, including Australia, Canada, Argentina and Brazil. Civilian versions are in use as fire bombers.

SPECIFICATIONS	
Country of origin:	United States
Type:	twin-engined carrier-based anti-submarine aircraft
Powerplant:	two 1135kW (1525hp) Wright R-1820-82 radial piston engines
Performance:	maximum speed 438km/h (272mph); range 1558km (968 miles); service ceiling 6949m (22,800ft)
Weights:	empty 7871kg (17,357lb); maximum 11,069kg (24,408lb)
Dimensions:	span 21m (69ft 8in); length 12.8m (42ft); height 4.9m (16ft 3in); wing area 45 sq m (485 sq ft)
Armament:	torpedoes, rockets, depth charges or one Mk 47 or Mk 101 nuclear depth bomb

US-2A Tracker

The US-2A was an S-2 with anti-submarine gear and weapons stripped out and replaced with seats or cargo space for utility duties. It could also tow targets. Japan acquired 47 Trackers for land-based use and converted a number into target tugs.

SPECIFICATIONS	
Country of origin:	United States
Type:	twin-engined carrier-based anti-submarine aircraft
Powerplant:	two 1135kW (1525hp) Wright R-1820-82 radial piston engines
Performance:	maximum speed 438km/h (272mph); range 1558km (968 miles); service ceiling 6949m (22,800ft)
Weights:	empty 7871kg (17,357lb); maximum 11,069kg (24,408lb)
Dimensions:	span 21m (69ft 8in); length 12.8m (42ft); height 4.9m (16ft 3in); wing area 45 sq m (485 sq ft)
Armament:	none

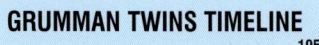   

GRUMMAN TWINS TIMELINE — 1952 — 1960

GRUMMAN CARRIER TWINS

S-2T Tracker

No real fixed-wing replacement exists for the Tracker, particularly for use on smaller carriers. Argentina, Brazil and Taiwan chose to modernize their aircraft in the 1980s. US firm Tracor fitted new turboprop engines and ASW equipment to create the S-2T Turbo Tracker.

SPECIFICATIONS

Country of origin:	United Kingdom
Type:	twin-engined carrier-based anti-submarine aircraft
Powerplant:	two 1141kW (1530hp) Garrett TPE331-15AW turboprop engines
Performance:	maximum speed 500 km/h (311 mph); range 1558km (968 miles); service ceiling 6949m (22,800ft)
Weights:	unknown
Dimensions:	span 21m (69ft 8in); length 12.8m (42ft); height 4.9m (16ft 3in); wing area 45 sq m (485 sq ft)
Armament:	torpedoes, rockets, bombs or depth charges

C-2A Greyhound

In service since 1966, the C-2 Greyhound is used in the carrier onboard delivery (or COD) role, transferring passengers, mail and urgent supplies such as aircraft engines from shore bases to carriers. It shares the wings, engines, landing gear and tail fins with the E-2 Hawkeye.

SPECIFICATIONS

Country of origin:	United States
Type:	twin-engined carrier-based cargo/passenger aircraft
Powerplant:	one 3400kW (4800hp) Allison T56-A-425 turboprop engines
Performance:	maximum speed 553km/h (343mph); range 2400km (1496 miles); service ceiling 10,210m (33,500ft)
Weights:	empty 15,310kg (33,746lb); maximum kg (lb)
Dimensions:	span 24.6m (80ft 7in); length 17.3m (56ft 10in); height 4.85m (15ft 11in); wing area 65 sq m (700 sq ft)
Armament:	none

E-2 Hawkeye

Replacing the E-1 Tracer, a derivative of the S-2 with a fixed radar housing on its back, the E-2 Hawkeye featured a rotating radar dish, or rotodome, mounted on a pylon. The Hawkeye was used to extend the radar range of the carrier battlegroup and direct fighters to intercept air threats.

SPECIFICATIONS

Country of origin:	United States
Type:	twin-engined carrier-based early warning aircraft (E-2C)
Powerplant:	one 3800kW (5100hp) Allison T56-A-427 turboprop engines
Performance:	maximum speed 604km/h (375mph); range 2583km (1605 miles); service ceiling 10,210m (33,500ft)
Weights:	empty 15,310kg (33,746lb); maximum 24,655kg (60,000lb)
Dimensions:	span 24.6m (80ft 7in); length 17.56m (57ft 7in); height 5.58m (18ft 4in); wing area 65 sq m (700 sq ft)
Armament:	none

1964    1989

COLD WAR MILITARY AIRCRAFT

Strategic Air Command

At its peak, the USAF's Strategic Air Command (SAC) had several thousand bombers and tankers as well as strategic reconnaissance aircraft and ballistic missiles. It even had its own fighter aircraft and transports. Under General Curtiss LeMay, SAC maintained an around-the-clock airborne alert of bombers ready to wreak destruction on the Soviet Union on the receipt of coded signals from the White House, an eventuality that never came to pass.

Convair B-36J Peacemaker

The extraordinary Convair B-36 had no fewer than 10 engines: four piston engines and six jets. It was designed to carry some of the extremely large atomic and hydrogen bombs of the time, but also had the capacity to drop up to 80 conventional bombs.

SPECIFICATIONS	
Country of origin:	United States
Type:	ten-engined strategic bomber
Powerplant:	six 2500kW (3800hp) Pratt & Whitney R-4360-53 radial and four 23kN (5200lb) thrust General Electric J47-GE-19 turbojet engines
Performance:	maximum speed 685 km/h (420 mph); range 10,945km (6800 miles); service ceiling 15,000m (48,000ft)
Weights:	empty 77,580kg (171,035lb); maximum 186,000kg (410,000lb)
Dimensions:	span 70.1m (230ft); length 49.40m (162ft 1in); height 14.25m (46ft 9in); wing area 443.3 sq m (4772 sq ft)
Armament:	16 20mm cannon; up to 39,010kg (86,000lb) of bombs

B-47E Stratojet

The B-47 was SAC's first all-jet bomber. Creating a large, high-performance, swept-wing jet with podded engines was a considerable challenge, but allowed a thinner wing than used on contemporary Soviet and British bombers. More than 1800 B-47s were built.

SPECIFICATIONS	
Country of origin:	United States
Type:	six-engined strategic bomber
Powerplant:	six 32.1kN (7200lb) thrust General Electric J47-GE-25 turbojet engines
Performance:	maximum speed 901km/h (560mph); range 5636km (3500 miles); service ceiling 11,978m (39,300ft)
Weights:	empty 35,867kg (79,074lb); maximum 102,512kg (226,000lb)
Dimensions:	span 35.36m (116ft); length 32.6m (107ft 1in); height 8.5m (28ft); wing area 132.7 sq m (1428 sq ft)
Armament:	two 20mm M2 cannon; up to 11,000kg (25,000lb) of conventional or nuclear bombs

STRATEGIC COMMAND TIMELINE
1947  1952  1956

STRATEGIC AIR COMMAND

B-52D Stratofortress

The B-52 has been in continuous service with Strategic Air Command in one form or another since 1955. Development of this remarkable warhorse, which started life as a turboprop-powered project, began in 1945. The aircraft had been designed to carry stand-off nuclear weapons, but in 1964 a rebuilding programme allowed it to carry 105 'iron bombs'.

SPECIFICATIONS	
Country of origin:	United States
Type:	long-range strategic bomber
Powerplant:	eight 44.5kN (10,000lb) Pratt & Whitney J57 turbojets
Performance:	max speed at 7315m (24,000ft) 1014km/h (630mph): service ceiling 13,720–16,765m (45,000–55,000ft); range 9978km (6200 miles)
Weights:	empty 77,200–87,100kg (171,000–193,000lb); loaded 204,120kg (450,000lb)
Dimensions:	wingspan 56.4m (185ft); length 48m (157ft 7in); height 14.75m (48ft 3in); wing area 371.60 sq m (4000 sq ft)
Armament:	remotely controlled tail mounting with four .5in MGs; internal bomb capacity 12,247kg (27,000lb) to 31,750kg (70,000lb)

KC-135A Stratotanker

SAC's fuel-thirsty jet bombers required an equally large fleet of tanker aircraft to support their patrols. The KC-135 was derived from the same prototype that parented the 707 airliner, and over 800 were built, the majority of them as KC-135As, illustrated.

SPECIFICATIONS	
Country of origin:	United States
Type:	four-engined aerial tanker
Powerplant:	four 244.7kN (55,000lb) Pratt & Whitney J57-59W thrust turbojet engines
Performance:	cruising speed 853km/h (530mph); range 4627km (2875 miles); service ceiling 10,980m (36,000ft)
Weights:	empty 44,665kg (98,465lb); maximum 134,720kg (29,000lb)
Dimensions:	wingspan 39.88m (130ft 10in); length 41.53m (136ft 3in); height 12.7m (41ft 8in); wing area 226.03 sq m (2,433 sq ft)
Armament:	none

Convair B-58 Hustler

The B-58 was the first supersonic bomber and the first to reach Mach 2. It was the first aircraft constructed mainly from a stainless-steel honeycomb sandwich, the first to have a slim body and fat payload pod so that when the load was dropped, the aircraft became slimmer and lighter. The first flight was made on 11 November 1956, and development continued for almost three years.

SPECIFICATIONS	
Country of origin:	United States
Type:	three-seat supersonic bomber
Powerplant:	four 69.3kN (15,600lb) General Electric J79-5B turbojets
Performance:	maximum speed 2125km/h (1385mph); service ceiling 19,500m (64,000ft); range on internal fuel 8248km (5125 miles)
Weights:	empty 25,200kg (55,560lb); maximum take-off 73,930kg (163,000lb)
Dimensions:	wingspan 17.31m (56ft 10in); length 29.5m (96ft 9in); height 9.6m (31ft 5in); wing area 145.32 sq m (1542 sq ft)
Armament:	one 20mm T171 Vulcan rotary cannon in radar-aimed tail barbette, nuclear or conventional weapons in disposable underfuselage pod

Dragon Lady: Lockheed U-2

The U-2 spyplane was initiated in the early 1950s with funding by the Central Intelligence Agency (CIA). The U-2 was designed by the 'Skunk Works' secret research division of Lockheed, led by 'Kelly' Johnson. The U-2 initially operated under the cover of a high-altitude research aircraft programme until CIA pilot Gary Powers was shot down on 1 May 1960 over Sverdlovsk. Later versions of the 'Dragon Lady' remain in service today.

U-2A

Under Project Aquatone, wearing civil registrations, the CIA's U-2s conducted reconnaissance overflights of the Soviet Union from forward bases in Turkey, Pakistan and Norway. They also flew over Cuba during the 1962 Missile Crisis and other hotspots until 1974. N803X was a U-2A.

SPECIFICATIONS	
Country of origin:	United States
Type:	high-altitude spyplane
Powerplant:	one 48.93kN (11,000lb) thrust Pratt & Whitney J75-P-37A turbojet engine
Performance:	maximum speed 795km/h (494mph); range 3542km (2200 miles); service ceiling 16,763m (55,000ft)
Weights:	empty 5306kg (11,700lb); maximum 9523kg (21,000lb)
Dimensions:	span 24.3m (80ft); length 15.1m (49ft 7in); height 3.9m (13ft); wing area 52.49 sq m (565 sq ft)
Armament:	none

U-2C

With a more powerful version of the J75 engine, the U-2C had a much higher altitude, putting it out of the range of most fighters and missiles. As extra insurance, it had an exhaust deflector beneath the tailpipe to mask the engine heat signature from infrared sensors.

SPECIFICATIONS	
Country of origin:	United States
Type:	high-altitude spyplane
Powerplant:	one 75.62kN (17,000lb) thrust Pratt & Whitney J75-P-13 turbojet engine
Performance:	maximum speed 850km/h (530mph); range 4830km (2610 miles)); service ceiling 25,930m (85,000ft)
Weights:	empty 5306kg (11,700lb); maximum 9523kg (21,000lb)
Dimensions:	span 24.3m (80ft); length 15.1m (49ft 7in); height 3.9m (13ft); wing area 52.49 sq m (565 sq ft)
Armament:	none

DRAGON LADY: LOCKHEED U-2

U-2CT

Designed along the principles of a glider, the U-2 was tricky to handle at high altitude and on landing. A number were lost in crashes with new pilots, so two U-2Cs were converted to U-2CT trainers with a second, raised cockpit. Only two of this version were built.

SPECIFICATIONS

Country of origin:	United States
Type:	high-altitude spyplane
Powerplant:	one 75.62kN (17,000lb) thrust Pratt & Whitney J75-P-13 turbojet engine
Performance:	maximum speed 850km/h (530mph); range 4830km (2610 miles); service ceiling 25,930m (85,000ft)
Weights:	unknown
Dimensions:	span 24.3m (80ft); length 15.1m (49ft 7in); height 3.9m (13ft); wing area 52.49 sq m (565 sq ft)
Armament:	none

U-2D

The U-2D was another two-seat version, although intended for operational use rather than training. Under Project Low Card, an optical spectrometer was fitted in a housing behind the cockpit. It was used to search for the plume of an ICBM launch. An operator was seated in the former camera bay.

SPECIFICATIONS

Country of origin:	United States
Type:	high-altitude spyplane
Powerplant:	one 48.93kN (11,000lb) thrust Pratt & Whitney J75-P-37A turbojet engine
Performance:	maximum speed 795km/h (494mph); range 3542km (2200 miles); service ceiling unknown
Weights:	unknown
Dimensions:	span 24.3m (80ft); length 15.1m (49ft 7in); height 3.9m (13ft); wing area 52.49 sq m (565 sq ft)
Armament:	none

U-2R

The U-2R, first flown in 1967, is significantly larger and more capable than the original aircraft. A distinguishing feature of these aircraft is the addition of a large instrumentation 'superpod' under each wing. It was designed for standoff tactical reconnaissance in Europe.

SPECIFICATIONS

Country of origin:	United States
Type:	single-seat high-altitude reconnaissance aircraft
Powerplant:	one 75.6kN (17,000lb) Pratt & Whitney J75-P-13B turbojet
Performance:	maximum cruising speed at more than 21,335m (70,000ft); operational ceiling 27,430m (90,000ft); maximum range 10,050km (6250 miles)
Weights:	empty 7031kg (15,500lb); maximum take-off weight 18,733kg (41,300lb)
Dimensions:	wingspan 31.39m (103ft); length 19.13m (62ft 9in); height 4.88m (16ft); wing area 92.9 sq m (1000 sq ft)
Armament:	none

Dassault Mirage III/5

Dassault's delta-winged Mirage III series began as the experimental Delta Mystère in 1955 and evolved into one of the most combat-proven supersonic fighters ever built. The first model, the Mirage IIIC interceptor, was improved as the Mirage IIIE, which had ground-attack capability and even more sales success. The Mirages 5 and 50 were dedicated export versions sold widely in Africa, the Middle East and South America. Further derivatives were produced by foreign users.

Dassault Mirage IIIC

The Mirage IIIC entered service in 1961, beginning over 25 years of service. The last Armée de l'Air unit to use them was EC 3/10 'Vexin', based at Djibouti-Ambouli until 1988. Contrasting with most of its unpainted metal home-based brethren, the African-based IIICs wore desert camouflage.

SPECIFICATIONS	
Country of origin:	France
Type:	single-engined interceptor fighter
Powerplant:	one 58.72kN (13,200lb) thrust SNECMA Atar 09B-3 afterburning turbojet engine and one 16.46kN (3700lb) thrust auxiliary SEFR 841 rocket motor
Performance:	maximum speed 2350km/h (1460mph); service ceiling 17,000m (55,755 ft); range 2012km (1250 miles)
Weights:	empty 6142kg (13,540lb); maximum 11,676kg (25,740lb)
Dimensions:	span 8.26m (27ft 2in); length 14.91m (48ft 10in); height 4.6m (14ft 10in); wing area 34.84 sq m (375 sq ft)
Armament:	two 30mm DEFA cannon; one Matra R.511 or R.530 AAM, up to 2295kg (5060lb) of bombs

Dassault Mirage IIIO

Australia considered purchasing a Mirage version powered by the Rolls-Royce Avon, but eventually chose the standard Atar engine for its IIIOs, which were generally similar to the IIIE model. This RAAF Mirage was with 77 Squadron, RAF, when based at Butterworth in Malaysia.

SPECIFICATIONS	
Country of origin:	France
Type:	single-engined interceptor fighter
Powerplant:	one 62.63kN (14,080lb) thrust SNECMA Atar 9B3 afterburning turbojet
Performance:	maximum speed 2350km/h (1460mph); range 1006km (625 miles); service ceiling 18,105m (59,400ft)
Weights:	empty 7035kg (15,510lb); maximum 13,671kg (30,140lb)
Dimensions:	span 8.26m (27ft 2in); length 14.91m (48ft 10in); height 4.6m (14ft 10in); wing area 34.84 sq m (375 sq ft)
Armament:	two 30mm DEFA 552A cannon; various AAMs, up to 2295kg (10,725lb) of bombs

Dassault Mirage IIIR

A reconnaissance version of the Mirage IIIE entered service in 1963 as the Mirage IIIR, replacing Republic RF-84F Thunderflashes in French service. This example served with EC 3/33 'Moselle' at Strasbourg-Entzheim. Export models were sold to South Africa, Switzerland, Abu Dhabi and Pakistan.

SPECIFICATIONS	
Country of origin:	France
Type:	single-engined reconnaissance aircraft
Powerplant:	one 58.72kN (13,200lb) thrust SNECMA Atar 09C afterburning turbojet engine
Performance:	maximum speed 1390km/h (864mph); range 1304km (810 miles); service ceiling 17,045m (55,921ft)
Weights:	empty 6608kg (14,569lb); maximum 13,718kg (30,242lb)
Dimensions:	span 8.26m (27ft 2in); length 15.54m (51ft); height 4.6m (14ft 10in); wing area 34.84 sq m (375 sq ft)
Armament:	none

Dassault Mirage 5M

In 1977, Zaire ordered 14 Mirage 5Ms, but took delivery of only eight. They served with 211 Escadrille of the Force Aérienne Zairoise at Kamina. The Mirage 5 was optimized for ground attack with more weapons pylons and a simpler radar system.

SPECIFICATIONS	
Country of origin:	France
Type:	single-engined fighter-bomber
Powerplant:	one 58.72kN (13,200lb) thrust SNECMA Atar 09C afterburning turbojet engine
Performance:	maximum speed 2350km/h (1460mph); range 1307km (812 miles); service ceiling 16,093m (52,800ft)
Weights:	empty 6586kg (14,520lb); maximum 13,671kg (30,140lb)
Dimensions:	span 8.26m (27ft 2in); length 15.65m (51ft 4in); height 2.87m (14ft 10in); wing area 34.84 sq m (375 sq ft)
Armament:	two 30mm DEFA 552A cannon; various AAMs, up to 3991kg (8800lb) of bombs

Dassault Mirage 5SDE

Saudi Arabia purchased about 30 Mirage 5s in 1972 on behalf of Egypt for diplomatic reasons. Although designated Mirage 5SDEs, they were essentially the same as Mirage IIIEs. Too late to participate in the Yom Kippur War with Israel, they saw combat in a short conflict with Libya in 1977.

SPECIFICATIONS	
Country of origin:	France
Type:	single-engined interceptor fighter
Powerplant:	one 62.63kN (14,080lb) thrust SNECMA Atar 9B3 afterburning turbojet
Performance:	maximum speed 2350km/h (1460mph); range 1006km (625 miles); service ceiling 18,105m (59,400ft)
Weights:	empty 7035kg (15,510lb); maximum 13,671kg (30,140lb)
Dimensions:	span 8.26m (27ft 2in); length 14.91m (48ft 10in); height 4.6m (14ft 10in); wing area 34.84 sq m (375 sq ft)
Armament:	two 30mm DEFA 552A cannon; various AAMs, up to 2295kg (10,725lb) of bombs

Mirage Modified

Denied the delivery of 50 Mirage 5Js that it had ordered and paid for, the Israelis sought to acquire plans of the Mirage and build it themselves. Secretly helped by Dassault, Israel Aircraft Industries was set up to produce a version called the *Nesher* (Eagle), which flew in 1969. The *Kfir* (Lion Cub), with the US J79 engine and canard foreplanes, followed. More recently, Israel has assisted South Africa with its Cheetah and Chile with the Pantera.

IAI Kfir

The original Mirage IIIC actually owes much of its inception to the close ties between Dassault and Israel. During the Six-Day War of 5–10 June 1967, this aircraft performed magnificently. Later, Israel Aircraft Industries set about devising an improved version of the Mirage III. The company adapted the airframe to take a General Electric J79 turbojet, under a programme dubbed Black Curtain. Some of these aircraft participated in the 1973 Yom Kippur War.

SPECIFICATIONS	
Country of origin:	Israel
Type:	single-seat interceptor
Powerplant:	one 79.6kN (17,900lb) General Electric J79-J1E turbojet
Performance:	maximum speed above 11,000m (36,090ft) 2445km/h (1520mph); service ceiling 17,680m (58,000ft); combat radius as interceptor 346km (215 miles)
Weights:	empty 7285kg (16,090lb); maximum take-off weight 16,200kg (35,715lb)
Dimensions:	wingspan 8.22m (26ft 11.5in); length 15.65m (51ft 4.25in); height 4.55m (14ft 11.25in); wing area 34.8 sq m (374.6 sq ft)
Armament:	one IAI (DEFA) 30mm cannon; nine external hardpoints with provision for up to 5775kg (12,732lb) of stores; for interception duties AIM-9 Sidewinder air-to-air missiles, or indigenously produced AAMs such as the Shafrir or Python

IAI Kfir C1 (F-21A)

The Kfir represented a significant improvement over the Mirage III. The installation of the J79 engine necessitated a redesign of the fuselage and the addition of a ram-cooling inlet ahead of the fin. The shorter engine resulted in a shorter rear fuselage, but the nose was lengthened to hold a comprehensive avionics suite.

SPECIFICATIONS	
Country of origin:	Israel
Type:	single-seat interceptor/ground attack aircraft
Powerplant:	one 79.6kN (17,900lb) General Electric J79-J1E turbojet
Performance:	maximum speed 2445km/h (1520mph); service ceiling 17,680m (58,000ft); combat radius as interceptor 346km (215 miles)
Weights:	empty 7285kg (16,090lb); maximum take-off weight 16,200kg (35,715lb)
Dimensions:	wingspan 8.22m (26ft 11.5in); length 15.65m (51ft 4.25in); height 4.55m (14ft 11.25in); wing area 34.8 sq m (374.6 sq ft)
Armament:	one IAI (DEFA) 30mm cannon; provision for 5775kg (12,732lb) of bombs, rockets, napalm tanks and missiles

MIRAGE MODIFIED

IAI Kfir TC2

The Kfir TC2 two-seat trainer first flew in 1982. It retained all the operational avionics of the C1, but had less fuel capacity. The long nose was angled down so as to not hamper the pilot's view during landing. Some TC2s were exported and the survivors were modified to TC7 standard.

SPECIFICATIONS

Country of origin:	Israel
Type:	single-seat interceptor
Powerplant:	one 79.6kN (17,900lb) General Electric J79-J1E turbojet
Performance:	maximum speed above 11,000m (36,090ft) 2445km/h (1520mph); service ceiling 17,680m (58,000ft); combat radius as interceptor 346km (215 miles)
Weights:	unknown
Dimensions:	wingspan 8.22m (26ft 11.5in); length 16.15m (53ft); height 4.55m (14ft 11.25in); wing area 34.8 sq m (374.60 sq ft)
Armament:	one 30mm cannon; provision for up to 5775kg (12,732lb) of stores; for interception duties AIM-9 Sidewinder air-to-air missiles, or Shafrir or Python AAMs

Atlas Cheetah D

The Atlas Cheetah is, in fact, the South African answer to an international arms embargo imposed on the country in 1977, which prevented the SAAF from importing a replacement for its ageing fleet of Mirage IIIs. The programme involved replacing nearly 50 per cent of the airframe. Production aircraft are modified from both single-seaters and twin seaters.

SPECIFICATIONS

Country of origin:	South Africa
Type:	one/two-seat combat and training aircraft
Powerplant:	one 70.6kN (15,873lb) SNECMA Atar 9K-50 turbojet
Performance:	maximum speed above 12,000m (39,370ft) 2337 km/h (1452mph); service ceiling 17,000m (55,775ft)
Weights:	not revealed
Dimensions:	wingspan 8.22m (26ft 11.5in); length 15.4m (50ft 6.5in); height 4.25m (13ft 11.5in); wing area 35 sq m (376.75sq ft)
Armament:	two 30mm DEFA cannon, Armscor V3B and V3C Kukri air-to-air missiles, provision for external stores such as cluster bombs, laser designator pods, and rockets

Dassault Mirage 50C (Pantera)

Chile purchased 17 Mirage 50s, including three two-seat 50DC models in 1982–3. Local firm ENAER began upgrading them as the *Pantera* (Panther) with Israeli help soon afterwards. The modernization package includes canard foreplanes and the ability to use a range of Israeli precison weapons.

SPECIFICATIONS

Country of origin:	France
Type:	single-engined fighter-bomber
Powerplant:	one 70.5kN (15,850lb) thrust SNECMA Atar 09K-50 afterburning turbojet engine
Performance:	maximum speed at high altitude 2350km/h (1460mph); service ceiling 18,105m (59,400ft); range 1408km (875 miles)
Weights:	empty 7136kg (15,730lb); maximum take-off 13,671kg (30,140lb)
Dimensions:	span 8.26m (27ft 2in); length 15.65m (51ft 4in); height 2.87m (14ft 10in); wing area 34.84 sq m (375 sq ft)
Armament:	two 30mm DEFA 552A cannon; various AAMs, up to 3991kg (8800lb) of bombs

Indo-Pakistani Wars

Partitioned at the time of independence in 1947, India and Pakistan fought major wars in 1965 and 1971, and have been involved in low-intensity conflict over the territory of Kashmir ever since. Air power played a vital role in the two major wars, and there were numerous air combats between Indian Air Force (IAF) and Pakistan Air Force (PAF) fighters, which were among the first battles involving supersonic aircraft.

North American F-86F Sabre

Pakistan's Sabre pilots made claims for 19 IAF aircraft in 1965, the majority of them Hunters. Fifteen of these claims match Indian losses. In 1971, they lost 28 Sabres to various causes. Pakistan received 120 Sabres, including Canadair Sabre Mk.6s (illustrated).

SPECIFICATIONS	
Country of origin:	United States/Canada
Type:	(CL-13B Sabre Mk. 6) single-engined jet fighter
Powerplant:	one 32.36kN (7275lb) thrust Avro Orenda Mark 14 turbojet engine
Performance:	maximum speed 965km/h (600mph); combat radius 530km (329 miles); service ceiling 14,600m (48,000ft)
Weights:	empty 4816 kg (10,500 lb); maximum 6628 kg (14,613 lb)
Dimensions:	span 11.58m (39ft); length 11.4m (37ft 6in); height 4.4m (14ft 8in); wing area 28.06 sq m (302 sq ft)
Armament:	six .5in (12.7mm) machine guns

Shenyang F-6 (MiG-19 'Farmer')

Pakistan had three squadrons of Shenyang F-6As (MiG-19s) in its inventory by 1971. Unusually for a non-Western type, they were equipped with AIM-9 Sidewinder missiles. In 1971, four F-6s were lost, one of them in air combat, but they claimed five Indian jets in return.

SPECIFICATIONS	
Country of origin:	USSR
Type:	single-seat all-weather interceptor
Powerplant:	two 31.9kN (7165lb) Klimov RD-9B turbojets
Performance:	max speed 1480km/h (920mph); service ceiling 17,900m (58,725ft); range 2200km (1367 miles)
Weights:	empty 5760kg (12,698lb); maximum take-off weight 9500kg (20,944lb)
Dimensions:	wingspan 9m (29ft 6.5in); length 13.58m (44ft 7in); height 4.02m (13ft 2.25in); wing area 25 sq m (269.11 sq ft)
Armament:	underwing pylons for four AA-1 Alkali air-to-air-missiles, or AA-2 Atoll

INDIA-PAKISTAN

Hawker Hunter F.56

Hunters were the most numerous combat aircraft in the IAF inventory in 1965, and equipped six squadrons in each war. Able to carry bombs or rockets as well as its heavy cannon, it was a powerful ground-attack platform. In one, 1971 battle Hunters were credited with destroying a Pakistani tank division.

SPECIFICATIONS	
Country of origin:	United Kingdom
Type:	single-seat fighter
Powerplant:	one 45.13kN (10,145lb) thrust Rolls-Royce Avon 207 turbojet engine
Performance:	maximum speed at sea level 1144km/h (710mph); service ceiling 15,240m (50,000ft); range on internal fuel 689km (490 miles)
Weights:	empty 6405kg (14,122lb) maximum 17,750kg (24,600lb)
Dimensions:	wingspan 10.26m (33ft 8in); length 13.98m (45ft 10.5in); height 4.02m (13ft 2in); wing area 32.42 sq m (349 sq ft)
Armament:	four 30mm Aden Cannon; up to 2722kg (6000lb) of bombs or rockets

Hawker Sea Hawk FB.Mk 3

India bought Sea Hawks from the United Kingdom and Germany in 1960 and used them in both wars. They flew from shore bases in 1965 and from the carrier INS *Vikrant* in 1971. Their most notable mission in 1971 was an attack on the port of Chittagong and the Pakistani shipping there.

SPECIFICATIONS	
Country of origin:	United Kingdom
Type:	single-seat carrier based fighter-bomber
Powerplant:	one 24kN (5400lb) Rolls-Royce Nene 103 turbojet
Performance:	maximum speed at sea level 969km/h (602mph); service ceiling 13,565m (44,500ft); combat radius (clean) 370km (230 miles)
Weights:	empty 4409kg (9720lb); maximum take-off weight 7348kg (16,200lb)
Dimensions:	wingspan 11.89m (39ft); length 12.09m (39ft 8in); height 2.64m (8ft 8in); wing area 25.83 sq m (278 sq ft)
Armament:	four 20mm Hispano cannon; plus underwing hardpoints with provision for four 227kg (500lb) bombs, or two 227kg (500lb) bombs and 20 three-inch or 16 five-inch rockets

English Electric Canberra B.66

India ordered 80 Canberras in 1957 and acquired more in later years, including 10 refurbished B(I).66 models as illustrated. In 1965, Indian Canberras bombed Pakistan's air bases by night, but also lost some of their number on the ground to PAF attacks.

SPECIFICATIONS	
Country of origin:	United Kingdom
Type:	twin-engined jet bomber
Powerplant:	two 33.23kN (7490lb) thrust turbojet engine Avon R.A.7 Mk,109 turbojet engines
Performance:	maximum speed 933km/h (580mph); range 5440km (3380 miles); service ceiling 15,000m (48,000ft)
Weights:	9820kg (21,650lb); maximum 24,948kg (55,000lb)
Dimensions:	span 19.51m (65ft 6in); length 19.96m (65ft 6in); height 4.77m (15ft 8in); wing area 88.19 sq m (960 sq ft)
Armament:	four 20mm cannon; two rocket pods or up to 2772kg (6000lb) of bombs

Lockheed F-104 Starfighter

The 'missile with a man in it' was one of many names for the Lockheed Starfighter. With its needle-like fuselage and tiny wings, the F-104 had exceptional speed and climb performance, but was not noted for its dogfight manoeuvrability. The US Air Force used it as an interceptor for a fairly short period, but in the so-called 'sale of the century', it was sold to seven European nations, who used it in the fighter and strike roles.

Lockheed F-104C Starfighter

The F-104 was dispatched to Vietnam in 1965, but was not a great success in the escort fighter and fighter-bomber roles. This F-104C of the 8th TFW was based at Udorn, Thailand, in 1966. It is fitted with a fixed refuelling probe to extend its range for low-level operations.

SPECIFICATIONS	
Country of origin:	United States
Type:	single-seat multi-mission strike fighter
Powerplant:	one 70.29kN (15,800lb) thrust General Electric J79-GE-7A afterburning turbojet engine
Performance:	maximum speed at 15,240m (50,000ft) 1845km/h (1146mph); service ceiling 15,240m (50,000ft); range 1740km (1081 miles)
Weights:	maximum 12,634kg (27,853lb)
Dimensions:	wingspan (excluding missiles) 6.36m (21ft 9in); length 16.66m (54ft 8in); height 4.09m (13ft 5in); wing area 18.22 sq m (196.10sq ft)
Armament:	one 20mm cannon, provision for AIM-9 Sidewinder on fuselage, under wings or on tips, and up to 908kg (2000lb) of bombs

Lockheed TF-104G Starfighter

West Germany undertook most initial training of their F-104 pilots in Arizona, using the two-seat TF-104G. A total of 220 examples of this combat-capable version were produced. Operating Starfighters at low level in European weather proved hazardous, and the Luftwaffe lost more than 200 F-104s.

SPECIFICATIONS	
Country of origin:	United States
Type:	single-seat multi-mission strike fighter
Powerplant:	one 69.4kN (15,600lb) General Electric J79-GE-11A turbojet
Performance:	max speed 1845km/h (1146mph); service ceiling 15,240m (50,000ft); range 1740km (1081 miles)
Weight:	empty 6348kg (13,995lb); maximum take-off 13,170kg (29,035lb)
Dimensions:	wingspan (excluding missiles) 6.36m (21ft 9in); length unreleased; height 4.09m (13ft 5in); wing area 18.22 sq m (196.10 sq ft)
Armament:	Provision for AIM-9 Sidewinder on fuselage, under wings or on tips, and/or stores up to a maximum of 1814kg (4000lb)

Lockheed F-104G Starfighter

The F-104G was designed for Germany, but was also exported to several nations, including the Republic of China, which used it from 1960 to 1998. Taiwanese F-104s destroyed two PLAAF J-6s in a skirmish over the Taiwan Straits in January 1967.

SPECIFICATIONS

Country of origin:	United States
Type:	single-seat multi-mission strike fighter
Powerplant:	one 69.4kN (15,600lb) General Electric J79-GE-11A turbojet engine
Performance:	maximum speed at 15,240m (50,000ft) 1845km/h (1146mph); service ceiling 15,240m (50,000ft); range 1740km (1081 miles)
Weights:	empty 6348kg (13,995lb); maximum take-off 13,170kg (29,035lb)
Dimensions:	wingspan (excluding missiles) 6.36m (21ft 9in); length 16.66m (54ft 8in); height 4.09m (13ft 5in); wing area 18.22 sq m (196.1 sq ft)
Armament:	one 20mm cannon, provision for AIM-9 Sidewinder on fuselage, under wings or on tips, and up to 1814kg (4000lb) stores

Lockheed CF-104G Starfighter

Denmark bought new F-104Gs from Lockheed in 1962 and added further ex-Canadian CF-104s in 1972. They frequently intercepted Soviet aircraft over the Baltic Sea. The RDAF retired its Starfighters in 1985, replacing them with F-16s.

SPECIFICATIONS

Country of origin:	United States
Type:	single-seat multi-mission strike fighter
Powerplant:	one 69.4kN (15,600lb) General Electric J79-GE-11A turbojet engine
Performance:	maximum speed at 15,240m (50,000ft) 1845km/h (1146mph); service ceiling 15,240m (50,000ft); range 1740km (1081 miles)
Weights:	empty 6348kg (13,995lb); maximum take-off 13,170kg (29,035lb)
Dimensions:	wingspan (excluding missiles) 6.36m (21ft 9in); length 16.66m (54ft 8in); height 4.09m (13ft 5in); wing area 18.22 sq m (196.10 sq ft)
Armament:	one 20mm M61A1 Vulcan cannon; AIM-9B/N Sidewinder AAMs; 2.75in (70mm) rocket pods

Aeritalia/Lockheed F-104 ASA Starfighter

Italy built nearly 250 F-104S Starfighters for itself and Turkey. The S designation denoted the ability to fire AIM-7 Sparrow missiles. The F-104 ASA (Aggiornamento Sistemi d'Arma) upgrade added a new radar and missiles in the 1990s. Italy retired the last military Starfighters in 2004.

SPECIFICATIONS

Country of origin:	Italy
Type:	single-seat multi-mission strike fighter
Powerplant:	one 79.6kN (17,900lb) thrust General Electric J79-GE-19 afterburning turbojet engine
Performance:	maximum speed 2330km/h (1450mph); service ceiling 17,680m (58,000ft); range 2920km (1815 miles)
Weights:	empty 6760kg (14,900lb), maximum 14,060kg (31,000lb)
Dimensions:	wingspan (excluding missiles) 6.36m (21ft 9in); length 16.66m (54ft 8in); height 4.09m (13ft 5in); wing area 18.22 sq m (196.10 sq ft)
Armament:	one 20mm M61A1 Vulcan cannon; AIM-9L Sidewinder, AIM-7 Sparrow or Selenia Aspide AAMs

COLD WAR MILITARY AIRCRAFT

Six-Day War

Following a long period of tension between Israel and its neighbours, Israel launched air attacks against Egypt on 5 June 1967, destroying most of the Egyptian Air Force on the ground. Iraqi and Syrian aircraft counterattacked, but most of their aircraft were lost in air combat or return raids by the Israeli Air Force. Israeli air losses were light, and within two days the Arab air forces had ceased to be a factor.

Sud Aviation SO.4050 Vautour

The Vautour IIA was a dedicated ground-attack version, equipped with heavy cannon as well as rockets and bombs. The IDF/AF's 110 Squadron operated a mix of IIA attackers and IIB bombers, while 119 Squadron flew IIN night fighters. Four Vautours were lost in the Six-Day War.

SPECIFICATIONS	
Country of origin:	France
Type:	two-seat medium bomber
Powerplant:	two 34.3kN (7716lb) SNECMA Atar 101E-3 turbojets
Performance:	maximum speed 1105km/h (687mph); service ceiling more than 15000m (49,210ft); range 5400km (3375 miles)
Weight:	maximum 21,000kg (46,300lb)
Dimensions:	wingspan 15.09m (49ft 6in); length 15.57m (51ft 1in); height 4.5m (14ft 9in)
Armament:	four 30mm DEFA cannon, pack containing 116 68mm (2.7in) rockets; up to 4400kg (9700lb) of bombs

Dassault Mystère IVA

Three squadrons of Mystère IVAs led the Israeli attack on Egypt. Once the Egyptian airfields had been neutralized, they turned to supporting ground forces in the Sinai. They were also involved in the controversial attack on the American spy ship USS Liberty.

SPECIFICATIONS	
Country of origin:	France
Type:	single-seat fighter bomber
Powerplant:	one 27.9kN (6,280lb) Hispano Suiza Tay 250A turbojet; or 3500kg (7716lb) Hispano Suiza Verdon 350 turbojet
Performance:	maximum speed 1120km/h (696mph); service ceiling 13,750m (45,000ft); range 1320km (820 miles)
Weights:	empty 5875kg (11,514lb); loaded 9500kg (20,950lb)
Dimensions:	span 11.1m (36ft 5.75in); length 12.9m (42ft 2in); height 4.4m (14ft 5in)
Armament:	two 30mm cannon with 150 rounds, provision for up to 907kg (2000lb) of stores, including tanks, rockets or bombs

SIX-DAY WAR

Dassault Mirage IIICJ

The most important IDF/AF fighter of the Six-Day War was the Mirage IIICJ Shahak (Heavens), which scored 56 aerial victories against the various Arab air forces. This aircraft served with 117 'First Jet' Squadron, which claimed 11 kills in 1967.

SPECIFICATIONS	
Country of origin:	France
Type:	single-engined interceptor fighter
Powerplant:	one 58.72kN (13,200lb) thrust SNECMA Atar 09B-3 afterburning turbojet engine and one 16.46kN (3700lb) thrust auxiliary SEFR 841 rocket motor
Performance:	maximum speed 2350km/h (1460mph); service ceiling 17,000m (55,755 ft); range 2012km (1250 miles)
Weights:	empty 6142kg (13,540lb); maximum 11,676kg (25,740lb)
Dimensions:	span 8.26m (27ft 2in); length 14.91m (48ft 10in); height 4.6m (14ft 10in); wing area 34.84 sq m (375 sq ft)
Armament:	two 30mm DEFA cannon; one Matra R.511 or R.530 AAM, up to 2295kg (5060lb) of bombs

Mikoyan-Gurevich MiG-17 'Fresco'

Egypt had four operational squadrons and a training unit of MiG-17s in 1967. Syria and Iraq also committed theirs to the fighting and lost a number in air combat. In all, the Arab air forces lost 89 MiG-15s and -17s, 90 per cent of them on the ground.

SPECIFICATIONS	
Country of origin:	USSR
Type:	single-seat fighter
Powerplant:	one 33kN (7,452lb) Klimov VK-1F turbojet
Performance:	maximum speed 1145km/h (711mph); service ceiling 16,600m (54,560ft); range at height with slipper tanks 1470km (913 miles)
Weights:	empty 4100kg (9040lb); maximum loaded 600kg (14,770lb)
Dimensions:	wingspan 9.45m (31ft); length 11.05m (36ft 3.75in); height 3.35m (11ft); wing area 20.6 sq m (221.74sq ft)
Armament:	one 37mm N-37 cannon and two 23mm NS-23 cannon, plus up to 500kg (1102lb) of mixed stores on underwing pylons

Mikoyan-Gurevich MiG-21 MF 'Fishbed'

Egypt lost an estimated 90 MiG-21s in the initial air attacks of July 1967. In the subsequent days, they were able to claim five Israeli aircraft destroyed and several more damaged. The combined Arab air forces had bought 235 MiG-21s but had many fewer operational in 1967.

SPECIFICATIONS	
Country of origin:	USSR
Type:	single-seat all-weather multi role fighter
Powerplant:	one 60.8kN (14,550lb) thrust Tumanskii R-13-300 afterburning turbojet
Performance:	maximum speed 2229km/h (1385mph); service ceiling 17,500m (57,400ft); range on internal fuel 1160km (721 miles)
Weights:	empty 5200kg (11,464lb); maximum take-off 10,400kg (22,925lb)
Dimensions:	wingspan 7.15m (23ft 5.5in); length (including probe) 15.76m (51ft 8.5in); height 4.1m (13ft 5.5in); wing area 23 sq m (247.58 sq ft)
Armament:	one 23mm cannon, provision for about 1500kg (3307lb) of stores, including air-to-air missiles, rocket pods, napalm tanks, or drop tanks

Yom Kippur War

On the Jewish holiday of Yom Kippur in October 1973, Syrian and Egyptian forces crossed Israel's borders. The initial days of fighting were the most desperate for Israel and the IDF/AF. Surface-to-air missiles played a much greater part in this war, causing most of Israel's losses. An American airlift of weapons, including aircraft and spare parts, is credited as a factor in preventing Israel's defeat.

Dassault Super Mystère B.2

Known as the 'Sambad', the Super Mystère B.2 was Israel's first supersonic fighter. By early 1973, the IDF/AF had upgraded its Super Mystère B.2s with the J52 engine of the Skyhawk, creating the *Sa'ar* (Tempest), which was used mainly on ground-attack missions.

SPECIFICATIONS	
Country of origin:	France
Type:	single-seat fighter bomber
Powerplant:	one 43.7kN (9833lb) SNECMA Atar 101 G-2/-3 turbojet
Performance:	maximum speed 1195km/h (743mph); service ceiling 17,000m (55,775ft); range 870km (540 miles)
Weights:	empty 6932kg (15,282lb); maximum take-off 10,000kg (22,046lb)
Dimensions:	wingspan 10.52m (34ft 6in); length 14.13m (46ft 4.25in); height 4.55m (14ft 11in); wing area 35 sq m (376.75 sq ft)
Armament:	two 30mm cannon, internal launcher for 35 68mm rockets, provision for up to 907kg (2000lb) of stores, including tanks, rockets or bombs

Douglas A-4N Skyhawk

Over 360 A-4 Skyhawks were delivered to Israel, where they were known as the *Ayit* (Vulture). Thirty examples of the A-4N, designed to Israel's specifications, had been delivered by October 1973. Over 50 A-4s of various sub-types were lost in 1973, mainly to SAMs.

SPECIFICATIONS	
Country of origin:	United States
Type:	single-engined fighter-bomber
Powerplant:	one 49.82kN (11,200lb) thrust Pratt & Whitney J52-P408 turbojet engine
Performance:	maximum speed 1038km/h (645mph); range 1090km (680 miles); service ceiling 11,705m (38,608ft)
Weights:	empty 4747kg (10,465lb); maximum 11,115kg (24,504lb)
Dimensions:	span 8.38m (27ft 6in); length 12.22m (40ft 2in); height 4.57m (15ft); wing area 24.15 sq m (260 sq ft)
Armament:	two 30mm DEFA 534 cannon

McDonnell Douglas F-4E Phantom II

The successes of the Israeli Defence Force/Air Force during the 1973 Yom Kippur War helped to seal the Phantom's reputation as the finest combat aircraft of its generation. Israel purchased 204 F-4Es during the early 1970s, and they remained in front-line operation for many years. Modifications included adoption of the indigenously produced Elta EL/M-2021 multi-mode radar.

SPECIFICATIONS	
Country of origin:	United States
Type:	two-seat all-weather fighter/attack aircraft
Powerplant:	two 79.6kN (17,900lb) General Electric J79-GE-17 turbojets
Performance:	maximum speed 2390km/h (1485mph); service ceiling 19,685m (60,000ft); range 817km (1750 miles)
Weights:	empty 12,700kg (28,000lb); maximum take-off 26,308kg (58,000lb)
Dimensions:	span 11.7m (38ft 5in); length 17.76m (58ft 3in); height 4.96m (16ft 3in); wing area 49.24 sq m (530 sq ft)
Armament:	one 20mm M61A1 Vulcan cannon and four AIM-7 Sparrow recessed under fuselage or other weapons up to 1370kg (3020lb) on centreline pylon; four wing pylons for two AIM-7, or four AIM-9

Mikoyan-Gurevich MiG-17F 'Fresco'

Syrian 'Frescoes' suffered at the hands of Israel's fighters in October 1973, particularly against the Mirages. One Syrian pilot made a claim for an F-4E, but otherwise their contribution was limited. By this time, the MiG-21 had become the predominant type in the Syrian AF inventory.

SPECIFICATIONS	
Country of origin:	USSR
Type:	single-seat fighter
Powerplant:	one 33.1kN (7452lb) Klimov VK-1F turbojet
Performance:	maximum speed 1145km/h (711mph); service ceiling 16,600m (54,560ft); range at height 1470km (913 miles)
Weights:	empty 4100kg (9040lb); maximum loaded 600kg (14,770lb)
Dimensions:	wingspan 9.45m (31ft); length 11.05m (36ft 3.75in); height 3.35m (11ft); wing area 20.6 sq m (221.74 sq ft)
Armament:	one 37mm N-37 cannon and two 23mm NS-23 cannon, plus up to 500kg (1102lb) of mixed stores on underwing pylons

Sukhoi Su-7BM 'Fitter-A'

The Su-7 'Fitter' was supplied to Syria, Iraq and Egypt (illustrated) before 1973. Although fast and with a good warload, the Su-7 had short range and an unreliable afterburner. Many were lost to IDF/AF fighters, although they inflicted considerable damage on Israeli ground forces.

SPECIFICATIONS	
Country of origin:	USSR
Type:	single-seat ground attack fighter
Powerplant:	one 110.3kN (24,802lb) Lyul'ka AL-21F-3 turbojet
Performance:	maximum speed approximately 2220km/h (1380mph); service ceiling 15,200m (49,865ft); combat radius 675km (419 miles)
Weights:	empty 9500kg (20,944lb); maximum take-off 19,500kg (42,990lb)
Dimensions:	wingspan 13.8m (45ft 3in) spread and 10m (32ft 10in) swept; length 18.75m (61ft 6in); height 5m (16ft 5in); wing area 40 sq m (430 sq ft)
Armament:	two 30mm cannon; provision for up to 4250kg (9370lb) of stores, including nuclear weapons, missiles, bombs and napalm tanks

McDonnell Douglas F-4 Phantom

The F-4 Phantom II was built to a US Navy requirement, but was quickly adopted by the USAF in a rare example for the 1950s of joint procurement of a weapons system. Able to carry a heavy ground attack load as well as eight air-to-air missiles, the F-4 also proved a hit on the export market. Phantoms were supplied to 10 nations. Production in the United States and Japan totalled 5195.

F4H-1 Phantom II

The Phantom II first appeared as the F4H-1F in the pre-1962 US Navy designation system. The Phantom made its maiden flight in May 1957 and the pre-production F4H-1s set a string of records for closed-circuit and point-to-point speed, and for climb to altitude.

SPECIFICATIONS	
Country of origin:	United States
Type:	(F4H-1F) twin-engined two-seat carrier-based fighter
Powerplant:	two 71.84kN (16,150lb) thrust General Electric J79-GE-2 turbojet engines
Performance:	maximum speed unknown; range unknown; service ceiling unknown
Weights:	unknown
Dimensions:	span 11.6m (38ft 4in); length 17.7m (58ft 4in); height 4.9m (16ft 3in); wing area 49.2 sq m (530 sq ft)
Armament:	four AIM-7 Sparrow under fuselage, provision for two AIM-7 or four AIM-9 Sidewinder; provision for 20mm M-61 cannon, and tanks and bombs to a weight of 6219kg (13,500lb)

F-4C Phantom II

The USAF adopted the Phantom in 1960, initially as the F-110A, but later redesignated F-4C. The first production model F-4s had larger radomes and raised rear canopies compared to the F4H-1. This F-4C served with the Michigan Air National Guard in the 1980s.

SPECIFICATIONS	
Country of origin:	United States
Type:	two seat all-weather fighter/attack aircraft
Powerplant:	two 75.6kN (17,000lb) General Electric J79-GE-15 turbojets
Performance:	maximum speed 2414km/h (1500mph); service ceiling 18,300m (60,000ft); range on internal fuel tanks 2817km (1750 miles)
Weights:	empty 12,700kg (28,000lb); maximum take-off 26,308kg (58,000lb)
Dimensions:	span 11.7m (38ft 5in); length 17.76m (58ft 3in); height 4.96m (16ft 3in); wing area 49.24 sq m (530 sq ft)
Armament:	four AIM-7 Sparrow; two wing pylons for two AIM-7, or four AIM-9 Sidewinder, provision for 20mm M-61 cannon; provision for bombs or other stores to a maximum weight of 6219kg (13,500lb)

MCDONNELL DOUGLAS F-4 PHANTOM

F-4S Phantom II

One of the lesser known of the F-4 variants, the F-4S was a development of the F-4J models constructed in small numbers for the US Navy. The F-4J had the AWG-10 pulse-doppler radar, drooping ailerons, slatted tail and J79-GE-10 engines. Also incorporated was an automatic carrier landing system. Production of carrier-based Phantoms lasted for a remarkable 17 years.

SPECIFICATIONS

Country of origin:	United States
Type:	two-seat all-weather fighter/attack carrier-borne aircraft
Powerplant:	two 79.6kN (17,900lb) General Electric J79-GE-10 turbojets
Performance:	maximum speed at high altitude 2414km/h (1500mph); service ceiling over 18,300m (60,000ft); range 2817km (1750 miles)
Weights:	empty 12,700kg (28,000lb); maximum take-off 26,308kg (58,000lb)
Dimensions:	span 11.7m (38ft 5in); length 17.76m (58ft 3in); height 4.96m (16ft 3in); wing area 49.24 sq m (530 sq ft)
Armament:	four AIM-7 Sparrow; two wing pylons for two AIM-7, or four AIM-9 Sidewinder, provision for 20mm cannon in external centreline pod; four wing pylons for tanks, bombs or other stores to a maximum weight of 6219kg (13,500lb)

Phantom FGR.2

The United Kingdom bought its own model of Phantom for the RAF and FAA. The FG.1 for the Navy and FGR.2 (Air Force, illustrated) had Rolls-Royce Spey engines, which required larger intakes and other airframe changes. When the aircraft carrier *Ark Royal* retired, its Phantoms went to the RAF.

SPECIFICATIONS

Country of origin:	United States
Type:	two-seat all-weather fighter/attack carrier-borne aircraft
Powerplant:	two 91.2kN (20,515lb) Rolls-Royce Spey 202 turbofans
Performance:	maximum speed 2230km/h (1386mph); service ceiling over 18,300m (60,000ft); range on internal fuel 2817km (1750 miles)
Weights:	empty 12,700kg (28,000lb); maximum take-off 26,308kg (58,000lb)
Dimensions:	span 11.7m (38ft 5in); length 17.55m (57ft 7in); height 4.96m (16ft 3in); wing area 49.24 sq m (530sq ft)
Armament:	four AIM-7 Sparrow recessed under fuselage; two wing pylons for two AIM-7, or four AIM-9 Sidewinder, provision for 20mm cannon; pylons for stores to a maximum weight of 7257kg (16,000lb)

RF-4EJ Phantom II

The Phantom proved easily adaptable into a high-speed tactical reconnaissance aircraft. The RF-4B (Marines) and RF-4C (USAF) preceded the RF-4E export version. RF-4Es were used by Germany, Iran, Greece, Turkey and Japan, whose version (illustrated) was the RF-4EJ.

SPECIFICATIONS

Country of origin:	United States
Type:	two-seat tactical reconnaissance aircraft
Powerplant:	two 79.6kN (17900lb) thrust General Electric J79-GE-17 afterburning turbojet engines
Performance:	maximum speed at 14,630m (48,000ft) 2390km/h (1485mph); service ceiling 18,900m (62,000ft); range 800km (500 miles)
Weights:	empty 13,768kg (30,328lb); maximum loaded 24,766kg (54,600lb)
Dimensions:	wingspan 11.7m (38ft 5in); length 18m (59ft); height 4.96m (16ft 3in); wing area 49.24 sq m (530 sq ft)
Armament:	none

Vietnam War: FACs and COIN

The Vietnam War began as a relatively low-level counter-insurgency (COIN) operation and grew to involve almost every US weapons system short of nuclear arms. In the South, it was a war of defending villages and military outposts, and interdicting supply routes from the North. Slow but agile prop-driven aircraft provided Forward Air Control (FAC) for fast jets that were vital for the close air support of troops.

Cessna O-2 Skymaster

The Cessna Model 337's unusual push-and-pull engine configuration gave it twin-engined power with single-engined drag and handling characteristics. As the O-2A, it made an effective FAC platform. This one was based at Nakhom Phanom, Thailand, in 1970.

SPECIFICATIONS

Country of origin:	United States
Type:	(O-2A) twin-engined Forward Air Control Aircraft
Powerplant:	two 157kW (210hp) Continental IO-360C flat-6 piston engines
Performance:	maximum speed 322km/h (200mph); range 2132KM (1325 miles); service ceiling 5940m (18,000ft)
Weights:	empty 1292kg (2848lb); loaded 2448kg (5400lb)
Dimensions:	span 11.63m (38ft 2in); length 9.07m (29ft 8in); height 2.79m (9ft 2in); wing area 18.8 sq m (202.5 sq ft)
Armament:	rocket pods for target marking

North American OV-10A Bronco

The purpose-built North American (later Rockwell) OV-10 Bronco replaced the O-2 with the USAF, but also served with the Navy and Marines as an FAC and light-attack aircraft. This OV-10A was assigned to the USMC's VMO-4 squadron at Quang Tri, Republic of Vietnam.

SPECIFICATIONS

Country of origin:	United States
Type:	twin-engined Forward Air Control Aircraft
Powerplant:	two 533kW (715hp) Garrett T76-G-410/412 turboprop engines
Performance:	maximum speed 452km/h (281mph); service ceiling 7315m (24,000ft); range 358km (576 miles)
Weights:	empty 3127kg (6893lb); maximum 6552kg (14,444lb)
Dimensions:	span 12.19m (40ft); length 12.67m (41ft 7in); height 4.62m (15ft 2in); wing area 64.57 sq m (695 sq ft)
Armament:	four 7.62mm M60C machine-guns; pods for 70mm (2.75in) or 125mm (5in) rockets; up to 125kg (500lb) of bombs

On Mark B-26K Counter Invader

SPECIFICATIONS	
Country of origin:	United States
Type:	Twin-engined attack aircraft
Powerplant:	two 1864kW (2500hp) Pratt & Whitney R-2800-52W radial piston engines
Performance:	maximum speed 520 km/h (323 mph); service ceiling 8717m (28,600ft); range 2382km (1480 miles)
Weights:	empty 11,399kg (25,130lb); maximum 17,804kg (39,250lb)
Dimensions:	span 21.79m (71ft 6in); length 15.71m (51ft 7in); height 5.79m (19ft); wing area 50.17 sq m (540 sq ft)
Armament:	14.5in (12.7mm) machine guns; up to 5443kg (12,000 lb) bombs

The On Mark Corporation converted Douglas B-26B and C Invaders into the B-26K Counter Invader for the USAF's Air Commando units in Vietnam. The structure was strengthened and wingtip fuel tanks added. They flew mainly night-interdiction missions over Vietnam and Laos.

Northrop F-5A Freedom Fighter

SPECIFICATIONS	
Country of origin:	United States
Type:	light tactical fighter
Powerplant:	two 18.1kN (4080lb) General Electric J85-GE-13 turbojets
Performance:	maximum speed at 10,975m (36,000ft) 1487km/h (924mph); service ceiling 15,390m (50,500ft); combat radius with maximum warload 314km (195 miles)
Weights:	empty 3667kg (8085lb); maximum take-off 9374kg (20,667lb)
Dimensions:	wingspan 7.7m (25ft 3in); length 14.38m (47ft 2in); height 4.01m (13ft 2in); wing area 15.79 sq m (170sq ft)
Armament:	two 20mm M39 cannon with 280rpg; provision for 1996kg (4400lb) of stores (missiles, bombs, cluster bombs, rocket launcher pods)

The USAF made relatively little use of the F-5, but sent a unit to Bien Hoa, Vietnam in late 1965 under an evaluation programme called 'Skoshi Tiger'. The results were mixed, but helped sell the F-5 as a light fighter and ground attacker to South Vietnam and many other Asian nations.

North American F-100F Super Sabre

SPECIFICATIONS	
Country of origin:	United States
Type:	single-seat fighter-bomber
Powerplant:	one 75.6kN (17,000lb) Pratt & Whitney J57-P-21A turbojet
Performance:	maximum speed at 10,670m (35,000ft) 1390km/h (864mph); service ceiling 14,020m (46,000ft); range with inernal fuel 966km (600 miles)
Weights:	maximum 17,745kg (39,122lb)
Dimensions:	wingspan 11.82m (38ft 9.5in); length 15.2m (50ft); height 4.95m (16ft 3in); wing area 35.77 sq m (385 sq ft)
Armament:	two 20mm Pontiac M39 cannon; up to 3402kg (7500lb) of rockets, bombs or cannon pods

The two-seat F-100F Super Sabre was used in a variety of roles in Vietnam, including close air support, anti-radar 'Wild Weasel' missions and as a fast FAC, able to reach a battlefield quickly and precisely mark defended targets for attack aircraft.

Vietnam War: Naval Air Power

From August 1964, the aircraft carriers on 'Yankee Station' off the coast of North Vietnam launched daily strike packages against military, industrial and transport targets. Increasingly sophisticated defences, including MiGs, SAMs and radar-controlled anti-aircraft guns of various calibres, took a heavy toll on carrier-based aircraft, with 530 aircraft lost and 620 aviators killed, missing or made prisoner.

Douglas A-4F Skyhawk

During its long career, the Skyhawk has proved to be one of the most versatile combat aircraft ever built, disproving those who argued that the small, lightweight machine would be outclassed by bigger, heavier aircraft. The aircraft pictured is an A-4F, the final attack version for the US Navy, distinguished by the dorsal hump carrying additional avionics and the J52-P-8A engine.

SPECIFICATIONS	
Country of origin:	United States
Type:	single-seat attack bomber
Powerplant:	one 41.3kN (9300lb) J52-8A turbojet
Performance:	maximum speed 1078km/h (670mph); service ceiling 14,935m (49,000ft); range with 4000lb load 1480km (920 miles)
Weights:	empty 4809kg (10,602lb); maximum take-off 12,437kg (27,420lb)
Dimensions:	wingspan 8.38m (27ft 6in); length excluding probe 12.22m (40ft 1.5in); height 4.66m (15ft 3in); wing area 24.15 sq m (260 sq ft)
Armament:	two 20mm Mk 12 cannon with 200rpg; five external hardpoints with provision for 3720kg (8200lb) of stores, including AGM-12 Bullpup air-to-surface missiles, AGM-45 Shrike anti-radar missiles, bombs, cluster bombs, dispenser weapons, rocket-launcher pods, cannon pods, drop tanks and ECM pods

McDonnell Douglas F-4G Phantom II

The Phantom was the Navy's principal fighter of the war. The F-4G illustrated here in an experimental camouflage scheme, was a version of the F-4B with an automatic carrier landing system. This version made one cruise with VF-213 on the USS *Kittyhawk*.

SPECIFICATIONS	
Country of origin:	United States
Type:	two-seat EW/radar-surpression aircraft
Powerplant:	two 54.2kN (17,000lb) thrust General Electric J79-GE-8A afterburning turbojet engines
Performance:	maximum speed at high altitude 2390km/h (1485mph); service ceiling over 18,975m (62,250ft); range 958km (595 miles)
Weights:	empty 13,300kg (29,321lb); maximum take-off 28,300kg (62,390lb)
Dimensions:	span 11.7m (38ft 5in); length 19.2m (63ft); height 5.02m (16ft 5.5in); wing area 49.24 sq m (530 sq ft)
Armament:	two AIM-7 Sparrow; wing pylons for radar suppression weapons

VIETNAM WAR: NAVAL AIR POWER

Douglas EKA-3 Skywarrior

The A3 Skywarrior is notable as the first carrier-based strategic nuclear bomber, designed to be operated from the deck of the Forrestal class of carriers that came into service in 1948. Both the outer wings and tail were designed to fold hydraulically and thus minimize the space occupied by the aircraft on deck. Deliveries began in March 1956 to the US Navy's VH-1 attack squadron. Later variants saw much service in Vietnam as Elint and ECM platforms.

SPECIFICATIONS	
Country of origin:	United States
Type:	carrier-based strategic bomber
Powerplant:	two 55.1kN (12,400lb) Pratt & Whitney turbojets
Performance:	maximum speed 982km/h (610mph); service ceiling 13,110m (43,000ft); range with maximum fuel 3220km (2000 miles)
Weights:	empty 17,875kg (39,409lb); maximum take-off 37,195kg (82,000lb)
Dimensions:	wing span 22.1m (72ft 6in); length 23.3m (76ft 4in); height 7.16m (23ft 6in); wing area 75.43 sq m (812 sq ft)
Armament:	two remotely controlled 20mm cannon in tail turret, plus provision for 5443kg (12,000lb) of conventional or nuclear weapons in internal bomb bay

Vought A-7B Corsair II

The A-7 entered combat with the Navy in Southeast Asia in December 1967, and replaced the A-4 with most of the light-attack squadrons by 1975. It brought higher performance and a sophisticated radar to the mission. This A-7B is in the markings of VA-46 'Clansmen', operating off the *John F. Kennedy*.

SPECIFICATIONS	
Country of origin:	United States
Type:	single-seat attack aircraft
Powerplant:	one 54.2kN (12,190lb) thrust Pratt & Whitney TF30-P-8 turbofan engine
Performance:	maximum speed at low-level 1123km/h (698mph); combat range with typical weapon load 1150km (4100 miles)
Weights:	empty 8972kg (19,781lb); maximum take-off 19,050kg (42,000lb)
Dimensions:	wingspan 11.8m (38ft 9in); length 14.06m (46ft 1.5in); height 4.9m (16ft 0.75in); wing area 34.84 sq m (375 sq ft)
Armament:	two 20mm Colt Mk 12 cannon; up to 6804kg (15,000lb) of bombs, air-to-surface missiles or other stores

Grumman A-6E Intruder

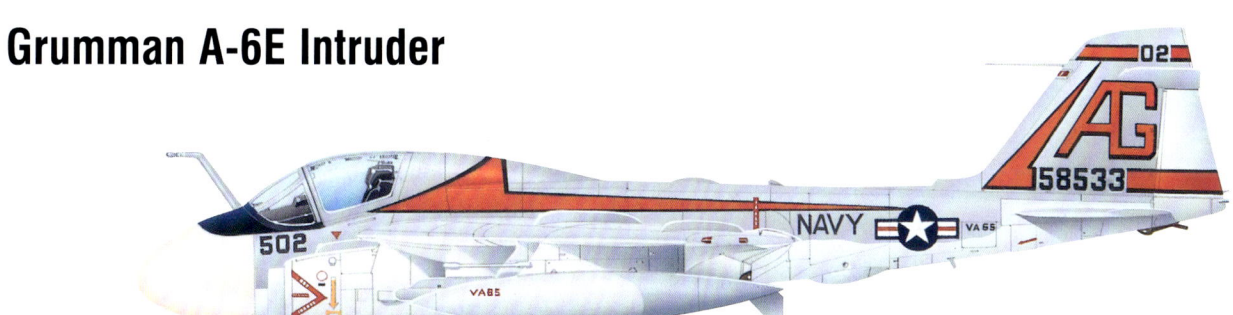

Entering service with the US Navy in February 1963, the Intruder was specifically planned for first pass blind attack on point surface targets at night or in any weather conditions. It was designed to be subsonic and is powered by two straight turbojets. Despite its considerable gross weight, the Intruder has excellent slow-flying qualities with full-span slats and flaps.

SPECIFICATIONS	
Country of origin:	United States
Type:	two-seat carrierborne and landbased all-weather strike aircraft
Powerplant:	two 41.3kN (9300lb) Pratt & Whitney J52-P-8A turbojets
Performance:	maximum speed at sea level 1043km/h (648mph); service ceiling 14,480m (47,500ft); range 1627km (1011 miles)
Weights:	empty 12,132kg (26,746lb); maximum take-off 26,581kg (58,600lb) for carrier launch or 27,397kg (60,400lb) for field take-off
Dimensions:	wingspan 16.15m (53ft); length 16.69; height 4.93m (16ft 2in); wing area 49.13 sq m (528.9 sq ft)
Armament:	five external hardpoints with provision for up to 8165kg (18,000lb) of stores, including nuclear weapons, bombs, missiles, and drop tanks

Vietnam War: Air Force Air Power

The US Air Force committed thousands of aircraft to Vietnam, from light observation types to strategic bombers. South Vietnam's airfields and airspace were so crowded that much of the effort had to be launched from Thailand or Guam. B-52 raids on Hanoi in 1972 helped bring the North to the negotiating table, but the millions of pounds of bombs dropped in the jungle did little to alter the final outcome.

Fairchild AC-119K Stinger

The gunship aircraft with an array of side-facing weapons was a concept originated by the USAF in Vietnam. Following the AC-47 'Spooky' was the AC-119G Shadow and AC-119K Stinger (illustrated). The Shadow was armed with four Miniguns and the Stinger had two Vulcan cannon.

SPECIFICATIONS	
Country of origin:	United States
Type:	Twin-engined gunship
Powerplant:	two 2610kW (3500hp) Wright R-3350-85 radial piston engines
Performance:	maximum speed 335km/h (210 mph); range 3219km (2000 miles); service ceiling 7300m (23,950ft)
Weights:	empty 18,200kg (40,125lb); maximum 28,100kg (62,000lb)
Dimensions:	span 33.3m (109ft 3in); length 26.37m (86ft 6in); height 8m (26ft 3in); wing area 134.43 sq m (1447 sq ft)
Armament:	two 20mm M61 Vulcan cannon

Lockheed AC-130 Spectre

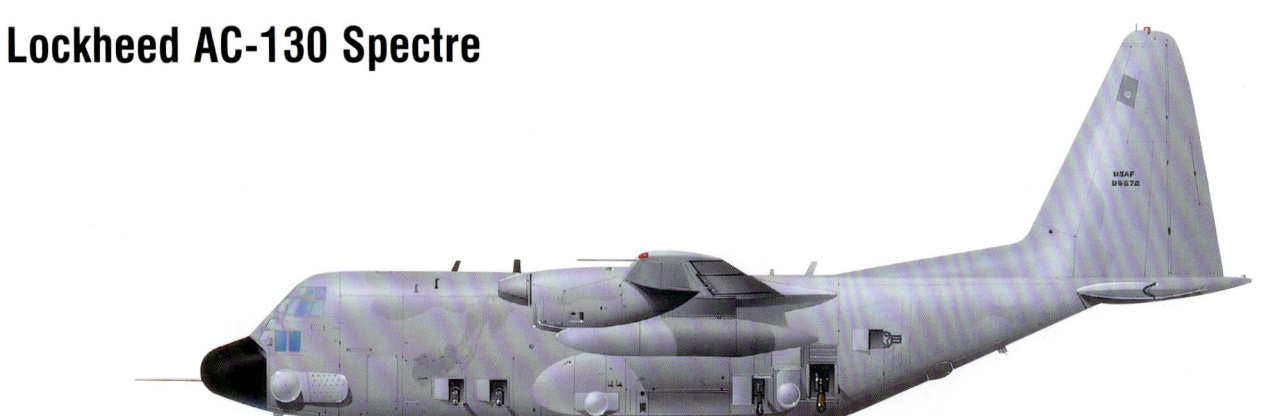

The definitive gunship is the AC-130, introduced into Vietnam in September 1967 and still in use in modernized form today. The AC-130A Spectre was used to hunt for night movements on the Ho Chi Minh Trail supply routes into the South. It also provided fire support for US bases under insurgent attack.

SPECIFICATIONS	
Country of origin:	United States
Type:	(AC-130A) four-engined gunship
Powerplant:	four 3661kW (4910hp) Allison T56-A-15 turboprop engines
Performance:	maximum speed 480km/h (300mph); range 4070km (2530 miles); service ceiling unknown
Weights:	empty unknown; maximum 69,750kg (155,000lb)
Dimensions:	span 40.4m (132ft 7in); length 29.8m (97ft 9in); height 11.7m (38ft 6in); wing area 162.2 sq m (1745.5 sq ft)
Armament:	four 7.62mm GAU-2/A Miniguns and four 20mm Vulcan cannon

Vietnam War: MiGs and Thuds

One of the most fiercely fought battles of the air war was that between the F-105 Thunderchief ('Thud') bombers and the North Vietnamese fighters defending the key industrial targets. A mountainous feature pointing towards Hanoi became known as 'Thud Ridge' after the F-105s that used its cover and fell there in large numbers. The North's MiGs claimed 22 F-105s up to 1970, when they were withdrawn.

Republic F-105B Thunderchief

The primary mission of Republic's replacement for the F-84F Thunderstreak was the delivery of nuclear and conventional weapons in all weathers, at high speeds and over long ranges. The production F-105B entered service in August 1958 with the USAF's 335th Tactical Fighter Squadron, three years later than planned. Seventy-five were completed before the aircraft was superseded by the F-105D.

SPECIFICATIONS

Country of origin:	United States
Type:	single-seat fighter-bomber
Powerplant:	one 104.5kN (23,500lb) Pratt & Whitney J75 turbojet
Performance:	maximum speed 2018km/h (1254mph); service ceiling 15,850m (52,000ft); combat radius with weapon load 370km (230 miles)
Weights:	empty 12,474kg (27,500lb); maximum take-off 18,144kg (40,000lb)
Dimensions:	wingspan 10.65m (34ft 11.25in); length 19.58m (64ft 3in); height 5.99m (19ft 8in); wing area 35.8 sq m (385 sq ft)
Armament:	one 20mm M61 cannon with 1029 rounds; internal bay with provision for up to 3629kg (8000lb) of bombs; five external pylons for additional load of 2722kg (6000lb)

Republic F-105F Thunderchief

In 1962, the USAF ordered 143 two-seat F-105F trainers. The aircraft were equipped with dual controls and full operational equipment. To incorporate the tandem cockpit, the fuselage was lengthened slightly. It was originally intended to use the aircraft for training and transition training, but the pressures of US involvement in the Vietnam conflict created an urgent requirement for these high-performance fighter-bombers, and many were used operationally in theatre.

SPECIFICATIONS

Country of origin:	United States
Type:	two-seat operational trainer
Powerplant:	one 108.9kN (24,500lb) Pratt & Whitney J75-19W turbojet
Performance:	maximum speed 2382km/h (1480mph); service ceiling 15,850m (52,000ft); combat radius with 16 750lb bombs 370km (230 miles)
Weights:	empty 12,890kg (28,393lb); maximum take-off 24,516kg (54,000lb)
Dimensions:	wingspan 10.65m (34ft 11.25in); length 21.21m (69ft 7.5in); height 6.15m (20ft 2in); wing area 35.8sq m (385sq ft)
Armament:	one 20mm M61 cannon with 1029 rounds; provision for up to 3629kg (8000lb) of bombs; pylons for additional load of 2722kg (6000lb)

VIETNAM WAR: AIR FORCE AIR POWER

Boeing B-52D

In Vietnam, the USAF found itself using strategic bombers designed for nuclear war to drop conventional bombs against suspected insurgents in thick jungle. These 'Arc Light' missions reportedly had a large psychological effect but did relatively little physical damage to the enemy. This B-52D of the 43rd Bomb Wing was based in Guam in 1972.

SPECIFICATIONS	
Country of origin:	United States
Type:	long-range strategic bomber
Powerplant:	eight 44.5kN (10,000lb) Pratt & Whitney J57 turbojets
Performance:	maximum speed at 7315m (24,000ft) 1014km/h (630mph): service ceiling 13,720–16,765m (45,000–55,000ft); standard range with maximum load 9978km (6200 miles)
Weights:	empty 77,200–87,100kg (171,000–193,000lb); loaded 204,120kg (450,000lb)
Dimensions:	wingspan 56.4m (185ft); length 48m (157ft 7in); height 14.75m (48ft 3in); wing area 371.6 sq m (4000 sq ft)
Armament:	remotely controlled tail mounting with four .5in machine guns; normal internal bomb capacity 12,247kg (27,000lb) including all SAC special weapons; 108 Mk 82 (500-pound) bombs internally and on underwing pylons

Douglas A-1H Skyraider

The outwardly anachronistic Skyraider was valued in Vietnam for its heavy weapons load and long loiter time over the target area. These attributes were most needed when covering the scene of a rescue operation, until the downed aircrew were aboard helicopters and on their way to safety.

SPECIFICATIONS	
Country of origin:	United States
Type:	single-engined attack aircraft
Powerplant:	one 2013kW (2700hp) Pratt & Whitney R-3350-26WA radial piston engine
Performance:	maximum speed 520km/h (320mph); range 2115km (1315mph); service ceiling 8660m (28,500ft)
Weights:	empty 5430kg (11,970lb); maximum 11,340kg (25,000lb)
Dimensions:	span 15.25m (50ft); length 11.84m (38ft 10in); height 4.78m (15ft 8in); wing area 37.19 sq m (400.31 sq ft)
Armament:	four 20mm cannon; up to 3600kg (8000lb) of bombs, rockets, or other stores

General Dynamics F-111

The variable geometry F-111 suffered a difficult gestation, earning it the unwelcome nickname Aardvark. It was developed to meet a bold Department of Defense edict for a common type of fighter to meet all future tactical needs of the US armed forces, and the first of 117 F-111s were eventually delivered into service in 1967. The Royal Australian Air Force bought the F-111C, the only export success for the aircraft.

SPECIFICATIONS	
Country of origin:	United States
Type:	two-seat multi-purpose attack aircraft
Powerplant:	two 11,1.6kN (25,100lb) Pratt & Whitney TF-30-P100
Performance:	maximum speed at optimum altitude 2655km/h (1650mph); service ceiling above 17,985m (59,000ft); range 4707km (2925 miles)
Weights:	empty 21,398kg (47,175lb); maximum take-off 45,359kg (100,000lb)
Dimensions:	wingspan unswept 19.2m (63ft); swept 9.74m (32ft 11.5in); length 22.4m (73ft 6in); height 5.22m (17ft 1.5in); wing area 48.77 sq m (525 sq ft) unswept
Armament:	one 20mm cannon and one 340kg (750lb) B43 bomb, or two B43 bombs in internal bay, provision for 14,290kg (31,000lb) of stores

Mikoyan-Gurevich MiG-17 'Fresco'

The North Vietnamese Air Force (NVAF) formed its first fighter unit, No 921 'Sao Do' (Red Star) Regiment, in 1964. It scored its first confirmed victories in April 1965 against F-105Ds. Most MiG-17 kills were scored with its 23mm and 37mm cannon.

SPECIFICATIONS	
Country of origin:	USSR
Type:	single-seat fighter
Powerplant:	one 33.1kN (7452lb) Klimov VK-1F turbojet
Performance:	maximum speed at 3000m (9840ft) 1145km/h (711mph); service ceiling 16,600m (54,560ft); range 1470km (913 miles)
Weights:	empty 4100kg (9040lb); maximum loaded 600kg (14,770lb)
Dimensions:	wingspan 9.45m (31ft); length 11.05m (36ft 3.75in); height 3.35m (11ft); wing area 20.6msq m (221.74sq ft)
Armament:	one 37mm N-37 cannon and two 23mm NS-23 cannon, plus up to 500kg (1102lb) of mixed stores on underwing pylons

Mikoyan-Gurevich MiG-19S 'Farmer'

The MiG-19, or Shenyang J-6, entered NVAF service in 1969 with the 925th Regiment. The small, agile MiGs proved hard to counter, particularly when US rules of engagement dictated their pilots get close enough to positively identify them, thus negating their medium-range missiles.

SPECIFICATIONS	
Country of origin:	USSR
Type:	single-seat all-weather interceptor
Powerplant:	two 31.9kN (7165lb) Klimov RD-9B turbojets
Performance:	maximum speed at 9080m (20,000ft) 1480km/h (920mph); service ceiling 17,900m (58,725ft); maximum range at high altitude with two drop tanks 2200km (1367 miles)
Weights:	empty 5760kg (12,698lb); maximum take-off weight 9500kg (20,944lb)
Dimensions:	wingspan 9m (29ft 6.5in); length 13.58m (44ft 7in); height 4.02m (13ft 2.25in); wing area 25 sq m (269.11 sq ft)
Armament:	underwing pylons for four AA-1 Alkali air-to-air-missiles, or AA-2 Atoll

Mikoyan-Gurevich MiG-21MF 'Fishbed'

The top five aces of the Vietnam War all flew MiG-21s, which entered the war in 1965. Nguyen Van Coc was the leading MiG-21 pilot, with seven USAF and USN aircraft and two unmanned aerial vehicles destroyed between April 1967 and December 1968.

SPECIFICATIONS	
Country of origin:	USSR
Type:	single-seat all-weather multi role fighter
Powerplant:	one 60.8kN (14,550lb) thrust Tumanskii R-13-300 afterburning turbojet
Performance:	maximum speed above 11,000m (36,090ft) 2229km/h (1385mph); service ceiling 17,500m (57,400ft); range on internal fuel 1160km (721 miles)
Weights:	empty 5200kg (11,464lb); maximum take-off weight 10,400kg (22,925lb)
Dimensions:	wingspan 7.15m (23ft 5.5in); length (including probe) 15.76m (51ft 8.5in); height 4.1m (13ft 5.5in); wing area 23sq m (247.58sq ft)
Armament:	one 23mm twin-barrell cannon in underbelly pack, four underwing pylons with provision for about 1500kg (3307kg) of stores

Lockheed P-3 Orion

Derived from the L-188 Electra airliner, the P-3 Orion has become the most widely used maritime patrol aircraft in the world. Designed primarily for anti-submarine work, the basic Orion can also undertake anti-shipping, search-and-rescue and overland reconnaissance missions. Specialized variants are used for airborne radar surveillance, electronic intelligence-gathering and scientific survey work.

P-3A Orion

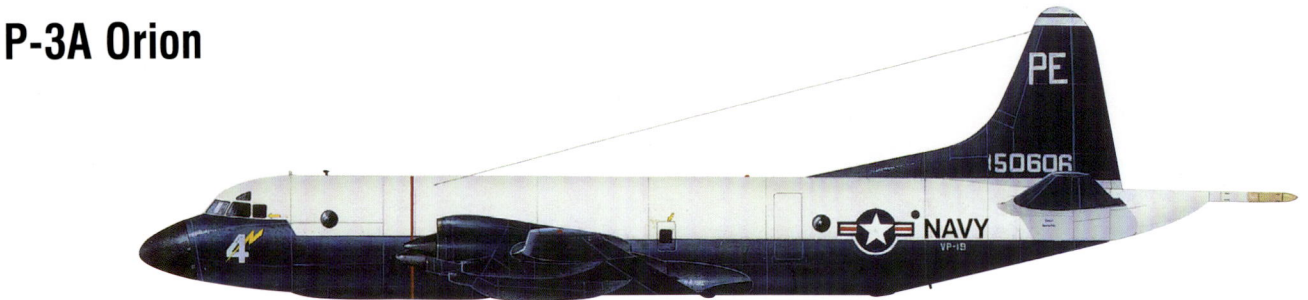

The P-3A entered service with the US Navy in 1962. In total, 157 of this model were built, all for domestic use. In 1975, patrol squadron VP-19 (illustrated) became the first to receive the P-3C with a revised internal layout, much-improved computers and navigation equipment.

SPECIFICATIONS	
Country of origin:	United States
Type:	four-engined maritime patrol aircraft
Powerplant:	four 3356kW (4500hp) Allison T56-A10W turboprop engines
Performance:	maximum speed 766km/h (476mph); range 4075km (2533 miles); service ceiling 8625m (28,300ft)
Weights:	empty 27,216kg (60,000lb); maximum 60,780kg (134,000lb)
Dimensions:	span 30.37m (99ft 8in); length 35.61m (116ft 9in); height 10.27m (33ft 8in); wing area 120.8 sq m (1300 sq ft)
Armament:	up to 9070kg (20,000lb) of ordnance, including bombs, mines and torpedoes

P-3B Orion

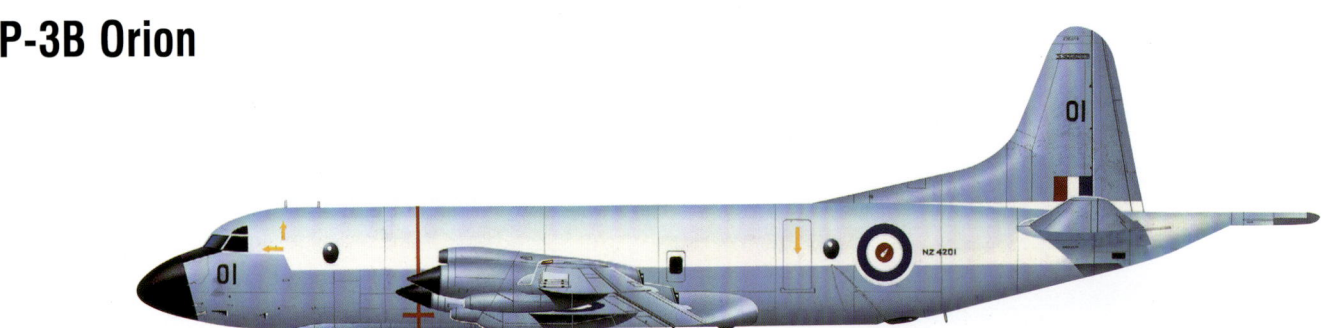

Since 1966, the Royal New Zealand Air Force has been an Orion operator. Its five P-3Bs were joined by an ex-Australian example in the 1980s, and all were updated to P-3K standard with avionics equivalent to the P-3C. Several structural upgrades have further extended their operational life.

SPECIFICATIONS	
Country of origin:	United States
Type:	four-engined maritime patrol aircraft
Powerplant:	four 3700kW (4600hp) Allison T56-A-14 turboprop engines
Performance:	maximum speed 766km/h (476mph); range 4075km (2533 miles); service ceiling 8625m (28,300ft)
Weights:	empty 27,216kg (60,000lb); maximum 60,780kg (134,000lb)
Dimensions:	span 30.37m (99ft 8in); length 35.61m (116ft 9in); height 10.27m (33ft 8in); wing area 120.8 sq m (1300 sq ft)
Armament:	up to 9070kg (20,000lb) of ordnance, including bombs, mines and torpedoes

EP-3E Aries II

SPECIFICATIONS	
Country of origin:	United States
Type:	four-engined maritime patrol aircraft
Powerplant:	four 3356kW (4500hp) Allison T56-A10W turboprop engines
Performance:	maximum speed 766km/h (476mph); range 4075km (2533 miles); service ceiling 8625m (28,300ft)
Weights:	empty 27,216kg (60,000lb); maximum 60,780kg (134,000lb)
Dimensions:	span 30.37m (99ft 8in); length 32.3m (105ft 11in); height 10.44m (34ft 3in); wing area 120.8 sq m (1300 sq ft)
Armament:	none

Based on the P-3A, the EP-3E has all its ASW mission equipment replaced with systems for antennas to gather electronic intelligence. This EP-3E Aries II served with VQ-1 'World Watchers', which gained worldwide attention in April 2001 when one of its aircraft was forced to land in China.

RP-3A Orion

SPECIFICATIONS	
Country of origin:	United States
Type:	four-engined maritime patrol aircraft
Powerplant:	four 3356kW (4500hp) Allison T56-A10W turboprop engines
Performance:	maximum speed 766km/h (476mph); range 4075km (2533 miles); service ceiling 8625m (28,300ft)
Weights:	empty 27,216kg (60,000lb); maximum 60,780kg (134,000lb)
Dimensions:	span 30.37m (99ft 8in); length 35.61m (116ft 9in); height 10.27m (33ft 8in); wing area 120.8 sq m (1300 sq ft)
Armament:	none

'El Coyote' was a modified P-3A used by the US Navy Oceanographic Office for Project Seascan, which measured sea temperatures, currents and salinity around the globe. This data had application for the tracking of submarines, as well as general environmental research.

P-3 Orion AEW&C

SPECIFICATIONS	
Country of origin:	United States
Type:	airborne warning and control aircraft
Powerplant:	four 3700kW (4600hp) Allison T56-A-14 turboprop engines
Performance:	maximum speed unknown; range unknown; service ceiling 8625m (28,300ft)
Weights:	unknown
Dimensions:	span 30.37m (99ft 8in); length 35.61m (116ft 9in); height 10.27m (33ft 8in); wing area 120.8 sq m (1300 sq ft)
Armament:	none

Developed as a private venture by Lockheed, the P-3 AEW&C (airborne early warning and control) aircraft was offered to countries unable to afford the E-3 Sentry. The only customer was the US Customs Service, which uses the 'Dome' to hunt drug-running boats and aircraft in the Caribbean.

COLD WAR MILITARY AIRCRAFT

Vertical Take-off

The certainty of fixed airfields being put out of action in the first hours of any major war led to efforts throughout the 1960s to create a viable vertical take-off and landing (VTOL) aircraft that could operate from positions such as forest clearings close to the battlefield. Most concepts performed poorly with the available thrust, but the P.1127 formed the basis of the Harrier, which has served the RAF since 1969.

Ryan XV-5A

SPECIFICATIONS	
Country of origin:	United States
Type:	Single-seat fighter
Powerplant:	two 11.79kN (2650lb-thrust) General Electric J85-GE-5 turbojet engines
Performance:	maximum speed 804km/h (498mph); range unknown; service ceiling 12,200m (40,028ft)
Weights:	loaded 5580kg (12,302lb)
Dimensions:	span 9.09m (29ft 10in); length 13.56m (44ft 6in); height 4.5m (14ft 9in)
Armament:	eight 7.7mm (.303in) Browning MGs

The XV-5 was developed for the US Army as a potential battlefield surveillance platform. The power system used two engines driving three fans to provide lift and pitch control. The first flight was made in May 1964, and the first vertical flight was achieved in July, but the prototype was destroyed in a 1965 accident. The army's test programme encompassed 338 flights up to 1967.

VFW-Fokker VAK-191B

SPECIFICATIONS	
Country of origin:	Germany/Holland
Type:	experimental V/STOL aircraft
Powerplant:	two Rolls-Royce R.B 162-81 lift jets and one Rolls-Royce/MTU R.B 193-12 vectored/thrust turbofan for forward propulsion.
Performance:	maximum speed (est) 1046km/h (650 mph); service ceiling (est) 15,250m (50,000ft); range with maximum fuel after vertical take-off 500km (311 miles)
Weights:	maximum vertical take-off 8000kg (17,625lb)
Dimensions:	wingspan 6.16m (20ft); length 13m (42ft 0.5in); height 4m (13ft)
Armament:	none

A West German-Dutch collaboration, the VAK 191B V/STOL was a reconnaissance-strike aircraft with a fairly conventional swept-wing monoplane configuration design. The first of three prototype aircraft flew in 1971, but the small wing hindered short take-off and landing and the project was terminated in the mid-1970s.

VERTICAL TAKE-OFF TIMELINE

 1960  1964

VERTICAL TAKE-OFF

Yakovlev Yak-38 'Forger-A'

SPECIFICATIONS	
Country of origin:	USSR (CIS)
Type:	V/STOL carrier-based fighter-bomber
Powerplant:	two 29.9kN (6724lb) Rybinsk RD-36-35VFR lift turbojets; one 6950kg (15,322lb) Tumanskii R-27V-300 vectored-thrust turbojet
Performance:	maximum speed at high altitude 1009km/h (627mph); service ceiling 12,000m (39,370ft); combat range on hi-lo-hi mission with maximum weapon load 370km (230 miles)
Weights:	empty 7485kg (16,502lb); maximum take-off weight 11,700kg (25,795lb)
Dimensions:	wingspan 7.32m (24ft); length 15.5m (50ft 10in); height 4.37m (14ft 4in); wing area 18.5 sq m (199.14sq ft)
Armament:	four external hardpoints with provision for 2000kg (4409lb) of stores, including missiles, bombs, pods and drop tanks

Apart from the Harrier, the Yak-38 was the only operational jet VTOL aircraft in the world, albeit a far less capable one. Operational service began in 1976. Unlike the Harrier, the Yak-38 used two fixed turbojets mounted in tandem behind the cockpit for lift, with auxiliary inlets on the top of the fuselage. These were augmented by a third vectoring thrust unit in the rear fuselage.

Hawker Siddeley P.1127

SPECIFICATIONS	
Country of origin:	United Kingdom
Type:	Experimental VTOL aircraft
Powerplant:	one 67kW (15,000lb) thrust Bristol Siddeley Pegasus 3 vectored-thrust turbofan engine
Performance:	maximum speed 878km/h (545mph); range unknown; service ceiling unknown
Weights:	empty 4500kg (10,000lb); maximum 17,000kg (7700lb)
Dimensions:	span 6.99m (22ft 11in); length 12.95m (42ft 6in); height 3.28m (10ft 9in); wing area 18.68 sq m (201 sq ft)
Armament:	none

The P.1127, the first of what would become the Harrier family, was flown in November 1960. A great deal of test flying in conventional mode and ground tests preceded the first full VTOL flight in September 1961. Five P.1127s were built, one of which made the first vertical carrier landing in 1963.

Hawker Siddeley Kestrel FGA.1

SPECIFICATIONS	
Country of origin:	United Kingdom
Type:	Experimental VTOL aircraft
Powerplant:	one 51.2kN (11,500lb) thrust Bristol Siddeley Pegasus 6 vectored-thrust turbofan engine
Performance:	maximum speed 878km/h (545mph); range unknown; service ceiling unknown
Weights:	empty 4500kg (10,000lb); maximum 17,000kg (7700lb)
Dimensions:	span 6.99m (22ft 11in); length 12.95m (42ft 6in); height 3.28m (10ft 9in); wing area 18.68 sq m (201 sq ft)
Armament:	none

The success of the experimental P.1127 led to the Kestrel for operational evaluation. A tri-national test unit consisting of the United Kingdom, West Germany and the United States (Army and Navy), called the Tri-partite Evaluation Squadron, flew the nine Kestrels on simulated missions in 1964–65.

1971

First-Generation Harriers

No 1 Squadron RAF introduced the first VTOL fighters to service in 1969. The aircraft soon attracted the attention of the US Marine Corps and an order for 110 AV-8As was made in 1971. Later, Spain and Thailand purchased AV-8As for use off their small carriers. The two-seat versions were produced to introduce pilots to the quirks of VTOL flying, and helped reduce the number of accidents suffered during hovering flight.

AV-8A Harrier

The US AV-8A was equivalent to the RAF's Harrier GR.1s with a few changes, such as AIM-9 Sidewinder capability. The Marines operated AV-8s from their amphibious carriers and shore bases in the close air-support role. The AV-8C designation was given to 47 rebuilt AV-8As.

SPECIFICATIONS	
Country of origin:	United Kingdom
Type:	V/STOL close support and reconnaissance aircraft
Powerplant:	one 91.2kN (20,500lb) thrust Rolls-Royce Pegasus 10 vectored-thrust turbofan engine
Performance:	maximum speed at low altitude over 1186km/h (737mph); service ceiling over 15,240m (50,000ft); range with one inflight refuelling 5560 km (3455 miles)
Weights:	basic operating empty 5579kg (12,300lb); maximum take-off weight 11,340kg (25,000lb)
Dimensions:	wingspan 7.7m (25ft 3in); length 13.87m (45ft 6in); height 3.45m (11ft 4in); wing area 18.68 sq m (201.1 sq ft)
Armament:	maximum of 2268kg (5000lb) of stores on underfuselage and underwing points; one 30mm Aden gun or similar gun, with 150 rounds, rockets, bombs

Harrier GR.3

The GR.3 Harrier is essentially the same as the GR.1, but with a retrofitted 9753kg (21,500lb) Rolls-Royce Pegasus 103 turbofan. Standard equipment on the GR.3 included inflight refuelling equipment, head-up display and a laser rangefinder. From 1970, the aircraft equipped one RAF squadron in the United Kingdom and three in Germany.

SPECIFICATIONS	
Country of origin:	United Kingdom
Type:	V/STOL close support and reconnaissance aircraft
Powerplant:	one 95.6kN (21,500lb) Rolls-Royce Pegasus vectored thrust turbofan
Performance:	maximum speed over 1186km/h (737mph); service ceiling over 15,240m (50,000ft); range with one inflight refuelling 5560km (3455 miles)
Weights:	empty 5579kg (12,300lb); maximum take-off weight 11,340kg (25,000lb)
Dimensions:	wingspan 7.7m (25ft 3in); length 13.87m (45ft 6in); height 3.45m (11ft 4in); wing area 18.68 sq m (201.1sq ft)
Armament:	maximum of 2268kg (5000lb) of stores; one 30mm Aden gun or similar gun, with 150 rounds, rockets, bombs

FIRST-GENERATION HARRIERS

Harrier T.4

The Harrier GR.1 for the RAF was followed by the T.2 conversion trainer. When the GR.3 was introduced, surviving T.2s were modified to T.4s by adding the laser rangefinder in the nose and the GR.3's other avionics. Twenty-five T.4s were built or converted from T.2s.

SPECIFICATIONS	
Country of origin:	United Kingdom
Type:	V/STOL close support and reconnaissance aircraft
Powerplant:	one 95.6kN (21,500lb) Rolls-Royce Pegasus vectored thrust turbofan
Performance:	maximum speed over 1186km/h (737mph); service ceiling over 15,240m (50,000ft); range with one inflight refuelling 5560km (3455 miles)
Weights:	empty 5579kg (12,300lb); maximum take-off weight 11,340kg (25,000lb)
Dimensions:	wingspan 7.7m (25ft 3in); length 13.87m (45ft 6in); height 3.45m (11ft 4in); wing area 18.68 sq m (201.1 sq ft)
Armament:	maximum of 2268kg (5000lb) of stores; one 30mm Aden gun or similar gun, with 150 rounds, rockets, bombs

AV-8S Matador

Spain's navy bought 12 early Harriers from the United States to operate from their World War II-era escort carrier *Dédalo* in 1976. These were known locally as AV-8S Matadors. Two were TAV-8S trainers. The Spanish later bought AV-8B Harrier IIs and sold its remaining AV-8Ss on to Thailand.

SPECIFICATIONS	
Country of origin:	United Kingdom
Type:	V/STOL close support and reconnaissance aircraft
Powerplant:	one 91.2kN (20,500lb) thrust Rolls-Royce Pegasus 11 vectored-thrust turbofan engine
Performance:	maximum speed over 1186km/h (737mph); service ceiling over 15,240m (50,000ft); range with one inflight refuelling 5560km (3455 miles)
Weights:	empty 5579kg (12,300lb); maximum take-off weight 11,340kg (25,000lb)
Dimensions:	wingspan 7.7m (25ft 3in); length 13.87m (45ft 6in); height 3.45m (11ft 4in); wing area 18.68 sq m (201.1sq ft)
Armament:	maximum of 2268kg (5000lb) of stores; one 30mm Aden gun or similar gun, with 150 rounds, rockets, bombs

Sea Harrier FRS.Mk 1

The Sea Harrier FRS.1 was ordered to equip the three Royal Navy 'through-deck cruisers' in the fighter, anti-submarine and surface-attack roles. Installing the Blue Fox radar meant lengthening the nose, and the cockpit was raised to accommodate a more substantial avionics suite and to afford the pilot a better all-round view. The aircraft proved an important asset during the Falklands War.

SPECIFICATIONS	
Country of origin:	United Kingdom
Type:	shipborne multi-role combat aircraft
Powerplant:	one 95.6kN (21,500lb) Rolls-Royce Pegasus vectored thrust turbofan
Performance:	maximum speed 1110km/h (690mph); service ceiling 15,545m (51,000ft); intercept radius 740km (460 miles)
Weights:	empty 5942kg (13,100lb); maximum take-off weight 11,884kg (26,200lb)
Dimensions:	wingspan 7.7m (25ft 3in); length 14.5m (47ft 7in); height 3.71m (12ft 2in); wing area 18.68sq m (201.1sq ft)
Armament:	two 30mm cannon, provision for AIM-9 Sidewinder or Matra Magic air-to-air missiles, and two Harpoon or Sea Eagle anti-shipping missiles, up to a total of 3629kg (8000lb) bombs

COLD WAR MILITARY AIRCRAFT

Last of the Flying Boats

For nations with significant sea areas to protect, large flying boats and amphibians continued to play a military role long after World War II. Patrol of sea lanes, anti-submarine warfare, mine-laying and combat search-and-rescue were all duties suited to aircraft flying low over water. Even today, Japan possesses a small fleet of Shin Meiwa amphibians, and Beriev has a jet flying boat in production, mainly for use as a firefighting aircraft.

Martin P5M Marlin

The Marlin was a post-war replacement for the Mariner. The US Navy used them throughout the Vietnam War to catch boats from the North supplying the Viet Cong. France leased 10 P5Ms in 1959 for use by the Aéronavale in West Africa.

SPECIFICATIONS	
Country of origin:	United States
Type:	(PFM-2) twin-engined flying boat
Powerplant:	two 2570kW (3450hp) Wright R-3350-32WA radial piston engines
Performance:	maximum speed 404km/h (251mph); range 3300km (2050 miles); service ceiling 7300m (24,000ft)
Weights:	empty 22,900kg (50,485lb); maximum 38,600kg (85,000lb)
Dimensions:	span 35.7m (117ft 2in); length 30.7m (100ft 7in); height 100m (32ft 9in); wing area 130.1 sq m (1406 sq ft)
Armament:	up to 8000lb of bombs, mines or torpedoes or one nuclear depth charge

Beriev Be-10 'Mallow'

Given the Nato reporting name 'Mallow', the Be-10 was one of a few jet-engined flying-boat types that were built in small numbers in the 1940s and 1950s. The Be-10 was displayed in public in 1961 and broke several records in its class, but did not enter large-scale service.

SPECIFICATIONS	
Country of origin:	USSR
Type:	twin-engined jet flying boat
Powerplant:	two 71.2kN (16,000lb) thrust Lyulka AL-7PB turbojet engines
Performance:	maximum speed 910km/h (565mph); service ceiling 12,500m (41,010ft); range 3150km (1957 miles)
Weights:	empty 27,600kg (60,900lb); maximum 48,500kg (106,900lb)
Dimensions:	span 28.6m (93ft 10in); length 31.45m (103ft 2in); height 10.7m (35ft 1in); wing area 130 sq m (1400 sq ft)
Armament:	four 23mm AM-23 cannon; up to 1360kg (3000lb) of bombs, torpedoes or mines

FLYING BOATS TIMELINE    1937    1948    1956

LAST OF THE FLYING BOATS

Beriev Be-12 'Mail'

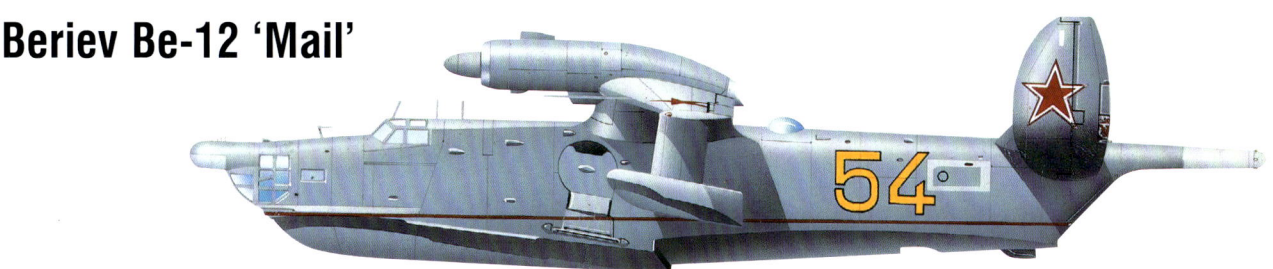

The Be-12 'Mail' amphibian was regarded as more worthy of development than the 'Mallow'. Derived from the piston-engined Be-6, it first flew in 1960 and was frequently encountered by Western pilots on patrol over the oceans of the world. A handful are still in Russian service.

SPECIFICATIONS

Country of origin:	USSR
Type:	twin-engined flying boat
Powerplant:	two 3864kW (5180hp) Ivchenko Progress AI-20D turboprop engines
Performance:	maximum speed 530km/h (330mph); range 3300km (2100 miles); service ceiling 8000m (26,247ft)
Weights:	empty 24,000kg (52,800lb); maximum 36,000kg (79,200lb)
Dimensions:	span 29.84m (97ft 11in); length 30.11m (98ft 9in); height 7.94m (26ft 1in); wing area 99 sq m (1065 sq ft)
Armament:	1500kg (3300lb) of bombs, depth-charges or torpedoes

Short Sunderland

The RNZAF used four Sunderland IIIs in wartime, and bought a further 16 in the early 1950s for patrol of New Zealand and the Pacific Islands. The last examples served with No 5 Squadron until fully replaced by P-3B Orions in 1967.

SPECIFICATIONS

Country of origin:	United Kingdom
Type:	(Sunderland Mk I) 10-seat maritime reconnaissance flying boat
Powerplant:	four 753kW (1010hp) Bristol Pegasus XXII nine-cylinder single-row radial engines
Performance:	maximum speed 336km/h (209mph); service ceiling 4570m (15,000ft); range 4023km (2500 miles)
Weights:	empty 13,875kg (30,589lb); maximum take-off weight 22,226kg (49,000lb)
Dimensions:	span 34.38m (112ft 9.5in); length 26.0m (85ft 3.5in); height 10.52m (34ft 6in)
Armament:	eight .303in machine guns in bow turret, in tail turret, in each beam position; internal bomb, depth charge and mine load of 907kg (2000lb)

Shin Meiwa PS-1

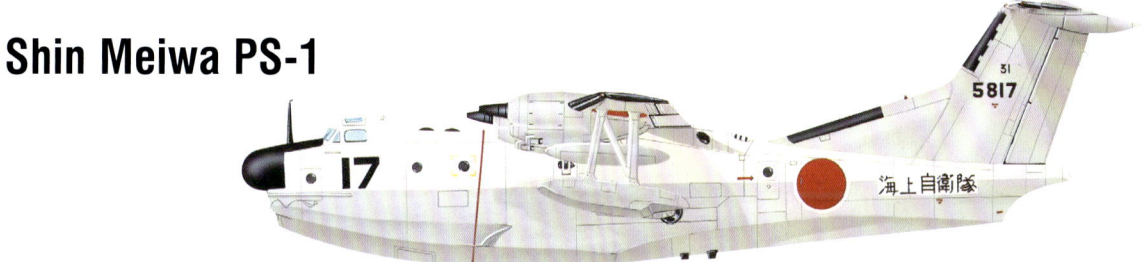

A successor to the wartime Kawanishi company, Shin Meiwa of Japan designed the PS-1 flying boat for the Japanese Maritime Self Defence Force in the late 1960s. The US-1A is an air-sea rescue variant with a retractable landing gear, and is still in service in small numbers.

SPECIFICATIONS

Country of origin:	Japan
Type:	four-engined flying boat
Powerplant:	four 2250kW (3017hp) General Electric T-64-IHI-10J turboprop engines
Performance:	maximum speed 545km/h (339mph); range 4700km (2921 miles); service ceiling 9000m (29,550ft)
Weights:	empty 26,300kg (57,982lb); maximum 43,000kg (94,799lb)
Dimensions:	span 33.1m (108ft 7in); length 33.5m (109ft 11in); height 9.7m (31ft 10in); wing area 135.8 sq m (1462 sq ft)
Armament:	bombs, torpedoes or depth charges

1960  1967

COLD WAR MILITARY AIRCRAFT

Convair Deltas

In 1948, Convair flew the world's first delta-wing aircraft, the XF-92A, which was part of a programme intended to lead to a supersonic fighter. This was terminated, but the US Air Force later issued a specification for an extremely advanced all-weather interceptor to carry the Hughes MX-1179 electronic control system. This effectively made the carrier aircraft subordinate to its avionics, a radical concept in the early 1950s. Convair was awarded the contract in September 1961.

F-102 Delta Dagger

Early flight trials of the F-102 prototype were disappointing, but once the design was right, 875 were delivered. In the search mode, the pilot flew with two control columns – the left-hand one was used to adjust the sweep angle and range of the radar.

SPECIFICATIONS	
Country of origin:	United States
Type:	supersonic all-weather single-seat fighter-interceptor
Powerplant:	one 76.5kN (17,200lb) Pratt & Whitney J57-P-23 turbojet
Performance:	maximum speed at 10,970m (36,000ft) 1328km/h (825mph); service ceiling 16,460m (54,000ft); range 2172km (1350 miles)
Weights:	empty 8630kg (19,050lb); maximum take-off weight 14,288kg (31,500lb)
Dimensions:	wingspan 11.62m (38ft 1.5in); length 20.84m (68ft 4.5in); height 6.46m (21ft 2.5in); wing area 61.45 sq m (661.5 sq ft)
Armament:	two AIM-26/26A Falcon missiles, or one AIM-26/26A plus two AIM-4A Falcons, or one AIM-26/26A plus two AIM-4C/Ds, or six AIM-4As, or six AIM-4C/Ds, some aircraft fitted with 12 2.75in folding-fin rockets

F-102 Delta Dagger

The only export customers for the Delta Dagger were Greece and Turkey. In 1974, when the countries fought over Cyprus, the two sides' F-102s did not endure any decisive engagements with each other. Each side had retired its 'Deuces' by 1980.

SPECIFICATIONS	
Country of origin:	United States
Type:	supersonic all-weather single-seat fighter-interceptor
Powerplant:	one 76.5kN (17,200lb) Pratt & Whitney J57-P-23 turbojet
Performance:	maximum speed at 10,970m (36,000ft) 1328km/h (825mph); service ceiling 16,460m (54,000ft); range 2172km (1350 miles)
Weights:	empty 8630kg (19,050lb); maximum take-off weight 14,288kg (31,500lb)
Dimensions:	wingspan 11.62m (38ft 1.5in); length 20.84m (68ft 4.5in); height 6.46m (21ft 2.5in); wing area 61.45 sq m (661.5 sq ft)
Armament:	Various combinations of AIM missiles, some aircraft fitted with 12 2.75in folding-fin rockets

TF-102 Delta Dagger

SPECIFICATIONS

Country of origin:	United States
Type:	supersonic all-weather single-seat fighter-interceptor
Powerplant:	one 76.5kN (17,200lb) Pratt & Whitney J57-P-23 turbojet
Performance:	maximum speed at 10,970m (36,000ft) 1328km/h (825mph); service ceiling 16,460m (54,000ft); range 2172km (1350 miles)
Weights:	empty 8630kg (19,050lb); maximum take-off weight 14,288kg (31,500lb)
Dimensions:	wingspan 11.62m (38ft 1.5in); length 20.84m (68ft 4.5in); height 6.46m (21ft 2.5in); wing area 61.45 sq m (661.5 sq ft)
Armament:	two AIM-26/26A Falcon missiles, or one AIM-26/26A plus two AIM-4A Falcons, or one AIM-26/26A plus two AIM-4C/Ds, or six AIM-4As, or six AIM-4C/Ds, some aircraft fitted with 12 2.75in folding-fin rockets

The two-seat TF-102A was unusual for a trainer version of a fighter in having a side-by-side seating arrangement. Although it retained the weapons capabilities of the F-102A, its performance was reduced. TF-102s were used for some missions in Vietnam, including as B-52 escorts.

F-106A Delta Dart

SPECIFICATIONS

Country of origin:	United States
Type:	supersonic all-weather single-seat fighter-interceptor
Powerplant:	one 76.5kN (17,200lb) Pratt & Whitney J57-P-23 turbojet
Performance:	maximum speed at 10,970m (36,000ft) 1328km/h (825mph); service ceiling 16,460m (54,000ft); range 2172km (1350 miles)
Weights:	empty 8630kg (19,050lb); maximum take-off weight 14,288kg (31,500lb)
Dimensions:	wingspan 11.62m (38ft 1.5in); length 20.84m (68ft 4.5in); height 6.46m (21ft 2.5in); wing area 61.45 sq m (661.5 sq ft)
Armament:	Various combinations of AIM missiles, some aircraft fitted with 12 2.75in folding-fin rockets

Although the F-102 was designed to carry the Hughes ECS, the avionics were then delivered in time and so were rescheduled for the F-106 programme. This was delayed by engine problems, and flight tests proved disappointing. The aircraft eventually entered service in October 1959 and remained active in updated versions until 1988.

F-106B Delta Dart

SPECIFICATIONS

Country of origin:	United States
Type:	supersonic all-weather single-seat fighter-interceptor
Powerplant:	one 76.5kN (17,200lb) Pratt & Whitney J57-P-23 turbojet
Performance:	maximum speed at 10,970m (36,000ft) 1328km/h (825mph); service ceiling 16,460m (54,000ft); range 2172km (1350 miles)
Weights:	empty 8630kg (19,050lb); maximum take-off weight 14,288kg (31,500lb)
Dimensions:	wingspan 11.62m (38ft 1.5in); length 20.84m (68ft 4.5in); height 6.46m (21ft 2.5in); wing area 61.45 sq m (661.5 sq ft)
Armament:	Various combinations of AIM missiles, some aircraft fitted with 12 2.75in folding-fin rockets

Unlike the TF-102, the trainer version of the Delta Dart, the F-106B – which was also combat capable – had its two seats arranged in tandem. Each F-106 unit, including two in the California Air National Guard, had a two-seater assigned.

Soviet Bombers

The Soviet Union's Long Range Aviation (Dalnaya Aviatsiya) force evolved from using Soviet copies of the B-29 Superfortress, and by the early 1960s it had become a powerful force of jet and turboprop nuclear-armed bombers. The Tupolev Design Bureau's 'Bear', 'Badger' and 'Blinder' bombers formed the backbone of the force. Unlike Strategic Air Command, Long Range Aviation did not mount airborne patrols and had far fewer tankers, relying on staging bases in the Arctic instead.

Ilyushin Il-28 'Beagle'

First appearing in prototype form as early as 1948, the Il-28 afforded Eastern Bloc armed forces the same degree of flexibility and duration of service as the Canberra did for Britain. The prototype was powered by two Soviet-built turbojets developed directly from the Rolls-Royce Nene, supplied by the British government in a fit of contrition.

SPECIFICATIONS	
Country of origin:	USSR
Type:	three seat bomber and ground attack/dual control trainer/torpedo carrier
Powerplant:	two 58.1kN (13,062lb) Klimov VK-1 turbojets
Performance:	maximum speed 902km/h (560mph); service ceiling 12,300m (40,355ft); range 2180km (1355 miles); with bomb load 1100km (684 miles)
Weights:	empty 12890kg (28,418lb); maximum take-off weight 21,200kg (46,738lb)
Dimensions:	wingspan 21.45m (70ft 4.5in); length 17.65m (57ft 10.75in); height 6.7m (21ft 11.8in); wing area 60.8 sq m (654.47 sq ft)
Armament:	four 23mm NR-23 cannon; internal bomb capacity up to 1000kg (2205lb), maximum bomb capacity 3000kg (6614lb); torpedo version had provision for two 400mm light torpedoes

Myasishchev M-4 'Bison C'

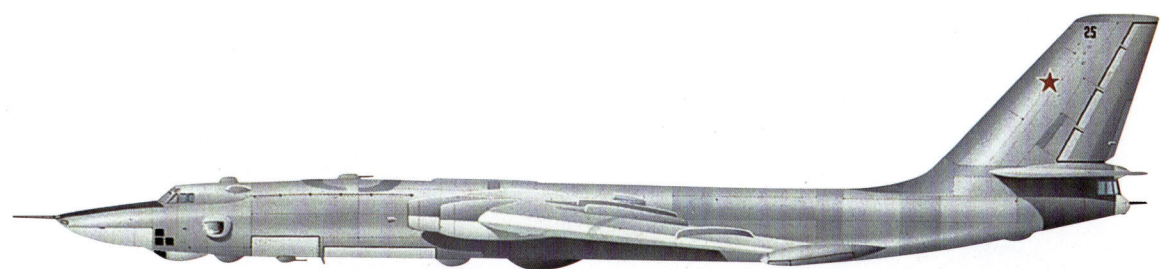

The M-4 was produced in some numbers as the 'Bison-A' strategic bomber, before being adapted for long-range strategic reconnaissance and ECM duties. In the 'Bison-C' sub-type, a large search radar was fitted inside a lengthened and modified nose. The 'C' model was most frequently encountered on high- and low-level missions over the Arctic, and the Atlantic and Pacific oceans.

SPECIFICATIONS	
Country of origin:	USSR
Type:	multi-role reconnaissance bomber
Powerplant:	four 127.4kN (28,660lb) Soloviev D-15 turbojets
Performance:	maximum speed 900km/h (560mph); service ceiling 15,000m (49,200ft); range 11,000km (6835 miles)
Weights:	empty 80,000kg (176,400lb); loaded 170,000kg (375,000lb)
Dimensions:	wing span 50.48m (165ft 7.5in); length 47.2m (154ft 10in); height 14.1m (46ft); wing area 309 sq m (3,326.16s q ft)
Armament:	six 23mm cannon in two forward turrets and tail turret; internal bay with provision over 4500kg (10,000lb) of stores

SOVIET BOMBERS

Tupolev Tu-22R 'Blinder-C'

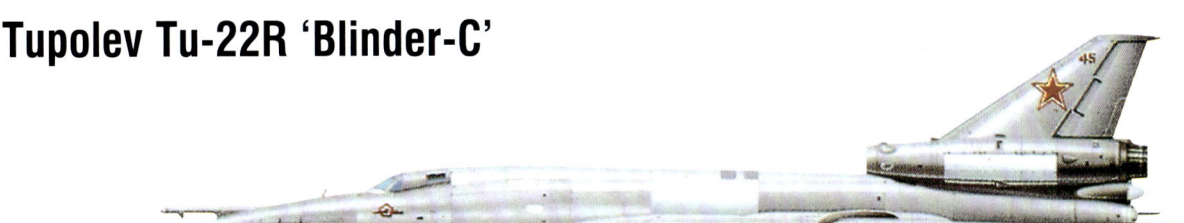

SPECIFICATIONS	
Country of origin:	USSR
Type:	long-range maritime reconnaissance/patrol aircraft
Powerplant:	two (estimated) 117.6kN (26,455lb) Koliesov VD-7 turbojets
Performance:	maximum speed 1487km/h (924mph); service ceiling 18,300m (60,040ft); combat radius with internal fuel 3100km (1926 miles)
Weights:	empty 40,000kg (88,185lb); maximum take-off 84,000kg (185,188lb)
Dimensions:	wingspan 23.75m (77ft 11in); length 40.53m (132ft 11.75in); height 10.67m (35ft); wing area 162 sq m (1722.28 sq ft)
Armament:	one 23mm NR-23 two-barrel cannon in radar-controlled tail barbette; internal weapons bay with provision for 12,000kg (26,455lb) of stores, including nuclear weapons and free-fall bombs, or one AS-4 carried semi-recessed under the fuselage

The Tu-22 was designed in the late 1950s as a replacement for the Tu-16, which was effectively rendered obsolete by a new generation of Western interceptors and missile systems. The Tu-22 'Blinder' was designed to penetrate hostile airspace at high speed and high altitude. The 'Blinder-C' was a dedicated maritime reconnaissance version with cameras and sensors in the weapons bay.

Tupolev Tu-16R 'Badger-K'

SPECIFICATIONS	
Country of origin:	USSR
Type:	medium bomber
Powerplant:	two 13.1kN (20,944lb) Mikulin RD-3M turbojets
Performance:	maximum speed at 6000m (19,685ft) 960km/h (507mph); service ceiling 15,000m (49,200ft); combat range with maximum weapon load 4800km (2983 miles)
Weights:	empty 40,300kg (88,846lb); maximum take-off 75,800kg (167,110lb)
Dimensions:	wingspan 32.99m (108ft 3in); length 36.5m (120ft); height 10.36m (34ft 2in); wing area 164.65 sq m (1772.3 sq ft)
Armament:	one forward and one rear ventral barbette each with two 23mm NR-23 cannon; two 23mm NR-23 cannon in radar-controlled tail position

The Tu-16R was a maritime/electronic reconnaissance version of the Tupolev medium bomber. 'Badger-E' had provision for a photoreconnaissance pallet in the weapons bay and passive Elint capability. 'Badger-F' was similar, but usually carried underwing ESM (Electronic Signal Monitoring) pods. Badger-K was based on Badger-F, but had enhanced Elint capability.

Tupolev Tu-95 'Bear'

SPECIFICATIONS	
Country of origin:	USSR
Type:	four-engined turboprop bomber
Powerplant:	four 11,000kW (14,800hp) Kuznetsov NK-12MV turboprop engines
Performance:	maximum speed 920km/h (575mph); range 15,000km (9400 miles); service ceiling 12,000m (39,000ft)
Weights:	empty 90,000kg (198,000lb); maximum 188,000kg (414,500lb)
Dimensions:	span 51.1m (167ft 8in); length 49.5m (162ft 5in); height 12.12m (39ft 9in); wing area 310 sq m (3330 sq ft)
Armament:	two 23mm AM-23 cannon; up to 15,000kg (33,000lb) of bombs or anti-surface or anti-shipping missiles

The Tu-95 'Bear' was unique in being a swept-wing turboprop bomber. Although slower than its counterparts, it had enormous range. Entering service in 1956, it remains in use by Russia. Later versions were equipped to carry variety of cruise missiles.

Fleet Air Arm

With the arrival of new fleet carriers like HMS *Hermes* and *Ark Royal*, and supersonic fighters and attack aircraft, the Fleet Air Arm was able to project airpower and deploy nuclear weapons over large parts of the world. Technologically, however, its aircraft were falling behind those of the United States, and the total number deployed was dropping due to budgetary pressures. Conventional fixed-wing naval aviation was abandoned in 1978 with the retirement of the *Ark Royal*.

Supermarine Scimitar F.1

The Scimitar had an extremely protracted gestation period. The first prototype, the Supermarine 508, was a thin, straight-winged design with a butterfly tail. Production aircraft were delivered from August 1957. A total of 76 were built, providing the Fleet Air Arm with a capable low-level supersonic attacker until the Scimitar was superseded by the Buccaneer in 1969.

SPECIFICATIONS

Country of origin:	United Kingdom
Type:	single-seat carrier-based multi-role aircraft
Powerplant:	two 50kN (11,250lb) Rolls-Royce Avon 202 turbojets
Performance:	maximum speed 1143km/h (710mph); service ceiling 15,240m (50,000ft); range, clean at height 966km (600 miles)
Weights:	empty 9525kg (21,000lb); maximum take-off weight 15,513kg (34,200lb)
Dimensions:	wingspan 11.33m (37ft 2in); length 16.87m (55ft 4in); height 4.65m (15ft 3in); wing area 45.06 sq m (485 sq ft)
Armament:	four 30mm Aden cannon, four 454kg (1000lb) bombs or four Bullpup air-to-ground missiles, or four sidewinder air-to-air missiles, or drop tanks

De Havilland Sea Vixen FAW.2

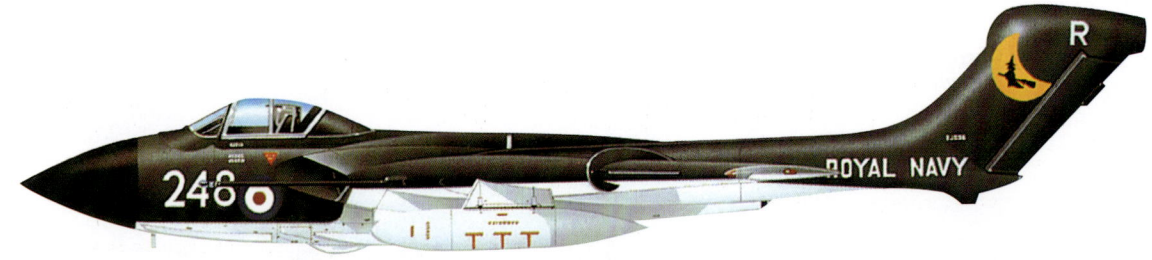

The Sea Vixen, like many of the aircraft operated by the Royal Navy, was originally designed to a 1946 RAF requirement for a land-based, all-weather interceptor. Mk 1s featured a hinged and pointed radome, power-folding wings and hydraulically steerable nosewheel. The FAW.2 had increased fuel capacity and provision for four Red Top missiles.

SPECIFICATIONS

Country of origin:	United Kingdom
Type:	two-seat all-weather strike fighter
Powerplant:	two 49.9kN (11,230lb) Rolls-Royce Avon 208 turbojets
Perfomance:	maximum speed 1110km/h (690mph); service ceiling 21,790m (48,000ft); range about 965.6km (600 miles) (FAW 1) and 1287.5km (800 miles) (FAW 2)
Weights:	empty weight about 22,000lb; maximum take-off 18,858kg (41,575lb)
Dimensions:	wingspan 15.54m (51ft); length 17.02 m (55ft 7in); height 3.28 m (10ft 9in); wing area 60.2 sq m (648 sq ft)
Armament:	four Red Top air-to-air missiles (FAW 2); on outer pylons 1000lb bombs, Bullpup air-to-surface missiles or equivalent stores

FLEET AIR ARM

Fairey Gannet AS.1

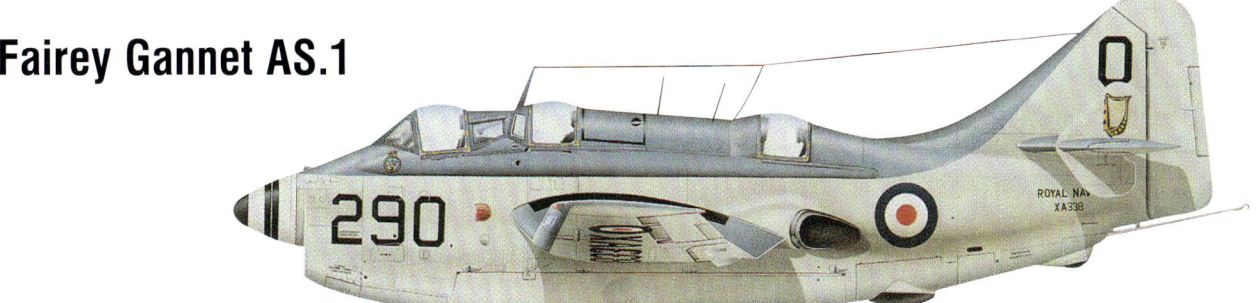

The Gannet AS.1 three-seat anti-submarine aircraft was the first of a family of variants that included early-warning, transport, trainer and electronic-warfare aircraft. The Double Mamba turboprop was essentially two coupled engines and drove counter-rotating propellers.

SPECIFICATIONS

Country of origin:	United Kingdom
Type:	carrier-based anti-submarine aircraft
Powerplant:	one 2200kW (2950hp) Double Mamba 100 turboprop engine
Performance:	maximum speed 499km/h (310mph); range 1111km (690 miles); service ceiling 7620m (25,000ft)
Weights:	empty 6835kg (15,069lb); maximum 8890kg (19,600lb)
Dimensions:	span 16.54m (54ft 4in); length 13.11m (43ft); height 4.18m (13ft 9in); wing area 44.9 sq m (483 sq ft)
Armament:	up to 908kg (2000lb) of torpedoes, mines, bombs or depth charges

Blackburn Buccaneer S.2

The Buccaneer was the first aircraft to be designed specifically for carrierborne strike operations at below radar level. The S.1 was marginal on power, but the greatly improved S.2 was a reliable and formidable aircraft. The first 84 were ordered by the Royal Navy, and after giving good service most were transferred to the Royal Air Force from 1969.

SPECIFICATIONS

Country of origin:	United Kingdom
Type:	two-seat attack aircraft
Powerplant:	two 50.4kN (11,255lb) Rolls-Royce RB.168 Spey Mk 101 turbofans
Performance:	maximum speed at 61m (200ft) 1040km/h (646mph); service ceiling over 12,190m (40,000ft); combat range with typical weapons load 3701km (2300 miles)
Weights:	empty 13,608kg (30,000lb); maximum take-off 28,123kg (62,000lb)
Dimensions:	wingspan 13.41m (44ft); length 19.33m (63ft 5in); height 4.97m (16ft 3in); wing area 47.82 sq m (514.7 sq ft)
Armament:	four 454kg (1000lb) bombs, fuel tank or reconnaissance pack, provision for 5443kg (12,000lb) of bombs or missiles

McDonnell Douglas Phantom FG.1

The Royal Navy's decision to buy the Phantom was governed by a requirement that the aircraft be equipped with British-built engines. To this end, an Anglicized version of the F-4J was produced, powered by two Rolls-Royce Spey turbofans; fitting these engines necessitated widening the fuselage. Twenty-eight aircraft were delivered to the Navy from 1964, and a further 20 to the Royal Air Force.

SPECIFICATIONS

Country of origin:	United States
Type:	two-seat all-weather fighter/attack carrier-borne aircraft
Powerplant:	two 91.2kN (20,515lb) Rolls-Royce Spey 202 turbofans
Performance:	maximum speed 2230km/h (1386mph); service ceiling over 18,300m (60,000ft); range with no weapon load 2817km (1750 miles)
Weights:	empty 12,700kg (28,000lb); maximum take-off 26,308kg (58,000lb)
Dimensions:	span 11.7m (38ft 5in); length 17.55m (57ft 7in); height 4.96m (16ft 3in); wing area 49.24 sq m (530 sq ft)
Armament:	four AIM-7 Sparrow; provision for 20mm M61A1 cannon; wing pylons for stores to a maximum weight of 7257kg (16,000lb)

F-5 Freedom Fighter/Tiger II

Northrop developed a lightweight combat aircraft under the designation N-156 in the late 1950s. The USAF used relatively few, but adopted a derivative as the T-38 Talon trainer. As the F-5 Freedom Fighter and the improved F-5E Tiger II, however, the design was a big export hit, used by 35 countries – most of whom still operate the aircraft today. Supersonic but simple, the F-5 was a good combat plane, but could also act as an adversary or aggressor in training missions.

T-38 Talon

The highly successful T-38 trainer aircraft was derived from a requirement issued by the US Government in the mid-1950s for a lightweight fighter to supply to friendly nations under the Military Assistance Program. Service began with the USAF in March 1961. The T-38 was used by the US Air Force Thunderbirds display team between 1974 and 1981.

SPECIFICATIONS	
Country of origin:	United States
Type:	two-seat supersonic basic trainer
Powerplant:	two 17.1kN (3850lb) General Electric J85-GE-5 turbojets
Performance:	maximum speed at 10,975m (36,000ft) 1381km/h (858mph); service ceiling 16,340m (53,600ft); range with internal fuel 1759km (1093 miles)
Weights:	empty 3254kg (7174lb); maximum take-off weight 5361kg (11,820lb)
Dimensions:	wingspan 7.7m (25ft 3in); length 14.14m (46ft 4.5in); height 3.92m (12ft 10.5in); wing area 15.79 sq m (170 sq ft)
Armament:	none

F-5A Freedom Fighter

The F-5A was a largely privately funded project by Northrup. In October 1962, the US Department of Defense decided to buy the aircraft in large numbers to supply friendly countries on advantageous terms. More than 1000 were supplied to Iran, Taiwan, Greece, South Korea, Phillipines, Turkey, Ethiopia, Morocco, Norway, Thailand, Libya and South Vietnam.

SPECIFICATIONS	
Country of origin:	United States
Type:	light tactical fighter
Powerplant:	two 18.1kN (4080lb) General Electric J85-GE-13 turbojets
Performance:	maximum speed at 10,975m (36,000ft) 1487km/h (924mph); service ceiling 15,390m (50,500ft); combat radius 314km (195 miles)
Weights:	empty 3667kg (8085lb); maximum take-off weight 9374kg (20,667lb)
Dimensions:	wingspan 7.7m (25ft 3in); length 14.38m (47ft 2in); height 4.01m (13ft 2in); wing area 15.79 sq m (170 sq ft)
Armament:	two 20mm M39 cannon; provision for 1996kg (4400lb) of stores on external pylons, including missiles, bombs, rocket launcher pods

F-5 FREEDOM FIGHTER/TIGER II

F-5B Freedom Fighter

SPECIFICATIONS	
Country of origin:	United States
Type:	light tactical fighter
Powerplant:	two 18.1kN (4080lb) General Electric J85-GE-13 turbojets
Performance:	maximum speed at 10,975m (36,000ft) 1487km/h (924mph); service ceiling 15,390m (50,500ft); combat radius 314km (195 miles)
Weights:	empty 3667kg (8085lb); maximum 8936kg (19,700lb)
Dimensions:	wingspan 7.7m (25ft 3in); length 14.14 m (46 ft 5 in); height 4.01m (13ft 2in); wing area 15.79 sq m (170 sq ft)
Armament:	provision for 1996kg (4400lb) of stores on external pylons, including air-to-air missiles, bombs, cluster bombs, rocket launcher pods

The two-seat F-5B is close in appearance to the T-38A, but is heavier, combat-capable and shares the F-5A's wing, braking parachute and other features. Over 200 were built, mainly for export. The example shown is from the Netherlands Air Force's 313 Squadron.

F-5E Tiger II

The F-5E Tiger II won a US industry competition in November 1970 for a follow-on International Fighter Aircraft to replace the F-5A. The improved aircraft was equipped with more powerful powerplants, extending nosegear to improve short field performance, extra fuel in a longer fuselage, new inlet ducts, widened fuselage and wing, root extensions and manoeuvring flaps. Deliveries began in 1972. The US Air Force still operates the aircraft for aggressor training.

SPECIFICATIONS	
Country of origin:	United States
Type:	light tactical fighter
Powerplant:	two 22.2kN (5000lb) General Electric J85-GE-21B turbojets
Performance:	maximum speed at 10,975m (36,000ft) 1741km/h (1082mph); service ceiling 15,790m (51,800ft); combat radius 306km (190 miles)
Weights:	empty 4410kg (9723lb); maximum take-off weight 11,214kg (24,722lb)
Dimensions:	wingspan 8.13m (26ft 8in); length 14.45m (47ft 4.75in); height 4.07m (13ft 4.25in); wing area 17.28 sq m (186 sq ft)
Armament:	two 20mm cannon; two air-to-air missiles, five external pylons with provision for 3175kg (7000lb) of stores, including missiles, bombs, ECM pods, cluster bombs, rocket launcher pods and drop tanks

RF-5E Tigereye

The RF-5E was a reconnaissance version of the F-5E Tiger, the improved version of the Freedom Fighter. The export success of this aircraft led to the development of a specialized tactical reconnaissance version, which first appeared at the Paris Air Show in 1978. Externally, the RF-5E is similar to the fighter, except for an extended 'chisel' nose housing camera equipment.

SPECIFICATIONS	
Country of origin:	United States
Type:	light tactical reconnaissance fighter
Powerplant:	two 22.2kN (5000lb) General Electric J85-GE-21B turbojets
Performance:	maximum speed at 10,975m (36,000ft) 1741km/h (1082mph); service ceiling 15,390m (50,500ft); combat radius 463km (288 miles)
Weights:	empty 4423kg (9750lb); maximum take-off weight 11,192kg (24,765lb)
Dimensions:	wingspan 8.13m (26ft 8in); length 14.65m (48ft 0.75in); height 4.07m (13ft 4.25in); wing area 17.28 sq m (186 sq ft)
Armament:	one 20mm cannon; two air-to-air missiles, five external pylons with provision for 3175kg (7000lb) of stores, including air-to-surface missiles

Maritime Patrol

Land-based maritime patrol aircraft have tended to be either converted airliner designs (such as the P-3 Orion and Il-38 'May') or purpose-built types like the Shackleton and Neptune. Advantages of the former include pressurized fuselages and greater internal space. Low-level flight over water causes stress on airframes and subjects them to salt corrosion, so maritime patrol aircraft need more frequent inspections and may require rebuilding to maintain their structural strength and to replace outdated avionics several times in their careers.

Avro Shackleton MR.2

Often said to have been derived from the Avro Lancaster, the Shackleton patrol aircraft owed more to the Lincoln, with which it shared wings, engines, tail surfaces and landing gear. Deliveries began in April 1951. The MR (maritime reconnaissance) Mk 2 had a longer fuselage, with the radar moved from the nose to a ventral 'dustbin' position, and a turret with two 20mm (.79in) cannon in the nose, rather than tail guns.

SPECIFICATIONS	
Country of origin:	United Kingdom
Type:	long-range maritime patrol aircraft
Powerplant:	four 1831kW (2455hp) Rolls-Royce Griffon 57A V-12 piston engines
Performance:	maximum speed 500km/h (311mph); service ceiling 6400m (21,000ft); range 5440km (3380 miles)
Weight:	39010kg (86,000lb) loaded
Dimensions:	wing span 36.58m (120ft); length 26.59m (87ft 3in); height 5.1m (16ft 9in)
Armament:	two Hispano No. 1 Mk 5 20mm (.79in) cannon in nose turret; up to 4536kg (10,000lb)

Kawasaki P-2J Neptune

SPECIFICATIONS
Country of origin:	Japan
Type:	four-engined maritime patrol aircraft
Powerplant:	two 2125kW (2850hp) General Electric T64-IHI-10 turboprop and two 13.7kN (3085kg) thrust IHI-JE turbojet engines
Performance:	maximum speed 1650km/h (403mph); service ceiling unknown; range 5633km (3500 miles)
Weights:	empty 19,278 kg (42,500 lb); maximum 34,020 kg (75,000 lb)
Dimensions:	span 30.9m (101ft 4in); length 27.9m (91ft 8in); height 8.9m (29ft 4in); wing area 93 sq m (1,000 sq ft)
Armament:	up to 16 5in rockets and up to 3628kg (8000lb) of bombs, depth charges or torpedoes

The last examples of the Lockheed Neptune series were the 89 produced by Kawasaki in Japan as the P-2J with turboprop main engines. The original piston-engined Neptunes served with many nations, including the United States, Netherlands, United Kingdom, France, Australia, Canada and Japan.

Ilyushin Il-38 'May'

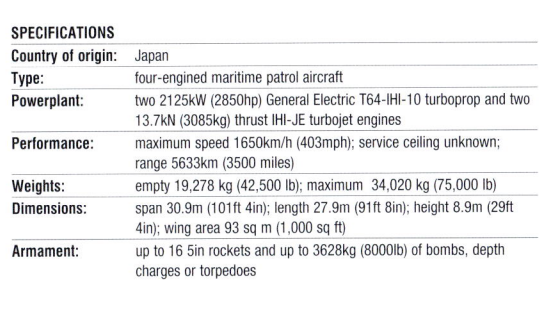

Like the P-3, the Il-38 had its origins as an airliner. The fuselage of the Il-18 'Coot' was stretched and ASW equipment fitted, including a Magnetic Anomaly Detector in a tail boom to create the Il-38 in 1967. About a quarter of the 176 built still serve in Russia, and India bought five.

SPECIFICATIONS
Country of origin:	USSR
Type:	four-engined maritime patrol aircraft
Powerplant:	3170kW (4250hp) Progress AI-20M turboprop engines
Performance:	maximum speed 650km/h (406mph); service ceiling 10,000m (32,800ft); range 9500km (5937 miles)
Weights:	empty 33,700kg (74,140lb); maximum 63,500kg (139,700lb)
Dimensions:	span 37.42m (122ft 9in); length 39.6m (129ft 11in); height 10.16m (33ft 4in); wing area 140 sq m (1506 sq ft)
Armament:	up to 5000kg (11,000lb) of depth-charges, torpedoes or missiles

Falklands: Argentine Navy

When Argentina invaded the Falkland Islands in April 1982, it did not expect the United Kingdom to dispatch a task force to recover the islands. The Argentine Navy was equipped with a conventional aircraft carrier and the islands were in range of land-based patrol aircraft, as well as air-refuelled aircraft equipped with anti-ship weapons such as the Exocet missile, which inflicted considerable damage.

Lockheed L-188PF Electra

The Lockheed Electras of the Argentine Navy were used as blockade-running transports between the mainland and Port Stanley on the Falklands. These flights were made at night and at low level. They also flew patrol missions using their weather radar to search for British ships.

SPECIFICATIONS	
Country of origin:	United States
Type:	four-engined transport
Powerplant:	one four 2796kW (3750hp) Allison 501D-13 turboprop engines
Performance:	maximum speed 652km/h (405mph); range 3541km (2200 miles); service ceiling 28,400ft (9500m)
Weights:	empty 27,895kg (61,500lb); maximum 52,664kg (116,000lb)
Dimensions:	span 30.18m (99m); length 31.81m (104ft 6in); height 10m (32ft 10in); wing area 120.8 sq m (1300 sq ft)
Armament:	none

Lockheed SP-2H Neptune

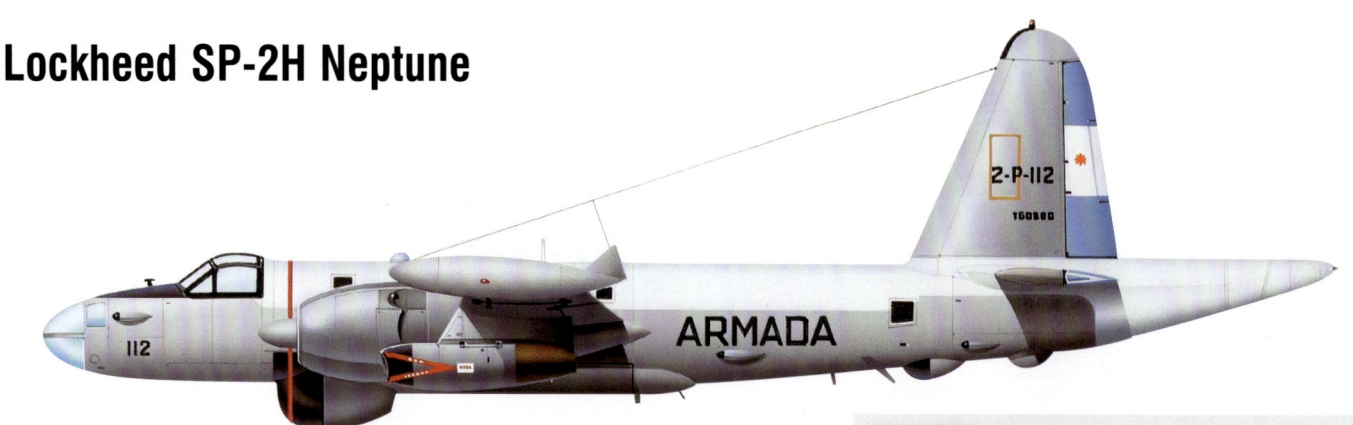

The Argentina Navy's Escuadrilla de Exploracion operated two SP-2H Neptunes in 1982. The aircraft illustrated was used to make radar searches for the task force, and guided the Super Etendards that struck HMS Sheffield with Exocets on 4 May 1982.

SPECIFICATIONS	
Country of origin:	United States
Type:	four-engined maritime patrol aircraft
Powerplant:	two 2759kW (3700hp) Wright R-3350-32W Cyclone radial and two 13.7kN (3085kg) thrust Westinghouse J-34-WE-36 turbojet engines
Performance:	maximum speed 586km/h (364mph); range 3540km (2200 miles); service ceiling 6827m (22,400ft)
Weights:	empty 49,935lb (22,650kg); maximum 79,895lb (35,240kg)
Dimensions:	span 31.65m (103ft 10in); length 27.9m (91ft 8in); height 8.9m (29ft 4in); wing area 93 sq m (1000 sq ft)
Armament:	up to 4540kg (10,000lb) of bombs, mines or torpedoes

FALKLANDS: ARGENTINE NAVY

Grumman S-2E Tracker

SPECIFICATIONS	
Country of origin:	United States
Type:	twin-engined carrier-based anti-submarine aircraft
Powerplant:	two 1135kW (1525hp) Wright R-1820-82WA radial piston engines
Performance:	maximum speed 438km/h (272mph); range 1558km (968 miles); service ceiling 6949m (22,800ft)
Weights:	empty 8505kg (18,750lb); maximum 13,222kg (29,150lb)
Dimensions:	span 21m (69ft 8in); length 12.8m (42ft); height 4.9m (16ft 3in); wing area 45 sq m (485 sq ft)
Armament:	torpedoes, rockets, depth charges

Based on the carrier *25 de Mayo*, the S-2E Trackers of the Escuadrilla Antisubmarina located the British fleet on 2 May but lost it again before a strike by the carrier's Skyhawks could be launched. They also carried out attacks against suspected submarines.

McDonnell Douglas A-4Q Skyhawk

SPECIFICATIONS	
Country of origin:	United States
Type:	single-seat attack bomber
Powerplant:	one 34.7kN (7800lb) J65-W-16A turbojet
Performance:	maximum speed 1078km/h (670mph); service ceiling 14,935m (49,000ft); range with 4000lb load 1480km (920 miles)
Weights:	empty 4809kg (10,602lb); maximum take-off weight 12,437kg (27,420lb)
Dimensions:	wingspan 8.38m (27ft 6in); length excluding probe 12.22m (40ft 1.5in); height 4.66m (15ft 3in); wing area 24.15 sq m (260 sq ft)
Armament:	two 20mm Mk 12 cannon; five external hardpoints with provision for 2268kg (5000lb) of stores, including air-to-surface missiles, bombs, rocket-launcher pods, cannon pods, drop tanks and ECM pods

Argentina has been one of the largest users of the Skyhawk, acquiring many ex-US Navy aircraft. During the late 1960s, it acquired 66 A-4B aircraft, which were refurbished and redesignated A-4P (air force) and A-4Q (navy). During the Falklands conflict, these were used extensively in attacks on British shipping, and despite suffering heavy losses at the hands of the Royal Navy Sea Harrier pilots, Argentine pilots inflicted some damage on the British fleet at San Carlos.

Dassault Super Etendard

SPECIFICATIONS	
Country of origin:	France
Type:	single-seat carrierborne strike/attack and interceptor aircraft
Powerplant:	one 49kN (11,023lb) SNECMA Atar 8K-50 turbojet
Performance:	maximum speed 1180km/h (733mph); service ceiling 13,700m (44,950ft); combat radius 850km (528 miles)
Weights:	empty 6500kg (14,330lb); maximum take-off 12,000kg (26,455lb)
Dimensions:	wingspan 9.6m (31ft 6in); length 14.31m (46ft 11.2in); height 3.86m (12ft 8in); wing area 28.4sq m (305.7sq ft)
Armament:	two 30mm cannon, provision for up to 2100kg (4630lb) of stores, including nuclear weapons and Exocet air-to-surface missiles

Dassault's super Etendard has a substantially redesigned structure, a more efficient engine, inertial navigation system and other upgraded avionics. The first prototype flew on 3 October 1975; deliveries to the Aeronavale began in June 1978. Fourteen Super Etendards were used by Argentina to great effect against British shipping during the Falklands War.

Falklands: Argentine Air Force

The few airfields on the Falklands were not capable of operating combat aircraft beyond the light Aermacchi MB.339 and Pucara attack aircraft. The Fuerza Aerea Argentina had to fly most of its missions from the mainland at the extremes of the range of its fighter-bombers. Only one pass at a target was usually possible, and there was no time for dogfighting over the islands.

Dassault Mirage IIIEA

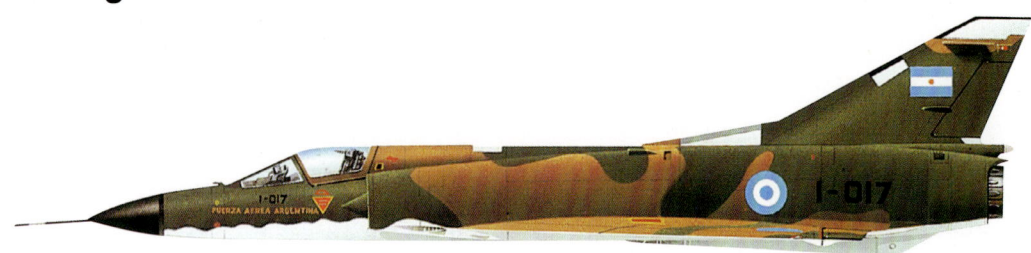

Responding to an Armée de l'Air light-interceptor specification of 1952, Dassault found the Mirage's initial powerplant insufficient and produced a larger, heavier and more powerful aircraft, the Mirage III. In October 1958, a pre-production Mirage IIIA-01 became the first West European fighter to attain Mach 2 in level flight. The longer and heavier IIIE was developed for a ground-attack role, with the Atar 9C turbojet and increased internal fuel.

SPECIFICATIONS	
Country of origin:	France
Type:	single-seat day visual fighter bomber
Powerplant:	one 60.8kN (13,668lb) SNECMA Atar 9C turbojet
Performance:	maximum speed at sea level 1390km/h (883mph); service ceiling 17,000m (55,755ft); combat radius at low level with 907kg (2000lb) load 1200km (745 miles)
Weights:	empty 7050kg (15,540lb); loaded 13,500kg (27,760lb)
Dimensions:	wingspan 8.22m (26ft 11.875in); length 16.5m (56ft); height 4.5m (14ft 9in); wing area 35 sq m (376.7 sq ft)
Armament:	two 30mm DEFA 552A cannon with 125rpg; three external pylons with provision for up to 3000kg (6612lb) of stores, including bombs, rockets, and gun pods

McDonnell Douglas A-4P Skyhawk

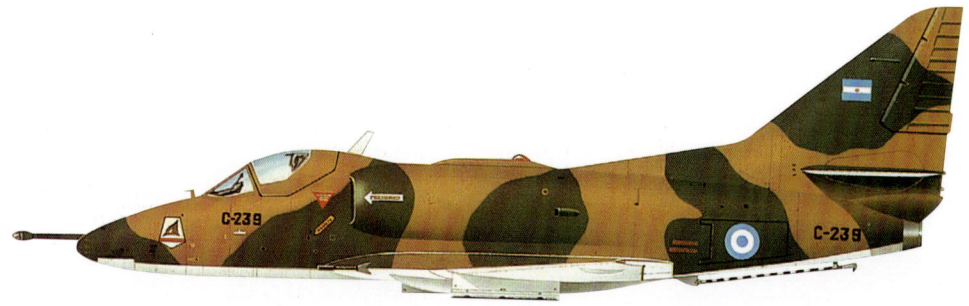

The McDonnell Douglas A-4P (later A-4B) had a strengthened rear fuselage, inflight refuelling equipment, provision for the Martin Bullpup air-to-surface missile, navigation and bombing computer, and the J65-W-16A turbojet. In total, 542 aircraft were built for the US Navy and US Marine Corps, 66 of which were rebuilt in the late 1960s for the Argentine Air Force and Navy as the A-4P. The A-4P was used extensively during the Falklands War.

SPECIFICATIONS	
Country of origin:	United States
Type:	single-seat attack bomber
Powerplant:	one 34.7kN (7800lb) J65-W-16A turbojet
Performance:	maximum speed 1078km/h (670mph); service ceiling 14,935m (49,000ft); range with 4000lb load 1480km (920 miles)
Weights:	empty 4809kg (10,602lb); maximum take-off weight 12,437kg (27,420lb)
Dimensions:	wingspan 8.38m (27ft 6in); length excluding probe 12.22m (40ft 1.5in); height 4.60m (15ft 3in); wing area 24.15 sq m (260 sq ft)
Armament:	two 20mm Mk 12 cannon with 200rpg; five external hardpoints with provision for 2268kg (5000lb) of stores

FALKLANDS: ARGENTINE AIR FORCE

Canberra B.2

Most of the 10 Canberras in service with the FAA's Grupo 2 de Bombardeo were deployed to Trelew Air Base for attacks against British troops on the islands. They also bombed a civilian tanker in error. Two were shot down, one by a Sea Harrier and another by a Sea Dart fired from HMS *Exeter*.

SPECIFICATIONS	
Country of origin:	United Kingdom
Type:	two-seat interdictor aircraft
Powerplant:	two 28.9kN (6500lb) Rolls Royce Avon Mk 101 turbojets
Performance:	maximum speed at 12,192m (40,000ft) 917km/h (570mph); service ceiling 14,630m (48,000ft); range 4274km (2656 miles)
Weights:	empty not published approx 11,790kg (26,000lb); maximum take-off weight 24,925kg (54,950lb)
Dimensions:	wingspan 29.49m (63ft 11in); length 19.96m (65ft 6in); height 4.78m (15ft 8in); wing area 97.08 sq m (1045 sq ft)
Armament:	internal bomb bay with provision for up to 2727kg (6000lb) of bombs, plus an additional 909kg (2000lb) of underwing pylons

FMA IA-58 Pucara

The Fábrica Militar de Aviones (FMA) Pucara was an indigenous counter-insurgency aircraft. Most of those deployed to the islands were destroyed in a Special Air Services raid. A-515 illustrated was captured intact and evaluated in the United Kingdom.

SPECIFICATIONS	
Country of origin:	United Kingdom
Type:	twin-engined light attack aircraft
Powerplant:	two 729kW (978hp) Turbomeca Astazou XVIG turboprop engines
Performance:	maximum speed 500km/h (310mph); range 3710km (2305 miles); service ceiling 10,000m (31,800ft)
Weights:	empty 4020kg (8862lb); maximum 6800kg (14,991lb)
Dimensions:	span 14.5m (47ft 6in); length 14.25m (46ft 9in); height 5.36m (17ft 7in); wing area 30.3 sq m (326 sq ft)
Armament:	two 20mm Hispano-Suiza HS.804 cannon and four 7.62mm FM M2-20 machine guns; up to 1500kg (3300lb) of bombs or rockets

C-130E Hercules

The C-130s of the FAA's Grupo 1 de Transporte Aereo Escuadron 1 flew transport missions to the islands, conducted aerial refuelling of Skyhawks and Super Etendards, and were even used as bombers with weapons racks under the wings. One was shot down on a reconnaissance mission by a Sea Harrier.

SPECIFICATIONS	
Country of origin:	United Kingdom
Type:	four-engined transport/tanker
Powerplant:	four 3021kW (4050hp) Allison T56-A-7A turboprop engines
Performance:	maximum speed 547km/h (340mph); range 3896km (2420 miles); service ceiling 7010m (23,000ft)
Weights:	empty 72,892lb (33,057kg); maximum 79,375kg (175,000lb)
Dimensions:	span 40.4m (132ft 7in); length 29.8m (97ft 9in); height 11.7m (38ft 6in); wing area 162.2 sq m (1745.5 sq ft)
Armament:	(as bomber) 12 500lb bombs on Multiple Ejector Racks

Falklands: British Forces

British fixed-wing airpower in the South Atlantic was mostly restricted to two small aircraft carriers and their Sea Harriers and Harrier GR.3s. The Exocet threat kept the ships to the east of the Falklands, restricting the ability to mount continuous patrols over the islands. The Avro Vulcan made its only combat missions here, flying from Ascension Island to bomb Stanley Airfield and attack radars on the Falklands.

Sea Harrier FRS.1

The Fleet Air Arm's Sea Harriers were assigned to No 809 Squadron on HMS *Invincible* and No 899 Squadron on *Hermes*. This example from the former unit was flown by Flight Lieutenant Dave Morgan when he shot down two Argentine Air Force A-4B Skyhawks over Choiseul Sound on 8 June 1982.

SPECIFICATIONS	
Country of origin:	United Kingdom
Type:	shipborne multi-role combat aircraft
Powerplant:	one 95.6kN (21,500lb) Rolls-Royce Pegasus Mk.104 vectored thrust turbofan
Performance:	maximum speed at sea level 1110km/h (690mph) with maximum AAM load; service ceiling 15,545m (51,000ft); intercept radius 740km (460 miles) on high level mission with full combat reserve
Weights:	empty 5942kg (13,100lb); maximum take-off 11,884kg (26,200lb)
Dimensions:	wingspan 7.7m (25ft 3in); length 14.5m (47ft 7in); height 3.71m (12ft 2in); wing area 18.68sq m (201.1sq ft)
Armament:	two 30mm Aden cannon, provision for AIM-9 Sidewinder air-to-air missiles, and two anti-shipping missiles, up to a total of 3629kg (8000lb)

Sea Harrier FRS.1

Sea Harrier XZ453 of No 899 Squadron damaged a Mirage III on 1 May 1982 over West Falkland with a sidewinder missile. The Argentine fighter was shot down by Argentine anti-aircraft guns while trying to make an emergency landing at Port Stanley. Sea Harriers destroyed 21 enemy aircraft in 1982.

SPECIFICATIONS	
Country of origin:	United Kingdom
Type:	shipborne multi-role combat aircraft
Powerplant:	one 95.6kN (21,500lb) Rolls-Royce Pegasus vectored thrust turbofan
Performance:	maximum speed at sea level 1110km/h (690mph) with maximum AAM load; service ceiling 15,545m (51,000ft); intercept radius 740km (460 miles) on high level mission with full combat reserve
Weights:	empty 5942kg (13,100lb); maximum take-off weight 11,884kg (26,200lb)
Dimensions:	wingspan 7.7m (25ft 3in); length 14.5m (47ft 7in); height 3.71m (12ft 2in); wing area 18.68 sq m (201.1 sq ft)
Armament:	two 30mm Aden cannon, provision for AIM-9 Sidewinder air-to-air missiles, and two anti-shipping missiles, up to a total of 3629kg (8000lb)

BAe Nimrod MR.2P

Hawker Siddeley began the design of the Nimrod in 1964, using the Comet 4C airliner as the basis for a new aircraft to replace the Avro Shackelton in the maritime patrol and anti-submarine warfare roles. Nimrods were very active during the Falklands War; inflight refuelling equipment was hastily added to a number of aircraft to allow them to operate from Ascension Island.

SPECIFICATIONS	
Country of origin:	United States
Type:	maritime patrol and anti-submarine warfare aircraft
Powerplant:	four 54kN (12,140lb) Rolls Royce Spey Mk 250 turbofans
Performance:	maximum speed 925km/h (575mph); service ceiling 12,800m (42,000ft); range on internal fuel 9262km (5,755 miles)
Weights:	empty 39,010kg (86,000lb); maximum take-off 87,090kg (192,000lb)
Dimensions:	wingspan 35m (114ft 10in); length 39.34m (129ft 1in); height 9.08m (29ft 9.5in); wing area 197.04 sq m (2,121 sq ft)
Armament:	internal bay with provision for 6123kg (13,500lb) of stores, including nine torpedoes and/or depth charges; underwing pylons for Harpoon anti-ship missiles or pairs of Sidewinder air-to-air missiles

Lockheed C-130K Hercules C.1

During the war, the RAF rapidly modified six C-130s with refuelling probes to extend their range. Flying from Ascension Island, Hercules made supply drops to the task force in mid-ocean. Several were later modified to refuel other aircraft, and were redesignated C.1Ks.

SPECIFICATIONS	
Country of origin:	United Kingdom
Type:	four-engined transport/tanker
Powerplant:	four 3021kW (4050hp) Allison T56-A-7A turboprop engines
Performance:	maximum speed 547km/h (340mph); range 3896km (2420 miles); service ceiling 7010m (23,000ft)
Weights:	empty 72,892lb (33,057kg); maximum 79,375kg (175,000lb)
Dimensions:	span 40.4m (132ft 7in); length 29.8m (97ft 9in); height 11.7m (38ft 6in); wing area 162.2 sq m (1745.5 sq ft)
Armament:	none

Avro Vulcan B.2

This Vulcan B.2a of No 101 Squadron is depicted in the late low-visibility colour scheme. The matt paint is paler than previous schemes, and Type B (two-colour) roundels and fin flash are also shown. The B.2A was optimized for low-level penetration missions. Vulcan B.2 XM607 of the Waddington Wing flew the first epic 'Black Buck' raid, flying from Ascension Island in mid-Atlantic with multiple aerial refuellings to bomb Stanley airfield. These raids convinced Argentina to keep back Mirage fighters to protect Buenos Aires.

SPECIFICATIONS	
Country of origin:	United Kingdom
Type:	low-level strategic bomber
Powerplant:	four 88.9kN (20,000lb) Olympus Mk.301 turbojets
Performance:	maximum speed 1038km/h (645mph) at high altitude; service ceiling 19,810m (65,000ft); range with normal bomb load about 7403km/h (4600 miles)
Weights:	maximum take-off weight 113,398kg (250,000lb)
Dimensions:	wingspan 33.83m (111ft); length 30.45m (99ft 11in); height 8.28m (27ft 2in); wing area 368.26 sq m (3,964 sq ft)
Armament:	internal weapon bay for up to 21,454kg (47,198lb) bombs

Spyplanes

Gathering intelligence on an enemy's disposition and intentions has now become more a matter of monitoring emissions across the radio spectrum than traditional observation and photography. The role of modern 'spyplanes' is grouped under the banner ISTAR (Intelligence, Surveillance, Target Acquisition and Reconnaissance), which includes photoreconnaissance, electronic intelligence gathering (Elint), signals intelligence (Sigint) and communications intelligence (Comint).

Boeing RC-135V

Although derived from the Boeing 707, the RC-135V bears little physical relation to the civilian aircraft. The RC-135V was the tenth of 12 variants, which have been tasked with electronic surveillance since the mid-1960s. As well as the cheek antennae fairings and sidewards-looking airborne radar (SLAR), the modified aircraft were fitted with a thimble nose and under-fuselage blade aerials.

SPECIFICATIONS	
Country of origin:	United States
Type:	electronic reconnaissance aircraft
Powerplant:	four 80kN (18,000lb) Pratt & Whitney TF33-P-9 turbojets
Performance:	maximum speed at 7620m (25,000ft) 991km/h (616mph); service ceiling 12,375m (40,600ft); range 4305km (2675 miles)
Weights:	empty 46,403kg (102,300lb) maximum take-off weight 124,965g (275,500lb)
Dimensions:	wingspan 39.88m (130ft 10in); length 41.53m (136ft 3in); height 12.7m (41ft 8in); wing area 226.03 sq m (2,433 sq ft)
Armament:	six 7.92mm (.3in) MGs; 1000kg (2205lb) bomb load

Lockheed SR-71A Blackbird

Deliveries of the SR-71 began in 1966, but it has the looks and performance of an aircraft of the twenty-first century. It was designed as a strategic reconnaissance aircraft to succeed the U-2. Although detailed design work began in 1959, the US Government did not formally acknowledge the existence of the SR-71 until 1964.

SPECIFICATIONS	
Country of origin:	United States
Type:	strategic reconnaissance aircraft
Powerplant:	two 144.5kN (32,500lb) Pratt & Whitney JT11D-20B bleed-turbojets
Performance:	maximum speed at 24,385m (80,000ft) more than 3219km/h (2000mph); ceiling in excess of 24,385m (80,000ft); standard range 4800km (2983 miles)
Weights:	empty 27,216kg (60,000lb); maximum take-off 77,111kg (170,000lb)
Dimensions:	wingspan 16.94m (55ft 7in), length 32.74m (107ft 5in); height 6.64m (18ft 6in); wing area 167.22 sq m (1800 sq ft)
Armament:	none

SPYPLANES

Lockheed TR-1A

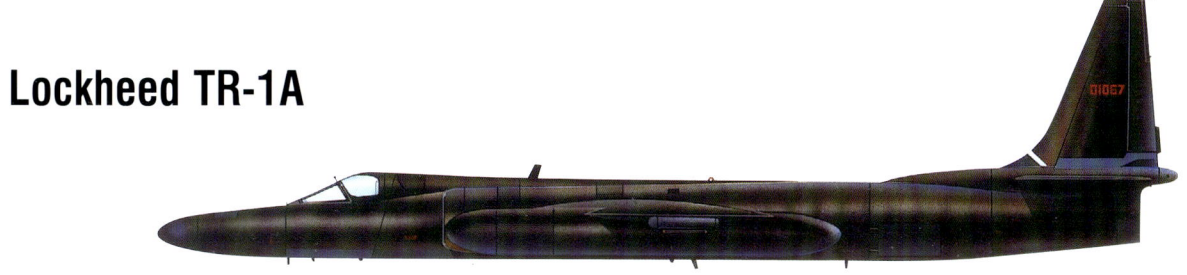

The first U-2s were deployed in England and Germany in 1956. Official reports announced that the glider-like aircraft were used for atmospheric research, when in fact they were overflying communist territory on reconnaissance missions. In 1978, the production line was reopened, and the first of 25 TR-1A aircraft followed. The TR-1A's primary role is that of tactical surveillance.

SPECIFICATIONS
Country of origin:	United States
Type:	single-seat high-altitude reconnaissance aircraft
Powerplant:	one 75.6kN (17,000lb) Pratt & Whitney J75-P-13B turbojet
Performance:	maximum cruising speed at more than 21,335m (70,000ft); operational ceiling 27,430m (90,000ft); maximum range 10,050km (6250 miles)
Weights:	empty 7031kg (15,500lb); maximum take-off weight 18,733kg (41,300lb)
Dimensions:	wingspan 31.39m (103ft); length 19.13m (62ft 9in); height 4.88m (16ft); wing area 92.9 sq m (1000 sq ft)
Armament:	none

Beech RC-12D

The US Army uses the RC-12D Guardrail for battlefield Comint and locating and jamming enemy radio transmitters. The data received is transmitted via a datalink for analysis by ground stations. The RC-12D is based on the airframe of the civilian Beech Super King Air.

SPECIFICATIONS
Country of origin:	United States
Type:	Tactical Communications intelligence aircraft
Powerplant:	two 634kW (850hp) Pratt & Whitney Canada PT6A-41 turboprop engines
Performance:	maximum speed 491km/h (306mph); endurance five hours; service ceiling 9449m (31,000ft)
Weights:	empty unknown; maximum 6412kg (14,136lb)
Dimensions:	span 16.92m (55ft 6in); length 13.34m (43ft 9in); height 4.57m (15ft); wing area 28.2 sq m (303 sq ft)
Armament:	none

BAe Nimrod R.1

The RAF's Nimrod fleet includes three specialized and secretive R.1 models serving with No 51 Squadron. With a similar role to the RC-135V, they are filled with sensors, recorders and language specialists. Communications such as mobile phone calls can be intercepted, monitored and analyzed in real time.

SPECIFICATIONS
Country of origin:	United Kingdom
Type:	electronic reconnaissance aircraft
Powerplant:	four 54.09kN (12,160lb thrust) Rolls-Royce Spey turbofans
Performance:	Maximum speed: 923km/h (575mph); operational ceiling 13,411m (44,000ft); maximum range 9,265km (5,755 miles)
Weights:	empty 39,009kg (86,000lb); maximum take-off weight 87,090kg (192,000lb)
Dimensions:	wingspan 35.0m (114ft 10in); length 38.65m (126ft 9in); height 9.14m (31ft); wing area 197.05m (2,121 sq ft)
Armament:	none

COLD WAR MILITARY AIRCRAFT

Heavy Lifters

For strategic airlift and some tactical missions, jet aircraft replaced 1950s-style propeller-driven transports in many of the world's larger air forces. Powerful high-bypass turbofans made possible super-heavy aircraft like the C-5 Galaxy (and the 747, which had its origins in the same competition). Ramp loading allowed vehicles as large as main battle tanks to be carried by air, and refuelling equipment gave the range to deploy whole infantry brigades anywhere in the world within hours.

Lockheed C-141B StarLifter

Designed in the early 1960s, the StarLifter was the most numerous of Military Airlift Command's strategic transport aircraft. The aircraft were delivered between April 1965 and February 1968. All 270 surviving C-141As were converted in the late 1970s to C-141B standard by stretching the fuselage by 7.11m (23ft 4in). The aircraft saw service in Vietnam, Grenada and in the 1991 Gulf War, before being retired in 2006.

SPECIFICATIONS	
Country of origin:	United States
Type:	heavy strategic transport
Powerplant:	four 93.4kN (21,000lb) Pratt & Whitney TF33-7 turbofans
Performance:	maximum speed 912km/h (567mph); range with maximum payload 4723km (2935 miles)
Weights:	empty 67,186kg (148,120lb); maximum take-off weight 155,582kg (343,000lb)
Dimensions:	wingspan 48.74m (159ft 11in); length 51.29m (168ft 3.5in); height 11.96m (39ft 3in); wing area 299.88 sq m (3228 sq ft)
Armament:	none

Lockheed C-5 Galaxy

For a time during the early 1970s, the giant C-5 Galaxy reigned as the world's largest aircraft, although it has now been overtaken by the Antonov An-225. Despite its huge size, the Galaxy can operate from rough airstrips. To this end, it has a high flotation landing gear with 28 wheels. The aircraft can carry complete missile systems and M1 Abrams tanks.

SPECIFICATIONS	
Country of origin:	United States
Type:	heavy strategic transport
Powerplant:	(C5A) four 82.3kN (41,000lb) General Electric TF39-1 turbofans
Performance:	maximum speed 919km/h (571mph); service ceiling at 272,910kg (615,000lb) 10,360m (34,000ft); range with maximum payload 100,228kg (220,967lb) 6033km (3749 miles)
Weights:	empty 147,528kg (325,244lb); maximum take-off weight 348,810kg (769,000lb)
Dimensions:	wingspan 67.88m (222ft 8.5in); length 75.54m (247ft 10in); height 19.85m (65ft 1.5in); wing area 575.98 sq m (6200 sq ft)
Armament:	none

HEAVY LIFTERS TIMELINE

 1963    1968    1970

HEAVY LIFTERS

Kawasaki C-1A

The C-1 was designed specifically to replace the Curtiss C-46 Commando transport aircraft in service in Japan. The first flight was made in November 1970; flight testing and evaluation led to a production order for 11 in 1972. The C-1 was designed with a short range suitable only for flights within Japan, in line with Japan's strict self-defence policy.

SPECIFICATIONS	
Country of origin:	Japan
Type:	short-range transport
Powerplant:	two 64.5kN (14,500lb) Mitsubishi (Pratt & Whitney) JT8-M-9 turbofans
Performance:	maximum speed at 7620m (25,000ft) 806km/h (501mph); service ceiling 11,580m (38,000ft); range 1300km (808 miles) with 7900kg (17,417lb) payload
Weights:	empty 23320kg (51,412lb); maximum take-off weight 45,000kg (99,208lb)
Dimensions:	wingspan 30.6m (100ft 4.75in); length 30.5m (100ft 4in); height 10m (32ft 9.3in); wing area 102.5 sq m (1297.09 sq ft)
Armament:	none

Lockheed TriStar K Mk 1

Since March 1986, the Royal Air Force has operated a converted version of the Lockheed Tristar jetliner as its primary tanker aircraft. Six of the 500 series aircraft were adapted for inflight refuelling operations. Four of the aircraft retained a commercial cabin configuration to allow passengers to be carried. The two other aircraft were fitted with a large cargo door on the port side.

SPECIFICATIONS	
Country of origin:	United Kingdom/United States
Type:	long-range strategic transport and inflight refuelling tanker
Powerplant:	three 222.3kN(50,000lb) Rolls-Royce RB.211-524B turbofans
Performance:	maximum cruising speed 964km/h (599mph) at 10,670m (35,000ft); service ceiling 13,105m (43,000ft); range on internal fuel with maximum payload 7783km (4836 miles)
Weights:	empty 110,163kg (242,684lb); maximum take-off weight 244,944kg (540,000lb)
Dimensions:	wingspan 50.09m (164ft 4in); length 50.05m (164ft 2.5in); height 16.87m (55ft 4in); wing area 329.96 sq m (3541 sq ft)
Armament:	none

Ilyushin Il-76 'Candid'

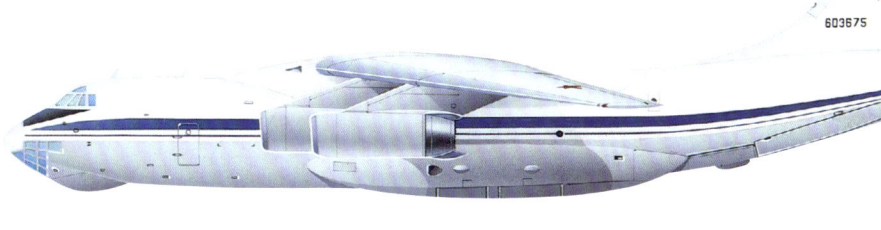

The Il-76 'Candid' (NATO reporting name) was first seen in the West at the 1971 Paris Air Salon. With a high cruising speed and intercontinental range, it was designed as a capable freighter that could carry large indivisible loads and operate from relatively poor and partially prepared airstrips. Aeroflot was the first operator, while India operates a fleet of 24.

SPECIFICATIONS	
Country of origin:	USSR (now CIS)
Type:	heavy freight transport
Powerplant:	four 117.6kN (26,455lb) Soloviev D-30KP-1 turbofans
Performance:	maximum speed at 11,000m (36,090ft) 850km/h (528mph); maximum cruising altitude 12,000m (39,370ft); range with 40,000kg (88,185lb) payload 5000km (3107 miles)
Weights:	empty about 75,000kg (165,347lb); maximum take-off weight 170,000kg (374,786lb)
Dimensions:	wingspan 50.5m (165ft 8.2in); length 46.59m (152ft 10.25in); height 14.76m (48ft 5in); wing area 300 sq m (3229.28 sq ft)
Armament:	provision for two 23mm cannon in tail

1971

COLD WAR MILITARY AIRCRAFT

Douglas A-4 Skyhawk

First flown in 1954 as a light bomber able to carry a single nuclear weapon, the A4D (later A-4) Skyhawk went on to be one of the longest-serving US Navy aircraft, finally retiring in 2003. Production for the United States and export also lasted a record time – 27 years – and the 'Bantam Bomber' was sold to eight nations, either new or second-hand. The Skyhawk remains in service with Argentina, Brazil, Israel, Singapore and several military contractors.

A-4C Skyhawk

The A-4C was the first model to be equipped with a fixed refuelling probe, and had an improved weapons delivery system compared to its predecessors. The 'Charlie' was used extensively in Vietnam, this example serving with VA-144 'Roadrunners' aboard the USS *Kitty Hawk*.

SPECIFICATIONS	
Country of origin:	United States
Type:	single-seat attack bomber
Powerplant:	one 34.7kN (7800lb) J65-W-16A turbojet
Performance:	maximum speed 1078km/h (670mph); service ceiling 14,935m (49,000ft); range with 4000lb load 1480km (920 miles)
Weights:	empty 4809kg (10,602lb); maximum take-off weight 12,437kg (27,420lb)
Dimensions:	wingspan 8.38m (27ft 6in); length excluding probe 12.22m (40ft 1.5in); height 4.66m (15ft 3in); wing area 24.15 sq m (260 sq ft)
Armament:	two 20mm Mk 12 cannon; provision for 2268kg (5000lb) of stores, including air-to-surface missiles, bombs, cluster bombs, dispenser weapons, rocket-launcher pods, cannon pods, drop tanks and ECM pods

A-4G Skyhawk

Australia purchased a version of the A-4F for use on its ex-British carrier HMAS *Melbourne*. The A-4Gs were supplemented by ex-USN A-4Fs and the survivors sold on to New Zealand when *Melbourne* was retired, supplementing that country's own fleet of A-4Ks.

SPECIFICATIONS	
Country of origin:	United States
Type:	single-seat attack bomber
Powerplant:	one 41.3kN (9300lb) J52-8A turbojet
Performance:	maximum speed 1078km/h (670mph); service ceiling 14,935m (49,000ft); range with 4000lb load 1480km (920 miles)
Weights:	empty 4809kg (10,602lb); maximum take-off weight 12,437kg (27,420lb)
Dimensions:	wingspan 8.38m (27ft 6in); length excluding probe 12.22m (40ft 1.5in); height 4.66m (15ft 3in); wing area 24.15 sq m (260 sq ft)
Armament:	two 20mm Mk 12 cannon; provision for 3720kg (8200lb) of stores, including AIM-9G Sidewinder AAMs, rocket-launcher pods and ECM pods

Douglas A-4M Skyhawk

SPECIFICATIONS
Country of origin:	United States
Type:	single-seat attack bomber
Powerplant:	one 49.82kN (11,500lb) thrust Pratt & Whitney J52-P408 turbojet engine
Performance:	maximum speed 1083km/h (673mph); service ceiling 14,935m (49,000ft); range 3310km (2200 miles)
Weights:	empty 4809kg (10,602lb); maximum take-off weight 12,437kg (27,420lb)
Dimensions:	wingspan 8.38m (27ft 6in); length excluding probe 12.22m (40ft 1.5in); height 4.66m (15ft 3in); wing area 24.15 sq m (260 sq ft)
Armament:	two 20mm Mk 12 cannon; provision for 3720kg (8200lb) of stores, including AIM-9G Sidewinder AAMs, rocket-launcher pods and ECM pods

The A-4M was a much-improved version for the US Marines, with an enlarged canopy, larger engine and revised avionics, including electronic counter measures (ECM) and laser-designation equipment. Some surplus A-4Ms were supplied to Argentina in the 1990s.

A-4PTM Skyhawk

SPECIFICATIONS
Country of origin:	United States
Type:	single-seat attack bomber
Powerplant:	one 34.7kN (7800lb) J65-W-16A turbojet
Performance:	maximum speed 1078km/h (670mph); service ceiling 14,935m (49,000ft); range with 4000lb load 1480km (920 miles)
Weights:	empty 4809kg (10,602lb); maximum take-off weight 12,437kg (27,420lb)
Dimensions:	wingspan 8.38m (27ft 6in); length excluding probe 12.22m (40ft 1.5in); height 4.66m (15ft 3in); wing area 24.15 sq m (260 sq ft)
Armament:	two 20mm Mk 12 cannon; provision for 2268kg (5000lb) of stores, including air-to-surface missiles, bombs, cluster bombs, drop tanks

Malaysia bought 40 ex-US Skyhawks in the mid-1980s and had Grumman update them from A-4C standard with modern avionics. The new designation was A-4PTM, which stands for *Persekutan Tanah Melayu*, or Federation of Malay States, although is sometimes said to mean 'Peculiar to Malaysia'.

TA-4J Skyhawk

SPECIFICATIONS
Country of origin:	United States
Type:	two-seat carrier trainer
Powerplant:	one 37.8kN (8500lb) J52-P-6 turbojet
Performance:	maximum speed 1084km/h (675mph); service ceiling 14,935m (49,000ft); range 1287km (800 miles)
Weights:	empty 4809kg (10,602lb); maximum take-off weight 11,113kg (24,500lb)
Dimensions:	wingspan 8.38m (27ft 6in); length excluding probe 12.98m (42ft 7.25in); height 4.66m (15ft 3in); wing area 24.15 sq m (260 sq ft)
Armament:	one 20mm cannon

Few people believed designer Ed Heinemann when he said he could build a jet attack bomber for the Navy at half the specified weight of 13,600kg (30,000lb). But ultimately his aircraft stayed in production for over 20 years, in a multiplicity of different versions. The TA-4J was a variant built for the US Navy, with the fuselage lengthened by .8m (2.5ft) to accommodate the instructor's tandem cockpit.

The Mighty Hercules

Lockheed's C-130 Hercules has been in production for over 50 years, with around 3000 aircraft built. It has become the standard military transport of the West, with few major nations not possessing any 'Herks'. The C-130 has been produced in versions for many tasks, including tankers, gunships, ski-planes and rescue variants. The C-130J, with new engines, propellers and electronics, is slowly emulating the export success of its predecessors.

C-130B Hercules

The C-130B entered service in 1959 and a number are still in use with air forces including South Africa, Turkey and Romania. It was the first model to have auxiliary fuel tanks under the wings, a feature of all subsequent Hercules prior to the C-130J.

SPECIFICATIONS	
Country of origin:	United States
Type:	four-engined transport aircraft
Powerplant:	four 3021kW (4050hp) Allison T56-A-7 turboprop engines
Performance:	maximum speed 547km/h (340mph); range 3896km (2420 miles); service ceiling 7010m (23,000ft)
Weights:	empty (34,686kg); maximum 79,375kg (175,000lb)
Dimensions:	span 40.4m (132ft 7in); length 29.8m (97ft 9in); height 11.7m (38ft 6in); wing area 162.2 sq m (1745.5 sq ft)
Armament:	none

EC-130E Hercules

Known as a 'flying broadcast station', the EC-130E Commando Solo is fitted with powerful transmitters to send programming to TV viewers and radio listeners in enemy territory during a conflict. These broadcasts can be used to influence public opinion or counter the enemy's own propaganda.

SPECIFICATIONS	
Country of origin:	United States
Type:	Psychological operations aircraft
Powerplant:	four 3660kW (4910hp) Allison T56-A-15 turboprop engines
Performance:	maximum speed 547km/h (340mph); range 3896km (2420 miles); service ceiling 7010m (23,000ft)
Weights:	empty unknown, maximum 70,300kg (155,000lb)
Dimensions:	span 40.4m (132ft 7in); length 29.8m (97ft 9in); height 11.7m (38ft 6in); wing area 162.2 sq m (1745.5 sq ft)
Armament:	none

EC-130Q Hercules

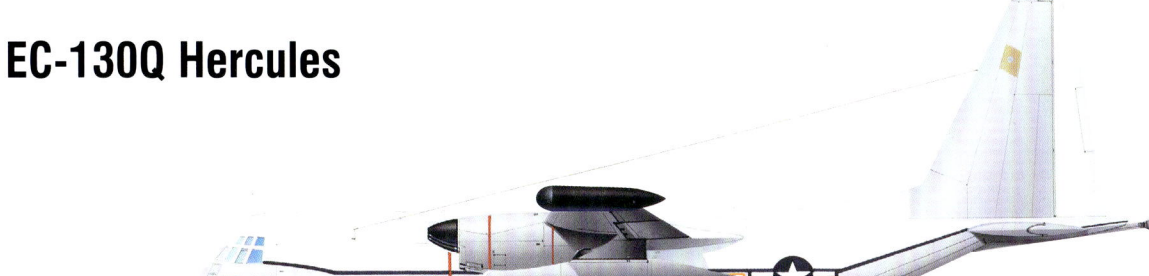

The EC-130Q was a version for the Navy's TACAMO (TAke Charge And Move Out) mission, which involved communication with submerged ballistic-missile submarines. The aircraft was fitted with a Very Low Frequency (VLF) radio transmitter and long trailing-wire antenna.

SPECIFICATIONS	
Country of origin:	United States
Type:	Communications relay aircraft
Powerplant:	four 3660kW (4910hp) Allison T56-A-15 turboprop engines
Performance:	maximum speed 547km/h (340mph); range 3896km (2420 miles); service ceiling 7010m (23,000ft)
Weights:	empty 72,892 lb (33,057 kg); maximum 79,375 kg (175,000 lb)
Dimensions:	span unknown; length 29.8m (97ft 9in); height 11.7m (38ft 6in); wing area 162.2 sq m (1745.5 sq ft)
Armament:	none

KC-130F Hercules

Originally introduced as the GV-1 in the pre-1962 designation system, the KC-130F became the US Navy and Marines' primary transport and tactical refuelling aircraft. The underwing fuel tanks were replaced with pods containing a hose reel and drogue. The 'Blue Angels' aerobatic team use KC-130Fs as support aircraft.

SPECIFICATIONS	
Country of origin:	United States
Type:	tanker/transport aircraft
Powerplant:	four 3660kW (4910hp) Allison T56-A-15 turboprop engines
Performance:	maximum speed 604km/h (374mph); range 3896km (2420 miles); service ceiling 10,058m (33,000ft)
Weights:	empty 72,892lb (33,057kg); maximum 70,306kg (155,000lb)
Dimensions:	span 40.4m (132ft 7in); length 29.8m (97ft 9in); height 11.7m (38ft 6in); wing area 162.2 sq m (1745.5 sq ft)
Armament:	none

WC-130H Hercules

The 'Hurricane Hunters' of the USAF's 53rd Weather Reconnaissance Squadron flew the WC-130H Hercules over the Atlantic and the Gulf of Mexico to detect hurricanes and gather data to predict their direction and strength. The unit was one of the first to get C-130J models.

SPECIFICATIONS	
Country of origin:	Germany
Type:	weather reconnaissance aircraft
Powerplant:	four 3660kW (4910hp) Allison T56-A-15 turboprop engines
Performance:	maximum speed 547km/h (340mph); range 3896km (2420 miles); service ceiling 10,058m (33,000ft)
Weights:	empty 72,892lb (33,057kg); maximum 79,375kg (175,000lb)
Dimensions:	span 40.4m (132ft 7in); length 29.8m (97ft 9in); height 11.7m (38ft 6in); wing area 162.2 sq m (1745.5 sq ft)
Armament:	(as bomber) 5670kg (12,500lb) bombs on Multiple Ejector Racks

COLD WAR MILITARY AIRCRAFT

One-offs

The exciting pace of development of aviation in the post-war years left many aircraft by the wayside. Some were technological dead-ends, such as rocket-powered interceptors. Others were too costly and ambitious. Two promising programmes, the Canadian Arrow and the British TSR.2, fell victim to the then-fashionable theory that missiles would make manned combat aircraft obsolete. In these cases, the national aerospace industries suffered and only American aircraft companies benefited.

Republic XF-91 Thunderceptor

The XF-91 was a bold attempt in 1946 to produce a high-altitude interceptor to the USAAF. Republic introduced unusual features such as a variable-incidence inverse tapered wing, with tandem-wheel main gears at the tips and the twin powerplant. The fairing for a Reaction Motors XLR-11-RM-9 rocket motor, visible under the tail, could be used to augment top speed for short periods.

SPECIFICATIONS	
Country of origin:	United States
Type:	experimental high-altitude interceptor
Powerplant:	one General Electric J47-GE-3 turbojet; Reaction Motors XLR-11-RM-9 rocket motor
Performance:	maximum speed attained 1812km/h (1126mph); ceiling (approximately) 15,250m (50,000ft)
Weights:	unknown
Dimensions:	unknown
Armament:	none

Saunders Roe SR.53

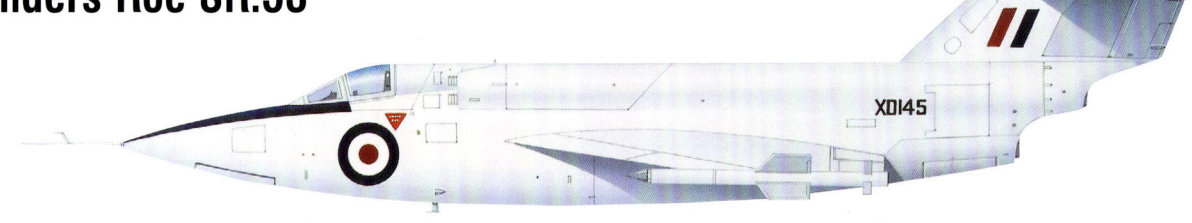

The SR.53 was intended as a pure rocket-powered interceptor, but a small turbojet was added to allow it to land under power after it had made its interception. One of the two SR.53s built crashed in 1958, but by then the UK Government had decided to cancel almost all manned fighter programmes.

SPECIFICATIONS	
Country of origin:	United Kingdom
Type:	experimental mixed-power interceptor
Powerplant:	one 7.3kN (1640lb) thrust Armstrong Siddeley Viper turbojet engine and one 35.6kN (8008lb) thrust Spectre Rocket engine
Performance:	maximum speed Mach 2.2; endurance: 7 minutes at full power; service ceiling 20,000m (65,600ft)
Weights:	loaded 8363kg (18,400lb)
Dimensions:	span 7.65m (25ft 1in); length 13.71m (10ft 5in); height 3.3m (10ft 10in); wing area 25.45 sq m (274 sq ft)
Armament:	two Firestreak or Blue Jay AAMs

ONE-OFFS TIMELINE

1949

1957

ONE-OFFS

Myasishchev M-50 'Bounder'

Vladimir M. Myasishchev's design for the M-50 was extremely advanced and was considered a significant potential threat when details of its capabilities first became known. Only ever built in prototype form, it featured a shoulder-mounted cropped delta wing, coupled with a conventional tail unit and all-swept surfaces. The fuselage was pressurized and incorporated a large weapons bay.

SPECIFICATIONS	
Country of origin:	USSR
Type:	prototype supersonic strategic bomber
Powerplant:	four wing-mounted 128.3kN (28,860lb) Soloviev D-15 turbojets
Performance:	(estimated) maximum speed at altitude 1950km/h (1,212mph)
Weights:	not released
Dimensions:	not released
Armament:	probably at least one cannon; internal bomb bay carrying stand-off nuclear weapons

Avro Arrow

The story of the Arrow bears a startling resemblance to that of the BAC TSR.2. Both projects showed great promise during the early stages of development in the mid-1950s, and both were destroyed by the decisions of politicians convinced that the days of the manned interceptor were numbered. The design incorporated a huge, high-set delta wing.

SPECIFICATIONS	
Country of origin:	Canada
Type:	two-seat all-weather long range supersonic interceptor
Powerplant:	two 104.5kN (23,500lb) Pratt and Whitney J75-P-3 turbojets
Performance:	Mach 2.3 recorded during tests
Weights:	empty 22,244kg (49,040lb); average take-off weight during trials 25,855kg (57,000lb)
Dimensions:	wingspan 15.24m (50ft); length 23.72m (77ft 9.75in); height 6.48m (21ft 3in); wing area 113.8 sq m (1225 sq ft)
Armament:	eight Sparrow air-to-air missiles in internal bay

BAC TSR.2

Conceived as a replacement for the English Electric Canberra, the cancellation of the TSR.2 programme was widely regarded within the aviation industry as the greatest disaster to befall the post-war British aviation industry. In retrospect, it is clear that much of the pioneering research carried out by the project team was of great benefit during the development of Concorde.

SPECIFICATIONS	
Country of origin:	United Kingdom
Type:	two-seat strike/reconnaissance aircraft
Powerplant:	two 136.1kN (30,610lb) thrust Bristol Siddeley Olympus 320 turbojets
Performance:	maximum speed at altitude 2390km/h (1485mph); operating ceiling 16,460m (54,000ft); range at low level 1287km (800 miles)
Weights:	average 36,287kg (80,000lb); maximum 43,545kg (96,000lb)
Dimensions:	wingspan 11.28m (37ft); length 27.13m (89ft); height 7.32m (24ft); wing area 65.03 sq m (700 sq ft)
Armament:	(planned) up to 2722kg (6000lb) of conventional or nuclear weapons internally; four underwing pylons for up to 1814kg (4000lb) of weapons

1958  1964

COLD WAR MILITARY AIRCRAFT

SAAB: Part 1

Neutral Sweden was surrounded by warring nations in 1939–45. Without modern aircraft, it had no way of preventing incursions across its borders or into its airspace. State aircraft maker Saab (Svenska Aeroplan Aktiebolaget), which had previously built aircraft under licence, created its first original design, the B 17, in 1940. It was followed by more innovative designs, including aircraft firsts: ejection seats; conversion of piston-engines to jet power; and, a first for Europe, swept-wing craft.

B 17

Saab hired American engineers to help it produce its first all-new design, the B 17, and the result bore more than a passing resemblance to the Curtiss Helldiver. The B 17 was built in dive-bomber and level-bomber variants and a dedicated photoreconnaissance version.

SPECIFICATIONS	
Country of origin:	Sweden
Type:	single-engined dive bomber
Powerplant:	one 882-kW (1,183-hp) Pratt and Whitney R-1830-S1C3G Twin Wasp radial piston engine
Performance:	maximum speed 435 km/h (270 mph)); range 1800km (1118 miles); service ceiling 8700m (28,543ft)
Weights:	empty 2600kg (5732lb); maximum 3605kg (7948lb)
Dimensions:	span 13.7m (45ft); length 9.8m (32ft 1in); height 4m (13ft 2in); wing area 28.5 sq m (307 sq ft)
Armament:	three 8mm machine guns, up to 500kg (1102lb) of bombs

91 Safir

The Safir was one of Saab's biggest export successes, selling to operators in over 20 countries. Sweden used it as its primary military trainer for many years, and Austria bought 24 in the mid-1960s, using them as primary trainers and navigation trainers.

SPECIFICATIONS	
Country of origin:	Sweden
Type:	two-seat trainer
Powerplant:	one 134-kW (180-hp) Avco Lycoming O-360-A1A piston engine
Performance:	maximum speed 266km/h (165mph); range 1000km (621 miles); service ceiling 5000m (16,400ft)
Weights:	empty 710kg (1565lb); maximum 1205kg (2657lb)
Dimensions:	span 10.6m (34ft 9in); length 7.95m (26ft 1in); height 2.2m (7ft 3in); wing area 13.6 sq m (146 sq ft)
Armament:	none

SAAB TIMELINE   1940   1943   1945

SAAB: PART 1

J 21

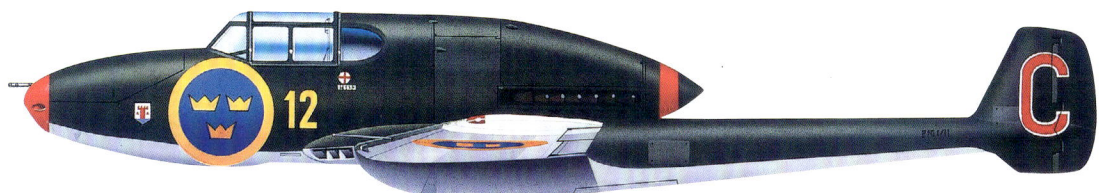

Inspired by the P-38 Lightning, but with a single pusher engine, the J 21 was one of the most unusual fighters to fly during the war years. It was equipped with an ejection seat so the pilot could avoid hitting the propeller if forced to bail out.

SPECIFICATIONS	
Country of origin:	Sweden
Type:	single-engined fighter-bomber
Powerplant:	one 1100kW (1475hp) Daimler-Benz DB 605B piston engine
Performance:	maximum speed 640 km/h (400 mph); range 1500km (930 miles); service ceiling 10,200m (33,450ft)
Weights:	empty 3350kg (7165lb); maximum 9730kg (4415lb)
Dimensions:	span 11.61m (38ft 1in); length 22.2m (38ft 1in); height 4m (13ft 2in); wing area 22.2 sq m (239 sq ft)
Armament:	one 20mm cannon and four 13.2mm machine-guns

J 29 Tunnan

The Tunnan (Barrel) was only the third swept-wing jet fighter to fly after the F-86 and MiG-15. Sweden took them on the only Flygvapnet (Air Force) combat deployment to date, flying ground-attack missions in support of UN operations in the Congo in 1961.

SPECIFICATIONS	
Country of origin:	Sweden
Type:	single-engined fighter
Powerplant:	27kN (6072.7lb) thrust Volvo RM 2B (de Havilland Ghost) afterburning turbojet engine
Performance:	maximum speed 1060km/h (659mph); range 1100km (684 miles); service ceiling 15,700m (51,000ft)
Weights:	empty 4845kg (10,680lb); maximum 8375kg (18,465lb)
Dimensions:	span 11m (36ft 1in); length 10.1m (33ft 2in); height 3.8m (12ft 6in); wing area 24 sq m (258 sq ft)
Armament:	four 20mm cannon, two AIM-9B Sidewinder AAMs

J 32B Lansen

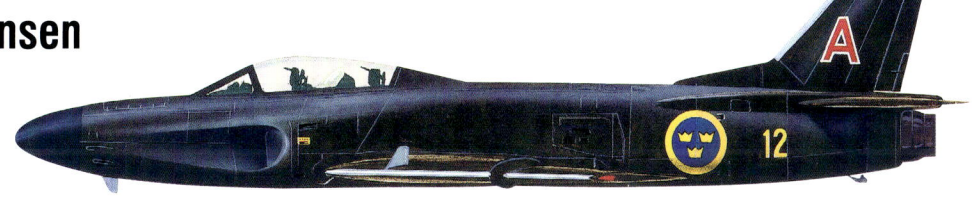

Designed to replace the Saab 18 light-bomber, the Type 32 was a large, all-swept machine of outstanding quality, designed and developed ahead of similar aircraft elsewhere in Western Europe. Entering service in 1953, the Type 32 served well into the 1990s as aggressor aircraft, target tugs and trials aircraft. The J 32B pictured had S6 radar fire control for lead/pursuit interception.

SPECIFICATIONS	
Country of origin:	Sweden
Type:	all-weather and night fighter
Powerplant:	one 67.5kN (15,190lb) Svenska Flygmotor (Rolls-Royce Avon) RM6A
Performance:	maximum speed 1114km/h (692mph); service ceiling 16,013m (52,500ft); range with external fuel 3220km (2000 miles)
Weights:	empty 7990kg (17,600lb); maximum loaded 13,529kg (29,800lb)
Dimensions:	wingspan 13m (42ft 7.75in); length 14.5m (47ft 6.75in); height 4.65m (15ft 3in); wing area 37.4 sq m (402.58 sq ft)
Armament:	four 30mm Aden M/55 cannon; four Rb324 (Sidewinder) air-to-air missiles or FFAR (Folding Fin Air-launched Rocket) pods

1948

1952

SAAB: Part Two

Since World War II, Saab has built most of Sweden's combat aircraft and trainers, continuing its innovation with the double-delta Draken (Dragon), the canard-wing Viggen (Thunderbolt) and the fly-by-wire Gripen (Griffin). Politics prevented fighter export sales to most countries likely to go to war with them, but more recently Saab has sold the Gripen to non-traditional customers.

MFI-15

A light utility aircraft and trainer, the civilian Saab MFI-15 Safari and military MFI-17 Supporter have been used by a number of air forces, including Norway, one of whose aircraft is shown. Pakistan has produced a version known locally as the Mushshak for its primary training needs.

SPECIFICATIONS	
Country of origin:	Sweden
Type:	utility aircraft/trainer
Powerplant:	one 149kW (200hp) Avco Lycoming IO-360-A1B6 piston engine
Performance:	maximum speed 235km/h (146mph); range unknown; service ceiling 4100m (13,450 ft)
Weights:	empty 690kg (1521 lb); maximum 1200kg (2646 lb)
Dimensions:	span 8.85m (29ft); length 7m (23ft); height 2.6m (8ft 6in); wing area 11.9 sq m (129 sq ft)
Armament:	none

J 35F Draken

The Draken was designed to a demanding specification for a single-seat interceptor that could operate from short air strips, had rapid time-to-height performance and supersonic performance. The 'double-delta' wing is an ingenious method of arranging items one behind the other to allow a long aircraft a small frontal area and correspondingly high aerodynamic efficiency.

SPECIFICATIONS	
Country of origin:	Sweden
Type:	single-seat all-weather interceptor
Powerplant:	one 76.1kN (17,110lb) Svenska Flygmotor RM6C turbojet
Performance:	maximum speed 2125km/h (1320mph); service ceiling 20,000m (65,000ft); range with maximum fuel 3250km (2020 miles)
Weights:	empty 7425kg (16,369lb); maximum take-off weight 16,000kg (35,274lb)
Dimensions:	wingspan 9.4m (30ft 10in); length 15.4m (50ft 4in); height 3.9m (12ft 9in); wing area 49.2 sq m (526.6 sq ft)
Armament:	one 30mm Aden M/55 cannon, air-to-air missiles, or up to 4082kg (9000lb) of bombs on attack mission

SAAB TIMELINE

 1955    1961    1963

SAAB: PART TWO

105/Sk 60

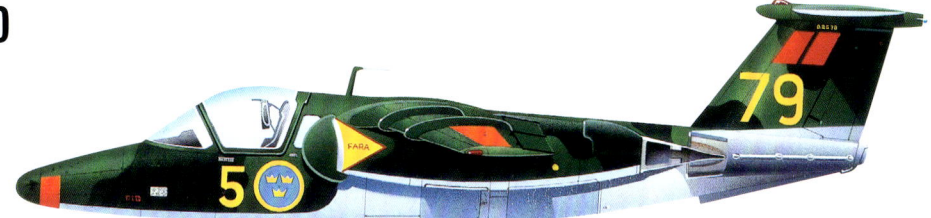

Having established its reputation with the Draken, Saab extended its range by developing the privately funded 105. This aircraft is a swept shoulder-wing monoplane with side-by-side cabin accommodation for either two or four crew. The first prototype flew in June 1963, and after successful evaluation by the Swedish Air Force, orders were placed for 150 production aircraft.

SPECIFICATIONS

Country of origin:	Sweden
Type:	training/liason aircraft with secondary attack capability
Powerplant:	two 73kN (1640lb) Turbomeca Aubisque turbofans
Performance:	maximum speed at 6095m (20,000ft) 770km/h (480mph); service ceiling 13,500m (44,290ft); range 1400km (870 miles)
Weights:	empty 2510kg (5534lb); maximum take-off weight 4050kg (8929lb)
Dimensions:	wingspan 9.5m (31ft 2in); length 10.5m (34ft 5.375in); height 2.7m (8ft 10.25in); wing area 16.3 sq m (175.46 sq ft)
Armament:	six external hardpoints with provision for up to 700kg (1543lb) of stores, including two Saab Rb05 air-to-surface missiles, or two 30mm cannon pods, or 12 135mm rockets, or bombs, cluster bombs and rocket launcher pods

SF 37 Viggen

The SF37 was a dedicated single-seat reconnaissance version intended to replace the S 35E in service with the Swedish air force. The first prototype flew in May 1973. Production aircraft were distinguished by a chisel nose containing seven cameras, which are often supplemented by surveillance pods on the shoulder hardpoints.

SPECIFICATIONS

Country of origin:	Sweden
Type:	single-seat all-weather attack aircraft
Powerplant:	one 115.7kN (26,015lb) Volvo Flygmotor RM8 turbofan
Performance:	maximum speed at high altitude 2124km/h (1320mph); service ceiling 18,290m (60,000ft); combat radius 1000km (621 miles)
Weights:	empty 11,800kg (26,015lb); maximum take-off weight 20,500kg (45,194lb)
Dimensions:	wingspan 10.6m (34ft 9.25in); length 16.3m (53ft 5.75in); height 5.6m (18ft 4.5in); wing area 46 sq m (495.16 sq ft)
Armament:	seven external hardpoints with provision for 6000kg (13,228lb) of stores, including cannon pods, rocket pods, missiles and bombs

JAS 39 Gripen

Saab has produced another excellent lightweight fighter in the form of the Gripen. It was conceived during the late 1970s as a replacement for the AJ, SH, SF and JA versions of the Saab 37 Viggen, and the configuration follows Saab's convention with an aft-mounted delta and swept-canard foreplanes. The flying surfaces are controlled via a fly-by-wire system.

SPECIFICATIONS

Country of origin:	Sweden
Type:	single-seat all-weather fighter, attack and reconnaissance aircraft
Powerplant:	one 80.5kN (18,100lb) Volvo Flygmotor RM12 turbofan
Performance:	maximum speed more than Mach 2; range on hi-lo-hi mission with external armament 3250km (2020 miles)
Weights:	empty 6622kg (14,600lb); maximum take-off weight 12,473kg (27,500lb)
Dimensions:	wingspan 8m (26ft 3in); length 14.1m (46ft 3in); height 4.7m (15ft 5in)
Armament:	one 27mm Mauser BK27 cannon, provision for air-to-air missiles, air-to-surface missiles, anti-ship missiles, bombs, cluster bombs, rocket-launcher pods, reconnaissance pods, drop tanks and ECM pods

1967

1988

Fouga Magister

One of the most successful and widely used trainer aircraft, the Magister was conceived and designed by Castello and Mauboussin for Fouga in 1950. It was the first purpose-built jet trainer. Despite the unusual butterfly-type tail, it proved a delight to fly. Total production of this and the version fitted with an arrestor hook (CM.75 Zephyr) was 437. In 1967, the Magister saw action with the Israeli Air Force during the Six-Day War.

CM.170 Magister

After prolonged testing, the Magister was put into production for L'Armée de l'Air. When Fouga was absorbed into the Potez company in 1958, Potez continued to produce a number of variants for international customers. Pictured is a version that flew with the 'Patrouille de France', the mount of the French national aerobatic team. The team now uses the Dassault/Dornier Alpha jet.

SPECIFICATIONS	
Country of origin:	France
Type:	two-seat trainer and light attack aircraft
Powerplant:	two 4kN (882lb) Turbomeca Marbore IIA turbojets
Performance:	maximum speed at 9150m (30,000ft) 715km/h (444mph); service ceiling 11,000m (36,090ft); range 925km (575 miles)
Weights:	empty equipped 2150kg (4740lb); maximum takeoff 3200kg (7055lb)
Dimensions:	over tip tanks 12.12m (39ft 10in); length 10.06m (33ft); height 2.8m (9ft 2in); wing area 17.3 sq m (186.1sq ft)
Armament:	two 7.5mm (.295in) or 7.62mm machine guns; rockets, bombs or Nord AS.11 missiles on underwing pylons

CM.170 Magister

Lebanon acquired Magisters in the 1960s and used them mainly as trainers. Their only combat operations were against the Palestine Liberation Organization (PLO) in refugee camps within Lebanon in 1973. Armed with 12.7mm machine guns, they attacked fortifications inside the camps.

SPECIFICATIONS	
Country of origin:	France
Type:	two-seat trainer and light attack aircraft
Powerplant:	two 4kN (882lb) Turbomeca Marbore IIA turbojets
Performance:	maximum speed at 9150m (30,000ft) 715km/h (444mph); service ceiling 11,000m (36,090ft); range 925km (575 miles)
Weights:	empty equipped 2150kg (4740lb); maximum takeoff 3200kg (7055lb)
Dimensions:	over tip tanks 12.12m (39ft 10in); length 10.06m (33ft); height 2.8m (9ft 2in); wing area 17.3 sq m (186.1sq ft)
Armament:	two 7.5mm (.295in) or 7.62mm machine guns; rockets, bombs or Nord AS.11 missiles on underwing pylons

CM.170 Magister

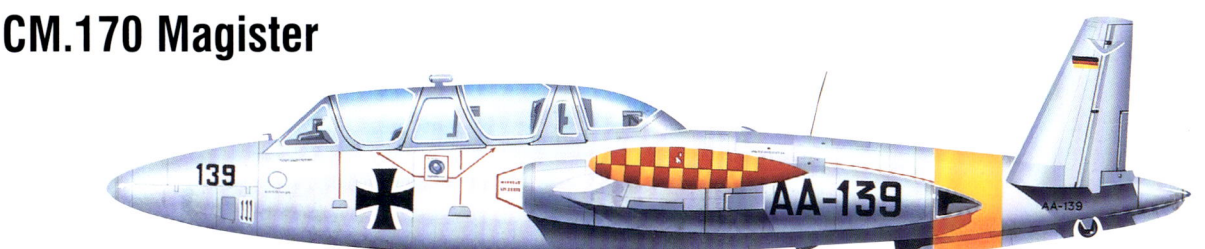

The Luftwaffe bought 40 Magisters, and Heinkel-Messerschmitt built 244 more under licence for use by the Luftwaffe and Marineflieger. This one belonged to the flight leader's school Flugzeug Fuhrerschule A, at Landsberg. The school had an aerobatic team called *Das Magister Team*.

SPECIFICATIONS	
Country of origin:	France
Type:	two-seat trainer and light attack aircraft
Powerplant:	two 4kN (882lb) Turbomeca Marbore IIA turbojets
Performance:	maximum speed at 9150m (30,000ft) 715km/h (444mph); service ceiling 11,000m (36,090ft); range 925km (575 miles)
Weights:	empty equipped 2150kg (4740lb); maximum takeoff 3200kg (7055lb)
Dimensions:	over tip tanks 12.12m (39ft 10in); length 10.06m (33ft); height 2.8m (9ft 2in); wing area 17.3 sq m (186.1sq ft)
Armament:	two 7.5mm (.295in) or 7.62mm machine guns; rockets, bombs or Nord AS.11 missiles on underwing pylons

CM.170 Magister

The Belgian Air Force aerobatic team *Les Diables Rouges* (Red Devils) flew the Magister from 1965 until 1977. In the 1990s and 2000s, the BAF revived the memory of the team with solo displays of a Magister dressed in *Les Diables Rouges* colours.

SPECIFICATIONS	
Country of origin:	France
Type:	two-seat trainer and light attack aircraft
Powerplant:	two 4kN (882lb) Turbomeca Marbore IIA turbojets
Performance:	maximum speed at 9150m (30,000ft) 715km/h (444mph); service ceiling 11,000m (36,090ft); range 925km (575 miles)
Weights:	empty equipped 2150kg (4740lb); maximum takeoff 3200kg (7055lb)
Dimensions:	over tip tanks 12.12m (39ft 10in); length 10.06m (33ft); height 2.8m (9ft 2in); wing area 17.3 sq m (186.1sq ft)
Armament:	two 7.5mm (.295in) or 7.62mm machine guns; rockets, bombs or Nord AS.11 missiles on underwing pylons

CM.170 Magister

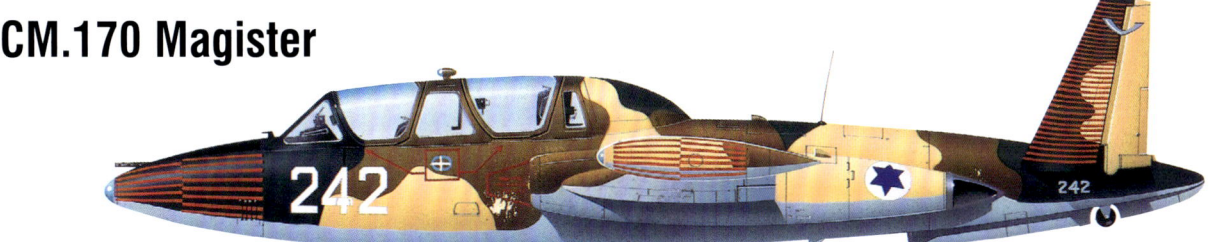

Israel has used the Magister since 1960 as a trainer and light-attack aircraft. Many were lost in attacks against Arab armour in 1967. Local industry upgraded 87 from 1980 as the *Tzukit* (Thrush or Merlin) and they remain in service, although they are due for replacement by the Beech Texan II turboprop.

SPECIFICATIONS	
Country of origin:	France
Type:	two-seat trainer and light attack aircraft
Powerplant:	two 4.7kN (1055lb) thrust Turbomeca Marboré IV turbojet engines
Performance:	maximum speed at 9150m (30,000ft) 715km/h (444mph); service ceiling 11,000m (36,090ft); range 925km (575 miles)
Weights:	empty equipped 2150kg (4740lb); maximum takeoff 3200kg (7055lb)
Dimensions:	over tip tanks 12.12m (39ft 10in); length 10.06m (33ft); height 2.8m (9ft 2in); wing area 17.3 sq m (186.1sq ft)
Armament:	two 7.5mm (.295in) or 7.62mm machine guns; rockets, bombs or Nord AS.11 missiles on underwing pylons

MiGs: Fagot, Fresco and Farmer

The first generation of Mikoyan-Gurevich MiG jet fighters—the MiG-15 'Fagot', MiG-17 'Fresco' and MiG-19 'Farmer'—filled the inventories of the Soviet Union and its allies for four decades. Even today, some Russian-built examples can be found in African and Asian air forces. Chinese derivatives are more common, with Pakistan and North Korea being particular bastions of the last Shenyang J-6s and Nanchang Q-5s, both descendants of the MiG-19 of 1955.

MiG-15bis 'Fagot' (S-103)

Czechoslovakia built the 'Fagot' under licence as the Avia S.102 (MiG-15) and S.103 (MiG-15bis). Single-seat MiG-15s were also produced in large numbers in Poland as the Lim-1 and Lim-2. Czech MiG-15s served the country's air force from 1951 until 1983.

SPECIFICATIONS	
Country of origin:	USSR
Type:	single-seat fighter
Powerplant:	one 26.5kN (5952lb) Klimov VK-1 turbojet
Performance:	maximum speed 1100km/h (684mph); service ceiling 15,545m (51,000ft); range at height with slipper tanks 1424km (885 miles)
Weights:	empty 4000kg (8820lb); maximum loaded 5700kg (12,566lb)
Dimensions:	wingspan 10.08m (33ft 0.75in); length 11.05m (36ft 3.75in); height 3.4m (11ft 1.75in); wing area 20.6 sq m (221.74 sq ft)
Armament:	one 37mm N-37 cannon and two 23mm NS-23 cannon, plus up to 500kg (1102lb) of mixed stores on underwing pylons

MiG-15UTI 'Midget'

There was no Soviet-built two-seat version of the MiG-17 or -19, but the MiG-15UTI 'Midget' served as the advanced trainer for thousands of pilots who went on to fly all models of MiG jet. Egypt was one of over 20 countries to operate the MiG-15 UTI.

SPECIFICATIONS	
Country of origin:	USSR
Type:	single-engined jet trainer
Powerplant:	one 26.5kN (5952lb) thrust Klimov RD-45FA turbojet engine
Performance:	maximum speed 1015km/h (631mph); service ceiling 15,545m (51,000ft); range 1054km (655 miles)
Weights:	empty 3724kg (8208lb); maximum loaded 5700kg (12,566lb)
Dimensions:	wingspan 10.08m (33ft 0.75in); length 11.05m (36ft 3.75in); height 3.7m (12ft 1.7in); wing area 20.6 sq m (221.74 sq ft)
Armament:	two 23mm NS-23 cannon

MiG-17 'Fresco'

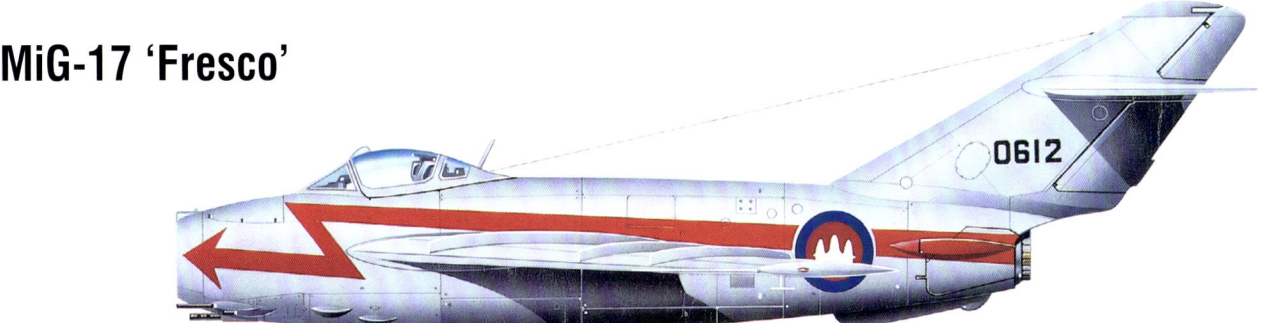

Although outwardly similar to the MiG-15, the -17 was in fact a completely different aircraft. The design began in 1949, its most important aspect being the new wing. Along with reduced thickness, a different section and platform, and three fences, this resulted in much-improved handling at high speed. Deliveries commenced in 1952; total production was more than 5000.

SPECIFICATIONS

Country of origin:	USSR
Type:	single-seat fighter
Powerplant:	one 33.1kN (7452lb) Klimov VK-1F turbojet
Performance:	maximum speed at 3000m (9,840ft) 1145km/h (711mph); service ceiling 16,600m (54,560ft); range 1470km (913 miles)
Weights:	empty 4100kg (9040lb); maximum loaded 600kg (14,770lb)
Dimensions:	wingspan 9.45m (31ft); length 11.05m (36ft 3.75in); height 3.35m (11ft); wing area 20.6 sq m (221.74 sq ft)
Armament:	one 37mm N-37 cannon and two 23mm NS-23 cannon, plus up to 500kg (1102lb) of mixed stores on underwing pylons

MiG-17PF 'Fresco-D'

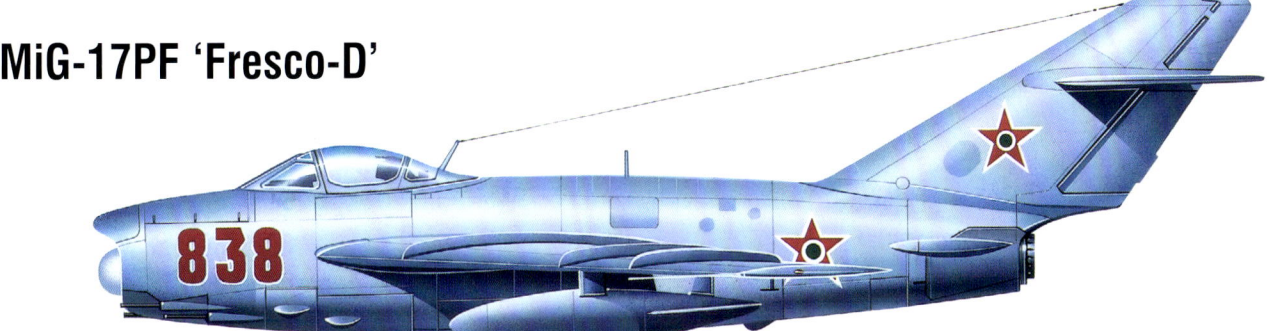

The MiG-17PF 'Fresco-D' was a night-fighter development with the RP-1 Izumrud or 'Scan Odd' radar mounted in the upper lip of the intake. The later MiG-17PFU could carry radar-guided missiles, but the PF was armed with cannon and bombs or unguided rockets only.

SPECIFICATIONS

Country of origin:	USSR
Type:	single-seat all-weather interceptor
Powerplant:	one 33.1kN (7452lb) thrust Klimov VK-1F afterburning turbojet
Performance:	maximum speed at 9080m (20,000ft) 1480km/h (920mph); service ceiling 17,900m (58,725ft); maximum range 2200km (1367 miles)
Weights:	empty 4182kg (9212lb); maximum 6350kg (14,000lb)
Dimensions:	wingspan 9m (29ft 6.5in); length 11.68m (38ft 4in); height 4.02m (13ft 2.25in); wing area 25 sq m (269.11 sq ft)
Armament:	three 23mm NS-23 cannon; up to 500kg (1102lb) of bombs or rockets

MiG-19 'Farmer'

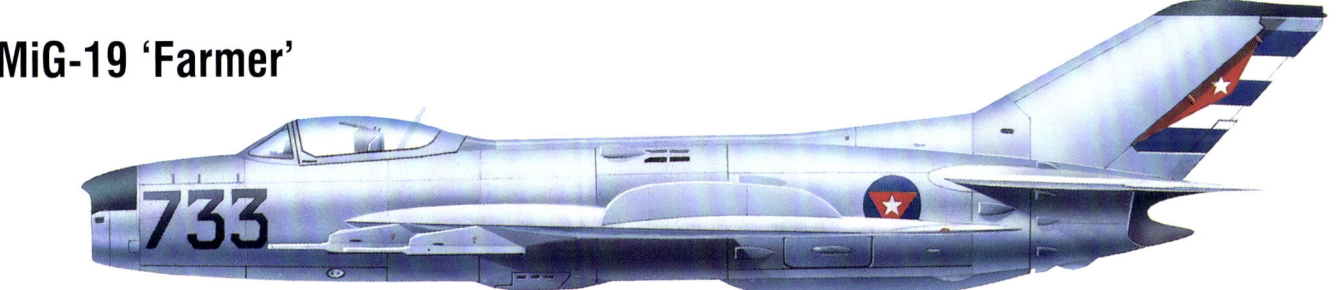

With the MiG-19, the Mikoyan-Gurevich bureau established itself at the forefront of the world's fighter design teams. It was first flown in September 1953, and steadily improved versions culminated in the MiG-19PM, with guns removed and pylons for four early beam-rider air-to-air missiles. In the late 1990s, some aircraft remained in service with training units.

SPECIFICATIONS

Country of origin:	USSR
Type:	single-seat all-weather interceptor
Powerplant:	two 31.9kN (7165lb) Klimov RD-9B turbojets
Performance:	maximum speed at 9080m (20,000ft) 1480km/h (920mph); service ceiling 17,900m (58,725ft); maximum range at high altitude with two drop tanks 2200km (1367 miles)
Weights:	empty 5760kg (12,698lb); maximum take-off weight 9500kg (20,944lb)
Dimensions:	wingspan 9m (29ft 6.5in); length 13.58m (44ft 7in); height 4.02m (13ft 2.25in); wing area 25 sq m (269.11 sq ft)
Armament:	underwing pylons for four AA-1 Alkali air-to-air-missiles, or AA-2 Atoll

Sukhoi Fitters

Less well-known in the West than MiG, the Sukhoi Design Bureau produced interceptors such as the Su-9 and Su-15, which were rarely seen outside the Soviet Union, and the 'Fitter' series, which was widely exported. The simple Su-7 with a fixed, swept wing was followed by the Su-17 (designated Su-20 for export) with variable-geometry ('swing') wings and the Su-22M with improved avionics and more fuel. Poland and Vietnam are among the few remaining Su-22 users.

Su-7BM 'Fitter-A'

Planned as a fighter to intercept the USAF's North American F-100 and F-101, the large swept-wing Sukhoi fighter in fact became the standard tactical fighter-bomber of the Soviet air forces. The Su-7B was ordered into production in 1958, and in a variety of sub-variants became the standard Soviet Bloc attack aircraft. Thousands were supplied to all Warsaw Pact nations, among other countries.

SPECIFICATIONS	
Country of origin:	USSR
Type:	ground-attack fighter
Powerplant:	one 88.2kN (19,842lb) Lyulka AL-7F turbojet
Performance:	maximum speed at 11,000m (36,090ft) approximately 1700km/h (1056mph); service ceiling 15,150m (49,700ft); typical combat radius 320km (199 miles)
Weights:	empty 8620kg (19,000lb); maximum take-off weight 13,500kg (29,750lb)
Dimensions:	wingspan 8.93m (29ft 3.5in); length 17.37m (57ft); height 4.7m (15ft 5in)
Armament:	two 30mm NR-30 cannon; four external pylons for two 750kg (1653lb) and two 500kg (1102lb) bombs, but with two tanks on fuselage pylons, total external weapon load is reduced to 1000kg (2205lb)

Su-7BMK 'Fitter-A'

The Su-7BMK was a version built largely for export between 1968 and 1971. Some served with the Soviet Air Force's Frontal Aviation arm, including the one illustrated, which was on the strength of a unit in the Trans-Baikal Military District in 1978.

SPECIFICATIONS	
Country of origin:	USSR
Type:	ground-attack fighter
Powerplant:	one 88.2kN (19,842lb) Lyulka AL-7F turbojet
Performance:	max speed at 11,000m (36,090ft) approximately 1700km/h (1056mph); service ceiling 15,150m (49,700ft); combat radius 320km (199 miles)
Weights:	empty 8620kg (19,000lb); maximum take-off weight 13,500kg (29,750lb)
Dimensions:	span 8.93m (29ft 3.5in); length 17.37m (57ft); height 4.7m (15ft 5in)
Armament:	two 30mm NR-30 cannon; four external pylons for two 750kg (1653lb) and two 500kg (1102lb) bombs, but with two tanks on fuselage pylons, total external weapon load is reduced to 1000kg (2205lb)

SUKHOI FITTERS

Su-7UM 'Moujik-A'

SPECIFICATIONS	
Country of origin:	USSR
Type:	ground-attack fighter
Powerplant:	one 94.1kN (21,164lb) thrust Lyulka AL-7F-1-250 afterburning turbojet engine
Performance:	max speed at 11,000m (36,090ft) approximately 1700km/h (1056mph); service ceiling 16,992m (55,760ft); range 1000km (621 miles)
Weights:	empty 8620kg (19,000lb); maximum take-off weight 13,500kg (29,750lb)
Dimensions:	span 8.93m (29ft 3.5in); length 17.37m (57ft); height 4.7m (15ft 5in)
Armament:	two 30mm NR-30 cannon; four external pylons for two 750kg (1653lb) and two 500kg (1102lb) bombs, but with two tanks on fuselage pylons, total external weapon load is reduced to 1000kg (2205lb)

Known as 'Moujik' to NATO, the Su-7UM was the two-seat transition trainer version of the Fitter-A, used by most operators, including Egypt. The installation of a second seat reduced the fuel capacity, and a periscope was needed by the instructor to have a forward view.

Su-17M-4 'Fitter-K'

SPECIFICATIONS	
Country of origin:	USSR
Type:	single-seat ground-attack fighter
Powerplant:	one 110.3kN (24,802lb) Lyul'ka AL-21F-3 turbojet
Performance:	max speed approximately 2220km/h (1380mph); service ceiling 15,200m (49,865ft); combat radius 675km (419 miles)
Weights:	empty 9500kg (20,944lb); maximum take-off weight 19,500kg (42,990lb)
Dimensions:	wingspan 13.8m (45ft 3in) spread and 10m (32ft 10in) swept; length 18.75m (61ft 6in); height 5m (16ft 5in); wing area 40 sq m (430 sq ft)
Armament:	two 30mm NR-30 cannon; nine external pylons with provision for up to 4250kg (9370lb) of stores, including tactical nuclear weapons

The Su-71G prototype with a variable-geometry wing was first flown in 1966. The new aircraft was found to have far superior performance than even the most developed Su-7, especially for short-field operations. Entering service in 1971, the ultimate development of the aircraft was the Su-17M-4, distinguishable by an airscoop for the cooling system on the leading edge of the tailfin root.

Su-20 'Fitter-C'

SPECIFICATIONS	
Country of origin:	USSR
Type:	single-seat ground-attack fighter
Powerplant:	one 110.3kN (24,802lb) Lyul'ka AL-21F-3 turbojet
Performance:	max speed approx 2220km/h (1380mph); service ceiling 15,200m (49,865ft); combat radius load 675km (419 miles)
Weights:	empty 9500kg (20,944lb); maximum take-off weight 19,500kg (42,990lb)
Dimensions:	span 13.8m (45ft 3in) spread, 10m (32ft 10in) swept; length 18.75m (61ft 6in); height 5m (16ft 5in); wing area 40 sq m (430 sq ft)
Armament:	two 30mm NR-30 cannon; nine external pylons with provision for up to 4250kg (9370lb) of stores, including tactical nuclear weapons

The first version of the variable-geometry wing Sukhoi Su-17 ground-attack aircraft made available for export was designated Su-20. Poland was the only country to receive the full-standard Fitter-C, but a reduced-equipment version was operated by Afghanistan, Algeria, Angola, Egypt, Iraq, North Korea and Vietnam.

MILITARY AIRCRAFT IN THE MODERN ERA

Military aircraft have matured to become 'systems platforms' as much as simply a means of taking weapons to the enemy.

Since the early 1990s, the previous generations of warplanes have been swept aside by aircraft with fly-by-wire control systems, data-linked sensors and the ability to travel at supersonic speeds. Composite materials such as carbon fibre make up a large part of aircraft structures. The increasing cost of such sophisticated warplanes has led to fierce competition by manufacturers for scarce orders.

Left: The Eurofighter Typhoon is an example of multinational cooperation to produce a fighter for the differing needs of four air forces.

Vought A-7 and F-8

The Vought Company (as Chance-Vought) had created the classic F4U Corsair during World War II, but stumbled in the early post-war years. It redeemed itself with the record-breaking F8U (F-8) Crusader, which became the US Navy's primary fighter and served with great distinction in Vietnam. The company reorganized as Ling-Temco Vought (LTV) in the 1960s. Its last aircraft, the A-7 Corsair II, successfully used the F-8's configuration for a compact attack aircraft.

Vought F-8D Crusader

In 1955, Vought began the development of a totally new Crusader. Designated XF8U-3 Crusader III, the aircraft was rejected in favour of the Phantom II. Vought steadily improved the aircraft, the most potent version being the F-8D, with J57-P-20 turbojet, extra fuel and new radar for a specially produced radar-homing AIM-9C Sidewinder air-to-air missile. A total of 152 F-8Ds were produced.

SPECIFICATIONS	
Country of origin:	United States
Type:	single-seat carrier-based fighter
Powerplant:	one 80kN (18,000lb) Pratt & Whitney J57-P-20 turbojet
Performance:	maximum speed at 12,192m (40,000ft) 1975km/h (1227mph); service ceiling about 17,983m (59,000ft); combat radius 966km (600 miles)
Weights:	empty 9038kg (19,925lb); maximum take-off weight 15,422g (34,000lb)
Dimensions:	wingspan 10.72m (35ft 2in); length 16.61m (54ft 6in); height 4.8m (15ft 9in)
Armament:	four 20mm Colt Mk 12 cannon with 144rpg, up to four Motorola AIM-9C Sidewinder air-to-air missiles; or two AGM-12A or AGM-12B Bullpup air-to-surface missiles

Vought RF-8G Crusader

The Crusader made a useful high-speed reconnaissance aircraft, and the RF-8A provided vital intelligence during the Cuban Missile Crisis and the Vietnam War. The RF-8G was rebuilt from older A models and became the last version in US service, finally retired from reserve units in 1987. It could carry four cameras in a fuselage bay, but no armament.

SPECIFICATIONS	
Country of origin:	United States
Type:	carrier-based reconnaissance aircraft
Powerplant:	one 80.1kN (18,000lb) thrust Pratt & Whitney J57-P-22 afterburning turbojet engine
Performance:	maximum speed at 12,192m (40,000ft) 1975km/h (1227mph); service ceiling about 17,983m (59,000ft); combat radius 966km (600 miles)
Weights:	empty 9038kg (19,925lb); maximum take-off weight 15,422g (34,000lb)
Dimensions:	wingspan 10.72m (35ft 2in); length 16.61m (54ft 6in); height 4.8m (15ft 9in);
Armament:	none

Vought F-8E(N) Crusader

Vought sold a version of the F-8E to the French Aéronavale, even though the carriers *Foch* and *Clemenceau* were thought too small for such aircraft. To create the F-8E (FN), Vought redesigned the wing and tail to provide greater lift and to improve low-speed handling. The first FN flew in June 1964, and in 1991, nearly 25 years after entering service, *Clemenceau's* aircraft were involved in the first Gulf War.

SPECIFICATIONS
Country of origin:	United States
Type:	single-seat carrier-borne interceptor and attack aircraft
Powerplant:	one 80kN (18,000lb) Pratt & Whitney J57-P-20A turbojet
Performance:	maximum speed at 10,975m (36,000ft) 1827km/h (1135mph); service ceiling 17,680m (58,000ft); combat radius 966km (600 miles)
Dimensions:	wingspan 10.87m (35ft 8in); length 16.61m (54ft 6in); height 4.8m (15ft 9in); wing area 32.51sq m (350sq ft)
Weights:	empty 9038kg (19,925lb); maximum take-off weight 15,420kg (34,000lb)
Armament:	four 20mm M39 cannon; provision for up to 2268kg (5000lb) of stores, including two Matra R530 air-to-air missiles or eight 5in rockets

Vought A-7D Corsair II

Though derived from the F-8 Crusader, the Corsair is a totally different aircraft. By restricting performance to high subsonic speed, it was possible to reduce structural weight, and correspondingly the range increased and weapon load multiplied by nearly four times. The first flight was made in September 1965. During the Vietnam War, more than 90,000 Corsair missions were flown.

SPECIFICATIONS
Country of origin:	United States
Type:	single-seat attack aircraft
Powerplant:	one 63.4kN (14,250lb) Allison TF41-1 (Rolls-Royce Spey) turbofan
Performance:	maximum speed at low-level 1123km/h (698mph); combat range with typical weapon load 1150km (4100 miles)
Weights:	empty 8972kg (19,781lb); maximum take-off weight 19,050kg (42,000lb)
Dimensions:	wingspan 11.8m (38ft 9in); length 14.06m (46ft 1.5in); height 4.9m (16ft 0.75in); wing area 34.84sq m (375sq ft)
Armament:	one 20mm M61 Vulcan, provision for up to 6804kg (15,000lb) of stores, including bombs, napalm tanks, air-to-surface missiles and drop tanks

Vought A-7H Corsair II

A number of nations expressed interest in the Vought A-7 at an early stage in the programme, but the first foreign nation to take delivery of the fighter was Greece. The A-7H aircraft are used in both ground-attack and air-defence roles, and may be equipped with AIM-9L Sidewinder.

SPECIFICATIONS
Country of origin:	United States
Type:	single-seat tactical fighter
Powerplant:	one 66.7kN (15,000lb) Allison TF-41-A-400 turbofan
Performance:	maximum speed at sea level 1112km/h (691mph); service ceiling 15,545m (51,000ft); range with typical load 1127km (700 miles)
Weights:	empty 8841kg (19,490lb); maximum take-off weight 19,051kg (42,000lb)
Dimensions:	wingspan 11.81m (38ft 9in); length 14.06m (46ft 1in); height 4.9m (16ft); wing area 34.84 sq m (375 sq ft)
Armament:	one 20-mm M61A1 multi-barrelled cannon; eight external pylons with provision for up to 6804kg (15,000lb) of stores, including bombs, cluster bombs, rocket pods and/or air-to-air missiles

Panavia Tornado

The Tornado was the result of a 1960s requirement for a strike and reconnaissance aircraft capable of carrying a heavy and varied weapons load and of penetrating Warsaw Pact defensive systems by day and night, at low level and in all weathers. To develop and build the aircraft, a consortium of European companies (primarily British, West German and Italian) was formed under the name of Panavia.

Tornado F.Mk 3

In the late 1960s, the RAF saw the need to replace its McDonnell Douglas Phantom II and BAe Lighting interceptors, and ordered the development of the Tornado ADV (Air Defence Variant), a dedicated air-defence aircraft with all-weather capability, based on the same airframe as the GR.1 ground-attack aircraft. Structural changes include a lengthened nose for the Foxhunter radar. The ADV was given the designations F.Mk 2 and F.Mk 3 in RAF service.

SPECIFICATIONS

Country of origin:	Germany/Italy/United Kingdom
Type:	all-weather air defence aircraft
Powerplant:	two 73.5kN (16,520lb) Turbo-Union RB.199-34R Mk 104 turbofans
Performance:	maximum speed above 11,000m (36,090ft) 2337km/h (1452mph); operational ceiling about 21,335m (70,000ft); intercept radius more than 1853km (1150 miles)
Weights:	empty 14,501kg (31,970lb); maximum take-off 27,987kg (61,700lb)
Dimensions:	wingspan 13.91m (45ft 7.75in) spread and 8.6m (28ft 2.5in) swept; length 18.68m (61ft 3in); height 5.95m (19ft 6.25in); wing area 26.6 sq m (286.3 sq ft)
Armament:	two 27mm IWKA-Mauser cannon with 180rpg, six external hardpoints with provision for up to 5806kg (12,800lb) of stores, including Sky Flash medium-range air-to-air missiles, AIM-9L Sidewinder short range air-to-air missiles and drop tanks

Tornado GR.1

SPECIFICATIONS

Country of origin:	United Kingdom/West Germany/Italy
Type:	tactical reconnaissance aircraft
Powerplant:	two 71.5kN (16,075lb) Turbo-Union RB.199-34R Mk 103 turbofans
Performance:	maximum speed 2337km/h (1452mph); service ceiling 15,240m (50,000ft); combat radius 1390km (864 miles)
Weights:	27,216kg (60,000lb) loaded
Dimensions:	wing span 13.91m (45ft 7in) spread and 8.6m (28ft 2.5in) swept; length: 16.72m (54ft 10in); height 5.95m (19ft 6.25in)
Armament:	up to 9000kg (19,840lb) of stores; Vinten Linescan infrared sensors and TIALD (Thermal Imaging and Laser Designator)

Tornado IDS

Italy's share in the Tornado programme comprised 100 Tornado IDS (interdictor/ strike) variants, equivalent to the RAF's GR.1. They later converted 15 to Tornado ECR (electronic combat and reconnaissance) standard, with the ability to launch HARM anti-radar missiles. A mid-life upgrade programme is underway for 80 Italian Tornadoes.

SPECIFICATIONS	
Country of origin:	United Kingdom/West Germany/Italy
Type:	multi-role combat aircraft
Powerplant:	two 71.5kN (16,075lb) Turbo-Union RB.199-34R Mk 103 turbofans
Performance:	maximum speed above 11,000m (36,090ft) 2337km/h (1452mph); service ceiling 15,240m (50,000ft); combat radius with weapon load on hi-lo-hi mission 1390km (864 miles)
Weights:	empty 14,091kg (31,065lb); maximum take-off 27,216kg (60,000lb)
Dimensions:	wingspan 13.91m (45ft 7in) spread; length 16.72m (54ft 10in); height 5.95m (19ft 6.25in); wing area 26.6 sq m (286.3 sq ft)
Armament:	two 27mm IWKA-Mauser cannon with 180rpg, seven external hard-points with provision for up to 9000kg (19,840lb) of stores, including ALARM anti-radiation missiles, air-to-air, air-to-surface and anti-ship missiles, conventional and guided bombs, cluster bombs, ECM pods and drop tanks TIALD (Thermal Imaging and Laser Designator)

The first Tornado GR.1s were delivered in July 1980. The RAF took delivery of 229 GR.1 strike aircraft, the Luftwaffe 212, the German Naval Air Arm 112, and the Aeronautica Militare Italiana (Italian Air Force) 100. RAF and Italian Tornados saw action in the 1991 Gulf War.

SEPECAT Jaguar

Developed jointly by BAC in Britain and Dassault-Breguet in France (Societé Européenne de Production de l'Avion Ecole de Combat at Appui Tactique), to meet a joint requirement of L'Armée de l'Air and the Royal Air Force, the Jaguar emerged as a far more powerful and effective aircraft than originally envisaged. It was planned as a light trainer and close-support machine, but as a single-seat tactical support aircraft, it formed the backbone of the French tactical nuclear strike force.

Jaguar A

The Jaguar A first flew in March 1969 and when service deliveries began in 1973, 160 aircraft were produced. Power was provided by a turbofan developed jointly from the Rolls-Royce RB.172 by Rolls-Royce and Turbomeca.

SPECIFICATIONS	
Country of origin:	France and United Kingdom
Type:	single-seat tactical support and strike aircraft
Powerplant:	two 32.5kN (7305lb) Rolls-Royce/Turbomeca Adour Mk 102 turbofans
Performance:	maximum speed at 11,000m (36,090ft) 1593km/h (990mph); combat radius on lo-lo-lo mission with internal fuel 557km (357 miles)
Weights:	empty 7000kg (15,432lb); maximum take-off weight 15,500kg (34,172lb)
Dimensions:	wingspan 8.69m (28ft 6in); length 16.83m (55ft 2.5in); height 4.89m (16ft 0.5in); wing area 24 sq m (258.34 sq ft)
Armament:	two 30mm DEFA cannon; provision for 4536kg (10,000lb) of stores, including tactical nuclear weapon or conventional loads, ASMs, drop tanks and rocket-launcher pods, and a reconnaissance pod

Jaguar GR.1

The RAF bought 165 single-seat Jaguar GR.1s from 1973, and they served with nine frontline squadrons over the years, including No 14 (illustrated), which operated in the strike role from RAF Bruggen in Germany. The Jaguar went through several upgrades before the last GR.3 models were retired in 2007.

SPECIFICATIONS	
Country of origin:	France and United Kingdom
Type:	single-seat tactical support and strike aircraft
Powerplant:	two 32.5kN (7305lb) Rolls-Royce/Turbomeca Adour Mk 102 turbofans
Performance:	maximum speed at 11,000m (36,090ft) 1593km/h (990mph); combat radius on lo-lo-lo mission with internal fuel 557km (357 miles)
Weights:	empty 7000kg (15,432lb); maximum take-off weight 15,500kg (34,172lb)
Dimensions:	wingspan 8.69m (28ft 6in); length 16.83m (55ft 2.5in); height 4.89m (16ft 0.5in); wing area 24 sq m (258.34 sq ft)
Armament:	two 30mm DEFA cannon; provision for 4536kg (10,000lb) of stores, including a nuclear weapon or conventional loads

SEPECAT JAGUAR

Jaguar International

The outstanding versatility of the Jaguar encouraged the Anglo-French SEPECAT company to develop a version for the export market, though by the mid-1990s only 169 had been ordered by four nations. The first Jaguar International took to the air in August 1976. The aircraft was optimized for anti-shipping, air-defence, ground-attack and reconnaissance roles. This example was used by Ecuador.

SPECIFICATIONS

Country of origin:	France and United Kingdom
Type:	single-seat tactical support and strike aircraft
Powerplant:	two 37.3kN (8400lb) Rolls-Royce/Turbomeca Adour Mk 811 turbofans
Performance:	maximum speed at 11,000m (36,090ft) 1699km/h (1056mph); combat radius on lo-lo-lo mission with internal fuel 537km (334 miles)
Weights:	empty 7700kg (16,976lb); maximum take-off weight 15,700kg (34,613lb)
Dimensions:	wingspan 8.69m (28ft 6in); length 16.83m (55ft 2.5in); height 4.89m (16ft 0.5in); wing area 24.18 sq m (260.28 sq ft)
Armament:	two 30mm Aden Mk.4 cannon; provision for 4763kg (10,500lb) of stores, including air-to-air missiles, anti-ship missiles, laser-guided or conventional bombs, napalm tanks, drop tanks and ECM pods

Jaguar T.2

The RAF version of the Jaguar E has the service designation Jaguar T.2. It retains full operational capability and is equipped to the same standard as the GR.1. The RAF received 38 T.2s, three more than originally planned. Updated aircraft are T.2A.

SPECIFICATIONS

Country of origin:	France and United Kingdom
Type:	single-seat tactical support and strike aircraft
Powerplant:	two 37.3kN (8040lb) Rolls-Royce/Turbomeca Adour Mk 104 turbofans
Performance:	maximum speed at 11,000m (36,090ft) 1593km/h (990mph); combat radius on lo-lo-lo mission with internal fuel 557km (357 miles)
Weights:	empty 7000kg (15,432lb); maximum take-off weight 15,500kg (34,172lb)
Dimensions:	wingspan 8.69m (28ft 6in); length 16.83m (55ft 2.5in); height 4.89m (16ft 0.5in); wing area 24 sq m (258.34 sq ft)
Armament:	two 30mm DEFA cannon; provision for 4536kg (10,000lb) of stores, including one tactical nuclear weapon or conventional loads

Jaguar IM (Shamsher)

Although most other users have retired their Jaguars, Hindustan Aeronautics Limited (HAL) continues to produce the Jaguar or Shamsher (Sword of Justice) for the Indian Air Force. A unique Indian variant is the Jaguar IM maritime strike aircraft, with an Agave radar and Sea Eagle missile capability.

SPECIFICATIONS

Country of origin:	France and United Kingdom
Type:	single-seat maritime attack aircraft
Powerplant:	two 37.3kN (8400lb) thrust Rolls-Royce/Turbomeca RT172-58 Adour Mk.811 afterburning turbofan engines
Performance:	maximum speed at 11,000m (36,090ft) 1699km/h (1056mph); combat radius on lo-lo-lo mission with internal fuel 537km (334 miles)
Weights:	empty 7700kg (16,976lb); maximum take-off weight 15,700kg (34,613lb)
Dimensions:	wingspan 8.69m (28ft 6in); length unknown; height 4.89m (16ft 0.5in); wing area 24.18 sq m (260.28 sq ft)
Armament:	two 30mm Aden Mk.4 cannon; provision for 4763kg (10,500lb) of stores

Grumman Intruder and Prowler

Selected from 11 competing designs in December 1957, the Intruder was specifically planned for first-pass blind-attack on point surface targets at night or in any weather conditions. The Intruder first came into service with the US Navy in February 1963; during the Vietnam War, the A-6A worked round-the-clock on precision bombing missions that no other aircraft was capable of undertaking until the introduction of the F-111.

A-6 Intruder

The A-6 was introduced in Vietnam and saw its last combat action over Iraq and Kuwait in 1991. This A-6E of VMA-533 was deployed to Bahrain for Operation Desert Storm in 1991. It is shown here carrying 227kg (500lb) Mk 82 bombs on underwing and centreline multiple ejector racks.

SPECIFICATIONS	
Country of origin:	United States
Type:	two-seat carrierborne and landbased all-weather strike aircraft
Powerplant:	two 41.4kN (9300lb) Pratt & Whitney J52-P-8A turbojets
Performance:	maximum speed at sea level 1043km/h (648mph); service ceiling 14,480m (47,500ft); range 1627km (1011 miles)
Weights:	empty 12,132kg (26,746lb); maximum take-off weight 26,581kg (58,600lb)
Dimensions:	wingspan 16.15m (53ft); length 16.69; height 4.93m (16ft 2in); wing area 49.13 sq m (528.9 sq ft)
Armament:	five external hardpoints with provision for up to 8165kg (18,000lb) of stores, including nuclear weapons, conventional and guided bombs, air-to-surface missiles and drop tanks

EA-6A Intruder

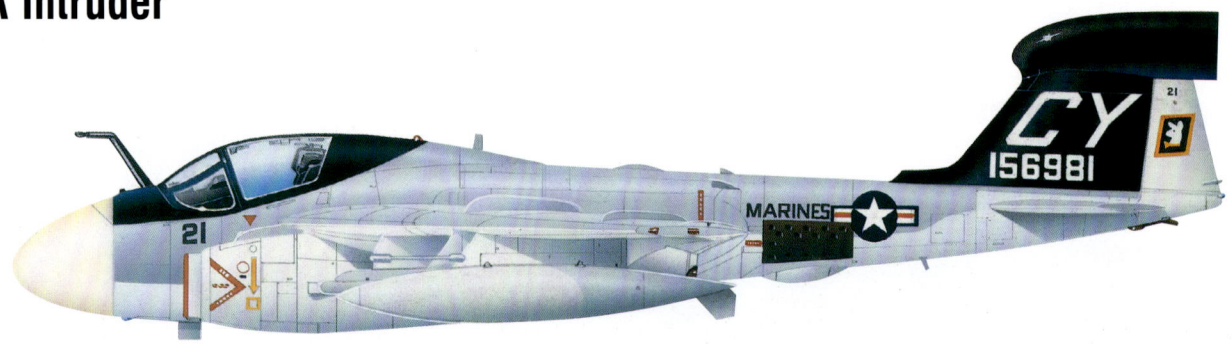

The EA-6A was a replacement for the elderly EF-10 Sky Knight electronic warfare aircraft. EA-6As were sent to Southeast Asia to deal with the increasing sophistication of North Vietnam's air-defence network. Able to jam and confuse radar, they had a rarely used capability to fire anti-radar Shrike missiles.

SPECIFICATIONS	
Country of origin:	United States
Type:	electronic warfare aircraft
Powerplant:	two 41.4kN (9300lb) Pratt & Whitney J52-P-8A turbojets
Performance:	maximum speed at sea level 1043km/h (648mph); service ceiling 14,480m (47,500ft); range 1627km (1011 miles)
Weights:	empty 12,132kg (26,746lb); maximum take-off weight 26,581kg (58,600lb)
Dimensions:	wingspan 16.15m (53ft); length 16.69; height unknown; wing area 49.13 sq m (528.9 sq ft)
Armament:	AGM-45 Shrike anti-radiation missiles

A-6E Intruder

The original A-6A model was replaced by the A-6E, with a new digital computer and other electronic improvements, from 1971. Later A-6Es had an undernose turret called TRAM (Target Recognition Attack Multi-sensor), which included an infrared sensor and a laser designator, which allowed the use of more precision weapons.

SPECIFICATIONS	
Country of origin:	United States
Type:	two-seat carrierborne and landbased all-weather strike aircraft
Powerplant:	two 41.4kN (9300lb) Pratt & Whitney J52-P-8A turbojets
Performance:	maximum speed at sea level 1043km/h (648mph); service ceiling 14,480m (47,500ft); range 1627km (1011 miles)
Weights:	empty 12,132kg (26,746lb); maximum take-off weight 26,581kg (58,600lb)
Dimensions:	wingspan 16.15m (53ft); length 16.69; height 4.93m (16ft 2in); wing area 49.13 sq m (528.9 sq ft)
Armament:	five external hardpoints with provision for up to 8165kg (18,000lb) of stores, including nuclear weapons, conventional and guided bombs, air-to-surface missiles and drop tanks

KA-6D Prowler

The KA-6D was a conversion of the A-6A modified to refuel carrier-based aircraft. Bombing equipment was removed and a hose-drum unit fitted in the fuselage. The weapons pylons were used to carry additional fuel tanks. This KA-6D also carries a 'buddy' refuelling pod on the centreline.

SPECIFICATIONS	
Country of origin:	United States
Type:	electronic countermeasures platform
Powerplant:	two 49.8kN (11,200lb) Pratt & Whitney J52-P-408 turbojets
Performance:	maximum speed at sea level 982km/h (610mph); service ceiling 11,580m (38,000ft); combat range with full external fuel 1769km (1099 miles)
Weights:	empty 14,588kg (32,162lb); maximum take-off weight 29,484kg (65,000lb)
Dimensions:	wingspan 16.15m (53ft); length 18.24m (59ft 10in); height 4.95m (16ft 3in); wing area 49.13 sq m (528.9 sq ft)
Armament:	none on early models, retrofitted with external hardpoints for four or six AGM-88 HARM air-to-surface anti-radar missiles

EA-6B Prowler

The US Navy rarely undertakes a strike mission without the protection offered by the EA-6 ECM. This aircraft was developed from the successful A-6 Intruder family. The cockpit provides seating for the pilot and three electronic warfare officers, who control the most sophisticated and advanced ECM equipment ever fitted to a tactical aircraft, including the ALQ-99 tactical jamming system.

SPECIFICATIONS	
Country of origin:	United States
Type:	electronic countermeasures platform
Powerplant:	two 49.8kN (11,200lb) Pratt & Whitney J52-P-408 turbojets
Performance:	maximum speed at sea level 982km/h (610mph); service ceiling 11,580m (38,000ft); combat range 1769km (1099 miles)
Weights:	empty 14,588kg (32,162lb); maximum take-off weight 29,484kg (65,000lb)
Dimensions:	wingspan 16.15m (53ft); length 18.24m (59ft 10in); height 4.95m (16ft 3in); wing area 49.13 sq m (528.9 sq ft)
Armament:	none on early models, retrofitted with external hardpoints for four or six AGM-88 HARM air-to-surface anti-radar missiles

USA vs Libya

Libya and the United States of America fought several actions in the 1980s over Libya's claim to parts of the Mediterranean Sea and its support for terrorism. These included the Gulf of Sidra Incident in 1981, which saw US Navy F-14s destroy Libyan Su-22s; and Operation El Dorado Canyon in 1986, which followed a terrorist attack in a Berlin nightclub targeted at American servicemen. The US blamed Libya, and launched an attack on targets in Tripoli and Benghazi on 15 April 1986.

Grumman F-14A Tomcat

While asserting the right to navigate in international waters, patrols from the carrier USS *Nimitz* encountered Libyan Arab Republic Air Force (LARAF) fighters on several occasions in August 1981. Responding to a Libyan missile launch, F-14As from VF-41 'Black Aces' destroyed two LARAF Su-22 'Fitters' with AIM-9 Sidewinders.

SPECIFICATIONS	
Country of origin:	United States
Type:	two-seat carrierborne fleet defence fighter
Powerplant:	two 92.9kN (20,900lb) Pratt & Whitney TF30-P-412A turbofans
Performance:	maximum speed at high altitude 2517km/h (1564mph); service ceiling 17,070m (56,000ft); range about 3220km (2000 miles)
Weights:	empty 18,191kg (40,104lb); maximum take-off 33,724kg (74,349lb)
Dimensions:	wingspan 19.55m (64ft 1.5in) unswept; 11.65m (38ft 2.5in) swept; length 19.1m (62ft 8in); height 4.88m (16ft); wing area 52.49sq m (565sq ft)
Armament:	one 20mm M61A1 Vulcan rotary cannon with 675 rounds; external pylons for a combination of AIM-7 Sparrow medium range air-to-air missiles, AIM-9 medium range air-to-air missiles, and AIM-54 Phoenix long range air-to-air missiles

General Dynamics F-111F

The main strike of El Dorado Canyon was launched from the UK by F-111s, including F-111Fs of the 48th Tactical Fighter Wing at Lakenheath (illustrated). One F-111F and crew was lost on the raid, having probably hit the sea while flying at low level.

SPECIFICATIONS	
Country of origin:	United States
Type:	two-seat attack aircraft
Powerplant:	111.7kN (25,100lb) thrust Pratt & Whitney TF30-P-100 afterburning turbojet engines
Performance:	maximum speed 2334km/h (1,452 mph); range 5851km (3634 miles); service ceiling 18,287m (60,000ft)
Weights:	empty 20,943kg (46,172lb); maximum 44,875kg (98,950lb)
Dimensions:	span 19.2m (63ft); length 22.4m (73ft 6in); height 5.2m (17ft 2in); wing area 61 sq m (657 sq ft)
Armament:	up to 11,250kg (25,000lb) of bombs, rockets, missiles or fuel tanks

USA VS LIBYA

Mikoyan-Gurevich MiG-23 'Flogger-E'

Libya and a number of other Arab countries purchased a much-simplified export version of the MiG-23M 'Flogger-B', designated MiG-23 'Flogger-E'. The aircraft retains the same basic airframe as its predecessor, but is powered by the 98kN (22,046lb) Tumanskii R-27F2M-300 turbojet.

SPECIFICATIONS	
Country of origin:	USSR
Type:	single-seat air combat fighter
Powerplant:	one 98kN (22,046lb) Tumanskii R-27F2M-300 turbojet
Performance:	maximum speed at altitude about 2445km/h (1520mph); service ceiling over 18,290m (60,000ft); combat radius 966km (600 miles)
Weights:	empty 10,400kg (22,932lb); maximum loaded 18,145kg (40,000lb)
Dimensions:	wingspan 13.97m (45ft 10in) spread and 7.78m (25ft 6.25in) swept; length (including probe) 16.71m (54ft 10in); height 4.82m (15ft 9.75in); wing area 37.25 sq m (402 sq ft) spread
Armament:	one 23mm GSh-23L cannon, six external hardpoints with provision for up to 3000kg (6614lb) of stores, including AA-2 Atoll air-to-air missiles, cannon pods, rocket launcher pods, large calibre rockets and bombs

Mikoyan-Gurevich MiG-25P 'Foxbat-A'

Designed as an interceptor to meet the challenge of the planned US B-70 long-range, high-speed strategic bomber in the late 1950s, the MiG-25 far outclassed any Western aircraft in terms of speed and height when it finally entered service in the early 1970s (and long after the B-70 programme was cancelled). This aircraft is also operated by Libya, Algeria, India, Iraq and Syria.

SPECIFICATIONS	
Country of origin:	USSR
Type:	single-seat interceptor
Powerplant:	two 100kN (22,487lb) Tumanskii R-15B-300 turbojets
Performance:	maximum speed at altitude about 2974km/h (1848mph); service ceiling over 24,385m (80,000ft); combat radius 1130km (702 miles)
Weights:	empty 20,000kg (44,092lb); maximum take-off 37,425kg (82,508lb)
Dimensions:	wingspan 14.02m (45ft 11.75in); length 23.82m (78ft 1.75in); height 6.1m (20ft 0.5in); wing area 61.4 sq m (660.9 sq ft)
Armament:	external pylons for four air-to-air missiles in the form of either two each of the IR- and radar-homing AA-6 'Acrid', or two AA-7 'Apex' and two AA-8 'Aphid' weapons

Ilyushin Il-76MD 'Candid'

F-111Fs and A-6E Intruders attacked several Libyan airfields in August 1986, destroying about 14 MiG-23s on the ground and up to five Il-76 'Candid' freighters belonging to the LARAF and state-owned airlines. One of them was this aircraft, Il-76MD 5A-DZZ of Libyan Arab Airlines.

SPECIFICATIONS	
Country of origin:	USSR
Type:	heavy freight transport
Powerplant:	four 117.6kN (26,455lb) Soloviev D-30KP-1 turbofans
Performance:	maximum speed at 11,000m (36,090ft) 850km/h (528mph); maximum cruising altitude 12,000m (39,370ft); range 5000km (3107 miles)
Weights:	empty about 75,000kg (165,347lb); maximum take-off weight 170,000kg (374,786lb)
Dimensions:	wingspan 50.5m (165ft 8.2in); length 46.59m (152ft 10.25in); height 14.76m (48ft 5in); wing area 300 sq m (3229.28 sq ft)
Armament:	provision for two 23mm cannon in tail

Airborne Early Warning

Experience in the Pacific in World War II showed the need for airborne radars that could detect attackers beyond the radar range of surface ships. Carrier-based airborne early warning (AEW) aircraft were soon joined by land-based radar 'pickets'. The mission evolved to encompass the control of strike aircraft and fighters on offensive operations, as well as a degree of overwater and overland surveillance.

Avro Shackleton AEW.2

The Shackleton AEW.2 was created from the MR.2 patrol aircraft to provide AEW coverage for the Royal Navy, which was phasing out the carrier-based Gannet, as well as for the RAF. Its radar was the Gannet's 1940s-vintage APS-20, and the 'Shack' was slow, low-flying and noisy. It lasted in service until 1990.

SPECIFICATIONS	
Country of origin:	United Kingdom
Type:	8–10 crew long-range maritime patrol aircraft
Powerplant:	four 1831kW (2455hp) Rolls-Royce Griffon 57A V-12 piston engines
Performance:	max speed 500km/h (311mph); service ceiling 6400m (21,000ft); range 5440km (3380 miles)
Weights:	39010kg (86,000lb) loaded
Dimensions:	wing span 36.58m (120ft); length 26.59m (87ft 3in); height 5.1m (16ft 9in)
Armament:	none

Grumman E-2C Hawkeye

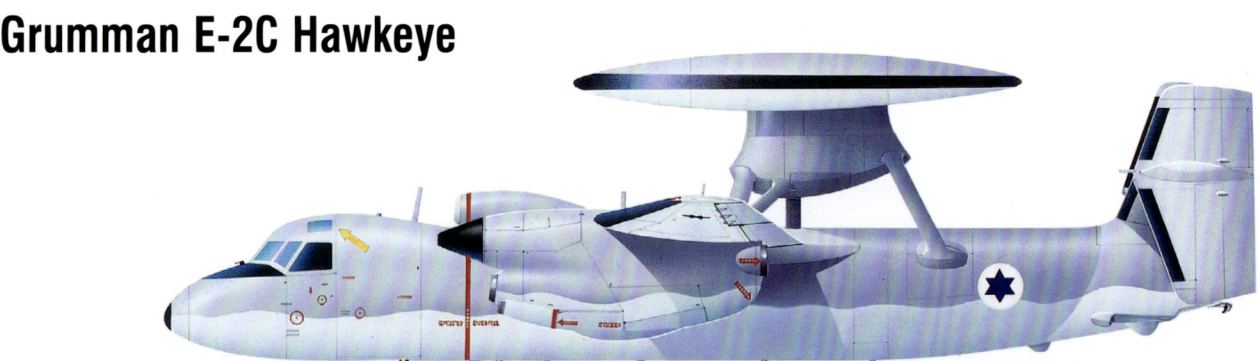

Israel was one of several operators without aircraft carriers to acquire the E-2C Hawkeye to patrol its airspace. Others included Singapore, Egypt and Taiwan. The IDF/AF Hawkeyes were retired in the 1990s, and three of the four purchased were later sold to the Mexican Navy.

SPECIFICATIONS	
Country of origin:	United States
Type:	twin-engined carrier-based early warning aircraft (E-2C)
Powerplant:	one 3800kW (5100hp) Allison T56-A-427 turboprop engines
Performance:	maximum speed 604km/h (375mph); range 2583km (1605 miles); service ceiling 10,210m (33,500ft)
Weights:	empty 15,310kg (33,746lb); maximum 24,655kg (60,000lb)
Dimensions:	span 24.6m (80ft 7in); length 17.56m (57ft 7in); height 5.58m (18ft 4in); wing area 65 sq m (700 sq ft)
Armament:	none

AIRBORNE EARLY WARNING

Hawker Siddeley Nimrod AEW.3

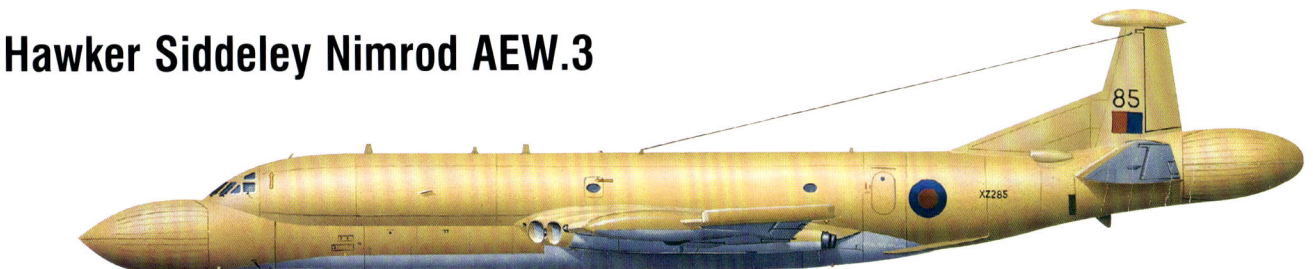

SPECIFICATIONS	
Country of origin:	United Kingdom
Type:	maritime patrol and anti-submarine warfare aircraft
Powerplant:	four 54kN (12,140lb) Rolls Royce Spey Mk 250 turbofans
Performance:	maximum speed 925km/h (575mph); service ceiling 12,800m (42,000ft); range on internal fuel 9262km (5,755 miles)
Weights:	maximum take-off weight 85,185kg (187,800lb)
Dimensions:	wingspan 35m (114ft 10in) excluding wingtip ESM pods; length 41.97m (37 ft 9 in); height 9.08m (29ft 9.5in); wing area 197.04 sq m (2121 sq ft)
Armament:	none

The Nimrod AEW.3 was an attempt to replace the Shackleton with a jet aircraft fitted with modern multi-mode radars. It proved very difficult to coordinate the pictures from the nose and tail radars, and the project went way over budget. It was cancelled and the RAF bought E-3 Sentries.

Boeing E-3A Sentry

SPECIFICATIONS	
Country of origin:	Germany
Type:	Airborne Warning and Control System platform
Powerplant:	four 93-kN (21,000-lb) thrust Pratt & Whitney TF33-PW-100A turbofan engines
Performance:	maximum speed 855km/h (530mph); service ceiling 12,500m (41,000ft); range 7400km (4598 miles)
Weights:	empty 73,480kg (162,000lb); maximum 147, 400kg (325,000lb)
Dimensions:	span 44.42m (145ft 9in); length 46.61m (152ft 11in); height 12.6m (41ft 4in); wing area 3050 sq m (283 sq ft)
Armament:	none

The Boeing E-3 Sentry introduced the term AWACS to common use, and became the platform of choice for the United States, Saudi Arabia, the United Kingdom, France and NATO. The airframe is based on that of the Boeing 707. The Westinghouse APY-1 radar in its giant rotodome can detect targets up to 650km (400 miles) away.

Beriev A-50 'Mainstay'

SPECIFICATIONS	
Country of origin:	Germany
Type:	AWACS aircraft
Powerplant:	four 157kN (35,200lb) thrust Aviadvigatel PS-90A turbofan engines
Performance:	maximum speed 750km/h (466mph); service ceiling 10,000m (32,800ft); range 7500km (4660 miles)
Weights:	loaded 190,000kg (418,880lb)
Dimensions:	span 50.5m (165ft 8in); length 46.59m (152ft 10in); height 14.76m (48ft 5in); wing area 300 sq m (3228 sq ft)
Armament:	none

Based on a stretched Ilyushin Il-76, the A-50 was developed by Beriev as the Soviet counterpart to the E-3. Known to NATO as the 'Mainstay', it entered service in 1989 and about 40 were produced. A programme to cooperate with China on an improved variant was halted, but India has ordered the type.

MILITARY AIRCRAFT IN THE MODERN ERA

McDonnell F-15 Eagle

To succeed the F-4 Phantom in US service, McDonnell Douglas produced the F-15 Eagle. Since its inception, this aircraft has assumed the crown as the world's greatest air-superiority fighter, although it has now been superseded by later F-15C and -B variants in US service. The first prototype of the F-15A, a single-seat, twin-turbofan, swept-wing aircraft, flew in July 1972. Impressive flying characteristics became immediately apparent during flight testing.

F-15A Eagle

The F-15's powerful Pratt & Whitney engines and extensive use of titanium in construction (more than 20 per cent of the airframe weight of production aircraft) enabled high sustained speeds (Mach 2.5 plus) at high altitude. Deliveries began in November 1974 and production continued until 1979, with 385 built.

SPECIFICATIONS	
Country of origin:	United States
Type:	single-seat air superiority fighter with secondary strike/attack role
Powerplant:	two 106kN (23,810lb) Pratt & Whitney F100-PW-100 turbofans
Performance:	maximum speed at high altitude 2655km/h (1650mph); initial climb rate over 15,240m (50,000ft)/min; ceiling 30,500m (100,000ft); range on internal fuel 1930km (1200 miles)
Weights:	empty 12,700kg (28,000lb); with maximum load 25,424kg (56,000lb)
Dimensions:	span 13.05m (42ft 9.75in); length 19.43in (63ft 9in); height 5.63m (18ft 5in); wing area 56.48 sq m (608 sq ft)
Armament:	one 20mm M61A1 cannon, pylons with provision for up to 7620kg (16,800lb) of stores

F-15B Eagle

The F-15 SMTD (Short take-off and Maneuver Technology Demonstrator) was a modification of an F-15B to test canard foreplanes, thrust-vectoring engines and advanced pilot-interface systems. Most aspects were not continued, but the thrust-vectoring research was useful to the F-22 programme.

SPECIFICATIONS	
Country of origin:	United States
Type:	Technology demonstrator
Powerplant:	two 106kN (23,810lb) thrust Pratt & Whitney F100-PW-229 IPE afterburning turbofan engines
Performance:	maximum speed unknown; service ceiling unknown; range unknown
Weights:	unknown
Dimensions:	span 13.05m (42ft 9.75in); length 19.43in (63ft 9in); height 5.63m (18ft 5in); wing area 56.48 sq m (608 sq ft)
Armament:	none

MCDONNELL F-15 EAGLE

F-15A Eagle

SPECIFICATIONS

Country of origin:	United States
Type:	single-seat fighter and strike aircraft
Powerplant:	two 71.1kN (16,000lb) General Electric F404-GE-400 turbofans
Performance:	maximum speed at 12,190m (40,000ft) 1912km/h (1183mph); combat ceiling 15,240m (50,000ft); combat radius 1065km (662 miles)
Weights:	empty 10,455kg (23,050lb); maximum take-off 25,401kg (56,000lb)
Dimensions:	wingspan 11.43m (37ft 6in); length 17.07m (56ft); height 4.66m (15ft 3.5in); wing area 37.16 sq m (400 sq ft)
Armament:	one 20mm M61A1 Vulcan rotary cannon; nine hardpoints with provision for up to 7711kg (17,000kg) of stores

The F-15A was supplied to Air Defense Command to replace its F-106 interceptors during the 1980s. One unit was the 5th Fighter Interceptor Squadron 'Spittin' Kittens' at Minot, North Dakota. The 5th FIS used the Eagle for only a few years, disbanding in 1988.

F-15J Eagle

SPECIFICATIONS

Country of origin:	United States/Japan
Type:	single-seat strike/attack aircraft and air superiority fighter
Powerplant:	two 105.7kN (23,770lb) Pratt & Whitney F100-PW-220 turbofans
Performance:	maximum speed at high altitude 2655km/h (1650mph); service ceiling 30,500m (100,000ft); range 5745km (3570 miles)
Weights:	empty 12,793kg (23,770lb); maximum take-off 30,844kg (68,000lb)
Dimensions:	wingspan 13.05m (42ft 9.75in); length 19.43in (63ft 9in); height 5.63m (18ft 5in); wing area 56.48 sq m (608 sq ft)
Armament:	one 20mm M61A1 cannon, provision for up to 10,705kg (23,600lb) of stores, including missiles, bombs, tanks, pods and rockets

By the late 1970s, the USAF had accepted the increasing tactical necessity for an interceptor that could provide top cover during long-range strike missions. Consequently, the F-15A was upgraded to the F-15C. The most obvious change is the provision for two low-drag conformal fuel tanks (CFTs). When built under license to Japan, the F-15C was designated the F-15J.

F-15B Strike Eagle

SPECIFICATIONS

Country of origin:	United States
Type:	Two seat strike aircraft demonstrator
Powerplant:	two 105.9kN (23,810lb) Pratt & Whitney F100-PW-229 turbofans
Performance:	max speed at high altitude 2655km/h (1650mph); service ceiling 30,500m (100,000ft); range with fuel tanks 5745km (3570 miles)
Weights:	empty 14,379kg (31,700lb); maximum take-off 36,741kg (81,000lb)
Dimensions:	wingspan 13.05m (42ft 9.75in); length 19.43in (63ft 9in); height 5.63m (18ft 5in); wing area 56.48sq m (608sq ft)
Armament:	one 20mm M61A1 cannon; provision for up to 11,100kg (24,500lb) of stores, air-to-air missiles and conventional and guided bombs

In 1980, under a private-venture programme called 'Strike Eagle', McDonnell Douglas converted an F-15B to a demonstrator of an advanced two-seat strike aircraft. It competed in the Dual-Role Fighter contest and was selected over the F-16XL. The production version became the F-15E.

Second-Generation Harriers

The AV-8B version of the Harrier was developed for the US Marine Corps, which had a requirement for a single-seat, close-support aircraft to supersede the AV-8A Harriers in service from the mid-1970s. The design resulted from a collaboration between the two companies, which had individually sought to improve on the Harrier design. The first of four full-scale development aircraft was flown on 5 November 1981 and entered service with the Marine Corps in January 1985.

McDonnell Douglas AV-8B Harrier II

The AV-8B was the first of the second-generation Harriers to enter service, replacing AV-8As and A-4M Skyhawks in the Marines' light-attack squadrons. Early AV-8Bs like this VMA-331 'Bumblebees' aircraft had day-attack capability only, with no radar or laser systems.

SPECIFICATIONS	
Country of origin:	United States and United Kingdom
Type:	V/STOL close-support aircraft
Powerplant:	one 105.8kN (23,800lb) Rolls-Royce Pegasus vectored thrust turbofan
Performance:	maximum speed at sea level 1065km/h (661mph); service ceiling more than 15,240m (50,000ft); combat radius with 2722kg (6000lb) bombload 277km (172 miles)
Weights:	empty 5936kg (13,086lb); maximum take-off weight 14,061kg (31,000lb)
Dimensions:	wingspan 9.25m (30ft 4in); length 14.12m (46ft 4in); height 3.55m (11ft 7.75in); wing area 21.37 sq m (230 sq ft)
Armament:	one 25mm GAU-12U cannon, six external hardpoints with provision for up to 7711kg (17,000lb) (Short take-off) or 3175kg (7000lb) (Vertical take-off) of stores

McDonnell Douglas TAV-8B Harrier

Built in small numbers for the USMC, Italy and Spain, the TAV-8B retained the attack capability of the AV-8B, but has rarely – if ever – been used on operations. The TAV-8B also served as the basis for the Harrier T.10 for the RAF.

SPECIFICATIONS	
Country of origin:	United States and United Kingdom
Type:	V/STOL close-support aircraft
Powerplant:	one 95.4kN (21,450lb) thrust Rolls-Royce F402-RR-406 Pegasus vectored-thrust turbofan engine
Performance:	maximum speed at sea level 1065km/h (661mph); service ceiling more than 15,240m (50,000ft); combat radius with 2722kg (6000lb) bombload 277km (172 miles)
Weights:	empty 6450kg (14,233lb); maximum 14,058kg (31,000lb)
Dimensions:	span 9.2m (30ft 4in); length 15.3m (50ft 3in); height 3.5m (11ft 7in); wing area 22.1 sq m (239 sq ft)
Armament:	none

SECOND-GENERATION HARRIERS

British Aerospace Harrier GR.5

SPECIFICATIONS

Country of origin:	United Kingdom and United States
Type:	V/STOL close-support aircraft
Powerplant:	one 95.4kN (21,450lb) thrust Rolls-Royce Pegasus 11-21/Mk. 105 vectored-thrust turbofan engine
Performance:	max speed 1065km/h (661mph); service ceiling more than 15,240m (50,000ft); combat radius 277km (172 miles)
Weights:	empty 7050kg (15,542lb); maximum take-off weight 14,061kg (31,000lb)
Dimensions:	wingspan 9.25m (30ft 4in); length 14.36m (47ft 1.5in); height 3.55m (11ft 7.75in); wing area 21.37 sq m (230 sq ft)
Armament:	two 25mm Aden cannon; provision for up to 4082kg (9000lb) (Short take-off) or 3175kg (7000lb) (Vertical take-off) of stores

First of the 'big wing' Harriers for the Royal Air Force, the GR.5 was something of an interim model, delayed for technical and budgetary reasons. Most were converted into GR.7s, with FLIR systems and night-vision goggle capability.

British Aerospace Sea Harrier FRS Mk 2

SPECIFICATIONS

Country of origin:	United Kingdom
Type:	shipborne multi-role combat aircraft
Powerplant:	one 95.6kN (21,500lb) Rolls-Royce Pegasus vectored thrust turbofan
Performance:	maximum speed 1185km/h (736mph); service ceiling 15,545m (51,000ft); intercept radius 185km (115 miles) on hi-hi-hi CAP with 90 minuted loiter on station
Weights:	empty 5942kg (13,100lb); maximum take-off weight 11,884kg (26,200lb)
Dimensions:	wingspan 7.7m (25ft 3in); length 14.17m (46ft 6in); height 3.71m (12ft 2in); wing area 18.68 sq m (201.1 sq ft)
Armament:	two 25mm Aden cannon, provision for missiles, up to a total of 3629kg 8000lb) rockets, pods and drop tanks

In 1985, BAe began modernizing the fleet of FRS Mk 1s. The primary aim was to give the Sea Harrier the ability to engage beyond-visual-range targets with the AIM-120 AMRAAM medium-range air-to-air missile. The most obvious difference is the shape of the forward fuselage, which accommodates the Ferranti Blue Vixen pulse-Doppler track-while-scan radar. Deliveries commenced in 1992.

British Aerospace Harrier GR.7

SPECIFICATIONS

Country of origin:	United Kingdom and United States
Type:	V/STOL close-support aircraft
Powerplant:	one 96.7kN (21,750lb) Rolls-Royce Pegasus vectored-thrust turbofan
Performance:	max speed 1065km/h (661mph); service ceiling more than 15,240m (50,000ft); combat radius 277km (172 miles)
Weights:	empty 7050kg (15,542lb); maximum take-off weight 14,061kg (31,000lb)
Dimensions:	wingspan 9.25m (30ft 4in); length 14.36m (47ft 1.5in); height 3.55m (11ft 7.75in); wing area 21.37 sq m (230 sq ft)
Armament:	provision for up to 4082kg (9000lb) (Short take-off) or 3175kg (7000lb) (Vertical take-off) of stores

The GR.7 became the definitive Harrier II for the RAF. They have been used over the Balkans, Iraq and Afghanistan, and fly from RN carriers as part of Joint Force Harrier. Conversion with new avionics and structural improvements has created the GR.9, able to carry a wider range of precision weapons.

MILITARY AIRCRAFT IN THE MODERN ERA

MiG-23/27 'Flogger'

The MiG-23 and related MiG-27 ground-attack aircraft superseded the MiG-21 as primary equipment for the Soviet tactical air forces and Voyska PVO home-defence interceptor force. The aircraft is still flown by all the former Warsaw Pact air forces, though it is now a little long in the tooth when compared to European and American aircraft. Most MiG-23MFs serve in the fighter role, and are configured for high performance with modest weapons loads.

MiG-23MF 'Flogger-B'

This aircraft served with the Czech Air Force and has the designation MiG-23MF. This was the major production version from 1978, with improved radar and an infrared sensor pod. The ventral fin folds prior to landing.

SPECIFICATIONS	
Country of origin:	USSR
Type:	single-seat air combat fighter
Powerplant:	one 98kN (22,046lb) Rumanskii R-27F2M-300
Performance:	max speed about 2445km/h (1520mph); service ceiling over 18,290m (60,000ft); combat radius on hi-lo-hi mission 966km (600 miles)
Weights:	empty 10,400kg (22,932lb); maximum loaded 18,145kg (40,000lb)
Dimensions:	wingspan 13.97m (45ft 10in) spread and 7.78m (25ft 6.25in) swept; length 16.71m (54ft 10in); height 4.82m (15ft 9.75in); wing area 37.25 sq m (402 sq ft) spread
Armament:	one 23mm GSh-23L cannon, underwing pylons for AA-3 Anab, AA-7 Apex, and/or AA-8 Aphid air-to-air missiles

MiG-23M 'Flogger-B'

By 1975, several hundred MiG-23s, including the attack and trainer versions, had been delivered to Warsaw Pact air forces. Production continued until the mid-1980s; by far the largest operator was the Soviet Union. The engine – one of the most powerful to be fitted to a combat aircraft – gives good short-field performance and high top speed.

SPECIFICATIONS	
Country of origin:	USSR
Type:	single-seat air combat fighter
Powerplant:	one 100kN (22,485lb) Khachaturov R-29-300 turbojet
Performance:	max speed about 2445km/h (1520mph); service ceiling over 18,290m (60,000ft); combat radius on hi-lo-hi mission 966km (600 miles)
Weights:	empty 10,400kg (22,932lb); maximum loaded 18,145kg (40,000lb)
Dimensions:	span 13.97m (45ft 10in) spread and 7.78m (25ft 6.25in) swept; length 16.71m (54ft 10in); height 4.82m (15ft 9.75in); wing area 37.25 sq m (402 sq ft) spread
Armament:	one 23mm GSh-23L cannon, underwing pylons for air-to-air missiles

MiG-23UB 'Flogger-C'

A two-seat version of the MiG-23 was produced for conversion training. The second cockpit, for the instructor, is to the rear of the standard cockpit. The seat is slightly raised and is provided with a retractable periscopic sight to give a more comprehensive forward view.

SPECIFICATIONS

Country of origin:	USSR
Type:	two-seat conversion trainer
Powerplant:	one 98kN (22,046lb) Tumanskii R-27F2M-300 turbojet
Performance:	max speed at altitude about 2445km/h (1520mph); service ceiling over 18,290m (60,000ft); operational radius about 966km(600 miles)
Weights:	empty 11,000kg (24,200lb); maximum loaded 18,145kg (40,000lb)
Dimensions:	wingspan 13.97m (45ft 10in) spread and 7.78m (25ft 6.25in) swept; length 16.71m (54ft 10in); height 4.82m (15ft 9.75in); wing area 37.25 sq m (402 sq ft) (spread)
Armament:	one 23mm GSh-23L cannon; provision for up to 3000kg (6614lb) of stores, including air-to-air missiles, pods and bombs

MiG-23BN 'Flogger F'

The MiG-23BN/BM 'Flogger-F' are basically fighter-bomber versions of the MiG-23 for the export market. The aircraft have similar nose shape, the same laser rangefinder, raised seat, cockpit external armour plate, and low-pressure tyres of the Soviet air forces' MiG-27 'Flogger-D', but retain the powerplant, variable geometry intakes and cannon armament of the MiG-23MF 'Flogger B'.

SPECIFICATIONS

Country of origin:	USSR
Type:	single-seat fighter bomber
Powerplant:	one 98kN (22,046lb) Tumanskii R-27F2M-300 turbojet
Performance:	max speed about 2445km/h (1520mph); service ceiling over 18,290m (60,000ft); combat radius on hi-lo-hi mission 966km (600 miles)
Weights:	empty 10,400kg (22,932lb); maximum loaded 18,145kg (40,000lb)
Dimensions:	span 13.97m (45ft 10in) spread, 7.78m (25ft 6.25in) swept; length 16.71m (54ft 10in); height 4.82m (15ft 9.75in); wing area 37.25 sq m (402 sq ft) spread
Armament:	one 23mm GSh-23L cannon, provision for 3000kg (6614lb) of stores

MiG-27 'Flogger-J'

The MiG-27 was a highly developed version of the MiG-23. The aircraft was designed from the outset as a dedicated ground-attack aircraft and is optimized for operations over the battlefield. The most obvious difference is the nose, which was designed to give the pilot an enhanced view of the ground during approaches. The aircraft began to enter service in the late 1970s.

SPECIFICATIONS

Country of origin:	USSR
Type:	single-seat ground attack aircraft
Powerplant:	one 103.4kN (23,353lb) Tumanskii R-29B-300 turbojet
Performance:	max speed 1885km/h (1170mph); service ceiling over 14,000m (45,900ft); combat radius on lo-lo-lo mission 540km (335 miles)
Weights:	empty 11,908kg (26,252lb); maximum loaded 20,300kg (44,750lb)
Dimensions:	span 13.97m (45ft 10in) spread, 7.78m (25ft 6.25in) swept; length 17.07m (56ft 0.75in); height 5m (16ft 5in); wing area 37.35 sq m (402 sq ft) spread
Armament:	one 23mm cannon, provision for up to 4000kg (8818lb) of stores

War over Lebanon

In June 1982, Israel launched an invasion of southern Lebanon to destroy Palestinian terrorist bases there. Syria became involved and large air battles were fought over the Bekaa, Lebanon's central valley. By this time, Israel had replaced the majority of its equipment of French origin with the latest aircraft from the United States, and Syria's aircraft were at best half a generation behind. It is estimated that 86 Syrian aircraft were destroyed for no Israeli losses in aerial combat.

General Dynamics F-16A Fighter

Israel's first 75 F-16s were delivered from 1980. Even before the Lebanon War, they destroyed four Syrian aircraft in skirmishes. In 1982, they claimed 44 Syrian jets and also destroyed numerous radar and SAM sites while acting in the 'Wild Weasel' role.

SPECIFICATIONS	
Country of origin:	United States
Type:	single-seat air combat and ground-attack fighter
Powerplant:	either one 105.7kN (23,770lb) Pratt & Whitney F100-PW-200 or one 128.9kN (28,984lb) General Electric F110-GE-100 turbofans
Performance:	maximum speed 2142km/h (1320mph); service ceiling above 15,240m (50,000ft); operational radius 925km (525 miles)
Weights:	empty 7070kg (15,586lb); maximum take-off weight 16,057kg (35,400lb)
Dimensions:	span 9.45m (31ft); length 15.09m (49ft 6in); height 5.09m (16ft 8in); wing area 27.87 sq m (300 sq ft)
Armament:	one General Electric M61A1 20mm multi-barrelled cannon, wingtip missile stations; provision for up to 9276kg (20,450lb) of stores

McDonnell Douglas F-15 Eagle

The first Israeli F-15s arrived in 1976, having been acquired as a counter to Syrian (and Soviet) MiG-25 incursions into Israel's airspace. In Operation Peace for Galilee, the 1982 invasion, they scored around 40 kills, mainly with Sidewinder and Python 3 short-range missiles.

SPECIFICATIONS	
Country of origin:	United States
Type:	single-seat air superiority fighter with secondary strike/attack role
Powerplant:	two 105.9kN (23,810lb) Pratt & Whitney F100-PW-100 turbofans
Performance:	max speed at high altitude 2655km/h (1650mph); service ceiling 30,500m (100,000ft); range on internal fuel 1930km (1200 miles)
Weights:	empty 12,700kg (28,000lb); with maximum load 25,424kg (56,000lb)
Dimensions:	span 13.05m (42ft 9.75in); length 19.43m (63ft 9in); height 5.63m (18ft 5in); wing area 56.48 sq m (608 sq ft)
Armament:	one 20mm M61A1 cannon, provision for up to 7620kg (16,800lb) of stores, including air-to-air missiles, conventional and guided bombs

WAR OVER LEBANON

Mikoyan-Gurevich MiG-23MS 'Flogger-E'

SPECIFICATIONS	
Country of origin:	USSR
Type:	single-seat air combat fighter
Powerplant:	one 98kN (22,046lb) Tumanskii R-27F2M-300 turbojet
Performance:	max speed about 2445km/h (1520mph); service ceiling over 18,290m (60,000ft); combat radius on hi-lo-hi mission 966km (600 miles)
Weights:	empty 10,400kg (22,932lb); maximum loaded 18,145kg (40,000lb)
Dimensions:	span 13.97m (45ft 10in) spread, 7.78m (25ft 6.25in) swept; length 16.71m (54ft 10in); height 4.82m (15ft 9.75in); wing area 37.25 sq m (402 sq ft) spread
Armament:	one 23mm GSh-23L cannon with 200 rounds, provision for up to 3000kg (6614lb) of stores

Syria's most important fighter in 1982 was the MiG-23. Against the Mirage or Phantom, it was a comparable opponent, but its missiles were the 1960s-era AA-2 'Atoll'. In combat with the F-15 and F-16, armed with modern all-aspect missiles, the Syrian MiG pilots had little chance.

Mikoyan-Gurevich MiG-23BN 'Flogger-F'

SPECIFICATIONS	
Country of origin:	USSR
Type:	single-seat fighter bomber
Powerplant:	one 98kN (22,046lb) Tumanskii R-27F2M-300 turbojet
Performance:	max speed about 2445km/h (1520mph); service ceiling over 18,290m (60,000ft); combat radius on hi-lo-hi mission 966km (600 miles)
Weights:	empty 10,400kg (22,932lb); maximum loaded 18,145kg (40,000lb)
Dimensions:	span 13.97m (45ft 10in) spread, 7.78m (25ft 6.25in) swept; length 16.71m (54ft 10in); height 4.82m (15ft 9.75in); wing area 37.25 sq m (402 sq ft) spread
Armament:	one 23mm GSh-23L cannon, provision for 3000kg (6614lb) of stores

The Syrians also flew the export model of the MiG-27, the MiG-23BN 'Flogger-F'. This had considerably poorer ECM systems than the versions in Soviet service. Israeli air-to-air victories in 1982 include a dozen MiG-23BNs amongst the various 'Floggers' claimed destroyed.

Hawker Hunter F.70

SPECIFICATIONS	
Country of origin:	United Kingdom
Type:	single-seat fighter
Powerplant:	one 45.13kN (10,145lb) thrust Rolls-Royce Avon 207 turbojet engine
Performance:	max speed 1144km/h (710mph); service ceiling 15,240m (50,000ft); range on internal fuel 689km (490 miles)
Weights:	empty 6405kg (14,122lb) maximum 17,750kg (24,600lb)
Dimensions:	span 10.26m (33ft 8in); length 13.98m (45ft 10.5in); height 4.02m (13ft 2in); wing area 32.42 sq m (349 sq ft)
Armament:	four 30mm Aden Cannon; up to 2722kg (6000lb) of bombs or rockets; AIM-9 Sidewinder AAMs or AGM-65 ASMs

Lebanon received 19 Hunters over a period of nearly 20 years. Although they fought with Israeli fighters in 1967 and were active in 1982–83, there were no combats in this period. After years of dormancy, Lebanon restored several of its Hunters to airworthy status in 2008.

Desert Storm: Iraq and Kuwait

Having fought a long and inconclusive war with Iran, Iraqi forces under Saddam Hussein invaded neighbouring Kuwait in August 1990. With a strength on paper of 550 frontline aircraft, Iraq had the sixth largest air force in the world, including some of the latest Soviet designs. Its many air bases were protected by Western-built hardened shelters, and its air-defence system was well integrated. However, Iraq's pilots were poorly trained and their officer corps weakened by Saddam's purges.

Mikoyan-Gurevich MiG-21PF 'Fishbed'

An estimated 204 MiG-21s were in the Iraqi inventory in 1990, making them the most numerous combat aircraft type facing the US-led coalition. They and their pilots were optimized for point defence under strict ground control, and were ineffective when faced by jamming and destruction of control centres.

SPECIFICATIONS	
Country of origin:	USSR
Type:	single-seat all-weather multi role fighter
Powerplant:	one 73.5kN (16,535lb) Tumanskii R-25 turbojets
Performance:	max speed 2229km/h (1385mph); service ceiling 17,500m (57,400ft); range on internal fuel 1160km (721 miles)
Weights:	empty 5200kg (11,464lb); maximum take-off 10,400kg (22,925lb)
Dimensions:	wingspan 7.15m (23ft 5.5in); length (including probe) 15.76m (51ft 8.5in); height 4.1m (13ft 5.5in); wing area 23 sq m (247.58 sq ft)
Armament:	one 23mm GSh-23 twin-barrell cannon, provision for about 1500kg (3307kg) of stores, including air-to-air missiles, rocket pods, drop tanks

Dassault Mirage F1 CK

Both Iraq and Kuwait were users of the Mirage F1 in 1990. Kuwait's Mirage F1CKs were less sophisticated and less numerous than Iraq's F1EQ models. At least two KAF Mirages were destroyed on the ground during the Iraqi invasion and others captured. The remainder fled to Saudi Arabia and took a limited part in the effort to recapture the country under the 'Free Kuwait Air Force' banner.

SPECIFICATIONS	
Country of origin:	France
Type:	single-seat multi-mission fighter attack aircraft
Powerplant:	one 100kN (15,873lb) SNECMA Atar 9K-50 turbojet
Performance:	max speed at high altitude 2350km/h (1460mph); service ceiling 20,000m (65,615ft); range with maximum load 900km (560 miles)
Weights:	empty 7400kg (16,314lb); maximum take-off 15,200kg (33,510lb)
Dimensions:	wingspan 8.4m (27ft 7in); length 15m (49ft 2.25in); height 4.5m (14ft 9in); wing area 25sq m (269 sq ft) spread
Armament:	two 30mm 553 DEFA cannon with 135 rpg, five external pylons with provision for up to 6300kg (13,889lb) of stores; AIM-9 Sidewinder and Matra 530 air-to-air missiles

Mikoyan-Gurevich MiG-23BN

SPECIFICATIONS
Country of origin:	USSR
Type:	single-seat fighter bomber
Powerplant:	one 98kN (22,046lb) Tumanskii R-27F2M-300 turbojet
Performance:	max speed about 2445km/h (1520mph); service ceiling over 18,290m (60,000ft); combat radius on hi-lo-hi mission 966km (600 miles)
Weights:	empty 10,400kg (22,932lb); maximum loaded 18,145kg (40,000lb)
Dimensions:	span 13.97m (45ft 10in) spread, 7.78m (25ft 6.25in) swept; length 16.71m (54ft 10in); height 4.82m (15ft 9.75in); wing area 37.25 sq m (402 sq ft) spread
Armament:	one 23mm GSh-23L cannon, provision for up to 3000kg (6614lb) of stores, including air-to-air missiles, AS-7 Kerry air-to-surface missiles, cannon pods, rocket launcher pods, large calibre rockets and bombs

About 70 MiG-23BNs were delivered to Iraq in the mid-1980s and some were later fitted with Mirage F.1-style refuelling probes. Four MiG-23BNs were among the 150 Iraqi aircraft that fled to Iran to escape destruction by coalition air forces near the end of the war.

Mikoyan-Gurevich MiG-29

SPECIFICATIONS
Country of origin:	USSR
Type:	single-seat air-superiority fighter with secondary ground attack capability
Powerplant:	two 81.4kN (18,298lb) Sarkisov RD-33 turbofans
Performance:	max speed 2443km/h (1518mph); service ceiling 17,000m (55,775ft); range with internal fuel 1500km (932 miles)
Weights:	empty 10,900kg (24,030lb); maximum take-off 18,500kg (40,785lb)
Dimensions:	wingspan 11.36m (37ft 3.75in); length (including probe) 17.32m (56ft 10in); height 7.78m (25ft 6.25in); wing area 35.2 sq m (378.9 sq ft)
Armament:	one 30mm GSh-30 cannon, provision for up to 4500kg (9921lb) of stores

Although Iraq had ordered over 130 MiG-29s by 1990, only about two squadrons' worth were in service before the invasion of Kuwait. They failed to score against the Kuwait Air Force, and although regarded as the main air threat to coalition forces, were no more effective in 1991.

McDonnell Douglas A-4KU Skyhawk

SPECIFICATIONS
Country of origin:	United States
Type:	single-seat attack bomber
Powerplant:	one 49.82kN (11,500lb) thrust Pratt & Whitney J52-P408 turbojet engine
Performance:	maximum speed 1083km/h (673mph); service ceiling 14,935m (49,000ft); range with 4000lb load 3310 km (2200 miles)
Weights:	empty 4809kg (10,602lb); maximum take-off 12,437kg (27,420lb)
Dimensions:	wingspan 8.38m (27ft 6in); length (excluding probe) 12.22m (40ft 1.5in); height 4.66m (15ft 3in); wing area 24.15 sq m (260 sq ft)
Armament:	two 20mm Mk 12 cannon with 200rpg; five external hardpoints with provision for 3720kg (8200lb) of stores

Kuwait's A-4KU Skyhawks are said to have destroyed several Iraqi helicopters during the initial invasion. When their base was shelled, they relocated to a highway strip and continued attacks, before retreating to Saudi Arabia and flying as the 'Free Kuwait Air Force'.

Desert Storm: Coalition Air Forces

Within days of the invasion, the first elements of an international coalition were sent to air bases in the region. The largest non-US contributor was the United Kingdom, but combat aircraft from France, Italy, Canada, Saudi Arabia, Bahrain, Qatar and exiled Kuwaiti units took part in Operation Desert Shield, which became Desert Storm when the air campaign began in January 1991. The RAF's Operation Granby was the largest air component, incorporating bombers, fighters and other aircraft.

Panavia Tornado GR.1

Based at Tabuk, Saudi Arabia, Tornado GR.1 'MiG Eater' flew 40 missions and destroyed a MiG-29 on the ground. The RAF lost six Tornados in combat during the initial low-level attack missions. They then mainly switched to striking bridges and hunting for mobile 'Scud' missile launchers.

SPECIFICATIONS	
Country of origin:	Germany/Italy/United Kingdom
Type:	multi-role combat aircraft
Powerplant:	two 71.5kN (16,075lb) Turbo-Union RB.199-34R Mk 103 turbofans
Performance:	max speed above 11,000m (36,090ft) 2337km/h (1452mph); service ceiling 15,240m (50,000ft); combat radius with weapon load on hi-lo-hi mission 1390km (864 miles)
Weights:	empty 14,091kg (31,065lb); maximum take-off 27,216kg (60,000lb)
Dimensions:	wingspan 13.91m (45ft 7in) spread and 8.6m (28ft 2.5in) swept; length 16.72m (54ft 10in); height 5.95m (19ft 6.25in); wing area 26.6 sq m (286.3 sq ft)
Armament:	two 27mm IWKA-Mauser cannon with 180rpg, seven external hardpoints with provision for up to 9000kg (19,840lb) of stores, including nuclear and JP233 runway denial weapon, ALARM anti-radiation missiles, air-to-air, air-to-surface and anti-ship missiles, conventional and guided bombs, cluster bombs, ECM pods and drop tanks

SEPECAT Jaguar GR.1A

The RAF's Gulf War Jaguars were fitted with overwing Sidewinder pylons for self-defence, although they never encountered an enemy fighter in the air. GR.1A 'Sadman' flew 47 missions over Iraq from its base at Muharraq, on the island of Bahrain.

SPECIFICATIONS	
Country of origin:	France and United Kingdom
Type:	single-seat tactical support and strike aircraft
Powerplant:	two 32.5kN (7305lb) Rolls-Royce/Turbomeca Adour Mk 102 turbofans
Performance:	maximum speed at 11,000m (36,090ft) 1593km/h (990mph); combat radius on lo-lo-lo mission with internal fuel 557km (357 miles)
Weights:	empty 7000kg (15,432lb); maximum take-off 15,500kg (34,172lb)
Dimensions:	wingspan 8.69m (28ft 6in); length 16.83m (55ft 2.5in); height 4.89m (16ft 0.5in); wing area 24 sq m (258.34 sq ft)
Armament:	two 30mm DEFA cannon with 150rpg; five external hardpoints with provision for 4536kg (10,000lb) of stores, including one tactical nuclear weapon, eight 454kg (1000lb) bombs

DESERT STORM: COALITION AIR FORCES

Hawker Siddeley Buccaneer S.2B

SPECIFICATIONS	
Country of origin:	United Kingdom
Type:	two-seat attack aircraft
Powerplant:	two 50kN (11,255lb) Rolls Royce RB.168 Spey Mk 101 turbofans
Performance:	max speed 1040km/h (646mph); service ceiling over 12,190m (40,000ft); combat range 3701km (2300 miles)
Weights:	empty 13,608kg (30,000lb); maximum take-off 28,123kg (62,000lb)
Dimensions:	wingspan 13.41 (44ft); length 19.33m (63ft 5in); height 4.97m (16ft 3in); wing area 47.82 sq m (514.7 sq ft)
Armament:	four 454kg (1000lb) bombs, fuel tank, or reconnaissance pack on inside of rotary bomb door, four underwing pylons with provision for up to 5443kg (12,000lb) of bombs or missiles, including Harpoon and Sea Eagle anti-shipping missiles, and Martel anti-radar missiles

Buccaneers from RAF Lossiemouth were dispatched to provide laser designation for the RAF's Tornadoes once the latter switched to medium-level operations. Later in the conflict, they carried out their own bombing missions. This Buccaneer bore the names 'Lynn', 'Glenfiddich' and 'Jaws'.

Dassault Mirage 2000C

SPECIFICATIONS	
Country of origin:	France
Type:	single-seat air-superiority and attack fighter
Powerplant:	one 97kN (21,834lb) SNECMA M53-P2 turbofan
Performance:	max speed 2338km/h (1453mph); service ceiling 18,000m (59,055ft); range with 1000kg (2205lb) load 1480km (920 miles)
Weights:	empty 7500kg (16,534lb); maximum take-off 17,000kg (37,480lb)
Dimensions:	wingspan 9.13m (29ft 11.5in); length 14.36m (47ft 1.25in); height 5.2m (17ft 0.75in); wing area 41 sq m (441.3 sq ft)
Armament:	two DEFA 554 cannon; nine external pylons with provision for up to 6300kg of stores, including air-to-air missiles, rocket launcher pods, and various attack loads, including 1000lb bombs

Under the codename Operation Daguet, France provided 12 Mirage 2000Cs to help in the defence of Saudi Arabia. France also sent Mirage F.1s and Jaguar As to participate, plus transports, helicopters and a considerable ground component.

McDonnell Douglas F-15C Eagle

SPECIFICATIONS	
Country of origin:	United States
Type:	single-seat air-superiority fighter with secondary strike/attack role
Powerplant:	two 101kN (23,830lb) thrust Pratt & Whitney F100-PW-220 afterburning turbofan engines
Performance:	maximum speed at high altitude 2655km/h (1650mph); service ceiling 19,812m (65,000ft); range 5552km (3450 miles)
Weights:	empty 12,247kg (27,000lb); maximum 29,937kg (66,000lb)
Armament:	one 20mm M61A1 cannon with 960 rounds, external pylons with provision for up to 10,705kg (23,600lb) of stores

The only non-American fighter kill of the war was scored by Saudi F-15C pilot Captain al-Shamrani of No 13 Squadron RSAF, who used AIM-9L Sidewinders to destroy two Iraqi Mirage F.1EQs that were heading to launch an Exocet attack on coalition warships.

Desert Storm: US Air Power

The vast bulk of missions against Iraq in 1991 were flown by US Air Force, Marine Corps and Navy aircraft. Over 12,600 strike sorties were flown against fixed strategic targets alone, and many more against battlefield targets and in support of these missions. Air supremacy was won quite quickly, and tactical airpower weakened or neutralized much of the Iraqi Army, but few – if any – SS-1 'Scud' mobile missile launchers were put out of action, and they continued to be fired throughout the war.

McDonnell Douglas F-15C MSIP

The F-15C was the dominant fighter of the war, credited with destroying 33 Iraqi Air Force and Army aircraft in flight. This aircraft of the 33rd Tactical Fighter Wing destroyed three of them, including a MiG-23 with an AIM-9 Sparrow missile.

SPECIFICATIONS	
Country of origin:	USA
Type:	single-seat air superiority fighter with secondary strike/attack role
Powerplant:	two Pratt and Whitney F100-PW-220 turbofans
Performance:	maximum speed at high altitude 2655km/h (1650mph); initial climb rate over 15,240m (50,000ft)/min; ceiling 30,500m (100,000ft); range 5745km (3570 miles)
Weights:	empty 12,700kg (28,000lb); maximum take-off 30,844kg (68,000lb)
Dimensions:	wingspan 13.05m (42ft 9in); length 19.43m (63ft 9in); height 5.63m (18ft 5in); wing area 56.48sq m (608sq ft)
Armament:	one 20mm M61A1 cannon with 960 rounds, external pylons with provision for up to 7620kg (16,800lb) of stores, for example four AIM-7 Sparrow air-to-air missiles and four AIM-9 Sidewinder AAMs; when configured for attack role conventional and guided bombs, rockets, air-to-air surface missiles; tanks and/or ECM pods.

Fairchild Republic A-10A

The Fairchild Republic A-10A grew out of the US Air Force's A-X programme, begun in 1967, to produce a highly battleproof, heavily armed close air-support aircraft to replace the A-1 Skyraider. The A-10A is dominated by the huge GAU-8/A cannon, but the range of weaponry that it can carry is devastating. This was proved during actions against Iraqi armour the 1991 Gulf War.

SPECIFICATIONS	
Country of origin:	United States
Type:	single-seat close support aircraft
Powerplant:	two 40.3kN (9065lb) General Electric TF34-GE-100 turbofans
Performance:	max speed 706km/h (439mph); combar radius 402km (250 miles) for a 2-hour loiter with 18 Mk82 bombs plus 750 rounds cannon ammunition
Weights:	empty 11,321kg (24,959lb); maximum take-off 22,680kg (50,000lb)
Dimensions:	wingspan 17.53m (57ft 6in); length 16.26m (53ft 4in); height 4.47m (14ft 8in); wing area 47.01 sq m (506 sq ft)
Armament:	one 30mm GAU-8/A rotary cannon with capacity for 1350 rounds of ammunition, eleven hardpoints with provision for up to 7528kg (16,000lb) of disposable stores

DESERT STORM: US AIR POWER

Northrop Grumman E-8A J-STARS

SPECIFICATIONS
Country of origin:	United States
Type:	four-engined surveillance aircraft
Powerplant:	four 85.5kN (19,200lb) thrust TF33-102C turbofan engines
Performance:	max speed 945km/h (587mph); endurance nine hours; service ceiling: 12,802 m (42,000 ft)
Weights:	empty 77,564kg (171,000lb); maximum 152,409kg (336,000lb)
Dimensions:	span 44.4m (145ft 9in); length 46.6m (152ft 11in); height 13m (42ft 6in)
Armament:	none

An all-new technology deployed to the Gulf was the Joint Surveillance Target Attack Radar System (J-STARS), which allowed constant monitoring of ground movements in real time. The two E-8A development aircraft, still under testing, were rushed to the region, where they provided invaluable intelligence.

McDonnell Douglas F/A-18D

SPECIFICATIONS
Country of origin:	United States
Type:	tandem-seat conversion trainer with combat capability
Powerplant:	two 71.1kN (16,000lb) General Electric F404-GE-400 turbofans
Performance:	maximum speed at 12,190m (40,000ft) 1912km/h (1183mph); combat ceiling about 15,240m (50,000ft); combat radius 1020km (634 miles) on attack mission
Weights:	empty 10,455kg (23,050lb); maximum take-off 25,401kg (56,000lb)
Dimensions:	wingspan 11.43m (37ft 6in); length 17.07m (56ft); height 4.66m (15ft 3.5in); wing area 37.16 sq m (400 sq ft)
Armament:	one 20mm M61A1 Vulcan six-barrell rotary cannon with 570 rounds, nine external hardpoints with provision for up to 7711kg (17,000lb) of stores

Desert Storm marked the combat debut of the F/A-18 Hornet. The type proved its ability to switch between fighter and bomber roles in the same mission. On the first day of the war, Navy Hornets destroyed two Iraqi MiG-21s and lost one of their number to a MiG-25. This is an F/A-18D of VMFA(AW)-121.

McDonnell Douglas-BAe AV-8B Harrier

SPECIFICATIONS
Country of origin:	United States and United Kingdom
Type:	V/STOL close-support aircraft
Powerplant:	one 105.8kN (23,800lb) Rolls Royce F402-RR-408 Pegasus vectored thrust turbofan
Performance:	max speed 1065km/h (661mph); service ceiling more than 15,240m (50,000ft); combat radius 277km (172 miles)
Weights:	empty 5936kg (13,086lb); maximum take-off 14,061kg (31,000lb)
Dimensions:	wingspan 9.25m (30ft 4in); length 14.12m (46ft 4in); height 3.55m (11ft 7.75in); wing area 21.37 sq m (230 sq ft)
Armament:	one 25mm GAU-12U cannon; provision for up to 7711kg (17,000lb) (short take-off) or 3175kg (7000lb) (vertical take-off) of stores

VMA-311's AV-8B Harrier IIs operated from King Abdul Aziz air base in Saudi Arabia and from austere forward operating locations in the desert. The Harriers provided effective close air support but proved vulnerable to IR-guided missiles due to their large underside heat signature.

MILITARY AIRCRAFT IN THE MODERN ERA

Grumman F-14 Tomcat

The F-14 was developed largely because of the failure of the F-111B fleet-fighter programme, yet has not enjoyed a trouble-free service life itself. Continuing problems led to escalating maintenance costs and a relatively high accident rate. Despite these problems, the Tomcat is widely regarded as the finest interceptor flying anywhere in the world. The F-14 succeeded the F-4 as the premier fleet-defence fighter. A total of 478 F-14As were supplied to the US Navy.

F-14A Tomcat

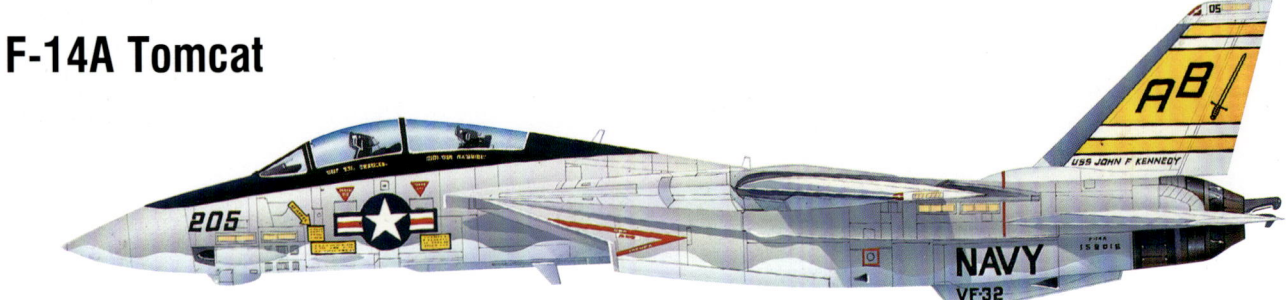

VF-32 'Swordsmen' was the first Atlantic Fleet squadron to form up with the Tomcat, making its first operational cruise in June 1975 on the USS *John F. Kennedy*. From the 1970s until the late 1980s, F-14s were mostly painted in gloss grey and white with colourful unit markings.

SPECIFICATIONS	
Country of origin:	United States
Type:	two-seat carrierborne fleet defence fighter
Powerplant:	two 92.9kN (20,900lb) Pratt & Whitney TF30-P-412A turbofans
Performance:	maximum speed at high altitude 2517km/h (1564mph); service ceiling 17,070m (56,000ft); range about 3220km (2000 miles)
Weights:	empty 18,191kg (40,104lb); maximum take-off 33,724kg (74,349lb)
Dimensions:	wingspan 19.55m (64ft 1.5in) unswept; 11.65m (38ft 2.5in) swept; length 19.1m (62ft 8in); height 4.88m (16ft); wing area 52.49 sq m (565 sq ft)
Armament:	one 20mm M61A1 Vulcan rotary cannon with 675 rounds; external pylons for a combination of AIM-7 Sparrow medium range air-to-air missiles, AIM-9 medium range air-to-air missiles, and AIM-54 Phoenix long range air-to-air missiles

F-14A Tomcat

In the 1980s, it was realized that brightly painted aircraft were more visible to infrared and other sensors, as well as to the naked eye. F-14 pilot C. J. Heatly and artist Keith Ferris were behind several experimental camouflage schemes, including this one, seen on an F-14A of VF-1.

SPECIFICATIONS	
Country of origin:	United States
Type:	two-seat carrierborne fleet defence fighter
Powerplant:	two 92.9kN (20,900lb) Pratt & Whitney TF30-P-412A turbofans
Performance:	maximum speed at high altitude 2517km/h (1564mph); service ceiling 17,070m (56,000ft); range about 3220km (2000 miles)
Weights:	empty 18,191kg (40,104lb); maximum take-off 33,724kg (74,349lb)
Dimensions:	wingspan 19.55m (64ft 1.5in) unswept; 11.65m (38ft 2.5in) swept; length 19.1m (62ft 8in); height 4.88m (16ft); wing area 52.49 sq m (565 sq ft)
Armament:	one 20mm M61A1 Vulcan rotary cannon with 675 rounds; external pylons for a combination of medium and long range air-to-air missiles

GRUMMAN F-14 TOMCAT

F-14A Tomcat

SPECIFICATIONS

Country of origin:	United States
Type:	two-seat carrierborne fleet defence fighter
Powerplant:	two 92.9kN (20,900lb) Pratt & Whitney TF30-P-412A turbofans
Performance:	maximum speed at high altitude 2517km/h (1564mph); service ceiling 17,070m (56,000ft); range about 3220km (2000 miles)
Weights:	empty 18,191kg (40,104lb); maximum take-off 33,724kg (74,349lb)
Dimensions:	wingspan 19.55m (64ft 1.5in) unswept; 11.65m (38ft 2.5in) swept; length 19.1m (62ft 8in); height 4.88m (16ft); wing area 52.49 sq m (565 sq ft)
Armament:	one 20mm M61A1 Vulcan rotary cannon with 675 rounds; external pylons for a combination of AIM-7 Sparrow medium range air-to-air missiles, AIM-9 medium range air-to-air missiles, and AIM-54 Phoenix long range air-to-air missiles

The only export customer for the Tomcat was the Imperial Iranian Air Force, which received 79 of 80 F-14As ordered before the Islamic revolution in 1979. Despite US sanctions, the Islamic Republic of Iran Air Force managed to keep about two dozen F-14s in service into the 2000s.

F-14B Tomcat

SPECIFICATIONS

Country of origin:	United States
Type:	two-seat carrierborne fleet defence fighter
Powerplant:	two 120kN (27,000lb) General Electric F110-GE-400 turbofans
Performance:	maximum speed at high altitude 1988km/h (1241mph); service ceiling 16,150m (53,000ft); range about 1994km (1239 miles) with full weapon load
Weights:	empty 18,951kg (41,780lb); maximum 33,724kg (74,349lb)
Dimensions:	span 19.55m (64ft 1.5in) unswept; 11.65m (38ft 2.5in) swept; length 19.1m (62ft 8in); height 4.88m (16ft); wing area 52.49 sq m (565 sq ft)
Armament:	one 20mm M61A1 Vulcan rotary cannon with 675 rounds; external pylons for a combination of medium and long range air-to-air missiles

The F-14B introduced the more powerful and less trouble-prone F110 engine, Martin-Baker ejection seats and other improvements. This F-14B of VF-74 'Bedevilers' illustrates the toned-down all-over grey colour schemes that became prevalent during the 1990s.

F-14D Tomcat

SPECIFICATIONS

Country of origin:	United States
Type:	two-seat carrierborne fleet defence fighter
Powerplant:	two 120kN (27,000lb) General Electric F110-GE-400 turbofans
Performance:	max speed 1988km/h (1241mph); service ceiling 16,150m (53,000ft); range about 1994km (1239 miles) with full weapon load
Weights:	empty 18,951kg (41,780lb); maximum take-off 33,724kg (74,349lb)
Dimensions:	span 19.55m (64ft 1.5in) unswept; 11.65m (38ft 2.5in) swept; length 19.1m (62ft 8in); height 4.88m (16ft); wing area 52.49 sq m (565 sq ft)
Armament:	one 20mm M61A1 Vulcan rotary cannon with 675 rounds; external pylons for a combination of AIM-7 Sparrow medium range air-to-air missiles, AIM-9 medium range air-to-air missiles, and AIM-54A/B/C Phoenix long range air-to-air missiles

In 1984, it was decided to develop an interim improved version of the F-14 with General Electric F110-GE-400, designated the F-14A (Plus). Thirty-two aircraft were converted and later designated F-14B. The F-14D project suffered a seemingly endless round of cancellations and reinstatements prior to the funding of 37 new-build aircraft and 18 rebuilds from F-14As.

MILITARY AIRCRAFT IN THE MODERN ERA

Afghan Wars

Throughout its history, Afghanistan has been in an almost constant state of war, either against outsiders or as a result of civil conflict. From 1979 to 1989, the Soviet Union attempted to exert its influence, launching an invasion to support the Communist government in Kabul. Over the following decade, various groups, many of them covertly backed by the United States, fought to expel Soviet forces. The USSR used many types of tactical aircraft and even strategic bombers.

MiG-17 'Fresco-C'

The Afghan Air Force was equipped with a variety of mostly Soviet-built types, including around 100 MiG-17s delivered from 1957. By 1985, they had an estimated 50 remaining. By mid-2001, what was left of the Air Force inventory was split amongst the Taliban and various Afghan factions.

SPECIFICATIONS	
Country of origin:	USSR
Type:	single-seat fighter
Powerplant:	one 33kN (7,452lb) Klimov VK-1F turbojet
Performance:	maximum speed at 3000m (9,840ft) 1145km/h (711mph); service ceiling 16,600m (54,560ft); range at height with slipper tanks 1470km (913 miles)
Weights:	empty 4100kg (9040lb); maximum loaded 600kg (14,770lb)
Dimensions:	wingspan 9.45m (31ft); length 11.05m (36ft 3.75in); height 3.35m (11ft); wing area 20.6 sq m (221.74 sq ft)
Armament:	one 37mm N-37 cannon and two 23mm NS-23 cannon, plus up to 500kg (1102lb) of mixed stores on underwing pylons

Sukhoi Su-25 'Frogfoot'

The prototype 'Frogfoot' first flew in 1975, and production of the single-seat close-support Su-25K (often compared to the Fairchild A-10 Thunderbolt II) began in 1978. A nose-mounted laser rangefinder and marked target seeker reportedly allows bombing accuracy to within 5m (16.4ft) over a stand-off range of 20km (12.5 miles). A trial unit was deployed to Afghanistan as early as 1980.

SPECIFICATIONS	
Country of origin:	USSR
Type:	single-seat close-support aircraft
Powerplant:	two 44.1kN (9921lb) Tumanskii R-195 turbojets
Performance:	maximum speed at sea level 975km/h (606mph); service ceiling 7000m (22,965ft); combat radius on lo-lo-lo mission 750km (466 miles)
Weights:	empty 9500kg (20,950lb); maximum take-off 17,600kg (38,800lb)
Dimensions:	wingspan 14.36m (47ft 1.5in); length 15.53m (50ft 11.5in); height 4.8m (15ft 9in); wing area 33.7 sq m (362.75 sq ft)
Armament:	one 30mm GSh-30-2 cannon with 250 rounds; eight external pylons with provision for up to 4400kg (9700lb) of stores, including AAMs, ASMs, ARMs, anti-tank missiles, guided bombs, cluster bombs

AFGHAN WARS

Mikoyan-Gurevich MiG-23MLD 'Flogger-K'

This MiG-23MLD was the commander's aircraft of the 120th Fighter Regiment, which deployed to Bagram in Afghanistan in 1986. The 'Floggers' were mainly used to escort Su-22 bombers on missions close to the Pakistan border. The white stars on the nose mark combat missions flown.

SPECIFICATIONS

Country of origin:	USSR
Type:	single-seat air combat fighter
Powerplant:	one 100kN (22,485lb) Khachaturov R-29-300 turbojet
Performance:	maximum speed at altitude about 2445km/h (1520mph); service ceiling over 18,290m (60,000ft); combat radius on hi-lo-hi mission 966km (600 miles)
Weights:	empty 10,400kg (22,932lb); maximum loaded 18,145kg (40,000lb)
Dimensions:	wingspan 13.97m (45ft 10in) spread and 7.78m (25ft 6.25in) swept; length (including probe) 16.71m (54ft 10in); height 4.82m (15ft 9.75in); wing area 37.25 sq m (402 sq ft) spread
Armament:	one 23mm GSh-23L cannon, underwing pylons for AA-3 Anab, AA-7 Apex, and/or AA-8 Aphid air-to-air missiles

Tupolev Tu-22PD 'Blinder-E'

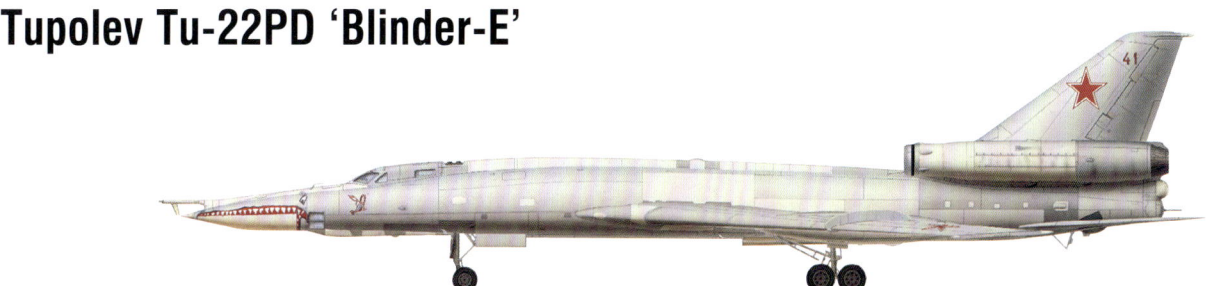

Used as a jammer and Elint aircraft, the Tu-22PD was used as an escort for the TU-22M-3 'Backfire' bombers used occasionally in Afghanistan, such as during the siege of Khost in 1987. Operating close to the Pakistan border, the 'Blinders' helped protect the 'Backfires' from interception.

SPECIFICATIONS

Country of origin:	USSR
Type:	electronic warfare aircraft
Powerplant:	two 161.9kN (36,376lb) thrust RD-7M2 afterburning turbojet engines
Performance:	maximum speed 1487km/h (924mph); service ceiling 18,300m (60,040ft); combat radius with internal fuel 3100km (1926 miles)
Weights:	empty 40,000kg (88,185lb); maximum take-off 84,000kg (185,188lb)
Dimensions:	wingspan 23.75m (77ft 11in); length 40.53m (132ft 11.75in); height 10.67m (35ft); wing area 162 sq m (1722.28 sq ft)
Armament:	one 23mm NR-23 two-barrel cannon

General Dynamics F-16A

This F-16A, flown by Squadron Leader Hameed Qadri of the Pakistan Air Force, destroyed two Soviet Su-22M-3s in an engagement over the border area in May 1986. A friendly-fire incident in April 1987 saw a Pakistani pilot accidentally shoot down his wingman over Afghanistan.

SPECIFICATIONS

Country of origin:	United States
Type:	single-seat air combat and ground attack fighter
Powerplant:	either one 105.7kN (23,770lb) Pratt & Whitney F100-PW-200 or one 128.9kN (28,984lb) General Electric F110-GE-100 turbofan
Performance:	maximum speed 2142km/h (1320mph); service ceiling above 15,240m (50,000ft); operational radius 925km (525 miles)
Weights:	empty 7070kg (15,586lb); maximum take-off 16,057kg (35,400lb)
Dimensions:	wingspan 9.45m (31ft); length 15.09m (49ft 6in); height 5.09m (16ft 8in); wing area 27.87 sq m (300 sq ft)
Armament:	one 20mm multi-barrelled cannon, wingtip missile stations; seven external hardpoints with provision for up to 9276kg (20,450lb) of stores

Stealth Attack Aircraft

The F-117 is probably the most important aircraft to enter service in the past two decades. It is likely that the secretive programme of stealth technology began in the wake of a number of radar-guided missile attacks on US-built F-4s during the 1973 Yom Kippur War. Delivered by Lockheed in 1982, Nighthawks really hit the headlines in the 1991 Gulf War, when pilots penetrated Iraqi air space undetected and delivered useful quantities of ordnance with pinpoint accuracy.

Lockheed F-117A Nighthawk

Both Lockheed and Northrop submitted proposals for the Experimental Stealth Technology requirement issued by the Department of Defense, with Lockheed's proposal being selected in 1977. The F-117 has been used several times in war. Its first mission was during the United States invasion of Panama in 1989. During that invasion two F-117A Nighthawks dropped two bombs on Rio Hato airfield. The F-117A can employ a variety of weapons and is equipped with sophisticated navigation and attack systems integrated into a state-of-the-art digital avionics suite that increases mission effectiveness and reduces pilot workload.

SPECIFICATIONS	
Country of origin:	United States
Type:	single-seat stealth attack aircraft
Powerplant:	two 48kN (10,800lb) General Electric F404-GE-F1D2 turbofans
Performance:	maximum speed about Mach 1 at high altitude; combat radius about 1112km (691 miles) with maximum payload
Weights:	empty about 13,608kg (30,000lb); maximum take-off 23,814kg (52,500lb)
Dimensions:	wingspan 13.2m (43ft 4in); length 20.08m (65ft 11in); height 3.78m (12ft 5in); wing area about 105.9 sq m (1140 sq ft)
Armament:	provision for 2268kg (5000lb) of stores on rotary dispenser in weapon bay; including the AGM-88 HARM anti-radiation missile; AGM-65 Maverick ASM, GBU-19 and GBU-27 optronically guided bombs, BLU-109 laser-guided bomb, and B61 free-fall nuclear bomb

STEALTH ATTACK AIRCRAFT

Lockheed XST 'Have Blue'

First flown secretly in December 1977 under the codename 'Have Blue', the Lockheed XST proved the aerodynamic shape intended for the F-117, although there were numerous differences, including the inwardly canted tailfins. Both prototypes crashed during testing.

SPECIFICATIONS	
Country of origin:	United States
Type:	stealth aircraft prototype
Powerplant:	two 12.4kN (2800lb) thrust General Electric CJ610 turbofan engines
Performance:	max speed unknown; range unknown; service ceiling unknown
Weights:	loaded 5440kg (11,993lb)
Dimensions:	span 6.86m (22ft 6in); length 11.58m (38ft 0in); height 2.29m (7ft 6in); wing area unknown
Armament:	none

Lockheed F-117A Nighthawk

Known as 'Spell Bound' during Operation Desert Storm, this F-117A flew eight or nine combat missions, the lowest number of those Nighthawks deployed. F-117s also saw action over Panama, Kosovo and Iraq in 2003. They were retired from USAF service in 2008.

SPECIFICATIONS	
Country of origin:	United States
Type:	single-seat stealth attack aircraft
Powerplant:	two 48kN (10,800lb) General Electric F404-GE-F1D2 turbofans
Performance:	maximum speed about Mach 1 at high altitude: combat radius about 1112km (691 miles) with maximum payload
Weights:	empty about 13,608kg (30,000lb); max take-off 23,814kg (52,500lb)
Dimensions:	wingspan 13.2m (43ft 4in); length 20.08m (65ft 11in); height 3.78m (12ft 5in); wing area about 105.9 sq m (1140 sq ft)
Armament:	provision for 2268kg (5000lb) of stores on rotary dispenser in weapon bay; including B61 free-fall nuclear bomb

MILITARY AIRCRAFT IN THE MODERN ERA

Warsaw Pact

By the late 1980s, the air forces of the Warsaw Pact, including the 16th Air Army, part of the Soviet Group of Forces in Germany, had largely re-equipped with the latest tactical aircraft. Each of the main types in NATO and United States Air Forces Europe had its equivalent on the other side of the Iron Curtain. Only the Su-27 was kept back for defence of the Motherland. After the edifice crumbled in 1989, the Soviets began to withdraw, leaving a legacy of air bases and aircraft.

Mikoyan-Gurevich MiG-21MF 'Fishbed-J'

The Air Force of the German Democratic Republic was given the best versions of the Soviet aircraft rather than the downgraded models available to other WarPac air arms. This MiG-21MF was with JG 8 at Marxwalde in 1985, a unit that disbanded in September 1990.

SPECIFICATIONS	
Country of origin:	USSR
Type:	single-seat all-weather multi role fighter
Powerplant:	one 60.8kN (14,550lb) thrust Tumanskii R-13-300 afterburning turbojet
Performance:	maximum speed 2229km/h (1385mph); service ceiling 17,500m (57,400ft); range on internal fuel 1160km (721 miles)
Weights:	empty 5200kg (11,464lb); maximum take-off 10,400kg (22,925lb)
Dimensions:	wingspan 7.15m (23ft 5.5in); length (including probe) 15.76m (51ft 8.5in); height 4.1m (13ft 5.5in); wing area 23 sq m (247.58 sq ft)
Armament:	one 23mm GSh-23 twin-barrell cannon in underbelly pack, four underwing pylons with provision for 1500kg (3307kg) of stores

Sukhoi SU-17M-4 'Fitter K'

Among the last Russian Air Force aircraft in Germany were the Su-17M-4s of the 20th Fighter-Bomber Regiment at Gross Dölln (Templin), which finally packed up and left in April 1994. This machine carries AS-14 'Kedge' anti-radar missiles under the fuselage.

SPECIFICATIONS	
Country of origin:	USSR
Type:	single-seat ground-attack fighter
Powerplant:	one 110.3kN (24,802lb) Lyul'ka AL-21F-3 turbojet
Performance:	max speed approximately 2220km/h (1380mph); service ceiling 15,200m (49,865ft); combat radius 675km (419 miles)
Weights:	empty 9,500kg (20,944lb); maximum take-off 19,500kg (42,990lb)
Dimensions:	wingspan 13.8m (45ft 3in) spread and 10m (32ft 10in) swept; length 18.75m (61ft 6in); height 5m (16ft 5in); wing area 40 sq m (430 sq ft)
Armament:	two 30mm NR-30 cannon; nine external pylons with provision for up to 4250kg (9370lb) of stores, including tactical nuclear weapons

Mikoyan-Gurevich MiG-23 BN 'Flogger-H'

The unified German Luftwaffe retained most of the former GDR's equipment for only a short time. Most of the older MiGs were retired quickly, but some, such as this MiG-23BN, were used by test unit WTD-61 for evaluation against Western types.

SPECIFICATIONS

Country of origin:	USSR
Type:	single-seat fighter bomber
Powerplant:	one 98kN (22,046lb) Tumanskii R-27F2M-300 turbojet
Performance:	maximum speed about 2445km/h (1520mph); service ceiling over 18,290m (60,000ft); combat radius on hi-lo-hi 966km (600 miles)
Weights:	empty 10,400kg (22,932lb); maximum loaded 18,145kg (40,000lb)
Dimensions:	wingspan 13.97m (45ft 10in) spread and 7.78m (25ft 6.25in) swept; length (including probe) 16.71m (54ft 10in); height 4.82m (15ft 9.75in); wing area 37.25 sq m (402 sq ft) spread
Armament:	one 23mm GSh-23L cannon with 200 rounds, six external hardpoints with provision for up to 3000kg (6614lb) of stores, including AA-2 Atoll air-to-air missiles, AS-7 Kerry air-to-surface missiles, cannon pods, rocket launcher pods, large calibre rockets and bombs

Mikoyan-Gurevich MiG-29 'Fulcrum A'

Following the break-up of the Soviet Union, the Czechoslovak Air Force was greatly reduced in size. The splitting of the country into Czech and Slovak Republics saw the one unit of MiG-29s transferred to the new Slovak AF. The nine aircraft have since been modernized and supplemented by others.

SPECIFICATIONS

Country of origin:	USSR
Type:	single-seat air-superiority fighter with secondary ground attack capability
Powerplant:	two 81.4kN (18,298lb) Sarkisov RD-33 turbofans
Performance:	max speed above 11000m (36,090ft) 2443km/h (1518mph); service ceiling 17,000m (55,775ft); range with internal fuel 1500km (932 miles)
Weights:	empty 10,900kg (24,030lb); maximum take-off 18,500kg (40,785lb)
Dimensions:	wingspan 11.36m (37ft 3.75in); length (including probe) 17.32m (56ft 10in); height 7.78m (25ft 6.25in); wing area 35.2 sq m (378.9 sq ft)
Armament:	one 30mm GSh-30 cannon with 150 rounds, eight external hardpoints with provision for up to 4500kg (9921lb) of stores, including infrared- or radar-guided air-to-air missiles

Sukhoi Su-24MR 'Fencer-E'

The 'Fencer E' is a version of the Su-24 strike and attack aircraft, designed for tactical reconnaissance. Approximately 65 Su-24MRs have been constructed with internal and external podded sensors of various types. Some of these sensors can transmit data to ground-based receivers for real-time surveillance. Service deliveries began in 1985.

SPECIFICATIONS

Country of origin:	USSR
Type:	two-seat maritime reconnaissance aircraft
Powerplant:	two 110.3kN (24,802lb) Lyul'ka AL-21F-3A turbojets
Performance:	maximum speed above 11,000m (36,090ft) approximately 2316km/h (1,439mph); service ceiling 17,500m (57,415ft); combat radius on hi-lo-hi mission with 3000kg (6614lb) load 1050km (650 miles)
Weights:	empty 19,00kg (41,888lb); maximum take-off 39,700kg (87,520lb)
Dimensions:	span 17.63m (57ft 10in) spread, 10.36m (34ft) swept; length 24.53m (80ft 5.75in); height 4.97m (16ft 0.75in); wing area 42 sq m (452.1 sq ft)
Armament:	(in secondary strike role) nine external pylons with provision for up to 8000kg (17,635lb) of stores, which may include air-to-air missiles

F-16 Fighting Falcon

The F-16 was undoubtedly one of the most important fighter aircraft of the twentieth century. It started fairly inauspiciously as a technology demonstrator to see to what degree it would be possible to build a useful fighter that was significantly smaller and cheaper than the F-15 Eagle. Interest from a number of America's NATO allies led to a revision of the programme and it was announced that the US Air Force would buy 650. General Dynamics' first production aircraft was flown in August 1978.

F-16A

The 8th Tactical Fighter Wing based at Kunsan in Korea was the first unit outside the United States to be equipped with F-16s, when it exchanged its last F-4s for Fighting Falcons in May 1981. This F-16A wears the codes and emblem of the 8th TFW, known as the 'Wolf Pack'.

SPECIFICATIONS

Country of origin:	United States
Type:	single-seat air combat and ground attack fighter
Powerplant:	either one 105.7kN (23,770lb) Pratt & Whitney F100-PW-200 or one 128.9kN (28,984lb) General Electric F110-GE-100 turbofan
Performance:	maximum speed 2142km/h (1320mph); service ceiling above 15,240m (50,000ft); operational radius 925km (525 miles)
Weights:	empty 7070kg (15,586lb); maximum take-off 16,057kg (35,400lb)
Dimensions:	wingspan 9.45m (31ft); length 15.09m (49ft 6in); height 5.09m (16ft 8in); wing area 27.87 sq m (300 sq ft)
Armament:	one General Electric M61A1 20mm multi-barrelled cannon, wingtip missile stations; seven external hardpoints with provision for up to 9276kg (20,450lb) of stores

F-16/79

The F-16/79 was an attempt to produce a less-sophisticated variant for the export market. An improved version of the J 79 engine, as used in the F-104 and F-4, was installed in two F-16s for demonstration purposes. The J 79 required heavy heat shielding and produced less thrust, and no one bought the F-16/79.

SPECIFICATIONS

Country of origin:	United States
Type:	single-seat air combat and ground attack fighter
Powerplant:	one 80.1kN (18,000lb) thrust General Electric J79-GE-17X afterburning turbojet engine
Performance:	Mach 2; service ceiling above 15,240m (50,000ft); radius 925km (525 miles)
Weights:	empty 7730kg (17,042lb); maximum 17,010kg (37,500lb)
Dimensions:	wingspan 9.45m (31ft); length 15.09m (49ft 6in); height 5.09m (16ft 8in); wing area 27.87 sq m (300 sq ft)
Armament:	one General Electric M61A1 20mm multi-barrelled cannon, wingtip missile stations; provision for up to 9276kg (20,450lb) of stores

F-16 FIGHTING FALCON

F-16N

To replace older F-5s and A-4s in the adversary role, the US Navy acquired a batch of 26 F-16Ns and two-seat TF-16Ns in the late 1980s. Armament was removed and the wing was strengthened for use in regular air-combat training. This F-16N served with VF-43 at NAS Oceana, Virginia.

SPECIFICATIONS

Country of origin:	United States
Type:	single-engined adversary fighter
Powerplant:	one 76.3kN (17,155lb) thrust General Electric F110-GE-100 afterburning turbofan engine
Performance:	maximum speed 2142km/h (1320mph); service ceiling above 15,240m (50,000ft); operational radius 925km (525 miles)
Weights:	unknown
Dimensions:	wingspan 9.45m (31ft); length 15.09m (49ft 6in); height 5.09m (16ft 8in); wing area 27.87 sq m (300 sq ft)
Armament:	none

F-16A Block 15 ADF

The F-16 ADF is a version optimized for air-defence interception with a modified radar, improved IFF equipment and a searchlight for identifying intruders at night. Most were supplied to Air National Guard units such as Puerto Rico's 198th Fighter Squadron.

SPECIFICATIONS

Country of origin:	United States
Type:	Air Defence Fighter
Powerplant:	one 105.7kN (23,770lb) thrust Pratt & Whitney F100-PW-220 afterburning turbofan engine
Performance:	maximum speed 2142km/h (1320mph); service ceiling 16,764m (55,000ft); range 3862km (2400 miles)
Weights:	empty 7387kg (16,285lb); maximum 17,010kg (37,500lb)
Dimensions:	wingspan 9.45m (31ft); length 15.09m (49ft 6in); height 5.09m (16ft 8in); wing area 27.87 sq m (300 sq ft)
Armament:	One 20mm M61A1 Vulcan cannon; AIM-9 Sidewinder and AIM-7 Sparrow or AIM-120 AMRAAM air-to-air missiles

F-16C Block 50D

The F-16C became the major production version after 1984. The Block 50/52 appeared in late 1990, offered with Pratt & Whitney F100 (Block 50) or General Electric F110 (Block 52) engines. One customer for many of the F-16C sub-variants has been Greece, one of whose Block 50s is illustrated.

SPECIFICATIONS

Country of origin:	United States
Type:	single-seat air combat and ground attack fighter
Powerplant:	one 126.7kN (28,500lb) thrust Pratt & Whitney F100-PW-229 afterburning turbofan engine
Performance:	maximum speed 2177km/h (1353mph); service ceiling 15,240m (49,000ft); range 3862km (2400 miles)
Weights:	empty 8273kg (18,238lb); maximum 19,187kg (42,300lb)
Dimensions:	wingspan 9.45m (31ft); length 15.09m (49ft 6in); height 5.09m (16ft 8in); wing area 27.87 sq m (300 sq ft)
Armament:	one General Electric M61A1 20mm multi-barrelled cannon, wingtip missile stations; seven external hardpoints with provision for up to 9276kg (20,450lb) of stores

Balkan Air Wars

MILITARY AIRCRAFT IN THE MODERN ERA

The state of Yugoslavia was broken into separate nations by a series of wars from 1991 to 1999. Aircraft in Federal Yugoslav armouries and air bases was appropriated by the various factions. Air power was mostly limited to sporadic ground-attack operations. Despite atrocities committed by most participants, the West was slow to take action. In 1993, a no-fly zone over Bosnia was imposed by NATO, in which four J-21s of the Republika Srpska Air Force were later shot down by USAF F-16s.

SOKO G-2A Galeb

In 1948, SOKO began licensed production of foreign designs before embarking on the design and construction of the G-2A Galeb trainer in 1957. This is a conventional low-wing monoplane of all-metal construction, retractable tricycle undercarriage and turbojet power. The crew are accommodated in tandem seats in a heated and air-conditioned cockpit.

SPECIFICATIONS	
Country of origin:	Yugoslavia
Type:	basic trainer
Powerplant:	one 11.1kN (2500lb) Rolls-Royce Viper 11 Mk 226 turbojet
Performance:	maximum speed at 6000m (19,685ft) 730km/h (454mph); service ceiling 12,000m (39,370ft); range with maximum standard fuel 1240km (771 miles)
Weights:	empty 2620kg (5776lb); maximum take-off weight 4300kg (9480lb)
Dimensions:	wingspan 9.73m (31ft 11in); length 10.34m (33ft 11in); height 3.28m (10ft 9in); wing area 19.43 sq m (209.15 sq ft)
Armament:	two 12.7mm machine guns with 80rpg; underwing racks for 150kg (331lb) bomblet containers, 100kg (220lb) bombs, 127mm rockets, and 55mm rocket-launcher pods

SOKO J-21 Jastreb

It was a relatively simple process for SOKO designers to convert the G-2A Galeb into a single-seat, light-attack aircraft. To improve weapons-carrying ability, an uprated version of the Viper engine was introduced, but apart from some local airframe strengthening, uprated wing hardpoints and the installation of a braking parachute, little was changed.

SPECIFICATIONS	
Country of origin:	Yugoslavia
Type:	single-seat light attack aircraft
Powerplant:	one 13.3kN (3000lb) Rolls-Royce Viper Mk 531 turbojet
Performance:	max speed 820km/h (510mph); service ceiling 12,000m (39,370ft); combat radius with standard fuel 1520km (944 miles)
Weights:	empty 2820kg (6217lb); maximum take-off weight 5100kg (11,244lb)
Dimensions:	wingspan 11.68m (38ft 3.75in); length 10.88m (38ft 8.25in); height 3.64m (11ft 11.25in); wing area 19.43 sq m (209.15 sq ft)
Armament:	three 12.7mm machine guns with 135 rpg; inboard hardpoints with provision for 500kg (1102lb) of stores

BALKAN AIR WARS

UTVA 75

SPECIFICATIONS	
Country of origin:	Yugoslavia
Type:	trainer/light-attack aircraft
Powerplant:	one 134kW (180hp) Lycoming IO-360-B1F 4 cylinder flat piston engine
Performance:	maximum speed 215km/h (134mph); range 800km (500 miles); service ceiling 4000m (13,100ft)
Weights:	empty 685kg (1510lb); maximum 970kg (2135lb)
Armament:	mountings for machine-gun pods, two-round rocket launchers or 200kg (441lb) of bombs
Dimensions:	span 9.73m (31ft 11in); length 7.11m (23ft 4in); height 3.15m (10ft 4in); wing area 14.63 sq m (158 sq ft)
Armament:	none

Another indigenous Yugoslav type that found itself in various hands after the break-up of the country, the UTVA-75 was a trainer with a light secondary ground-attack capability. Many were camouflaged and used as such by Croatia. The survivors serve as trainers in the modern Croatian Air Force.

Mikoyan-Gurevich MiG-21-bis 'Fishbed'

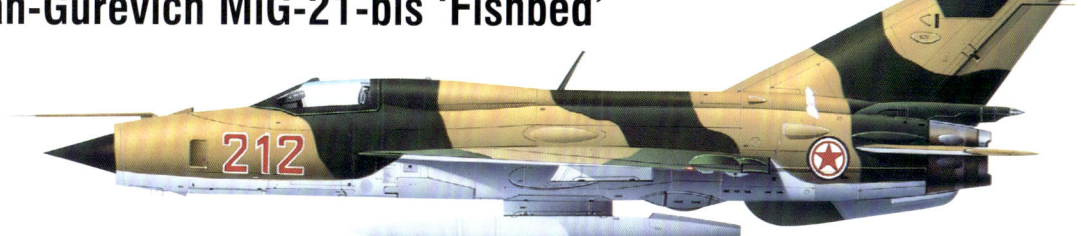

SPECIFICATIONS	
Country of origin:	USSR
Type:	single-seat all-weather multi role fighter
Powerplant:	one 73.5kN (16,535lb) Tumanskii R-25 turbojets
Performance:	max speed 2229km/h (1385mph); service ceiling 17,500m (57,400ft); range on internal fuel 1160km (721 miles)
Weights:	empty 5200kg (11,464lb); maximum take-off weight 10,400kg (22,925lb)
Dimensions:	wingspan 7.15m (23ft 5.5in); length (including probe) 15.76m (51ft 8.5in); height 4.1m (13ft 5.5in); wing area 23 sq m (247.58 sq ft)
Armament:	one 23mm GSh-23 twin-barrell cannon in underbelly pack, provision for about 1500kg (3307kg) of stores

Yugoslavia operated a large number of MiG-21s before 1991, including MiG-21bis, MF and PFM models, but their reliability fell to a very low level during the civil wars. With the loss of many MiG-29s in 1999 and the poor condition of the survivors, the MiG-21 again became Serbia's main fighter.

Mikoyan-Gurevich MiG-29B 'Fulcrum'

SPECIFICATIONS	
Country of origin:	USSR
Type:	single-seat air-superiority fighter with secondary ground attack capability
Powerplant:	two 81.4kN (18,298lb) Sarkisov RD-33 turbofans
Performance:	maximum speed above 11000m (36,090ft) 2443km/h (1518mph); service ceiling 17,000m (55,775ft); range with internal fuel 1500km (932 miles)
Weights:	empty 10,900kg (24,030lb); maximum take-off 18,500kg (40,785lb)
Dimensions:	wingspan 11.36m (37ft 3.75in); length (including probe) 17.32m (56ft 10in); height 7.78m (25ft 6.25in); wing area 35.2sq m (378.9sq ft)
Armament:	one 30mm GSh-30 cannon with 150 rounds, eight external hardpoints with provision for up to 4500kg (9921lb) of stores.

Serbia's MiG-29s made only about a dozen sorties in the Kosovo War in 1999. Six 'Fulcrums' were shot down, two of them by USAF 493rd Fighter Squadron F-15 pilot Jeffery Hwang, including the MiG-29A illustrated. A MiG-29 shot down by a Dutch F-16 was the first non-US NATO victory.

Balkan Air Wars: NATO

NATO's first military action was to shoot down Bosnian Serb aircraft in 1994 under Operation Deny Flight. In April 1995, Operation Deliberate Force targeted the Army of Republika Srpska to prevent further massacres, but the major action was Operation Allied Force in 1999, which ended the conflict over Kosovo.

BAe Sea Harrier FA.2

Flying from Royal Navy carriers in the Adriatic, FAA Sea Harriers enforced the no-fly zone and provided close air support for UN troops. One Sea Harrier FRS.1 was shot down by a Serbian SAM in 1994, but this is a later FA.2, flying from HMS *Illustrious* during 'Allied Force' in 1999.

SPECIFICATIONS	
Country of origin:	United Kingdom
Type:	shipborne multi-role combat aircraft
Powerplant:	one 95.6kN (21,500lb) Rolls-Royce Pegasus Mk 106 vectored thrust turbofan
Performance:	maximum speed at sea level with maximum AAM load 1185km/h (736mph); service ceiling 15,545m (51,000ft); intercept radius 185km (115 miles) on hi-hi-hi CAP with 90 minuted loiter on station
Weights:	empty 5942kg (13,100lb); maximum take-off weight 11,884kg (26,200lb)
Dimensions:	wingspan 7.7m (25ft 3in); length 14.17m (46ft 6in); height 3.71m (12ft 2in); wing area 18.68 sq m (201.1 sq ft)
Armament:	two 25mm Aden cannon with 150 rounds, five external pylons with provision for AIM-9 Sidewinder, AIM-120 AMRAAM, and two Harpoon or Sea Eagle anti-shipping missiles, up to a total of 3629kg (8000lb)

BAe/McDonnell Douglas Harrier GR.7

RAF Harrier GR.7s saw their first active combat during Allied Force, although their LGB-dropping missions were initially hampered by cloud and bad weather. This led to a switch to conventional 'dumb' munitions such as BL-755 cluster bombs and eventually to the development of British GPS-guided bombs.

SPECIFICATIONS	
Country of origin:	United Kingdom/United States
Type:	V/STOL close-support aircraft
Powerplant:	one 96.7kN (21,750lb) Rolls-Royce Mk 105 Pegasus vectored-thrust turbofan
Performance:	max speed 1065km/h (661mph); service ceiling more than 15,240m (50,000ft); combat radius 277km (172 miles)
Weights:	empty 7050kg (15,542lb); maximum take-off weight 14,061kg (31,000lb)
Dimensions:	wingspan 9.25m (30ft 4in); length 14.36m (47ft 1.5in); height 3.55m (11ft 7.75in); wing area 21.37 sq m (230 sq ft)
Armament:	two 25mm Aden cannon with 100rpg; six external hardpoints with provision for up to 4082kg (9000lb) (short take-off) or 3175kg (7000lb) (vertical take-off) of stores

BALKAN AIR WARS: NATO

McDonnell Douglas CF-18A Hornet

SPECIFICATIONS	
Country of origin:	United States
Type:	single-seat multi-mission fighter
Powerplant:	two 71.1kN (16,000lb) General Electric F404-GE-400 turbofans
Performance:	maximum speed at 12,190m (40,000ft) 1912km/h (1183mph); combat ceiling about 15,240m (50,000ft); combat radius 740km (460 miles) on escort mission or 1065km (662 miles) in attack role
Weights:	empty 10,455kg (23,050lb); maximum take-off 25,401kg (56,000lb)
Dimensions:	wingspan 11.43m (37ft 6in); length 17.07m (56ft); height 4.66m (15ft 3.5in); wing area 37.16 sq m (400 sq ft)
Armament:	one 20mm M61A1 Vulcan six-barrel rotary cannon with 570 rounds, nine external hardpoints with provision for up to 7711kg (17,000lb) of stores

Canada contributed CF-18A Hornets to Allied Force with a unit designated Task Force Aviano. The CAF Hornets were able to fly in night and bad weather, and could strike with laser-guided bombs aided by Nighthawk targeting pods. Spanish Air Force, USN and USMC Hornets also took part.

Boeing E-3D Sentry AEW.1

SPECIFICATIONS	
Country of origin:	United States
Type:	Airborne Warning and Control System platform
Powerplant:	four 106.8kN (24,000lb) thrust CFM56-2A-3 turbofan engines
Performance:	maximum speed 852km/h (529mph); range 3200km (1988 miles); service ceiling 10,668m (35,000ft)
Weight:	loaded 147,000kg (324,000lb)
Dimensions:	span 44.98m (147ft 7in); length 46.68m (153ft); height 12.6m (41ft 4in); wing area 3050 sq m (283 sq ft)
Armament:	none

During Allied intervention over the Balkans, E-3 Sentries from the United States, NATO and the RAF were vital in guiding fighters to interceptions, controlling rescue missions and warning of enemy air activity. The E-3Ds of the RAF's Nos 8 and 23 Squadrons assisted in several successful fighter engagements.

Northrop Grumman B-2A

SPECIFICATIONS	
Country of origin:	United States
Type:	strategic bomber and missile-launch platform
Powerplant:	four 84.5kN (19,000lb) General Electric F118-GE-110 turbofans
Performance:	maximum speed at high altitude 764km/h (475mph); service ceiling 15,240m (50,000ft); range on high level mission with standard fuel and 16,919kg (37,300lb) warload 11,675km (7255 miles)
Weights:	empty 45,360kg (100,000lb); max take-off 181,437kg (400,000lb)
Dimensions:	wingspan 52.43m (172ft); length 21.03m (69ft); height 5.18m (17ft); wing area more than 464.5 sq m (5000 sq ft)
Armament:	two internal bomb bays, provision for up to 22,680kg (50,000lb) of stores; each bay can carry 16 1.1 megaton thermonuclear free-fall bombs, 22 680kg (1500lb) bombs, or 80 227kg (500lb) free-fall bombs

The B-2A Spirit 'stealth bomber' made its combat debut over Kosovo, Serbia and Montenegro in March 1999. Flying non-stop missions of up to 44 hours from Whiteman AFB, Missouri, the B-2s flew the longest-duration bombing raids in history and employed GPS-guided bombs for the first time.

Sukhoi Fencer and Frogfoot

In the 1980s, Soviet forces began to field more modern fighter, ground-attack and strike aircraft equivalent to those in the West. The Sukhoi Su-24 'Fencer' swing-wing bomber matched the F-111, and the Su-25 'Frogfoot' was a heavily armoured close air-support platform along the same lines as the A-10. Both types were widely exported, or left in the hands of former Soviet Republics, and have seen combat in Iraq, Chechnya, Georgia and other post Cold-War hotspots.

Su-25 'Frogfoot-A'

The single-seat Su-25 became fully operational in 1984, serving extensively in Afghanistan. As a testament to the longevity of the type, the Republic of Macedonia Air Force used Su-25s in their fight against Albanian insurgents in 2001, and in 2008 Georgia and Russia were both reported to be employing the Su-25 in South Ossetia.

SPECIFICATIONS	
Country of origin:	USSR
Type:	single-seat close-support aircraft
Powerplant:	two 44.1kN (9921lb) Tumanskii R-195 turbojets
Performance:	max speed 975km/h (606mph); service ceiling 7000m (22,965ft); combat radius on lo-lo-lo mission 750km (466 miles)
Weights:	empty 9500kg (20,950lb); maximum take-off weight 17,600kg (38,800lb)
Dimensions:	wingspan 14.36m (47ft 1.5in); length 15.53m (50ft 11.5in); height 4.8m (15ft 9in); wing area 33.7 sq m (362.75 sq ft)
Armament:	one 30mm GSh-30-2 cannon with 250 rounds; eight external pylons with provision for up to 4400kg (9700lb) of stores

Su-25K 'Frogfoot-A'

The Su-25K was an export variant of the basic 'Frogfoot' built from the mid-1980s. The Czechoslovak Air Force had 36 Su-25Ks at the time of the 'Velvet Divorce' between the Czech and Slovak republics in 1992–93. The Czech Air Force retained 24 of them but retired the type from service in 2000.

SPECIFICATIONS	
Country of origin:	USSR
Type:	single-seat close-support aircraft
Powerplant:	two 44.1kN (9921lb) Tumanskii R-195 turbojets
Performance:	max speed 975km/h (606mph); service ceiling 7000m (22,965ft); combat radius on lo-lo-lo mission 750km (466 miles)
Weights:	empty 9500kg (20,950lb); maximum take-off weight 17,600kg (38,800lb)
Dimensions:	wingspan 14.36m (47ft 1.5in); length 15.53m (50ft 11.5in); height 4.8m (15ft 9in); wing area 33.7 sq m (362.75 sq ft)
Armament:	one 30mm GSh-30-2 cannon with 250 rounds; eight external pylons with provision for up to 4400kg (9700lb) of stores

Su-25UTG 'Frogfoot B'

The Su-25UB 'Frogfoot-B' two-seat trainer has a longer forward fuselage to accommodate a second cockpit. Production of a navalized version, the Su-25UTG, began in the late 1980s with strengthened undercarriage and arrestor gear. The aircraft pictured passed to the Ukrainian Air Force after the dissolution of the Soviet Union.

SPECIFICATIONS	
Country of origin:	USSR
Type:	two-seat carrier-training aircraft
Powerplant:	two 44.1kN (9921lb) Tumanskii R-195 turbojets
Performance:	maximum speed at sea level 950km/h (590mph); service ceiling 10,000m (32,810ft); combat radius on lo-lo-lo mission with 4400kg (9700lb) load 4000km (248 miles)
Weights:	empty 9500kg (20,950lb); maximum take-off 17,600kg (38,800lb)
Dimensions:	wingspan 14.36m (47ft 1.5in); length 15.53m (50ft 11.5in); height 4.8m (15ft 9in); wing area 33.7 sq m (362.75 sq ft)
Armament:	one 30mm GSh-30-2 cannon with 250 rounds; eight external pylons with provision for up to 4400kg (9700lb) of stores

Su-24 'Fencer-B'

The 'Fencer-B' was a variant of the initial-production Su-24, identified by its brake parachute housing and revised shape around the rear fuselage. This example probably served with a Guards regiment based at Osla in Poland in the early 1990s.

SPECIFICATIONS	
Country of origin:	USSR
Type:	two-seat strike and attack aircraft
Powerplant:	two 110.3kN (24,802lb) Lyul'ka AL-21F-3A turbojets
Performance:	max speed approximately 2316km/h (1439mph); service ceiling 17,500m (57,415ft); combat radius 1050km (650 miles)
Weights:	empty 19,000kg (41,888lb); maximum take-off 39,700kg (87,520lb)
Dimensions:	span 17.63m (57ft 10in) spread, 10.36m (34ft) swept; length 24.53m (80ft 5in); height 4.97m (16ft 0.75in); wing area 42 sq m (452.1 sq ft)
Armament:	one 23mm GSh-23-6 six-barrelled cannon; nine external pylons with provision for up to 8000kg (17,635lb) of stores

Su-24M 'Fencer-D'

In 1965, the Soviet government prompted Sukhoi to begin designing a new Soviet variable-geometry attack aircraft comparable in performance to the F-111. One of the requirements was the ability to penetrate radar defences at very low level and at supersonic speeds. Service deliveries of the 'Fencer A' began in 1974, with the improved 'Fencer D' (Su-24M) entering service in 1986.

SPECIFICATIONS	
Country of origin:	USSR
Type:	two-seat strike and attack aircraft
Powerplant:	two 110.3kN (24,802lb) Lyul'ka AL-21F-3A turbojets
Performance:	max speed approximately 2316km/h (1439mph); service ceiling 17,500m (57,415ft); combat radius 1050km (650 miles)
Weights:	empty 19,000kg (41,888lb); maximum take-off 39,700kg (87,520lb)
Dimensions:	span 17.63m (57ft 10in) spread, 10.36m (34ft) swept; length 24.53m (80ft 5in); height 4.97m (16ft 0.75in); wing area 42 sq m (452.1 sq ft)
Armament:	one 23mm GSh-23-6 six-barrelled cannon; nine external pylons with provision for up to 8000kg (17,635lb) of stores

SEAD and Jamming

Suppression of Enemy Air Defences (SEAD) is a role that evolved from World War II-era electronic jamming and 'Wild Weasel' missions in Vietnam. It is defined by NATO as 'that activity which neutralizes, temporarily degrades or destroys enemy air defences by destructive and/or disruptive means'. This can be in the form of either a 'soft-kill' method such as jamming or spoofing radars, or by 'hard-kill' anti-radiation missiles or other ordnance.

Grumman EA-6B Prowler

The Prowler was unarmed until the Improved Capability II (ICAP II) upgrade in the 1980s, which allowed it to carry the AGM-88 high-speed anti-radiation missile (HARM). This missile can reach a radar before it has time to shut down or use inbuilt memory to strike a target that has stopped emitting.

SPECIFICATIONS	
Country of origin:	United States
Type:	electronic countermeasures platform
Powerplant:	two 49.8kN (11,200lb) Pratt & Whitney J52-P-408 turbojets
Performance:	max speed at sea level 982km/h (610mph); service ceiling 11,580m (38,000ft); combat range with full external fuel 1769km (1099 miles)
Weights:	empty 14,588kg (32,162lb); maximum take-off 29,484kg (65,000lb)
Dimensions:	wingspan 16.15m (53ft); length 18.24m (59ft 10in); height 4.95m (16ft 3in); wing area 49.13 sq m (528.9 sq ft)
Armament:	none on early models, retrofitted with external hardpoints for four or six AGM-88 HARM air-to-surface anti-radar missiles

Mikoyan-Gurevich MiG-25BM 'Foxbat F'

Russia's SEAD aircraft have included unmodified Su-25s and variants of the MiG-25 'Foxbat'. The MiG-25BM 'Foxbat-F' was based on the airframe of the MiG-25RB reconnaissance model. Its normal armament for the SEAD role was the Raduga Kh-58 (AS-11 'Kilter') anti-radiation missiles.

SPECIFICATIONS	
Country of origin:	USSR
Type:	SEAD aircraft
Powerplant:	two 109.8kN (24,691lb) Tumanskii R-15BD-300 turbojets
Performance:	maximum speed at altitude about 3339km/h (2112mph); service ceiling 27,000m (88,585ft); operational radius 900km (559 miles)
Weights:	empty 19,600kg (43,211lb); maximum take-off 33,400kg (73,634lb)
Dimensions:	wingspan 13.42m (44ft 0.75in); length 23.82m (78ft 1.75in); height 6.1m (20ft 0.5in); wing area not disclosed
Armament:	four Raduga Kh-58 (AS-11 'Kilter') anti-radiation missiles

SEAD AND JAMMING

Grumman (General Dynamics) EF-111A

The biggest threat to US aircraft in Vietnam proved to be ground-based radar-guided missiles, supplied by the USSR to NVA forces. A development programme was begun and Grumman's adapted F-111A entered service in 1981, its most recognizable feature being the fin-tip pod that houses the jamming system's receiver and antenna.

SPECIFICATIONS	
Country of origin:	United States
Type:	two-seat ECM tactical jamming aircraft
Powerplant:	two 82.3kN (18,500lb) Pratt & Whitney TF-30-P3 turbofans
Performance:	maximum speed at optimum altitude 2272km/h (1412mph); service ceiling above 13,715m (45,000ft); range with internal fuel 1495km (929 miles)
Weights:	empty 25,072kg (55,275lb); maximum take-off 40,346kg (88,948lb)
Dimensions:	wingspan unswept 19.2m (63ft); swept 9.74m (32ft 11.5in); length 23.16m (76ft); height 6.1m (20ft); wing area 48.77 sq m (525 sq ft) unswept

McDonnell Douglas F-4G 'Phantom II'

The F-4G was designed and built specifically for the radar-suppression role in the wake of significant USAF losses to Soviet-supplied SA-2 'Guideline' SAMs over Vietnam. By 1972, about 12 F-4C 'Wild Weasels' had been introduced to service. The F-4G was the result of a much more extensive modification programme, and was produced by modifying F-4Es.

SPECIFICATIONS	
Country of origin:	United States
Type:	two-seat EW/radar-surpression aircraft
Powerplant:	two 79.6kN (17,900lb) General Electric J79-GE-17 turbojets
Performance:	maximum speed at high altitude 2390km/h (1485mph); service ceiling over 18,975m (62,250ft); range on internal fuel with weapon load 958km (595 miles)
Weights:	empty 13,300kg (29,321lb); maximum take-off weight 28,300kg (62,390lb)
Dimensions:	span 11.7m (38ft 5in); length 19.2m (63ft); height 5.02m (16ft 5.5in); wing area 49.24 sq m (530 sq ft)
Armament:	two AIM-7 Sparrow recessed under rear fuselage; wing pylons for radar suppression weapons such as AGM-45 Shrike, AGM-78 Standard or AGM-88 HARM anti-radiation missiles

Tornado ECR

Germany and Italy collaborated on the Tornado ECR electronic combat and reconnaissance version of the standard ground attack Tornado IDS. The Luftwaffe accepted 35 ECRs equipped with an emitter locator and armed with the AGM-88 HARM anti-radar missile.

SPECIFICATIONS	
Country of origin:	Germany, Italy and UK
Type:	multi-role combat aircraft
Powerplant:	two 7292kg (16,075lb) Turbo-Union RB.199-34r Mk 103 turbofans
Performance:	maximum speed above 11,000m (36,090ft) 2337km/h (1452mph); service ceiling 15,240m (50,000ft); combat radius wuth weapon load on hi-lo-mission 1390km (864 miles)
Weights:	empty 14091kg (31,065lb); maximum loaded 27,216kg (60,000lb)
Dimensions:	wingspan 13.91m (45ft 7in) spread and 8.6m (28ft 2.5in) swept; length 16.72m (54ft 10in); height 5.95m (19ft 6.25in); wing area 26.6 sq m (286.3 sq ft)
Armament:	two 27mm IWKA-Mauser cannon with 180 rpg, seven enternal hard points with provision for up to 9000kg (19,840lb) of stores; including AGM-88 HARM anti-radar missile

Dassault Mirage 2000

Early research and experience had shown that the delta-wing configuration carried some notable disadvantages, not least a lack of low-speed manoeuvrability. With the development of fly-by-wire technology during the late 1960s and early 1970s, it was possible for airframe designers to overcome some of these problems, when coupled with advances in aerodynamics. The 2000C was designed by Dassault to be a single-seat interceptor to replace the F.1.

Mirage 2000-01

First flown in March 1978, the Mirage 2000 built on the success of the delta-winged Mirage III and IV with the addition of fly-by-wire controls. Four prototypes were built, all of them single-seaters. The test programme was quick by later standards and the first production aircraft flew in 1982.

SPECIFICATIONS	
Country of origin:	France
Type:	single-seat air-superiority and attack fighter
Powerplant:	one 83.36kN (18,839lb) thrust SNECMA M53-2 afterburning turbofan engine
Performance:	maximum speed at high altitude 2338km/h (1453mph); service ceiling 18,000m (59,055ft); range 1480km (920 miles)
Weights:	empty 7500kg (16,534lb); maximum take-off weight 17,000kg (37,480lb)
Dimensions:	wingspan 9.13m (29ft 11.5in); length 14.36m (47ft 1.25in); height 5.2m (17ft 0.75in); wing area 41 sq m (441.3 sq ft)
Armament:	two DEFA 554 cannon with 125rpg; nine external pylons with provision for up to 6300kg of stores, including R.530 air-to-air missiles, AS.30 or A.30L missiles, rocket launcher pods

Mirage 2000C

The 2000C was adopted by the French Government in December 1975 as the primary combat aircraft of the French air force, and was developed initially under contract as an interceptor and air-superiority fighter. Deliveries to L'Armée de l'Air began in July 1984: early production examples were fitted with the SNEMCA M53-5; aircraft built after that date have the more powerful M53-P2.

SPECIFICATIONS	
Country of origin:	France
Type:	single-seat air-superiority and attack fighter
Powerplant:	one 97.1kN (21,834lb) SNECMA M53-P2 turbofan
Performance:	maximum speed at high altitude 2338km/h (1453mph); service ceiling 18,000m (59,055ft); range 1480km (920 miles)
Weights:	empty 7500kg (16,534lb); maximum take-off weight 17,000kg (37,480lb)
Dimensions:	wingspan 9.13m (29ft 11.5in); length 14.36m (47ft 1.25in); height 5.2m (17ft 0.75in); wing area 41 sq m (441.3 sq ft)
Armament:	two DEFA 554 cannon with 125rpg; nine external pylons with provision for up to 6300kg of stores, including R.530 air-to-air missiles, AS.30 or A.30L missiles, rocket launcher pods

DASSAULT MIRAGE 2000

Mirage 2000P

SPECIFICATIONS	
Country of origin:	France
Type:	single-seat air-superiority and attack fighter
Powerplant:	one 97.1kN (21,834lb) SNECMA M53-P2 turbofan
Performance:	maximum speed at high altitude 2338km/h (1,453mph); service ceiling 18,000m (59,055ft); range with 1000kg (2205lb) load 1,480km (920 miles)
Weights:	empty 7500kg (16,534lb); maximum take-off weight 17,000kg (37,480lb)
Dimensions:	wingspan 9.13m (29ft 11.5in); length 14.36m (47ft 1.25in); height 5.2m (17ft 0.75in); wing area 41 sq m (441.3 sq ft)
Armament:	two DEFA 554 cannon with 125rpg; nine external pylons with provision for up to 6300kg of stores, including R.530 air-to-air missiles, AS.30 or A.30L missiles, rocket launcher pods, and various attack loads, including 1000lb bombs. For air defence weapon training, the Cubic Corpn AIS (airborne instrumentation subsystem)

Peru was the first Latin-American export customer for the Mirage 2000, buying 10 Mirage 2000Ps and two 2000DP two-seaters, although it had originally ordered 26 in total. The export-standard radar allowed a smaller range of weapons options than was available for aircraft of L'Armée de l'Air.

Mirage 2000B

SPECIFICATIONS	
Country of origin:	France
Type:	dual-seat jet trainer with operational capability
Powerplant:	one 97.1kN (21,034lb) SNECMA M53-P2 turbofan
Performance:	maximum speed 2338km/h (1453mph); service ceiling 18,000m (59,055ft); range with two 1700 litre (374 Imp gal) drop tanks 1850km (1150miles)
Weights:	empty 7600kg (16,755lb); maximum take-off weight 17,000kg (37,480lb)
Dimensions:	wingspan 9.13m (29ft 11.5in); length 14.55m (47ft 9in); height 5.15m (16ft 10.75in); wing area 41 sq m (441.3 sq ft)
Armament:	seven external pylons with provision for R.530 air-to-air missiles, AS.30 or A.30L missiles and 1000lb bombs

Because of the complexity of the third-generation Mirage 2000, the French air force decided to pursue a programme of development for a two-seat trainer to run concurrently with the single-seat 2000C. The fifth Mirage 2000 prototype was flown in this format as the 2000B in October 1980. Production aircraft are distinguished by a slightly longer fuselage.

Mirage 2000N

SPECIFICATIONS	
Country of origin:	France
Type:	single-seat strike aircraft
Powerplant:	one 97.1kN (21,834lb) SNECMA M53-P2 turbofan
Performance:	maximum speed at high altitude 2338km/h (1453mph); service ceiling 18,000m (59,055ft); range with two 1700 litre (374 Imp gal) drop tanks 1850km (1150 miles)
Weights:	empty 7600kg (16,755lb); maximum take-off weight 17,000kg (37,480lb)
Dimensions:	wingspan 9.13m (29ft 11.5in); length 14.55m (47ft 9in); height 5.15m (16ft 10.75in); wing area 41 sq m (441.3 sq ft)
Armament:	one ASMP stand off nuclear missile, seven external pylons with provision for air-to-air missiles and various attack loads, including 1000lb bombs

The Mirage 2000N is France's primary nuclear-strike aircraft. Although also capable of conventional attack, its main armament is the Mach-2 ASMP (*Air-Sol Moyenne Portée*) missile with a range of about 250km (155 miles) from high altitude, and a nuclear yield of 150 or 300 kilotons.

A-10 Thunderbolt II

The Fairchild Republic A-10A grew out of the US Air Force's A-X programme, begun in 1967, to produce a highly battleproof, heavily armed, close air-support aircraft to replace the A-1 Skyraider. In December 1970, three companies were chosen to build prototypes for evaluation, and Fairchild's YA-10A emerged as the winner in January 1973. The A-10A is dominated by the huge GAU-8/A cannon, but the range of weaponry that it can carry is devastating, as demonstrated during the 1991 Gulf War.

Fairchild-Republic YA-10A Thunderbolt II

The YA-10A won a competitive fly-off against the Northrop YA-9A in 1973, and went on to a period of evaluation for the USAF. Because the GAU-8 cannon was not ready, the two prototypes (this is the second) were armed with the M61 Vulcan used by most other USAF tactical aircraft.

SPECIFICATIONS	
Country of origin:	United States
Type:	prototype close air support aircraft
Powerplant:	two 40.3kN (9065lb) General Electric TF34-GE-100 turbofans
Performance:	maximum speed at sea level 706km/h (439mph); combat radius 402km (250 miles)
Weights:	unknown
Dimensions:	wingspan 17.53m (57ft 6in); length 16.26m (53ft 4in); height 4.47m (14ft 8in); wing area 47.01 sq m (506 sq ft)
Armament:	one 20mm M61A1 Vulcan cannon

Fairchild-Republic A-10A Thunderbolt II

With its intended role being low-level tank-hunting in regions such as Western Europe and Korea, the question of a suitable camouflage scheme for the A-10 became important. This is one pattern trialled during the USAF/Army Joint Attack Weapons Systems (JAWS) tests in 1977.

SPECIFICATIONS	
Country of origin:	United States
Type:	single-seat close support aircraft
Powerplant:	two 40.3kN (9065lb) General Electric TF34-GE-100 turbofans
Performance:	maximum speed at sea level 706km/h (439mph); combat radius 402km (250 miles)
Weights:	empty 11,321kg (24,959lb); maximum take-off 22,680kg (50,000lb)
Dimensions:	wingspan 17.53m (57ft 6in); length 16.26m (53ft 4in); height 4.47m (14ft 8in); wing area 47.01 sq m (506 sq ft)
Armament:	one 30mm rotary cannon with 1350 rounds, 11 hardpoints with provision for up to 7528kg (16,000lb) of disposable stores

FAIRCHILD-REPUBLIC A-10 THUNDERBOLT II

Fairchild-Republic A-10A N/AW Thunderbolt II

The simple A-10A had no avionics for night or bad-weather flying. The first production A-10A was converted with a second seat, a podded weather radar and other avionics to create the A-10A N/AW (Night/Adverse Weather). A proposed A-10B based on this design was ordered but later cancelled.

SPECIFICATIONS

Country of origin:	United States
Type:	two-seat night/adverse weather attack aircraft
Powerplant:	two 40.3kN (9065lb) General Electric TF34-GE-100 turbofans
Performance:	maximum speed at sea level 706km/h (439mph); combat radius 402km (250 miles) for a 2-hour loiter with 18 Mk82 bombs plus 750 rounds cannon ammunition
Weights:	empty 11,321kg (24,959lb); maximum take-off 22,680kg (50,000lb)
Dimensions:	wingspan 17.53m (57ft 6in); length 16.26m (53ft 4in); height unknown; wing area 47.01 sq m (506 sq ft)
Armament:	one 30mm GAU-8/A rotary cannon with capacity for 1350 rounds of ammunition, eleven hardpoints with provision for up to 7528kg (16,000lb) of disposable stores; weapons include conventional bombs, incendiary bombs, Rockeye cluster bombs

Fairchild-Republic A-10A Thunderbolt II

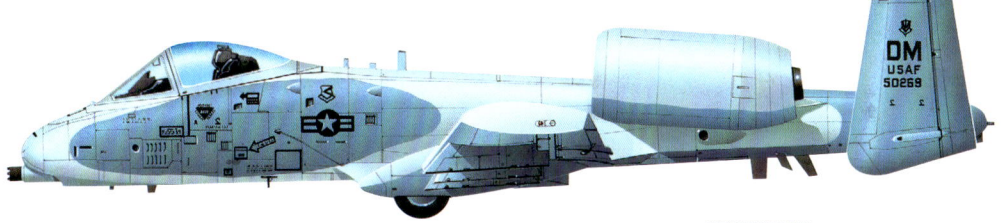

Although A-10s were fielded for many years in the 'European One' or 'lizard' colour scheme, this gave way in time to light greys similar to that of other USAF tactical aircraft. This A-10 is seen in an earlier grey/grey camouflage that saw limited use with the 355th TFW at Davis Monthan AFB, Arizona.

SPECIFICATIONS

Country of origin:	United States
Type:	single-seat close support aircraft
Powerplant:	two 40.3kN (9065lb) General Electric TF34-GE-100 turbofans
Performance:	maximum speed at sea level 706km/h (439mph); combat radius 402km (250 miles) for a 2-hour loiter with 18 Mk82 bombs plus 750 rounds cannon ammunition
Weights:	empty 11,321kg (24,959lb); maximum take-off 22,680kg (50,000lb)
Dimensions:	wingspan 17.53m (57ft 6in); length 16.26m (53ft 4in); height 4.47m (14ft 8in); wing area 47.01 sq m (506 sq ft)
Armament:	one 30mm GAU-8/A rotary cannon with capacity for 1350 rounds of ammunition, 11 hardpoints with provision for up to 7528kg (16,000lb) of disposable stores

Fairchild-Republic OA-10A

Conflict over the role of the Air Force versus the Army in providing close air support nearly caused the A-10's retirement several times. As part of this debate, several A-10 units, including that of the Pennsylvania ANG, were given the FAC role and their aircraft renamed OA-1As, despite being unchanged.

SPECIFICATIONS

Country of origin:	United States
Type:	single-seat close support aircraft
Powerplant:	two 40.3kN (9065lb) General Electric TF34-GE-100 turbofans
Performance:	maximum speed at sea level 706km/h (439mph); combat radius 402km (250 miles) loaded
Weights:	empty 11,321kg (24,959lb); maximum take-off 22,680kg (50,000lb)
Dimensions:	wingspan 17.53m (57ft 6in); length 16.26m (53ft 4in); height 4.47m (14ft 8in); wing area 47.01 sq m (506 sq ft)
Armament:	one 30mm GAU-8/A rotary cannon with capacity for 1350 rounds of ammunition, 11 hardpoints with provision for up to 7528kg (16,000lb) of stores, pods for 70mm (2.75in) target-making rockets

General Dynamics F-111

The variable-geometry General Dynamics F-111 suffered a difficult gestation, earning it the unwelcome nickname 'Aardvark'. Developed to meet a bold Department of Defense edict that a common type of fighter should be developed to meet all future tactical needs of the US armed forces, the F-111 seemed at the outset both a success and a great failure. After a troubled development process, the first of 117 aircraft, designated F-111As, were eventually delivered in 1967.

F-111A/TACT

The F-111A suffered several accidents and mysterious disappearances in Vietnam, but eventually overcame most of its numerous teething troubles. Under the Transonic Aircraft Technology (TACT) and other programmes, NASA tested this F-111 with new 'supercritical' wing sections.

SPECIFICATIONS	
Country of origin:	United States
Type:	two-seat attack aircraft
Powerplant:	two 82.29kN (18,500lb) thrust Pratt & Whitney TF30-P-3 afterburning turbofan engines
Performance:	maximum speed 2338km/h (1453mph); range 5094km (3165 miles); service ceiling 17,678m (58,000ft)
Weights:	empty 20,943kg (46,172lb); maximum 44,838kg (98,850lb)
Dimensions:	span 19.2m (63ft); length 22.4m (73ft 6in); height 5.33m (17ft 6in); wing area 48.77 sq m (525 sq ft)
Armament:	one 20mm M61A1 Vulcan rotary cannon; up to 13,608kg (30,000lb) of bombs, missiles or fuel tanks

FB-111A

Strategic Air Command adopted its own model of the F-111 as a replacement for the B-58 Hustler and some B-52s. With a longer wing and more powerful engines, the FB-111 could carry six AGM-69 Short-Range Attack Missiles (SRAMs) under the wings and in its internal bomb bay.

SPECIFICATIONS	
Country of origin:	United States
Type:	two-seat attack aircraft
Powerplant:	two 90.52kN (20,350lb) thrust Pratt & Whitney TF30-P-7 afterburning turbofan engines
Performance:	maximum speed 2338km/h (1453mph); range 7702km (4786 miles); service ceiling 15,320m (50,263ft)
Weights:	empty 21,763kg (47,980lb); maximum 54,091kg (119,250lb)
Dimensions:	span 21.34m (70ft); length 22.4m (73ft 6in); height 5.33m (17ft 6in); wing area 51.1 sq m (550 sq ft)
Armament:	six Boeing AGM-69 SRAM nuclear missiles; up to 17,010kg (37,500lb) of bombs, missiles or fuel tanks

GENERAL DYNAMICS F-111

F-111E

The F-111E was an improved F-111A with more efficient intakes and better navigation and electronic-warfare equipment. All the Es were allocated to the 20th TFW at Upper Heyford, England. They had less precision-bombing capability than the F-111F, but were used against Libya in 1986.

SPECIFICATIONS

Country of origin:	United States
Type:	two-seat attack aircraft
Powerplant:	two 90.52kN (20,350lb) thrust Pratt & Whitney TF30-P-7 afterburning turbofan engines
Performance:	maximum speed 2338km/h (1453mph); range 7702km (4786 miles); service ceiling 15,320m (50,263ft)
Weights:	empty 21,763kg (47,980lb); maximum 54,091kg (119,250lb)
Dimensions:	span 21.34m (70ft); length 22.4m (73ft 6in); height 5.33m (17ft 6in); wing area 51.1 sq m (550 sq ft)
Armament:	six Boeing AGM-69 SRAM nuclear missiles; up to 17,010kg (37,500lb) of bombs, missiles or fuel tanks

F-111C

Australia was the only export customer for the F-111, acquiring 24 F-111Fs in the 1970s and later adding F-111Gs to the fleet. The F-111C had the longer wings of the FB-111 and the ability to use weapons such as the Harpoon anti-shipping missile. They are the last F-111s remaining in service.

SPECIFICATIONS

Country of origin:	United States
Type:	two-seat attack aircraft
Powerplant:	two 92.70kN (20,840lb) thrust Pratt & Whitney TF30-P-109RA afterburning turbofan engines
Performance:	maximum speed 2338km/h (1453mph); range 7702km (4786 miles); service ceiling 15,320m (50,263ft)
Weights:	empty 20,943kg (46,172lb); maximum 41,414kg (91,300lb)
Dimensions:	span 21.34m (70ft); length 22.4m (73ft 6in); height 5.33m (17ft 6in); wing area 51.1 sq m (550 sq ft)
Armament:	up to 13,608kg (30,000lb) of bombs or missiles, including AGM-84 Harpoon anti-ship and AGM-88 HARM anti-radiation missiles

F-111G

The FB-111 was retired from SAC as the B-1B Lancer came into service. A number were modified with increased conventional weapons capability and new intakes. Redesignated as F-111Gs, they served with the 27th TFW at Cannon AFB, New Mexico.

SPECIFICATIONS

Country of origin:	United States
Type:	two-seat attack aircraft
Powerplant:	two 90.52kN (20,350lb) thrust Pratt & Whitney TF30-P-7 afterburning turbofan engines
Performance:	maximum speed 2338km/h (1453mph); range 7702km (4786 miles); service ceiling 15,320m (50,263ft)
Weights:	empty 21,763kg (47,980lb); maximum 54,091kg (119,250lb)
Dimensions:	span 21.34m (70ft); length 22.40m (73ft 6in); height 5.33m (17ft 6in); wing area 51.1 sq m (550 sq ft)
Armament:	six Boeing AGM-69 SRAM nuclear missiles; up to 17,010kg (37,500lb) of bombs, missiles or fuel tanks

Tactical Transports

The market for tactical transport aircraft able to move troops and equipment around within a region or combat theatre, and to operate from short runways there, is one largely ignored by US manufacturers but enthusiastically taken up elsewhere. Turboprop or turbofan engines, multiwheel landing gears and high-lift wings, as well as other devices, allow operation from austere strips in 'hot-and-high' conditions, where C-130s and larger aircraft cannot go or struggle to carry a useful load.

Aeritalia G.222

The Aeritalia (now Alenia) G.222 has a wide fuselage and particularly good short take-off performance. The Italian Air Force bought nearly 50, but new examples were exported in small numbers. The C-27J Spartan, based on the G.222 with C-130J technology, has had some notable sales successes.

SPECIFICATIONS	
Country of origin:	Italy
Type:	twin-engined tactical transport
Powerplant:	two 2535kW (3400hp) General Electric T64-GE-P4D turboprop engines
Performance:	maximum speed 540km/h (336mph); range 4685km (2910 miles); service ceiling 7620m (25,000ft)
Weights:	empty 11,940kg (26,320lb); maximum 31,800kg (70,107lb)
Dimensions:	span 28.7m (94ft 2in); length 22.7m (74ft 6in); height 9.8m (32ft 2in); wing area 82 sq m (893 sq ft)
Armament:	none

CASA 212 Aviocar

Able to take off in as little as 400m (1312ft), the CASA 212 Aviocar was particularly popular with African and Latin-American air forces. In the Middle East, Saudi Arabia and Jordan were customers. The Royal Jordanian Air Force bought four 212-100s in the mid-1970s, one of which is shown.

SPECIFICATIONS	
Country of origin:	Spain
Type:	twin-engined tactical transport
Powerplant:	two 671kW (900hp) Garrett TPE331-10R-513C turboprop engines
Performance:	maximum speed 370km/h (230mph); range 2680km (1665 miles); service ceiling 7925m (26,000ft)
Weights:	empty 4400kg (9700lb); maximum 8000kg (17,637lb)
Dimensions:	span 20.28m (66ft 7in); length 15.18m (49ft 9in); height 6.6m (21ft 8in); wing area 41 sq m (441 sq ft)
Armament:	none

TACTICAL TRANSPORTS

Antonov An-72 'Coaler'

The design of the An-72 'Coaler' is optimized for STOL capability, with a variety of high-lift features to permit short-field operation. The most noticeable of these is the positioning of the twin powerplants, at a position high up and well forward on the wing. When the inboard flaps are deployed, the engine exhaust is deflected over them, producing greatly increased lift.

SPECIFICATIONS

Country of origin:	USSR (Ukraine)
Type:	STOL transport
Powerplant:	two 63.7kN (14,330lb) Zaporozhye/Lotarev D-36 turbofans
Performance:	maximum speed 705km/h (438mph) at 10,000m (32,810ft); service ceiling 11,800m (38,715ft); range 800km (497 miles) with maximum payload
Weights:	empty 19,050kg (41,997lb); maximum take-off weight 34,500kg (76,059lb)
Dimensions:	wingspan 31.89m (104ft 7.5in); length 28.07m (92ft 1in); height 8.65m (28ft 4.5in); wing area 98.62 sq m (1062 sq ft)
Armament:	none

Transall C.160

Built by the Franco-German Transport Allianz consortium over a 20-year period, the Transall C-160 became the main airlifter for L'Armée de l'Air and the Luftwaffe. Other military users were Turkey and South Africa, and total production numbered over 200 aircraft.

SPECIFICATIONS

Country of origin:	France and Germany
Type:	twin-engined transport
Powerplant:	two 4225kW (5565hp) Rolls-Royce Tyne 22 turboprops engines
Performance:	maximum speed 513km/h (319mph); range 1850km (1150 miles); service ceiling 8230m (30,000ft)
Weights:	empty 30,000kg (62,700lb); maximum 46,000kg (103,400lb)
Dimensions:	span 40m (131ft 3in); length 32.4m (106ft 4in); height 12.36m (38ft 3in); wing area 160 sq m (1721 sq ft)
Armament:	none

Airtech CN-235

CASA of Spain and IPTN in Indonesia joined forces under the Airtech name in the early 1980s to produce the CN-235 military airlifter and civil airliner. The Saudi Air Force Royal Flight is one user, operating four CN-235M-10 models. The CN-235 spawned the larger CN-295, which also remains in production.

SPECIFICATIONS

Country of origin:	Spain and Indonesia
Type:	twin-engined transport
Powerplant:	two 1395-kW (1,750-hp) General Electric CT7C turboprop engines
Performance:	maximum speed 509km/h (317mph); range 5003km (3108 miles); service ceiling 9145m (30,000ft)
Weights:	empty 9800kg (21,605lb); maximum 15,100kg (33,290lb)
Dimensions:	span 25.81m (84ft 8in); length 21.4m (70ft 3in); height 8.18m (26ft 10in); wing area 59.1 sq m (636 sq ft)
Armament:	none

British Aerospace Hawk

The Hawk has been one of the truly outstanding successes of the British aerospace industry in the past three decades. Much of this success is due to the exceptional service life of the airframe, low maintenance requirements, the relatively inexpensive purchase price when originally offered for export, large optional payload, and its ability to operate in the medium range attack and air-superiority role for a fraction of the cost of more powerful types. The first operational aircraft were delivered in 1976.

BAe Hawk T.Mk 1

The Royal Air Force aerobatic team, the 'Red Arrows', flew its first season with the Hawk T.Mk 1 in 1980, replacing the Folland Gnat. The team uses nine Hawks, modified with tanks and piping able to make trails of red, white and blue smoke to enhance the display.

SPECIFICATIONS	
Country of origin:	United Kingdom
Type:	two-seat basic and advanced jet trainer
Powerplant:	one 23.1kN (5200lb) Rolls Royce/Turbomeca Adour Mk 151 turbofan
Performance:	maximum speed 1038km/h (645mph); service ceiling 15,240m (50,000ft); endurance 4 hours
Weights:	empty 3647kg (8040lb); maximum take-off weight 7750kg (17,085lb)
Dimensions:	wingspan 9.39m (30ft 9.75in); length 11.17m (36ft 7.75in); height 3.99m (13ft 1.75in); wing area 16.69 sq m (179.64sq ft)
Armament:	none

BAe Hawk T.Mk 1A

The RAF also operates the T.Mk 1 for weapons instruction. The T.Mk 1A has three pylons; the central one is normally occupied by a 30mm (1.2in) Aden cannon, the two underwing pylons can be fitted with a wide combination of weapons, including Matra rocket pods. This aircraft is carrying a centreline drop tank and rocket pods for weapons training.

SPECIFICATIONS	
Country of origin:	United Kingdom
Type:	two-seat weapons training aircraft
Powerplant:	one 23.1kN (5200lb) Rolls-Royce/Turbomeca Adour Mk 151 turbofan
Performance:	maximum speed 1038km/h (645mph); service ceiling 15,240m (50,000ft); endurance 4 hours
Weights:	empty 3647kg (8040lb); maximum take-off weight 7750kg (17,085lb)
Dimensions:	wingspan 9.39m (30ft 9.75in); length 11.17m (36ft 7.75in); height 3.99m (13ft 1.75in); wing area 16.69 sq m (179.64 sq ft)
Armament:	underfuselage/wing hardpoints with provision for up to 2567kg (5660lb) of stores, wingtip mounted air-to-air missiles

BAe Hawk T.52

The Hawk is the most successful British combat aircraft export of recent years, with sales to 15 countries, including Indonesia. The Indonesians eventually bought a total of 20 Hawk T.53s in four batches and later added further Hawk 100 and 200 ground-attack models.

SPECIFICATIONS

Country of origin:	United Kingdom
Type:	two-seat weapons training aircraft
Powerplant:	one 23.1kN (5200lb) Rolls-Royce/Turbomeca Adour Mk 151 turbofan
Performance:	maximum speed 1038km/h (645mph); service ceiling 15,240m (50,000ft); endurance 4 hours
Weights:	empty 3647kg (8040lb); maximum take-off weight 7750kg (17,085lb)
Dimensions:	wingspan 9.39m (30ft 9.75in); length 11.17m (36ft 7.75in); height 3.99m (13ft 1.75in); wing area 16.69 sq m (179.64 sq ft)
Armament:	underfuselage/wing hardpoints with provision for up to 2567kg (5660lb) of stores, wingtip mounted air-to-air missiles

BAe Hawk T.51

Finland was the Hawk's first export customer, buying 48 T.51s, the majority of which were built under licence by Valmet. As well as use as advanced trainers, they had a secondary air-defence role, armed with a machine-gun pod and two R-60 air-to-air missiles.

SPECIFICATIONS

Country of origin:	United Kingdom
Type:	two-seat weapons training aircraft
Powerplant:	one 23.1kN (5200lb) Rolls-Royce/Turbomeca Adour Mk 151 turbofan
Performance:	maximum speed 1038km/h (645mph); service ceiling 15,240m (50,000ft); endurance 4 hours
Weights:	empty 3647kg (8040lb); maximum take-off weight 7750kg (17,085lb)
Dimensions:	wingspan 9.39m (30ft 9.75in); length 11.17m (36ft 7.75in); height 3.99m (13ft 1.75in); wing area 16.69 sq m (179.64 sq ft)
Armament:	one 12.7mm (.5in) VKT machine-gun pod and two R-60 (AA-8 'Aphid' IR-homing AAMs

McDonnell Douglas T-45A Goshawk

The Goshawk is a development of the highly successful BAe (HS) Hawk trainer for the US Navy. A joint McDonnell Douglas/BAe venture that entered service in 1990, the aircraft is significantly different from the Hawk, with strong twin-wheel nose gear, strengthened long-stroke main gear legs, an arrestor hook and twin lateral airbrakes.

SPECIFICATIONS

Country of origin:	United States
Type:	tandem-seat carrier-equipped naval pilot trainer
Powerplant:	one 26kN (5845lb) Rolls Royce/Turbomeca F-405-RR-401 turbofan
Performance:	maximum speed at 2440m (8000ft) 997km/h (620mph); service ceiling 12,875m (42,250ft); range on internal fuel 1850km (1150 miles)
Weights:	empty 4263kg (9399lb); maximum take-off weight 5787kg (12,758lb)
Dimensions:	wingspan 9.39m (30ft 9.75in); length 11.97m (39ft 3.125in); height 4.27m (14ft); wing area 16.69 sq m (179.6 sq ft)
Armament:	none

MILITARY AIRCRAFT IN THE MODERN ERA

America's Strategic Bombers

The conclusion of the Cold War saw enormous changes to US strategic air power. In 1991, SAC bombers were taken off 24-hour alert. A drawdown in strategic assets saw the retirement of the B-52G and FB-111 and the purchase of fewer B-2s than planned. The B-1B later lost its nuclear role. The heavy bombers, however, found new roles in the 'War on Terror'. With GPS-guided weapons, the former nuclear bombers have conducted close air-support missions in Iraq and Afghanistan.

Boeing B-52G

The B-52G introduced a host of significant improvements, including a wet wing that housed far more fuel, more powerful turbojets, a shortened fin of increased chord, and a remote-controlled rear turret. It was first flown in October 1958, and a total of 193 B-52G models were built, the last in 1960. Some 173 of these were later converted to carry 12 Boeing AGM-86B Air Launched Cruise Missiles.

SPECIFICATIONS	
Country of origin:	United States
Type:	long-range strategic bomber
Powerplant:	eight 61.1kN (13,750lb) Pratt & Whitney J57-P-43W turbojets
Performance:	maximum speed 1014km/h (630mph): service ceiling 16,765m (55,000ft); standard range with maximum load 13,680km (8500 miles)
Weights:	empty 77,200–87,100kg (171,000–193,000lb); loaded 221,500kg (448,000lb)
Dimensions:	wingspan 56.4m (185ft); length 48m (157ft 7in); height 12.4m (40ft 8in); wing area 371.6 sq m (4000 sq ft)
Armament:	remotely controlled tail mounting with four .5in machine guns; normal internal bomb capacity 12,247kg (27,000lb), including all SAC special weapons; external pylons for two Hound Dog missiles

Boeing B-52H

The turbofan-engined Boeing B-52H became the only version in service when the B-52G was retired in 1992. The main visual difference was a 20mm (.8in) cannon in the tail rather than four machine guns, although this was later removed. B-52Hs have served in every recent major conflict.

SPECIFICATIONS	
Country of origin:	United States
Type:	long-range strategic bomber
Powerplant:	four 76kN (17,000lb) Pratt & Whitney thrust turbofan engines
Performance:	maximum speed at 7315m (24,000ft) 1014km/h (630mph): service ceiling 16,765m (55,000ft); range 13,680km (8500 miles)
Weights:	185,000lb (83,250kg); loaded 221,500kg (448,000lb)
Dimensions:	wingspan 56.4m (185ft); length 48.5m (159ft 4in); height 12.4m (40ft 8in); wing area 371.60 sq m (4000 sq ft)
Armament:	remotely controlled tail mounting with one 20mm M61A1 Vulcan cannon (later removed); up to 31,500kg (70,000lb) of bombs, mines or air-to-surface missiles

AMERICA'S STRATEGIC BOMBERS

Rockwell B-1A

SPECIFICATIONS	
Country of origin:	United States
Type:	four-engined strategic bomber prototype
Powerplant:	four 136.9kN (30,618lb) thrust General Electric F101-GE-101 afterburning turbofan engines
Performance:	maximum speed 2351km/h (1458mph); range 9915km (6085 miles); service ceiling 12,000m (39,360ft)
Weights:	maximum 176,810kg (388,982lb)
Dimensions:	span (unswept) 41.67m (136ft 8in); length 45.78m (150ft 2in); height 10.24m (33ft 7in); wing area 191 sq m (1950 sq ft)
Armament:	up to 52,160kg (114,752lb) of conventional or nuclear bombs, AGM-69A SRAMs or AGM-86 air-launched cruise missiles

The B-1A was intended to replace the B-52 in SAC. It was designed as a Mach 2 high-altitude bomber armed with bombs or short-range missiles. Doubts about its abililty to survive this approach, and the increasing cost, caused its cancellation in 1977, although two B-1As flew on as test aircraft.

Rockwell B-1B

SPECIFICATIONS	
Country of origin:	United States
Type:	long-range multi-role strategic bomber
Powerplant:	four 136.9kN (30,780lb) General Electric F101-GE-102 turbofans
Performance:	maximum speed at high altitude 1328km/h (825mph); service ceiling 15,240m (50,000ft); range on internal fuel 12,000km (7455 miles)
Weights:	empty 87,090kg (192,000lb); maximum take-off 216,634kg (477,000lb)
Dimensions:	wingspan 41.67m (136ft 8.5in) unswept, 23.84m (78ft 2.5in) swept; length 44.81m (147ft); height 10.36m (34ft); wing area 181.16 sq m (1950 sq ft)
Armament:	three internal bays with provision for up to 34,019kg (75,000lb) of weapons, plus eight underfuselage stations with a capacity of 26,762kg (59,000lb)

In the early days after its service entry in 1985, several B-1B aircraft were lost after engine failures. Very low-level penetration missions are dependent on state-of-the-art avionics, including a satellite communications link, Doppler radar altimeter, forward-looking and terrain-following radars, and a defensive suite weighing over a ton.

Northrop B-2

SPECIFICATIONS	
Country of origin:	United States
Type:	strategic bomber and missile-launch platform
Powerplant:	four 84.5kN (19,000lb) General Electric F118-GE-110 turbofans
Performance:	max speed 764km/h (475mph); service ceiling 15,240m (50,000ft); range on high-level mission loaded 11,675km (7255 miles)
Weights:	empty 45,360kg (100,000lb); maximum take-off 181,437kg (400,000lb)
Dimensions:	wingspan 52.43m (172ft); length 21.03m (69ft); height 5.18m (17ft); wing area more than 464.50 sq m (5,000 sq ft)
Armament:	two internal bomb bays each carrying one eight-round Boeing Rotary launcher for a total of 16 1.1 megaton B83 thermonuclear free-fall bombs

Since 1978, the B-2 has been developed to a US Air Force requirement for a strategic penetration bomber to complement and replace the Rockwell B-1 Lancer and the Boeing B-52 Stratofortress. The B-2's radar reflectivity is very low because of smooth blended surfaces and the use of radiation-absorbent materials. This image shows the first B-2A, Spirit of America.

China

The People's Liberation Army Air Force (PLAAF) has almost exclusively been equipped with aircraft of Soviet origin since its formation in 1949. When relations broke down between the two powers in 1960, Chinese state aircraft factories kept producing the same designs, in many cases long after production had ended in Russia. China supplied many fighters to Pakistan, North Korea and other nations. Today, China assembles Russian as well as indigenous designs.

Shenyang FT-6

The Shenyang FT-6 was the Chinese-produced version of the MiG-19UTI advanced trainer. Known as the JJ-6 in China, it replaced the FT-5 (JJ-5) in service with Egypt and Pakistan. Egypt's FT-6s served with 20 and 21 Squadrons of the EAF's 221 fighter ground-attack brigade.

SPECIFICATIONS	
Country of origin:	China
Type:	two-seat conversion trainer
Powerplant:	two 36.78kN (8267lb) Liming Wopen-6A (Tumansky RD-9B) afterburning turbojet engines
Performance:	maximum speed 1540km/h (957mph); service ceiling 17,900m (58,725ft); range with internal fuel 1390km (864 miles)
Weights:	empty 5760kg (12,699lb); maximum take-off weight 10,000kg (22,046lb)
Dimensions:	wingspan 9.2m (30ft 2.25in); length (without probe) 13.3m (43ft 8in); height 3.88m (12ft 8.75in); wing area 25 sq m (269.11 sq ft)
Armament:	one 30mm NR-30 cannon, four external hardpoints with provision for up to 500kg (1102lb) of stores

Harbin H-5

The Harbin Aircraft Manufacturing Corporation began as an aircraft-repair plant but went on to produce thousands of the H-5 (H standing for *Hongzha* or bomber), from 1965 to 1984. The H-5 was a reverse-engineered Il-28 'Beagle' and served past 2000 in China. The last user is probably North Korea.

SPECIFICATIONS	
Country of origin:	USSR
Type:	three-seat bomber & ground attack/dual control trainer/torpedo carrier
Powerplant:	two 26.3kN (5952lb) Klimov VK-1 turbojets
Performance:	maximum speed 902 km/h (560mph); service ceiling 12,300m (40,355ft); range 2180km (1355 miles); with bomb load 1100km (684 miles)
Weights:	empty 12890kg (28,418lb); maximum take-off weight 21,200kg (46,738lb)
Dimensions:	wingspan 21.45sq m (70ft 4.5in); length 17.65m (57ft 10.75in); height 6.7m (21ft 11.8in); wing area 60.8 sq m (654.47 sq ft)
Armament:	four 23mm NR-23 cannon (in nose and tail turret); internal bomb capacity of up to 1000kg (2205lb), maximum bomb capacity 3000kg (6614lb); torpedo version had provision for two 400mm light torpedoes

Shenyang J-6

When Sino-Soviet relations cooled in 1960, locally manufactured components were used to assemble Mikoyan-Gurevich MiG-19Ss. The Chinese-built MiG-19S was designated J-6 and entered service in mid-1962, becoming its standard day fighter. Production numbered in the thousands.

SPECIFICATIONS	
Country of origin:	China
Type:	single-seat day fighter
Powerplant:	two 31.9kN (7165lb) Shenyang WP-6 turbojets
Performance:	maximum speed 1540km/h (957mph); service ceiling 17,900m (58,725ft); range with internal fuel 1390km (864 miles)
Weights:	empty 5760kg (12,699lb); maximum take-off 10,000kg (22,046lb)
Dimensions:	wingspan 9.2m (30ft 2.25in); length 14.9m (48ft 10.5in); height 3.88m (12ft 8.75in); wing area 25 sq m (269.11 sq ft)
Armament:	three 30mm NR-30 cannon; four external hardpoints with provision for up to 500kg (1102lb) of stores, including air-to-air missiles, 250kg bombs, 55mm rocket-launcher pods, 212mm rockets or drop tanks

Shenyang F-6

Pakistan was a major user of the F-6 (Chinese J-6), and they were extensively used in the 1971 war with India, acquitting themselves well against the IAF's Su-7s and Hunters. This example served with No 11 Squadron, Pakistan Air Force, in around 1970.

SPECIFICATIONS	
Country of origin:	China
Type:	single-seat day fighter
Powerplant:	two 31.9kN (7165lb) Shenyang WP-6 turbojets
Performance:	maximum speed 1540km/h (957mph); service ceiling 17,900m (58,725ft); range with internal fuel 1390km (864 miles)
Weights:	empty 5760kg (12,699lb); maximum take-off 10,000kg (22,046lb)
Dimensions:	wingspan 9.2m (30ft 2.25in); length 14.9m (48ft 10.5in); height 3.88m (12ft 8.75in); wing area 25 sq m (269.11 sq ft)
Armament:	three 30mm NR-30 cannon; four external hardpoints with provision for up to 500kg (1102lb) of stores, including air-to-air missiles, 250kg bombs, 55mm rocket-launcher pods, 212mm rockets or drop tanks

Chengdu F-7P Airguard

The Pakistan Air Force operates many Chengdu F-7s – Chinese-produced MiG-21s – and made technical contributions to the F-7M, a modernized version. It also commissioned the F-7P Airguard with new radar and weapons. F-7Ps are being further upgraded with an Italian radar.

SPECIFICATIONS	
Country of origin:	China
Type:	single-engined fighter
Powerplant:	one 59.8kN (13,448lb) thrust Liyang afterburning turbojet engine
Performance:	maximum speed 2175km/h (1350mph); range 1740km (1081 miles); service ceiling 18,200m (59,720ft)
Weights:	empty 5275kg (11,629lb); loaded 7531kg (16,603lb)
Dimensions:	span 7.15m (23ft 6in); length 13.95m (45ft 9in); height 4.11m (13ft 6in); wing area 23 sq m (248 sq ft)
Armament:	two 30mm Type 30-1 cannon; pods for 57mm or 90mm rockets; up to 1300kg (2866lb) of bombs

Eurofighter Typhoon

In May 1988, an agreement was signed between the United Kingdom, West Germany and Italy to develop the Eurofighter. Spain joined in November of that year. The aircraft was designed ostensibly for the air-to-air role, with secondary air-to-surface capability. With its canard design and fly-by-wire control system, the aircraft is supremely manoeuvrable and has achieved considerable export success, with examples in service with Austria and Saudi Arabia as well as the four partner nations.

Typhoon T.1

The two-seat Typhoon has the same combat capability as the single-seater. This Typhoon T.1 of No. 29 Squadron, RAF is depicted dropping a Paveway IV laser/GPS-guided bomb. It is fitted with a PIRATE (Passive InfraRed Airborne Tracking Equipment) sensor forward of the cockpit.

SPECIFICATIONS	
Country of origin:	Germany/Italy/Spain/United Kingdom
Type:	twin-seat fighter/trainer
Powerplant:	two 90kN (20,250lb) Eurojet EJ200 turbofans
Performance:	maximum speed at 11,000m (36,090ft) 2125km/h (1321mph); combat radius about 463 and 556km
Weights:	empty 10,000 kg (22,044 lb); maximum take-off weight 23,000 kg (50,705 lb)
Dimensions:	wingspan 10.5m (34ft 5.5in); length 16.0m (52 ft 6 in); height 4m (13ft 1.5in); wing area 52.4 sq m (564.05 sq ft)
Armament:	one 27mm Mauser cannon; thirteen fuselage hardpoints for a wide variety of stores

Eurofighter Typhoon DA-2

The first British Typhoon was the second development aircraft DA-2. Flown initially with Turbo-Union RB.119 engines, it later received the Eurojet EJ 200 engines intended for the production aircraft. The Royal Air Force plans to acquire 232 Typhoons in three main tranches.

SPECIFICATIONS	
Country of origin:	Germany/Italy/Spain/United Kingdom
Type:	multi-role fighter
Powerplant:	two 9185kg (20,250lb) Eurojet EJ200 turbofans
Performance:	maximum speed at 11,000m (36,090ft) 2125km/h (1321mph); combat radius about 463 and 556km
Weights:	empty 9750kg (21,495lb); maximum take-off weight 21,000kg (46,297lb)
Dimensions:	wingspan 10.5m (34ft 5.5in); length 14.5m (47ft 4in); height 4m (13ft 1.5in); wing area 52.4 sq m (564.05 sq ft)
Armament:	one 27mm Mauser cannon; thirteen fuselage hardpoints for a wide variety of stores

EUROFIGHTER TYPHOON

BAe EAP

SPECIFICATIONS

Country of origin:	United Kingdom
Type:	experimental aircraft prototype
Powerplant:	two 71.3kN (16,000lb) thrust Turbo-Union RB199-104 turbofan engines
Performance:	maximum speed 2414km/h (1500mph); range unknown; service ceiling unknown
Weights:	empty 9935kg (21,900lb); maximum 18,145kg (40,000lb)
Dimensions:	span 10.5m (34ft 5in); length 16.8m (55ft 1in); height 5.8m (19ft); wing area 50 sq m (538 sq ft)
Armament:	none

The BAe EAP (Experimental Aircraft Prototype) was built to test concepts for the proposed European Fighter Aircraft (EFA), which later became the Typhoon. It incorporated many parts from the Tornado, including its engines and tailfin. Nearly 200 test flights were made from 1986.

MILITARY AIRCRAFT IN THE MODERN ERA

MiG-29 'Fulcrum'

In 1972, the Soviet Air Force began seeking a replacement for the MiG-21, -23, Sukhoi Su-15, and -17 fleets then in service. The MiG bureau submitted the winning entry, and flight testing of the new fighter – designated 'Ram L' (later 'Fulcrum') by Western intelligence – began in October 1977. First deliveries of the aircraft were made to Soviet Frontal Aviation units in 1983 and the type became operational in 1985. More than 600 of the first production model, the 'Fulcrum-A', were delivered.

Mikoyan-Gurevich MiG-29 'Fulcrum-A'

The MiG-29 was first seen clearly in the West when a squadron from Kubinka Air Base, including the aircraft illustrated, visited Finland's Kuopio-Rissala Air Base in July 1986. Many assumptions about Soviet aircraft, including poor build quality, were dispelled. The first MiG-29s visited Farnborough in 1988.

SPECIFICATIONS
Country of origin:	USSR
Type:	single-seat air-superiority fighter with secondary ground-attack capability
Powerplant:	two 81.4kN (18,298lb) Sarkisov RD-33 turbofans
Performance:	maximum speed 2443km/h (1518mph); service ceiling 17,000m (55,775ft); range with internal fuel 1500km (932 miles)
Weights:	empty 10,900kg (24,030lb); maximum take-off 18,500kg (40,785lb)
Dimensions:	wingspan 11.36m (37ft 3.75in); length (including probe) 17.32m (56ft 10in); height 7.78m (25ft 6.25in); wing area 35.2 sq m (378.9 sq ft)
Armament:	one 30mm GSh-30 cannon with 150 rounds, eight external hardpoints with provision for up to 4500kg (9921lb) of stores

Mikoyan-Gurevich MiG-29UB 'Fulcrum-B'

Although capable of carrying the same weapons load as the single-seat MiG-29, the MiG-29UB's offensive capacity is limited by its lack of radar. The infrared search-and-track (IRST) sensor was retained, allowing cueing of short-range missiles. Romania has now retired its 'Fulcrums'.

SPECIFICATIONS
Country of origin:	USSR
Type:	single-seat air-superiority fighter with secondary ground-attack capability
Powerplant:	two 81.4kN (18,298lb) Sarkisov RD-33 turbofans
Performance:	maximum speed 2232km/h (1387mph); service ceiling 17,762m (58,275ft); range 1835km (1140 miles)
Weights:	empty 15,300kg (33,731lb); maximum 19,700kg (43,431lb)
Dimensions:	wingspan 11.36m (37ft 3.75in); length 17.42m (57ft 1in); height 7.78m (25ft 6.25in); wing area 35.2 sq m (378.9 sq ft)
Armament:	one 30mm GSh-30 cannon with 150 rounds, eight external hardpoints with provision for up to 4500kg (9921lb) of stores

Mikoyan-Gurevich MiG-29 'Fulcrum-A'

Iran received anything from 24 to 40 MiG-29s from the Soviet Union, and may also have integrated a number of Iraqi aircraft into its inventory that fled in 1991. Only two squadrons are known to be equipped with them, the 11th Fighter Squadron at Tehran-Mehrabad and the 23rd at Tabriz.

SPECIFICATIONS

Country of origin:	USSR
Type:	single-seat air-superiority fighter with secondary ground-attack capability
Powerplant:	two 81.4kN (18,298lb) Sarkisov RD-33 turbofans
Performance:	maximum speed 2443km/h (1518mph); service ceiling 17,000m (55,775ft); range with internal fuel 1500km (932 miles)
Weights:	empty 10,900kg (24,030lb); maximum take-off 18,500kg (40,785lb)
Dimensions:	wingspan 11.36m (37ft 3.75in); length (including probe) 17.32m (56ft 10in); height 7.78m (25ft 6.25in); wing area 35.2 sq m (378.9 sq ft)
Armament:	one 30mm GSh-30 cannon with 150 rounds, eight external hardpoints with provision for up to 4500kg (9921lb) of stores

Mikoyan-Gurevich MiG-29M 'Fulcrum D'

Work commenced on advanced versions of the MiG-29 at the end of the 1970s, with attention paid to improving its range and versatility. One of the most significant changes was the incorporation of an advanced analogue fly-by-wire control system. Physical appearance is similar, although the MiG-29M has an extended chord tailplane and a recontoured dorsal fairing.

SPECIFICATIONS

Country of origin:	USSR
Type:	single-seat air-superiority fighter with secondary ground-attack capability
Powerplant:	two 92.1kN (20,725lb) Sarkisov RD-33K turbofans
Performance:	max speed 2300km/h (1430mph); service ceiling 17,000m (55,775ft); range with internal fuel 1500km (932 miles)
Weights:	empty 10,900kg (24,030lb); maximum take-off 18,500kg (40,785lb)
Dimensions:	wingspan 11.36m (37ft 3.75in); length (including probe) 17.32m (56ft 10in); height 7.78m (25ft 6.25in); wing area 35.2 sq m (378.9 sq ft)
Armament:	one 30mm GSh-30 cannon with 150 rounds, six external hardpoints with provision for up to 3000kg (6614lb) of stores

Mikoyan-Gurevich MiG-29K

The carrier-capable MiG-29K was tested in the early 1990s, but development stalled due to financial issues. This aircraft made the first landing of a conventional fixed-wing aircraft on a Soviet carrier in September 1990. A modernized version of the MiG-29K has been sold to India.

SPECIFICATIONS

Country of origin:	USSR
Type:	single-seat air-superiority fighter with secondary ground-attack capability
Powerplant:	two 92.1kN (20,725lb) Sarkisov RD-33K turbofans
Performance:	maximum speed above 11000m (36,090ft) 2300km/h (1430mph); service ceiling 17,000m (55,775ft); range 2900km (1802 miles)
Weights:	maximum 22,400kg (49,340lb)
Dimensions:	span 12m (39ft 4in); length 17.27m (56ft 8in); height 4.73m (15ft 6in); wing area 41.6 sq m (448 sq ft)
Armament:	one 30mm GSh-30 cannon with 150 rounds, six external hardpoints with provision for up to 3000kg (6614lb) of stores

India

The modern Indian Air Force (Bharatiya Vayu Sena) has always chosen its combat aircraft from a variety of international suppliers, in keeping with India's non-aligned status. Until recently, the United States has made little headway in India, and Russia, the United Kingdom, France and various European consortiums have provided most equipment. Hindustan Aeronautics Ltd (HAL) has long assembled foreign designs and built some indigenous trainers, but it is now designing combat aircraft locally.

Hawker Siddeley/HAL 748M

The Hawker Siddeley (BAe) 748 was just one of the many types produced under licence by HAL, which built 69 aircraft beginning in 1964. About half are still in service for communication and transport training duties. One was tested as an AEW platform with a large rotodome on a dorsal mount.

SPECIFICATIONS	
Country of origin:	United Kingdom and India
Type:	twin-engined military transport
Powerplant:	two 1700kW (2280hp) Rolls-Royce Dart RDa 7 Mk 536-2 turboprop engines
Performance:	maximum speed 452km/h (281mph); range 2630km (1645 miles); service ceiling 7620m (25,000ft)
Weights:	empty 11,671kg (25,730lb); maximum 29,092kg (46,500lb)
Dimensions:	span 31.23m (102ft 6in); length 20.42m (67ft); height 7.57m (24ft 10in); wing area 77 sq m (829 sq ft)
Armament:	none

Ilyushin Il-76TD Gajaraj 'Candid'

Known as the *Gajraj* (King Elephant), the Il-76 is India's main heavy airlifter, with nearly 30 in service. As well as aircraft bought from the Soviet Union in the mid-1980s, a further six Il-78 tanker versions were obtained second-hand from Uzbekistan after 2001.

SPECIFICATIONS	
Country of origin:	USSR (now CIS)
Type:	heavy freight transport
Powerplant:	four 117.6kN (26,455lb) Soloviev D-30KP-1 turbofans
Performance:	maximum speed 850km/h (528mph); maximum cruising altitude 12,000m (39,370ft); range 5000km (3107 miles)
Weights:	empty about 75,000kg (165,347lb); maximum take-off weight 170,000kg (374,786lb)
Dimensions:	wingspan 50.5m (165ft 8.2in); length 46.59m (152ft 10.25in); height 14.76m (48ft 5in); wing area 300 sq m (3229.28 sq ft)
Armament:	provision for two 23mm cannon in tail

Mikoyan-Gurevich MiG-21FL 'Fishbed E'

The MiG-21FL was the first major variant of the 'Fishbed' to enter Indian service and was flown by 10 squadrons. The MiG-21 remains in IAF service with the MiG-21bis 'Bison', an upgraded variant with a modernized cockpit, improved radar warning systems and beyond visual range missiles.

SPECIFICATIONS
Country of origin:	USSR
Type:	single-seat all-weather multi role fighter
Powerplant:	one 73.5kN (16,535lb) Tumanskii R-25 turbojets
Performance:	maximum speed above 11,000m (36,090ft) 2229km/h (1385mph); service ceiling 17,500m (57,400ft); range 1160km (721 miles)
Weights:	empty 5200kg (11,464lb); maximum take-off weight 10,400kg (22,925lb)
Dimensions:	wingspan 7.15m (23ft 5.5in); length (including probe) 15.76m (51ft 8.5in); height 4.1m (13ft 5.5in); wing area 23 sq m (247.58 sq ft)
Armament:	one 23mm GSh-23 twin-barrell cannon in underbelly pack, four underwing pylons with provision for about 1500kg (3307kg) of stores

SEPECAT Jaguar

This Jaguar IS (Indian Single-seater), of No 5 Squadron 'The Tuskers', was one of 35 supplied as pattern aircraft from the United Kingdom before local production by HAL began. At least 95 Jaguars in strike, maritime-attack and trainer variants have been built in India.

SPECIFICATIONS
Country of origin:	France and United Kingdom
Type:	single-seat tactical support and strike aircraft
Powerplant:	one 56.4kN (12,676lb) thrust Tumansky R-11 afterburning turbojet engine
Performance:	maximum speed at 11,000m (36,090ft) 1699km/h (1056mph); combat radius on lo-lo-lo mission with internal fuel 537km (334 miles)
Weights:	empty 7700kg (16,976lb); maximum take-off weight 15,700kg (34,613lb)
Dimensions:	wingspan 8.69m (28ft 6in); length 16.83m (55ft 2.5in); height 4.89m (16ft 0.5in); wing area 24.18 sq m (260.28 sq ft)
Armament:	two 30mm Aden Mk.4 cannon with 150rpg; seven external hardpoints with provision for 4763kg (10,500lb) of stores

Dassault Mirage 2000H Vajra

Export contracts for the agile and capable 2000C were plentiful, and by 1990 Dassault had received orders from Abu Dhabi, Egypt, Greece, India and Peru. The Indian aircraft pictured is one of 40 ordered in October 1982 that carry the designation 2000H. Final delivery was made in September 1984. Vajra means 'Thunder'. A follow-on order for a further nine aircraft was made in March 1986.

SPECIFICATIONS
Country of origin:	France
Type:	single-seat air-superiority and attack fighter
Powerplant:	one 97.1kN (21,834lb) SNECMA M53-P2 turbofan
Performance:	maximum speed 2338km/h (1453mph); service ceiling 18,000m (59,055ft); range with 1000kg (2205lb) load 1480km (920 miles)
Weights:	empty 7500kg (16,534lb); maximum take-off weight 17,000kg (37,480lb)
Dimensions:	wingspan 9.13m (29ft 11.5in); length 14.36m (47ft 1.25in); height 5.2m (17ft 0.75in); wing area 41 sq m (441.3 sq ft)
Armament:	two DEFA 554 cannon with 125rpg; nine external pylons with provision for up to 6300kg of stores

Dassault Rafale

The Rafale has been designed and built to replace the fleet of SEPECAT Jaguars with L'Armée de l'Air, and to form part of the new French nuclear-carrier force's air wing. The first flight took place on 4 July 1986. The airframe is largely constructed of composite materials, with a fly-by-wire control system. Early flight trials were particularly encouraging, with the aircraft achieving Mach 1.8 on only its second flight. Original production orders have been cut since the end of the Cold War.

Rafale M

Rafale is produced in three versions: the Rafale C single-seat, multi-role aircraft for the French air force; the two-seat Rafale B; and the navalized Rafale M, a production version of which is shown here.

SPECIFICATIONS	
Country of origin:	France
Type:	two-seat multi-role combat aircraft
Powerplant:	two 73kN (16,424lb) SNECMA M88-2 turbofans
Performance:	max speed 2130km/h (1324mph); service ceiling classified; combat radius 1854km (1152 miles)
Weights:	19,500kg (42,990lb) loaded
Dimensions:	wing span 10.9m (35ft 9in); length 15.3m (50ft 2in); height 5.34m (17ft 6in)
Armament:	one 30mm DEFA 791B cannon, up to 6000kg (13,228lb) of external stores

Dassault Rafale A

SPECIFICATIONS

Country of origin:	France
Type:	prototype combat aircraft
Powerplant:	two 72.96kN (16,402lb) thrust SNECMA M88-2 afterburning turbofan engines
Performance:	maximum speed 2125km/h (1321mph); range unknown; service ceiling unknown
Weights:	empty 9500kg (20,944lb); maximum 20,000kg (44,092lb)
Dimensions:	span 11.2m (36ft 9in); length 15.8m (51ft 10in); height unknown; wing area 47 sq m (506 sq ft)
Armament:	none

Under the Avion de Combat Experimental (ACX) programme, Dassault built the Rafale A demonstrator, which at the time met French air force and navy requirements for lighter aircraft better than the Eurofighter 2000. The Rafale A was slightly larger than the production Rafale models.

Although both L'Armée de l'Air and the French navy considered the Eurofighter to replace the SEPECAT Jaguar, the smaller and lighter Rafale was chosen. This is a Rafale-M navalized fighter.

Rafale M

SPECIFICATIONS

Country of origin:	France
Type:	carrier-based multi-role combat aircraft
Powerplant:	two 73kN (16,424lb) SNECMA M88-2 turbofans
Performance:	maximum speed at high altitude 2130km/h (1324mph); combat radius air-to-air mission 1853km (1152 miles)
Weights:	empty equipped 9800kg; maximum take-off weight 19,500kg (42,990lb)
Dimensions:	wingspan 10.9m (35ft 9.175in); length 15.3m (50ft 2.5in); height 5.34m (17ft 6.25in); wing area 46 sq m (495.1 sq ft)
Armament:	one 30mm DEFA 791B cannon, 14 external hardpoints with provision for up to 6000kg (13,228lb) of stores

MILITARY AIRCRAFT IN THE MODERN ERA

Tiltrotors

From the early 1950s, aircraft manufacturers sought to overcome the speed and range limitations of helicopters by creating hybrid aircraft that could switch between hovering and wingborne flight. Various concepts were tested over the years, including tilt-wings, tilt-props and tilt-fans. The best of these was the tiltrotor, which used fixed wings and tilting engine nacelles to turn propellers into rotors, and vice versa.

Dornier Do 29

The 1950s and 1960s saw an explosion of research into vertical and short take-off and landing aircraft in Germany. The smallest and lightest was the Dornier Do 29, derived from the Do 27. The Do 29, which first flew in December 1958, replaced the Do 27's single nose-mounted engine with two pusher turboprops that could be tilted downwards as much as 90°.

SPECIFICATIONS	
Country of origin:	Germany
Type:	single-seat V/STOL aircraft
Powerplant:	two 201kW (270hp) Lycoming GO-480-B1A6 six-cylinder piston engines
Performance:	cruising speed 290km/h (180mph); service ceiling unknown
Weights:	maximum 2500kg (5511lb)
Dimensions:	wingpsan 13.2m (43ft 4in); length 9.5m (31ft 2in); height unknown
Armament:	none

Canadair CL-84-1

Canada's Dynavert was a private-venture attempt to create a tilt-wing tactical transport and gunship for the export market. The CL-84 began hovering trials in May 1965 and was well advanced in its test programme when it crashed two years later. Improvements and further trials were made, but the Canadian military showed little interest in the project and development ceased.

SPECIFICATIONS	
Country of origin:	Canada
Type:	tilt-wing tactical transport plane and gunship
Powerplant:	two 1118kW (1500hp) Lycoming LTC1K-4A turboprop engines
Performance:	maximum speed 517km/h (321mph); service ceiling unknown
Weights:	maximum 6577kg (14,500lb)
Dimensions:	span 10.56m (34ft 8in); length 16.34m (53ft 8in); height (wing horizontal) 4.34m (14ft 3in)
Armament:	none

TILTROTORS TIMELINE			
	1958 	1965 	1977 

TILTROTORS

Bell XV-15

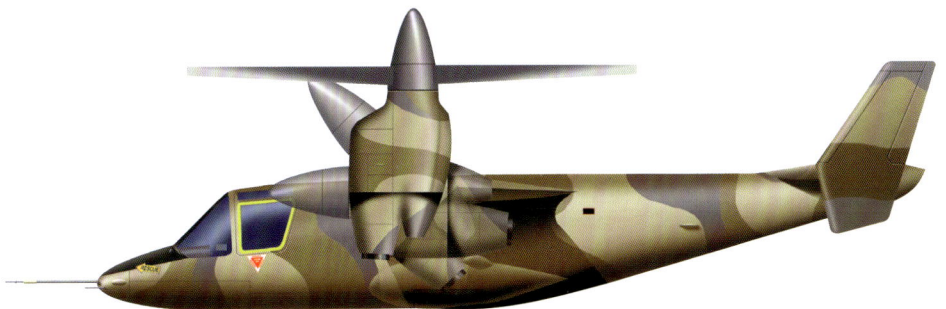

To prove its concept of tilting rotors, Bell produced the XV-15 in 1977. A relatively simple craft, its successful testing paved the way for the V-22. The XV-15 also led to the Bell-Agusta BA.609 of similar size, which is being offered for civil and military use, including potentially as a V-22 escort.

SPECIFICATIONS

Country of origin:	United States
Type:	Tiltrotor demonstrator
Powerplant:	two 1156kW (1550hp) Avco Lycoming LTC1K-4K turboshaft engines
Performance:	maximum speed 557km/h (350mph); range 825km (515 miles); service ceiling 8840m (29,500ft)
Weights:	empty 4574kg (10,083lb); maximum 6009kg (13,248lb)
Dimensions:	span 17.42m (57ft 2in); length 12.83m (42ft 1in); height 3.86m (12ft 8in); rotor diameter 7.62m (25ft)
Armament:	none

Bell/Boeing V-22 Osprey

Turning the tiltrotor concept into a useful military aircraft took much longer and was much costlier than anticipated. Its complex systems and unusual flight characteristics contributed to several accidents. This is the second Full-Scale Development (FSD) aircraft, which first flew in August 1989.

SPECIFICATIONS

Country of origin:	United States
Type:	Tiltrotor transport prototype
Powerplant:	two 4586kW (6150hp) Allison T406-A0-400 turboshaft engines
Performance:	maximum speed 584km/h (363mph); range 3892km (2418 miles); service ceiling 7925m (26,000ft)
Weights:	empty 14,433kg (31,820lb); maximum 21,546kg (47,500lb)
Dimensions:	span 25.55m (84ft 6in); length 17.47m (57ft 4in); height 6.63m (21ft 9in); rotor diameter 11.58m (38ft)
Armament:	none

Bell/Boeing MV-22 Osprey

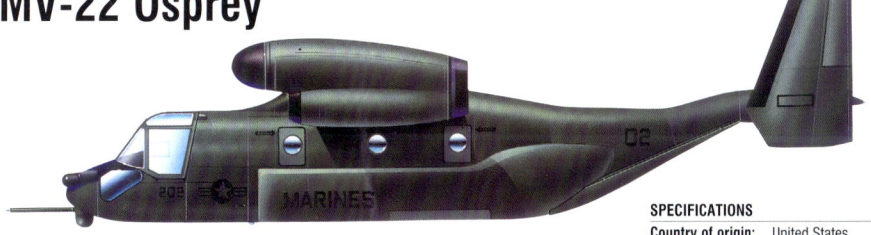

The Osprey is in service with the US Marines as the MV-22B assault transport (illustrated) and the USAF as the CV-22 for special-forces use. The much modified production aircraft first saw operational service in Iraq in 2008 with the Marines, proving superior to helicopters for most missions.

SPECIFICATIONS

Country of origin:	United States
Type:	Tiltrotor transport
Powerplant:	Powerplant: two 4586kW (6150hp) Rolls-Royce AE1107C turboshaft engines
Performance:	maximum speed 584km/h (363mph); range 3892km (2418 miles); service ceiling 7925m (26,000ft)
Weights:	empty 14,433kg (31,820lb); maximum 23,495kg (52,600lb)
Dimensions:	span 25.55m (84ft 6in); length 17.47m (57ft 4in); height 6.73m (22ft 1in); rotor diameter 11.58m (38ft)
Armament:	one 7.62mm M240 machine gun on rear ramp

1989

2002

MILITARY AIRCRAFT IN THE MODERN ERA

From Cobra to Super Hornet

The F/A-18 Hornet was originated by Northrop, developed and manufactured by McDonnell Douglas, and is now a Boeing product. Along the way, it has evolved from a lightweight fighter into a replacement for the F-14 Tomcat, A-6 Intruder and F-4 Phantom with the Navy and Marine Corps. The F/A-18A to D 'Legacy' Hornets had considerable export success, which the FA-18E/F 'Super Bug' looks likely to emulate.

Northrop YF-17

The YF-17 competed in a 'fly-off' evaluation for the USAF's Lightweight Fighter competition in the mid-1970s. The contest was won by the YF-16, but Northrop collaborated with McDonnell Douglas on a production version that became the substantially different F/A-18 Hornet.

SPECIFICATIONS	
Country of origin:	United States
Type:	twin-engined fighter prototype
Powerplant:	two 64.08kN (14,414lb) thrust General Electric YJ101-GE-100 turbojet engines
Performance:	maximum speed 2124km/h (1316mph); range 4500km (2790 miles); service ceiling 18,288m (59,800ft)
Weights:	empty 9527kg (20,960lb); maximum 13,894kg (30,567lb)
Dimensions:	span 10.67m (35ft); length 16.92m (55ft 6in); height 4.42m (14ft 6in); wing area 32.51 sq m (350 sq ft)
Armament:	one 20mm M61A1 Vulcan cannon; two AIM-9 Sidewinder AAMs

McDonnell Douglas F/A-18A Hornet

Although the Hornet was originally to have been produced in both fighter and attack versions, service aircraft are easily adapted to either role. Deliveries to the US Navy began in May 1980 and were completed in 1987.

SPECIFICATIONS	
Country of origin:	United States
Type:	single-seat fighter and strike aircraft
Powerplant:	two 71.1kN (16,000lb) General Electric F404-GE-400 turbofans
Performance:	maximum speed at 12,190m (40,000ft) 1912km/h (1183mph); combat ceiling 15,240m (50,000ft); combat radius 1065km (662 miles)
Weights:	empty 10,455kg (23,050lb); maximum take-off 25,401kg (56,000lb)
Dimensions:	wingspan 11.43m (37ft 6in); length 17.07m (56ft); height 4.66m (15ft 3.5in); wing area 37.16 sq m (400 sq ft)
Armament:	one 20mm M61A1 Vulcan rotary cannon; nine external hardpoints with provision for up to 7711kg (17,000kg) of stores

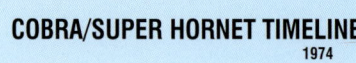
COBRA/SUPER HORNET TIMELINE
1974

1978

1979

McDonnell Douglas F/A-18B Hornet

The combat-capable, tandem-seat trainer Hornet is designated F/A-18B. The aircraft is produced with the same navigation/attack systems as the single-seat variant, although internal fuel capacity has been reduced due to the inclusion of a second seat under a longer canopy. Performance is similar to the single-seat variant, with the exception of range.

SPECIFICATIONS	
Country of origin:	United States
Type:	tandem-seat conversion trainer with combat capability
Powerplant:	two 71.1kN (16,000lb) General Electric F404-GE-400 turbofans
Performance:	maximum speed at 12,190m (40,000ft) 1912km/h (1183mph); combat ceiling about 15,240m (50,000ft); combat radius 1020km (634 miles) on attack mission
Weights:	empty 10,455kg (23,050lb); maximum take-off 25,401kg (56,000lb)
Dimensions:	wingspan 11.43m (37ft 6in); length 17.07m (56ft); height 4.66m (15ft 3.5in); wing area 37.16 sq m (400 sq ft)
Armament:	one 20mm M61A1 Vulcan rotary cannon with 570 rounds; nine external hardpoints with provision for up to 7711kg (17,000kg) of stores

McDonnell Douglas F/A-18C Hornet

The F/A-18A was updated in the late 1980s with new radar and avionics, and the ability to carry new weapons such as the AMRAAM missile. After the Gulf War, Kuwait took delivery of F/A-18Cs to replace its A-4KU Skyhawks. Other F/A-18C operators include Switzerland and Finland.

SPECIFICATIONS	
Country of origin:	United States
Type:	twin-engined fighter/attack aircraft
Powerplant:	two 79.2kN (17,750lb) thrust General Electric F404-GE-402 turbofan engines
Performance:	maximum speed 1915km/h (1190mph); service ceiling 15,000m (50,000ft); combat radius 1065km (662 miles)
Weights:	empty 11,200kg (24,700lb); maximum 23,400kg (51,500lb)
Dimensions:	wingspan 11.43m (37ft 6in); length 17.07m (56ft); height 4.66m (15ft 3.5in); wing area 37.16 sq m (400 sq ft)
Armament:	one 20mm M61A1 Vulcan six-barrel rotary cannon with 570 rounds, nine external hardpoints with provision for up to 7711kg (17,000lb) of stores

Boeing F/A-18E Super Hornet

The Super Hornet, built in single-seat F/A-18E and two-seat F/A-18F versions, has very little in common with its forebears except overall configuration. As well as fighter and attack missions, the F/A-18E can be used as a tanker aircraft with a centreline 'buddy' refuelling pod.

SPECIFICATIONS	
Country of origin:	United States
Type:	carrier-based single-seat fighter/attack aircraft
Powerplant:	two 97.90kN (22,000lb) thrust General Electric F414-GE-400 afterburning turbofan engines
Performance:	maximum speed 1190 km/h (1190mph); combat radius 722km (449 miles); service ceiling 15,000m (50,000ft)
Weights:	empty 13,900kg (30,600lb); maximum 29,900kg (66,000lb)
Dimensions:	span 13.62m (60ft 1in); length 13.62m (44ft 9in); height 4.88m (16ft); wing area 46.45 sq m (500 sq ft)
Armament:	one 20mm M61A1 Vulcan cannon; 11 external hardpoints for up to 8050kg (17,750lb) of stores

1987

1995

Japan

Japan's post-war constitution prohibits the establishment of armed forces, but nonetheless, the nation maintains air, ground and maritime 'self-defence forces' that are amongst the most powerful in Asia. Although forbidden to export weapons systems, Japan's aviation industry produces indigenously designed trainers, transports, patrol aircraft and some combat aircraft for its own needs. Japan has also manufactured American-designed fighters, from the F-86 Sabre to the F-15 Eagle.

McDonnell Douglas F-4EJ Kai Phantom

The EJ is a licence-built air-defence version of the Phantom F-4E. The original F-4E(J) model was built by McDonnell Douglas, and the remainder under licence by Mitsubishi with Kawasaki as a subcontractor. The last was delivered in May 1981. The original batch of 45 was then updated to F-4EJ Kai standard with improved weapon and avionics systems such as digital displays.

SPECIFICATIONS	
Country of origin:	United States
Type:	two-seat all-weather fighter/attack aircraft
Powerplant:	two 79.6kN (17,900lb) General Electric J79-GE-17 turbojets
Performance:	maximum speed 2390km/h (1485mph); service ceiling 19,685m (60,000ft); range on internal fuel with no weapon load 2817km (1750 miles)
Weights:	empty 12,700kg (28,000lb); maximum take-off 26,308kg (58,000lb)
Dimensions:	span 11.7m (38ft 5in); length 17.76m (58ft 3in); height 4.96m (16ft 3in); wing area 49.24 sq m (530 sq ft)
Armament:	one 20mm M61A1 Vulcan cannon and four AIM-7 Sparrow recessed under fuselage or other weapons up to 1370kg (3020lb) on centreline pylon; four wing pylons for stores to a maximum weight of 5888kg (12,980lb)

McDonnell Douglas F-15DJ Eagle

The F-15DJ is the two-seat version of the F-15C (the upgraded version of the F-15A and the principal production version) for the Japanese Air Self-Defence Force. This aircraft is configured to carry conformal fuel tanks that fit flush with the fuselage, leaving all store hardpoints available for the carriage of weapons. Twelve were delivered.

SPECIFICATIONS	
Country of origin:	United States
Type:	twin-seat air superiority fighter trainer with secondary strike/attack role
Powerplant:	two 105.4kN (23,700lb) Pratt & Whitney F100-PW-220 turbofans
Performance:	maximum speed at high altitude 2655km/h (1650mph); ceiling 30,500m (100,000ft); range on internal fuel 4631km (2878 miles)
Weights:	empty 13,336kg (29,400lb); maximum take-off 30,844kg (68,000lb)
Dimensions:	wingspan 13.05m (42ft 9.75in); length 19.43in (63ft 9in); height 5.63m (18ft 5in); wing area 56.48 sq m (608 sq ft)
Armament:	one 20mm M61A1 cannon with 960 rounds, external pylons with provision for up to 10,705kg (23,600lb) of stores

JAPAN TIMELINE

1975      1984      1985   

Kawasaki EC-1

First flying in 1970, the C-1 was designed specifically to replace the Curtiss C-46 Commando transport aircraft in service with the Japanese Air Self-Defence Force. This aircraft differs from standard models by distinctive radomes on the nose and tail, an ALQ-5 ECM system and antennae beneath the fuselage.

SPECIFICATIONS	
Country of origin:	Japan
Type:	ECM trainer aircraft
Powerplant:	two 64.5kN (14,500lb) Mitsubishi (Pratt & Whitney) JT8-M-9 turbofans
Performance:	maximum speed at 7620m (25,000ft) 806km/h (501mph); service ceiling 11,580m (38,000ft); range 1300km (808 miles) with 7900kg (17,417lb) payload
Weights:	empty 23320kg (51,412lb); maximum take-off weight 45,000kg (99,208lb)
Dimensions:	wingspan 30.6m (100ft 4.75in); length 30.5m (100ft 4in); height 10m (32ft 9.3in); wing area 102.5 sq m (1297.09 sq ft)
Armament:	none

Kawasaki T-4

The Kawasaki T-4 replaced the elderly Lockheed T-33 and Fuji T-1 in service in the training role with the Japan Air Self-Defence Force. Most of the airframe is built by Fuji, with the nose section and final assembly the responsibility of Kawasaki. The JASDF's 'Blue Impulse' team flies the T-4.

SPECIFICATIONS	
Country of origin:	Japan
Type:	twin-engined trainer
Powerplant:	two 32.56kN (7320lb) Ishikawajima-Harima F3-IHI-30 turbofan engines
Performance:	max speed 1038km/h (645mph); range 1668km (1036 miles); service ceiling 14,815m (48,606 ft)
Weights:	empty 3790kg (8356lb); maximum 7500kg (16,535lb)
Dimensions:	span 9.94 m (32 ft 7 in); length 13m (42 ft 8 in); height 4.6 m (15 ft 1 in); wing area 21 sq m (226 sq ft)
Armament:	two hardpoints for training bombs or rocket launchers

Mitsubishi XF-2B

Japan's requirement for a replacement for the Mitsubishi F-1 anti-ship attack aircraft led to a collaboration with General Dynamics to develop a version of the F-16, to be produced in Japan. The resulting F-2 was larger and heavier, and considerably more expensive, leading to a reduction in planned numbers.

SPECIFICATIONS	
Country of origin:	Japan and United States
Type:	single-engined fighter prototype
Powerplant:	one 131.7kN (29,607lb) thrust General Electric F110-GE-129 turbofan
Performance:	maximum speed 2125km/h (1321mph); combat radius 834km (518 miles); service ceiling 20,000m (65,555ft)
Weights:	empty 9527kg (21,000lb); maximum 22,000kg (48,500lb)
Dimensions:	span 11.13m (36ft 6in); length 15.52m (50ft 11in); height 4.69m (15ft 5in); wing area 34.84 sq m (375 sq ft)
Armament:	up to 9000kg (19,840lb) of stores, including Mitsubishi AAM-3 air-to-air missiles and ASM-2 anti-ship missiles

1988  1995

MILITARY AIRCRAFT IN THE MODERN ERA

Foxbat and Foxhound

The Mikoyan-Gurevich MiG-25 caused a sensation in the West with its Mach 3 performance, which was unmatched by any other production combat aircraft. As an interceptor and reconnaissance aircraft, it was not noted for its manoeuvrability, but its speed and altitude made it invulnerable to most contemporary weapons systems. The MiG-31 was an all-new aircraft, designed along the same lines but with two seats and a 'look-down shoot-down' capability.

Mikoyan-Gurevich MiG-25 'Foxbat'

It was designed to counter the planned B-70 bomber, but when the BO-70 programme was cancelled, the MiG-25 'Foxbat' was left in search of a role. It entered service in 1970 as an interceptor, its role now defined as being capable of countering all air targets in all weather conditions.

FOXBAT AND FOXHOUND

Mikoyan-Gurevich MiG-25R 'Foxbat'

The reconnaissance version of the 'Foxbat', the MiG-25R provided the Indian Air Force with a strategic reconnaissance capability, using two powerful cameras and SLAR radar. Known as the Garuda in IAF service, they saw the majority of their service with No 102 Squadron 'Trisonics'.

SPECIFICATIONS

Country of origin:	USSR
Type:	single-seat reconnaissance aircraft
Powerplant:	two 109.8kN (24,691lb) Tumanskii R-15BD-300 turbojets
Performance:	maximum speed at altitude about 3339km/h (2112mph); service ceiling 27,000m (88,585ft); operational radius 900km (559 miles)
Weights:	empty 19,600kg (43,211lb); maximum take-off weight 33,400kg (73,634lb)
Dimensions:	wingspan 13.42m (44ft 0.75in); length 23.82m (78ft 1.75in); height 6.10m (20ft 0.5in); wing area not disclosed
Armament:	six external pylons for six 500kg (1102lb) bombs

SPECIFICATIONS

Country of origin:	USSR
Type:	interceptor
Powerplant:	two 100kN (22,487lb) thrust Tumanskii R-15B-300 turbojets
Performance:	maximum speed 2974km/h (1848mph); range 1130km (702 miles); service ceiling 24,383m (80,000ft)
Weights:	loaded 37,425kg (82,508lb)
Dimensions:	span 14.02m (45ft 11in); length 23.82m (78ft 1in); height 6.1m (20ft)
Armament:	four underwing pylons for various combinations of air-to-air missiles

Mikoyan-Gurevich MiG-31 'Foxhound A'

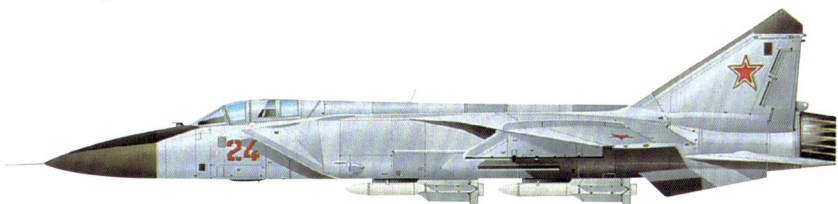

The MiG-31 was developed during the 1970s from the impressive MiG-25 'Foxbat' to counter the threat from low-flying cruise missiles and bombers. The MiG-31 was a vast improvement over the 'Foxbat', with tandem seat cockpit, IR search and tracking sensor, and the Zaslon 'Flash Dance' pulse-Doppler radar providing genuine fire-and-forget engagement capability against multiple targets flying at lower altitudes.

SPECIFICATIONS

Country of origin:	USSR
Type:	two-seat all weather interceptor and ECM aircraft
Powerplant:	two 151.9kN (34,171lb) Soloviev D-30F6 turbofans
Performance:	maximum speed at 17,500m (57,400ft) 3000km/h (1865mph); service ceiling 20,600m (67,600ft); combat radius 1400km (840 miles)
Weights:	empty 21,825kg (48,415lb); maximum take-off weight 46,200kg (101,850lb)
Dimensions:	wingspan 13.46m (44ft 2in); length 22.68m (74ft 5.25in); height 6.15m (20ft 2.25in); wing area 61.6 sq m (663 sq ft)
Armament:	one 23mm GSh-23-6 cannon with 260 rounds, eight external hardpoints with provision for air-to-air missiles, ECM pods or drop tanks

MILITARY AIRCRAFT IN THE MODERN ERA

Sweden

Neutral Sweden has long practised self-reliance in most fields of defence procurement, to include not only locally designed combat aircraft, but the development of radars, datalinks, missiles and systems. Where components such as engines have been sourced from abroad, Swedish engineers have worked to improve them. Despite post-Cold War cutbacks, the Swedish Air Force (Flygvapent) remains one of the best-equipped and technologically advanced in Europe.

Saab A 32A Lansen N

The Lansen first flew in 1952 and has had a very long career, with a couple of examples still in use in the radiological detection role. They were used mainly as ground attack aircraft, but also as reconnaissance aircraft, radar jammers and as airborne testbeds. This particular example was heavily modified to test the Ericsson PS 37 radar for the Viggen.

SPECIFICATIONS	
Country of origin:	Sweden
Type:	all-weather and night fighter
Powerplant:	one 67.5kN (15,190lb) Svenska Flygmotor (Rolls-Royce Avon) RM6A
Performance:	maximum speed 1114km/h (692mph); service ceiling 16,013m (52,500ft); range with external fuel 3220km (2000 miles)
Weights:	unknown
Dimensions:	wingspan 13m (42ft 7.75in); length unknown; height 4.65m (15ft 3in); wing area 37.4 sq m (402.58 sq ft)
Armament:	four 30mm Aden M/55 cannon; four Rb324 (Sidewinder) air-to-air missiles or FFAR (Folding Fin Air-launched Rocket) pods

Saab JA 37 Viggen

The interceptor version of the Viggen, and an integral part of the System 37 series, was the single-seat JA 37. Externally, the aircraft closely resembles the attack AJ 37, although the fin is slightly taller and the interceptor has four elevon actuators under the wing instead of three as on other versions. Production of the JA 37 totalled 149 aircraft, with the last delivered in June 1990.

SPECIFICATIONS	
Country of origin:	Sweden
Type:	single-seat all-weather interceptor aircraft with secondary attack capability
Powerplant:	one 125kN (28,109lb) Volvo Flygmotor RM8B turbofan
Performance:	maximum speed at high altitude 2124km/h (1320mph); service ceiling 18,290m (60,000ft); combat radius on lo-lo-lo mission with external armament 500km (311 miles)
Weights:	empty 15,000kg (33,060lb); maximum take-off 20,500kg (45,194lb)
Dimensions:	wingspan 10.6m (34ft 9.25in); length 16.3m (53ft 5.75in); height 5.9m (19ft 4.25in); wing area 46 sq m (495.16 sq ft)
Armament:	one 30mm Oerlikon KCA cannon with 150 rounds; six external hardpoints with provision for 6000kg (13,228lb) of stores

SWEDEN TIMELINE

 1953    1973    1978

SWEDEN

Saab Sk 37 Viggen

The two-seat conversion trainer of the Viggen, the SK 37, had an unusual seating arrangement, with a second cockpit and canopy for the instructor. Periscopes were built into the framing to allow a forward view. The last Viggens in service were SK 37E 'Eriks' used for electronic-warfare training.

SPECIFICATIONS	
Country of origin:	Sweden
Type:	two-seat conversion trainer
Powerplant:	one 115.7kN (26,015lb) Volvo Flygmotor RM8 turbofan
Performance:	maximum speed at high altitude 2124km/h (1320mph); service ceiling 18,290m (60,000ft); range unknown
Weights:	empty 11,800kg (26,015lb); maximum take-off 20,500kg (45,194lb)
Dimensions:	wingspan 10.6m (34ft 9.25in); length 16.3m (53ft 5.75in); height 5.16m (18ft 9in); wing area 46 sq m (495.16 sq ft)
Armament:	none

North American Sabreliner TP 86

Sweden's military test organisation (*Forsokcentralen*) uses two Rockwell Sabreliners, known as TP 86s, for evaluation of avionics systems. These modified Sabreliner 40 business jets have been used to test synthetic aperture radars, GPS receivers and meteorological research equipment.

SPECIFICATIONS	
Country of origin:	United States
Type:	twin-engined avionics testbed
Powerplant:	two 14.7kN (3307lb) Pratt & Whitney JT12A-8 turbojet engines
Performance:	maximum speed 885km/h (550mph); range 4020km (2500mph); service ceiling 12,200m (40,000ft)
Weights:	empty 4199kg (9257lb); maximum 8500kg (lb)
Dimensions:	span 13.61 m (44ft 8in); length 13.76m (45ft 1in); height 4.88m (16ft); wing area 31.79 sq m (342 sq ft)
Armament:	none

Saab JAS 39A Gripen

The JAS 39 Gripen has now replaced the Viggen in Flygvapnet service. This JAS 39A wears the markings of Flygflotilj 7 at Såtenäs, the first unit to be equipped with the Gripen. Although just over 200 were ordered, Sweden is reducing its fighter force to 100 improved JAS 39C/D versions.

SPECIFICATIONS	
Country of origin:	Sweden
Type:	single-seat all-weather fighter, attack and reconnaissance aircraft
Powerplant:	one 80.5kN (18,100lb) Volvo Flygmotor RM12 turbofan
Performance:	maximum speed more than Mach 2; range on hi-lo-hi mission with external armament 3250km (2020 miles)
Weights:	empty 6622kg (14,600lb); maximum take-off 12,473kg (27,500lb)
Dimensions:	wingspan 8m (26ft 3in); length 14.1m (46ft 3in); height 4.7m (15ft 5in)
Armament:	one 27mm Mauser BK27 cannon with 90 rounds, six external hardpoints with provision for Rb71 Sky Flash and Rb24 Sidewinder air-to-air missiles, Maverick air-to-surface missiles, Rb15F anti-ship missiles, bombs, cluster bombs, rocket-launcher pods, reconnaissance pods, drop tanks and ECM pods

1981  1988

F-22 Raptor

Designed in collaboration between Lockheed and Boeing, the YF-22 won a fly-off against Northrop and McDonnell Douglas's YF-23. The production F-22A Raptor is touted as an 'air-dominance' fighter, able to outperform all current and anticipated opponents. Modern technology allows greater stealth than the F-117 without the faceted shape. The powerful engines allow the Raptor to 'supercruise', or remain supersonic without using afterburner. The engine nozzles are moveable, allowing their use to help manoeuvre the aircraft in pitch and roll.

Lockheed Martin F-22A

The first F-22A Raptor development aircraft flew in 1997 and underwent testing at Edwards Air Force Base. The first squadron was declared operational in 2007. Plans for a force of 383 Raptors were scaled back to 187 due to the 137 million-dollar cost per aircraft.

SPECIFICATIONS	
Country of origin:	United States
Type:	Stealth fighter
Powerplant:	two 160kN (35,000lb) thrust Pratt & Whitney F119-PW-100 thrust-vectoring afterburning turbofan engines
Performance:	maximum speed 2410km/h (1500mph); service ceiling 15,524m (50,000ft); combat radius 2977km (1850 miles)
Weights:	empty 19,700kg (43,340lb); maximum take-off 38,000kg (83,500lb)
Dimensions:	wingspan 13.6m (44ft 6in); length 18.9m (62ft 1in); height 5.1m (16ft 8in); wing area 78.04 m2 (840 sq ft)
Armament:	one 20mm M61A2 Vulcan cannon; internal weapons bays for two AIM-9 Sidewinder and six AIM-120 ASRAAM air-to-air missiles

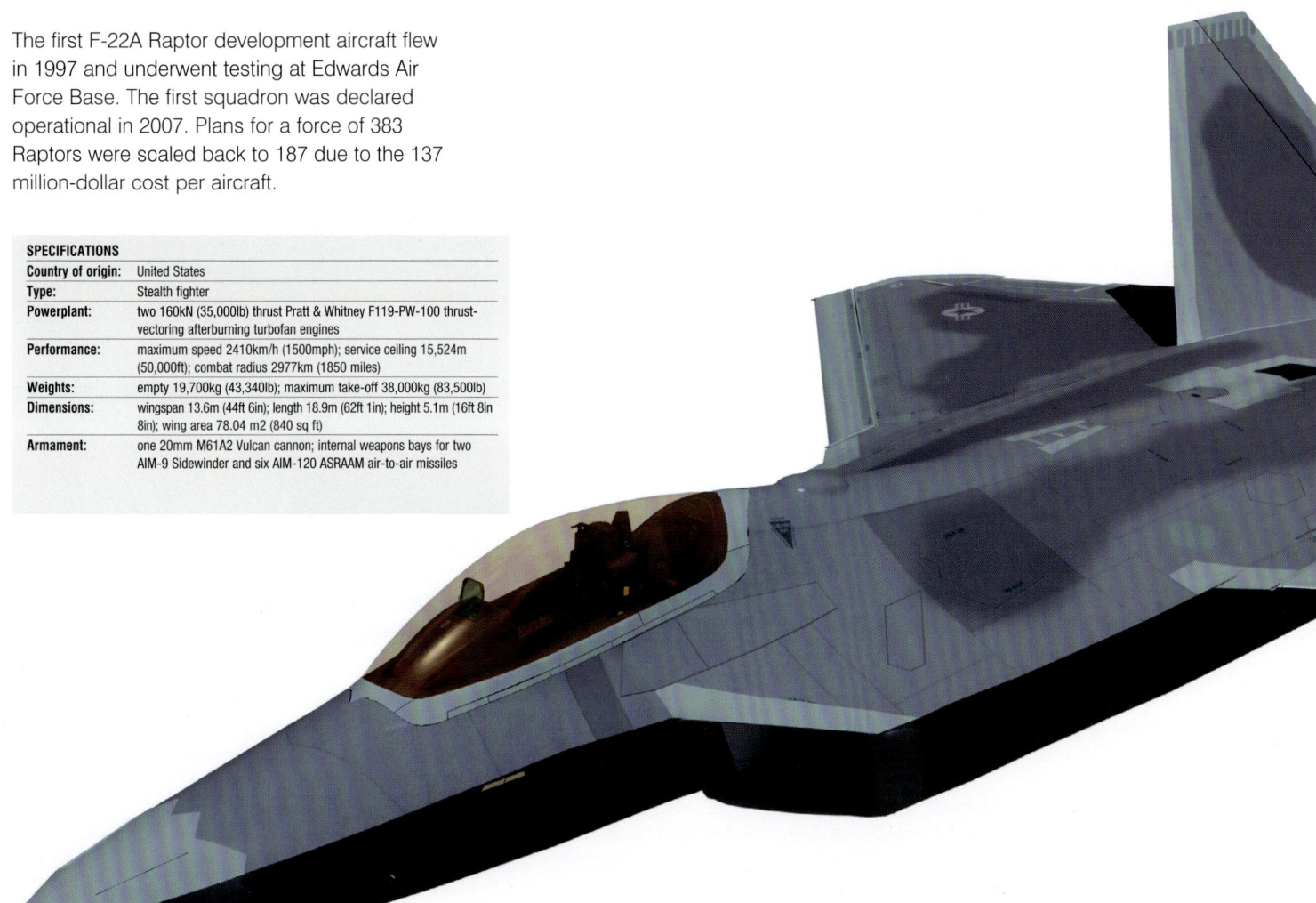

Northrop YF-23A

The Northrop design incorporated many of the stealth features seen on the B-2 Spirit bomber, and the first of two prototype aircraft – dubbed the 'Grey Ghost' – flew in August 1990. However, after a successful flight-testing programme involving both YF-23A prototypes (designated PAV-1 and -2), the aircraft was rejected in favour of the Lockheed YF-22. The two Northrop aircraft were subsequently placed in secure storage at Edwards Air Force Base. It is interesting to compare Northrop's approach to stealth with Lockheed's F-117 Nighthawk.

SPECIFICATIONS

Country of origin:	United States
Type:	single-seat tactical fighter
Powerplant:	one aircraft with two 155.6kN (35,000lb) Pratt & Whitney YF119-PW-100 turbofans; one with General Electric YF120-GE-100 turbofans
Performance:	maximum speed approximately Mach 2; service ceiling 19,812m (65,000ft); range on internal fuel 1200km (750 miles)
Weights:	empty 16,783kg (37,000lb); combat take-off 29,030kg (64,000lb)
Dimensions:	wingspan 13.2m (43ft 7in); length 20.5m (67ft 4in); height 4.2m (13ft 10in); wing area 87.8 sq m (945.07 sq ft)
Armament:	(planned) one 20mm M61 cannon, internal bay for AIM-9 Sidewinder air-to-air missiles and AIM-120 AMRAAMS, 'Have Dash 2' AAMs and 'Have Slick' air-to-surface missiles

Lockheed Martin YF-22 Raptor

The first F-22A Raptor development aircraft flew in 1997 and underwent testing at Edwards Air Force Base. The first squadron was declared operational in 2007. Plans for a force of 383 Raptors were scaled back to 187 due to the cost – $137 million per aircraft.

SPECIFICATIONS

Country of origin:	United States
Type:	Stealth fighter
Powerplant:	two 155.6kN (35,000lb) thrust Pratt & Whitney F119-PW-100 thrust-vectoring afterburning turbofan engines
Performance:	maximum speed 2410km/h (1500mph); range 2977km (1850 miles); service ceiling 15,524m (50,000ft)
Weights:	empty 19,700kg (43,340lb); maximum 38,000kg (83,500lb)
Dimensions:	span 13.6m (44ft 6in); length 18.9m (62ft 1in); height 5.1m (16ft 8in); wing area 78.04 sq m (840 sq ft)
Armament:	one 20mm M61A2 Vulcan cannon; internal weapons bays for two AIM-9 Sidewinder and six AIM-120 ASRAAM air-to-air missiles

Sukhoi 'Flankers'

Development of the Su-27 began in the mid-1970s, with the aim of producing a combat aircraft for Soviet forces comparable to the McDonnell Douglas F-15 Eagle. Given this seemingly daunting design brief, Sukhoi proceeded with impressive haste, and by the end of May 1977 the prototype Su-27 had flown. Development from prototype stage was somewhat longer and involved some fundamental design changes, necessitated by poor structural strength, flutter and excessive weight.

Su-27 T-10-1 'Flanker-A'

The prototype of the Su-27 series, the T-10-1 first flew in May 1977. It suffered from excessive drag, a weak structure and other problems. Even before the second crash among the four prototypes, it was deemed inferior to the F-15, and a total redesign was ordered.

SPECIFICATIONS	
Country of origin:	USSR
Type:	Prototype twin-engined fighter
Powerplant:	two 106kN (24,000lb) thrust Lyul'ka AL-21FZAI afterburning turbofan engines
Performance:	maximum speed unknown; range unknown; service ceiling unknown
Weights:	unknown
Dimensions:	unknown
Armament:	unknown

Su-27UB 'Flanker-C'

It was not until 1980 that full-scale production of the Su-27 began, and service entry started in 1984. The aircraft represents a significant advance over previous generations of Soviet aircraft. The Su-27UB 'Flanker-C', the first variant produced, was a tandem-seat trainer.

SPECIFICATIONS	
Country of origin:	USSR
Type:	tandem-seat operational conversion trainer
Powerplant:	two 122.5kN (27,557lb) Lyul'ka AL-31M turbofans
Performance:	maximum speed 2150km/h (1335mph); service ceiling 17,500m (57,400ft); combat radius 1500km (930 miles)
Weights:	maximum take-offweight 30,000kg (66,138lb)
Dimensions:	wingspan 14.7m (48ft 2.75in); length 21.94m (71ft 11.5in); height 6.36m (20ft 10.25in); wing area 46.5 sq m (500 sq ft)
Armament:	one 30mm GSh-3101 cannon with 149 rounds; 10 external hardpoints with provision for 6000kg (13,228kg) of stores

SUKHOI 'FLANKERS' TIMELINE — 1977 — 1985 — 1988

Su-33/Su-27K 'Flanker-D'

Experiments with Su-27s modified for carrier operations began as early as 1982, leading to what NATO calls the Su-27K 'Flanker-D' and Sukhoi calls the Su-33, with folding wings and a strengthened landing gear. The 1st Squadron of the Russian Navy occasionally flies them from the carrier *Admiral Kuznetsov*.

SPECIFICATIONS

Country of origin:	USSR
Type:	twin-engined carrier-based fighter
Powerplant:	two 130.4kN (29,321lb) thrust Lyul'ka AL-31K afterburning turbofan engines
Performance:	maximum speed 2300km/h (1429mph); range 1864 miles (3000km); service ceiling 17,000m (55,750ft)
Weights:	empty 18,400kg (40,600lb); maximum 33,000kg (72,753lb)
Dimensions:	span 14.7m (48ft 3in); length 21.15m (69ft 5in); height 5.85m (19ft 2in); wing area 67.8 sq m (730 sq ft)
Armament:	one 30mm GSh-301 cannon; 12 hardpoints for up to 6500kg (14,330lb) of bombs, air-to-surface missiles, rockets or air-to-air missiles

Su-27IB/Su-34 'Fullback'

Known variously during its development as the Su-27IB and the Su-32FN, the Su-34 'Fullback' has been chosen as Russia's future strike aircraft and is slowly entering squadron service. The side-by-side cockpit has provision for a foldaway galley and toilet to improve crew comfort on long-range missions.

SPECIFICATIONS

Country of origin:	United States
Type:	twin-engined attack aircraft
Powerplant:	two 137.2kN (30,845lb) thrust Lyul'ka AL-35F afterburning turbofans
Performance:	maximum speed 1900km/h (1180mph); range 4000km (2490 miles); service ceiling 15,000m (49,200ft)
Weights:	empty 22,000kg (48,502lb); maximum 45,100kg (99,425lb)
Dimensions:	span 14.7m (48ft 3in); length 23.34m (72ft 2in); height 6.09m (19ft 5in); wing area 62 sq m (666 sq ft)
Armament:	one SPPU-22 23-mm six-barrelled cannon with 140 rounds; 10 hardpoints for up to 8000kg (17,630lb) of stores

Su-35 (Su-27M)

One of the ongoing developments of the Su-27 is the single-seat Su-35 all-weather air-superiority fighter (derived from the 'Flanker-B'). This aircraft, which has similar powerplant and configuration to the Su-27, is an attempt to provide a second-generation Su-27 with improved agility and operational capability. It first flew in 1988.

SPECIFICATIONS

Country of origin:	USSR
Type:	single-seat all-weather air superiority fighter
Powerplant:	two 122.5kN (27,557lb) Lyul'ka AL-31M turbofans
Performance:	maximum speed at high altitude 2500km/h (1500mph); service ceiling 18,000m (59,055ft); combat radius 1500km (930 miles)
Weights:	maximum take-off weight 30,000kg (66,138lb)
Dimensions:	wingspan 14.7m (48ft 2.75in); length 21.94m (71ft 11.5in); height 6.36m (20ft 10.25in); wing area 46.5 sq m (500 sq ft)
Armament:	one 30mm GSh-3101 cannon with 149 rounds; 10 external hardpoints with provision for 6000kg (13,228kg) of stores

1990

1994

COLD WAR TANKS AND AFVS

Elements of the wartime German heavy tank designs were reflected in post-war armoured fighting vehicles like the British Centurion and the American M48 main battle tanks.

When the Centurion began to be phased out of British service in 1960, other nations were keen to acquire it. So good a design was the Centurion that it continued to serve in some numbers into the 1990s. Together with the American M48, it wrought havoc on Egypt's Russian-built tanks in the 1967 Arab–Israeli War.

Left: The American M26 Pershing heavy tank entered operational service in 1945, just too late to have any impact on World War II.

Centurion A41

The Centurion Main Battle Tank had its origins in World War II, when it was developed as a cruiser tank. The first prototype appeared in 1945 and the tank entered production shortly after. The Centurion saw action in Korea, Vietnam, Pakistan and the Middle East. Nearly 4500 were built before production ceased.

Centurion A41

The Centurion is one of the most successful tank designs in history. Indeed, its capacity for upgrading kept it in service beyond the 1960s with some armies. Variants include an armoured recovery vehicle; an amphibious recovery vehicle, used in the Falklands conflict in 1982; a bridge-layer and an AVRE.

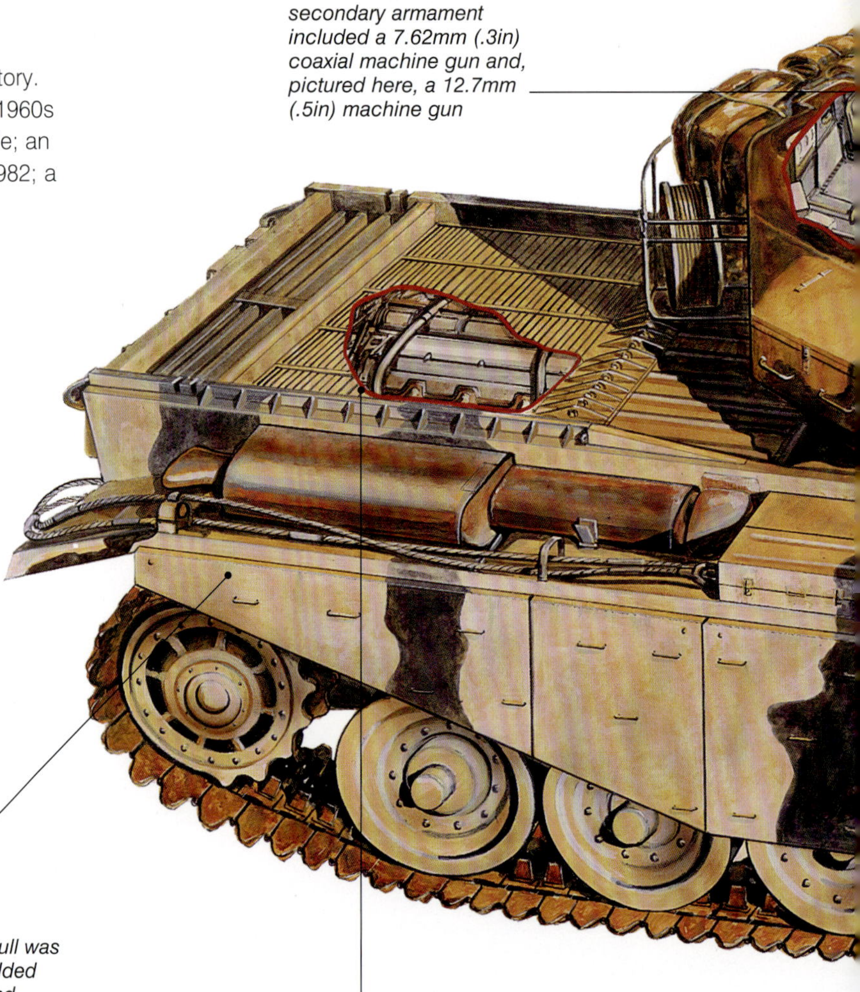

SECONDARY ARMAMENT
The Centurion's secondary armament included a 7.62mm (.3in) coaxial machine gun and, pictured here, a 12.7mm (.5in) machine gun

SPECIFICATIONS	
Country of origin:	United Kingdom
Crew:	4
Weight:	51,723kg (113,792lb)
Dimensions:	Length: 9.854m (32ft 4in); width: 3.39m (11ft 1.5in); height: 3.009m (9ft 10.5in)
Range:	205km (127 miles)
Armour:	51–152mm (2–6in)
Armament:	1 x 105mm (4.1in) gun; 2 x 7.62mm (.3in) MGs; 1 x 12.7mm (.5in) HMG
Powerplant:	1 x Rolls-Royce Meteor Mk IVB V-12 petrol engine, developing 485kW (650hp)
Performance:	Maximum road speed: 43km/h (27mph); fording: 1.45m (4ft 9in); vertical obstacle: .91m (3ft); trench: 3.352m (11ft)

HULL
The Centurion's hull was designed with welded sloped armour, and featured a partially cast turret mounting the main and secondary weapons. The sloped hull was also a feature of tanks like the German Panther

POWERPLANT
All through its career in the British Army, the Centurion used the standard Rolls-Royce Meteor petrol engine, which was a development of the Merlin aeroengine

Centurion A41

ARMAMENT
The Centurion was originally armed with a 20-pounder gun, but this was used only for a short time before it was replaced by a very effective 105mm (4.1in) weapon

SUSPENSION
The Centurion's designers extended the long-travel 5-wheel suspension used on the Comet with the addition of a sixth wheel and an extended spacing between the second and third wheels

Postwar Tanks

Elements of the best German tank designs were reflected in postwar armoured fighting vehicles like the British Centurion. But in the postwar era, tanks faced a very different scenario from the operational requirements that had led to their development in World War II – the prospect of fighting in a nuclear battleground.

M103 Heavy Tank

Prototypes of the M103 showed deficiencies in both the turret and gun control equipment. It was modified, but remained hard to employ because of its size (which made concealment difficult), short range and poor reliability. It was phased out during the 1960s.

SPECIFICATIONS	
Country of origin:	USA
Crew:	5
Weight:	56,610kg (124,544lb)
Dimensions:	Length: 11.3m (37ft 1.5in); width: 3.8m (12ft 4in); height: 2.9m (9ft 5.3in)
Range:	130km (80 miles)
Armour:	12.7–178mm (.5–7in)
Armament:	1 x 120mm (4.7in) rifled gun; 1 x 7.62mm (.3in) coaxial MG; 1 x 12.7mm (.5in) anti-aircraft HMG
Powerplant:	1 x Continental AV-1790-5B or 7C V-12 petrol engine developing 604kW (810hp)
Performance:	Maximum road speed: 34km/h (21mph); vertical obstacle: .91m (3ft); trench: 2.29m (7ft 6in)

T-10 Heavy Tank

The T-10 was was reserved for the domestic market and was designed to provide long-range fire-support for the T-54/55s, and to act as a spearhead for thrusts through heavily defended areas. Its cramped confines required the use of separate-loading ammunition.

SPECIFICATIONS	
Country of origin:	Soviet Union
Crew:	4
Weight:	49,890kg (109,760lb)
Dimensions:	Length (including gun): 9.875m (32ft 4.75in); length (hull): 7.04m (23ft 1in); width: 3.566m (11ft 8.5in); height: 2.25m (7ft 4.5in)
Range:	250km (155 miles)
Armour:	20–250mm (.79–9.84in)
Armament:	1 x 122mm (4.8in) gun; 2 x 12.7mm (.5in) machine guns (one coaxial, one anti-aircraft)
Powerplant:	1 x V-12 diesel engine developing 522kW (700hp)
Performance:	Maximum road speed: 42km/h (26mph); vertical obstacle: .9m (35.5in); trench: 3m (9ft 10in)

TIMELINE			
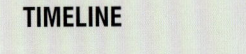	1945 	1952 	

Conqueror Heavy Tank

Based on the the Centurion, the FV214 Conqueror was cumbersome and hard to maintain. Its advantage over the Centurion was limited to a longer-range gun, and when the latter was upgunned the Conqueror had no role. It was phased out in the 1960s.

SPECIFICATIONS	
Country of origin:	United Kingdom
Crew:	4
Weight:	64,858kg (142,688lb)
Dimensions:	Length (gun forwards): 11.58m (38ft); length (hull): 7.72m (25ft 4in); width: 3.99m (13ft 1in); height: 3.35m (11ft)
Range:	155km (95 miles)
Armour:	17–178mm (.66–7in)
Armament:	1 x 120mm (4.7in) rifled gun; 1 x 7.62mm (.3in) coaxial MG
Powerplant:	1 x 12-cylinder petrol engine developing 604kW (810hp)
Performance:	Maximum road speed: 34km/h (21.3mph); vertical obstacle: .91m (3ft); trench: 3.35m (11ft)

M47 Patton I Medium Tank

The M47 Patton tank was an interim AFV, built while the T42 was under development. It was made up of a T42 turret on an M46 chassis. Very successful, it was widely used by many of America's NATO allies and by a further 18 countries.

SPECIFICATIONS	
Country of origin:	USA
Crew:	5
Weight:	46,165kg (45.4 tons)
Dimensions:	Length (over gun): 8.56m (28ft 1in); width: 3.2m (10ft 6in); height: 3.35m (11ft)
Range:	129km (80 miles)
Armour:	115mm (4.53in)
Armament:	1 x M36 90mm (3.54in) gun; two 12.7mm (.5in) MGs; one 7.62mm (.3in) MG
Powerplant:	1 x Continental AVDS-1790-5B V-12 petrol, 604.5kW (810hp)
Performance:	Maximum speed: 48km/h (30mph)

Type 59

Early models of the Type 59 lacked gun stabilization and night vision equipment. A laser rangefinder was later mounted on the front of the turret, where it was vulnerable to small-arms fire. The tank made a smoke screen by injecting diesel fuel into the exhaust pipe.

SPECIFICATIONS	
Country of origin:	China
Crew:	4
Weight:	36,000kg (79,200lb)
Dimensions:	Length: 9m (27ft 6in); width: 3.27m (10ft); height: 2.59m (7ft 8in)
Range:	600km (375 miles)
Armour:	39–203mm (1.5–8in)
Armament:	1 x 100mm (3.9in) gun; 2 x 7.62mm (.3in) MGs; 1 x 12.7mm (.5in) HMG
Powerplant:	1 x Model 12150L V-12 diesel engine developing 388kW (520hp)
Performance:	Maximum road speed: 50km/h (31.3mph); fording: 1.4m (4ft 7in); vertical obstacle: .79m (2ft 7in); trench: 2.7m (8ft 10in)

1955

T-54/55 Main Battle Tank

No other tank has been produced in such quantity as the T-54/55 series. It was developed as the Soviet Union's main battle tank in 1947, and remained in service with some countries 60 years later. The T-54 was based on the T-44, but with a 100mm (3.9in) gun. In service the tank was progressively updated.

T-54/55 Main Battle Tank

CREW
The low turret offered a small silhouette to the enemy, but this was achieved only by cramming the commander, gunner and loader into a very small and uncomfortable space

ARMAMENT
The T-54/55's main armament was originally the D-10T 100mm (3.9in) rifled gun, a mighty weapon in its heyday but one soon rendered inadequate by the pace of modern warfare

T-54/55 MAIN BATTLE TANK

T-54/55

AMMUNITION
The basic armour-piercing ammunition types for the D-10T gun lacked the penetrating power to destroy the latest generation of Western tanks except at very close range

The T-55 was a T-54 modified for operations on the nuclear battlefield. It had a thicker turret casting, more powerful engine and primitive NBC protection. The T-55B was the first model to incorporate infra-red night vision equipment and two-axis stabilization for the main gun.

SPECIFICATIONS	
Country of origin:	Soviet Union
Crew:	4
Weight:	36,000kg (35.42 tons)
Dimensions:	Length (hull): 6.45m (21ft 2in); width: 3.27m (10ft 9in); height: 2.4m (7ft 10in)
Range:	400km (250 miles)
Armour:	Up to 203mm (8in)
Armament:	1 x 100mm (3.94in) D-10T gun; 2 x 7.62mm (.3in) DT MGs; 1 x 12.7mm (.5in) DShK AA MG
Powerplant:	1 x V-54 12-cylinder diesel engine developing 388kW (520bhp) at 2000rpm
Performance:	Maximum speed: 48km/h (30mph)

POWERPLANT
The engines on Soviet-built T-54/55s were often so badly made that they self-destructed when oil lines became blocked by loose metal filings. Made of magnesium alloy, the engine caught fire easily

DIFFERENCES
Later model T-54/55s had numerous detail differences, including infra-red night vision and deep-fording equipment, automatic fire extinguishers and air-conditioning

COLD WAR TANKS AND AFVS

M48

Christened the Patton after the World War II US tank general, the M48 was designed hurriedly and rushed into production in 1953, after the Korean War had exposed the US Army's alarming shortage of modern tanks. Despite teething troubles, it became one of the most widely used medium tanks in the world.

M48

TURRET
A single-piececasting, the M48's turret is not ballistically well shaped. It is 120mm (4.7in) thick at the front and only 76mm (3in) thick along the sides

EXTRA ARMOUR
Many tank crews fix sections of track on the turret sides as an extra defence against RPG-7 anti-tank rockets

M48

LOADER
The loader in an M48 has a reasonable amount of room and can use his right hand to pick up the shells

The M48 was used in combat during the Vietnam War, where, despite misgivings that South East Asia didn't have the terrain suitable for heavy armour, it served successfully in the infantry support role. It was first used in tank-versus-tank combat during the Indo-Pakistan War of 1965, where it suffered heavy losses, particularly in action against the Indian Army's Centurions.

SPECIFICATIONS

Country of origin:	USA
Crew:	4
Weight:	48,987kg (107,998lb)
Dimensions:	Length (over gun): 9.31m (30ft 6in); width: 3.63m (11ft 11in); height: 3.01m (10ft 1in)
Range:	499km (310 miles)
Armour:	Up to 180mm (7in)
Armament:	1 x 105mm (4.13in) L7 gun; 3 x 7.62mm (.3in) MGs
Powerplant:	1 x Continental AVDS-1790-2 12-cylinder supercharged diesel engine developing 559.7kW (750hp)
Performance:	Maximum road speed: 48km/h (30mph)

MAIN GUN
The 105mm (4.13in) gun can easily be distinguished by the prominent blast deflector. Firing fixed ammunition, it is roughly equivalent in hitting power to the 100mm (3.94in) gun in the T-54/55

ESCAPE HATCH
Underneath the hull is a single escape hatch through which the crew can bail out

Medium Tanks

The limitations of the heavy tank became apparent during World War II, and the Soviet Union was the first to recognize that the key to success in armoured warfare was to deploy fast medium tanks, combining mobility with effective sloping armour. It was the medium tank that continued to make its mark in the postwar world.

M48A1

The M48 Patton II was designed to replace the earlier Sherman and M47 Patton tanks. The M48A1 had the commander's cupola of the earlier M1, allowing the M2HB 12.7mm (.5in) machine gun to be operated and reloaded from within the vehicle.

SPECIFICATIONS	
Country of origin:	USA
Crew:	4
Weight:	47,273kg (104,219lb)
Dimensions:	Length: 7.3m (23ft 11in); Width: 3.6m (11ft 11in); Height: 3.1m (11ft 10in)
Range:	216km (134 miles)
Armour:	Not revealed
Armament:	1 x 105mm (4.13in) gun; 3 x 7.62mm (.3in) MGs
Powerplant:	M48A1 AV1790-7C V-12 air-cooled petrol engine, developing 810hp (604kW)
Performance:	Maximum road speed: 42km/h (26mph)

M48A2

The M48A2 had an improved powerplant and transmission, redesigned rear plate, and improved turret control. The M48A2C variant had an improved rangefinder, new ballistic drive and bore evacuator for the main gun.

SPECIFICATIONS	
Country of origin:	
Crew:	4
Weight:	Not revealed
Dimensions:	Length: 6.95m (22ft 10in); width: 3.63m (11ft 11in); height: 3.27m (10ft 9in)
Range:	Not revealed
Armour:	Not revealed
Armament:	1 x 105mm (4.13in) gun; 3 x 7.62mm (.3in) MGs
Powerplant:	1 x General Dynamics Land Systems AVDS-1790 series diesel engine
Performance:	Maximum speed: not revealed

TIMELINE

 1950    1952    1953

MEDIUM TANKS

M48A5

The M48A5 was the first Patton variant to be fitted with the powerful 105mm (4.1in) M68 gun. Many kits were supplied to other operators to enable them to upgrade their existing Pattons to M48A5 standard.

SPECIFICATIONS
Country of origin:	USA
Crew:	4
Weight:	49,090kg (108,225lb)
Dimensions:	Length: 9.47m (11ft 1in); Width: 3.63m (10ft 11in); Height: 3.29m (10ft 10in)
Range:	500km (311 miles)
Armour:	Not revealed
Armament:	1 X 105mm M48 gun
Powerplant:	1 X M48A1 AV1790-7C V-12 air-cooled petrol engine, developing 810hp (604kW); 1 x AVDS-1790 2A RISE model engine, developing 750hp (559 kW)
Performance:	Maximum road speed: 48lm/h (30mph)

Centurion Mk 5

The Centurion Mk 5 was one of 13 sub-variants of the basic Centurion design. The tank was widely exported, some overseas customers making their own modifications to what was already an outstanding design in a bid to improve performance.

SPECIFICATIONS
Country of origin:	United Kingdom
Crew:	4
Weight:	50,728kg (111,836lb)
Dimensions:	Length: 9.8m (32ft 2in); Width: 3.39m (11ft 1in); Height: 2.94m (9ft 7in)
Range:	102km (63 miles)
Armour:	Not revealed
Armament:	Not revealed
Powerplant:	Rolls-Royce Mk IVB 12-cylinder liquid-cooled engine, developing 650hp (485kW)
Performance:	Maximum road speed: 34.6km/h (21.5 mph))

Magach (Israeli M48)

The Magach series of tanks was the Israeli Defence Force's name for the M48/M60. These were armed with a 90mm (3.54in) gun and used during the Six-Day War of June 1967. The Magach was upgraded by the Israelis, receiving the British 105mm (4.1in).

SPECIFICATIONS
Country of origin:	Israel
Crew:	4
Weight:	Not available
Dimensions:	Length: 6.95m (22ft 10in); width: 3.63m (11ft 11in); height: 3.27m (10ft 9in)
Range:	Not available
Armour:	Not available
Armament:	1 x 105mm (4.1in) gun; 3 x 7.62mm (.3in) MGs
Powerplant:	1 x General Dynamics Land Systems AVDS-1790 series diesel engine
Performance:	Not available

 1954    1967

COLD WAR TANKS AND AFVS

PT-76

Developed in the late 1940s, the Soviet PT-76 has proved to be an excellent design, as illustrated by the fact that it is still in service in many parts of the world. Fully amphibious, it is propelled through the water with the help of two water jets at the rear of the hull.

PT-76

MAIN GUN
Based on the Red Army's World War II tank gun, the PT-76's main gun, a D-56, fires APHE, FRAG-HE, HEAT and HVAP ammunition, but the sights are marked for only the first two

FRONT HULL ARMOUR
The glacis is protected by just 11mm (.43in) of armour sloped at 80°. This will keep out rifle rounds and most heavy machine-gun fire, but nothing else

PT-76

The PT-76 has also been in combat in the Arab-Israeli wars, and has been used in the counter-insurgency role in countries such as Indonesia. Still in use in the twenty-first century, the PT-76 is reported to have seen service with the Russian Army in Chechnya.

SCHNORKEL
The schnorkel fits over a ventilator in the turret rear when swimming, but can suck exhaust gases into the crew compartment

SPECIFICATIONS	
Country of origin:	Soviet Union
Crew:	3
Weight:	14,000kg (30,800lb)
Dimensions:	Length: 7.65m (25ft .25in); width: 3.14m (10ft 3.7in); height: 2.26m (7ft 4.75in)
Range:	260km (160 miles)
Armour:	5–17mm (.19–.66in)
Armament:	1 x 76mm (3in) gun; 1 x coaxial 7.62mm (.3in) MG; 1 x 12.7mm (.5in) AA HMG
Powerplant:	1 x V-6 6-cylinder diesel engine developing 179kW (240hp)
Performance:	Maximum road speed: 44km/h (27mph); fording: amphibious; vertical obstacle: 1.1m (3ft 7.3in); trench: 2.8m (9ft 2in)

EXTRA FUEL TANK
Flat fuel tanks, similar to those on the T-54/55 series tanks, can be mounted on the hull rear. Up to two can be carried on either side

DRIVER
When the vehicle is in the water, the driver cannot see much and relies on steering instructions from the vehicle commander

AMX-13

The AMX-13 was designed immediately after the end of World War II. Production began in 1952 and continued until the 1980s. Its design included an automatic loader in the turret bustle which included two revolver-type magazines, each holding six rounds of ammuntion. The tank was widely exported.

AMX-13

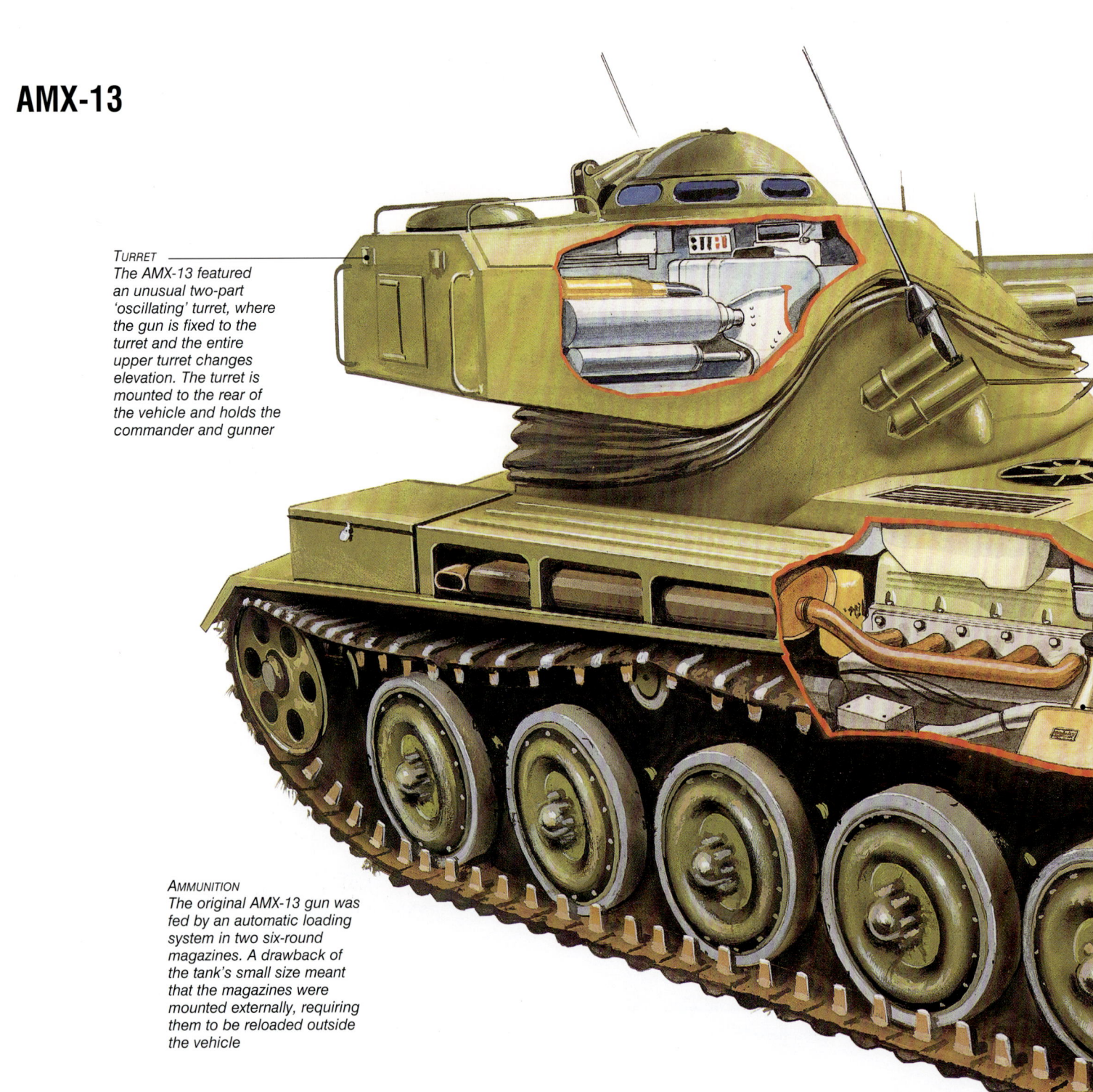

Turret
The AMX-13 featured an unusual two-part 'oscillating' turret, where the gun is fixed to the turret and the entire upper turret changes elevation. The turret is mounted to the rear of the vehicle and holds the commander and gunner

Ammunition
The original AMX-13 gun was fed by an automatic loading system in two six-round magazines. A drawback of the tank's small size meant that the magazines were mounted externally, requiring them to be reloaded outside the vehicle

AMX-13

ARMAMENT
The 75mm (2.95in) gun of the original AMX-13 production model was modelled on the German L/71 Panther gun, this being replaced by a 90mm (3.54in) weapon in 1966

A notable feature on the AMX-13 is the oscillating turret: the gun was fixed in the upper part, which pivoted on the lower part. The AMX-13 was extensively modified and used for a family of vehicles, from self-propelled guns and howitzers to engineer vehicles, recovery vehicles, bridgelayers and infantry fighting vehicles.

SPECIFICATIONS	
Country of origin:	France
Crew:	3
Weight:	15,000kg (33,000lb)
Dimensions:	Length: 6.36m (20ft 10.3in); width: 2.5m (8ft 2.5in); height: 2.3m (7ft 6.5in)
Range:	400km (250 miles)
Armour:	10–40mm (.4–1.57in)
Armament:	1 x 75mm (2.95in) gun; one 7.62mm (.3in) MG
Powerplant:	1 x SOFAM eight-cylinder petrol engine developing 186kW (250hp)
Performance:	Maximum road speed: 60km/h (37mph); fording: .6m (1ft 11.7in); vertical obstacle: .65m (2ft 1.7in); trench: 1.6m (5ft 3in)

CHASSIS
The chassis of the AMX-13, intensively modified in many cases, was used as the basis for one of the most complete families of AFV ever developed

POWERPLANT
The basic AMX-13 had a petrol engine offering an operating range up to 400km (250 miles). Later, Creusot-Loire offered a package to replace this with a Detroit Diesel unit

Postwar Self-Propelled Guns

In the postwar years, artillery exchanges became faster and more accurate. New systems enabled gunners to detect enemy artillery with the minimum delay and lay down swift counter-battery fire. The self-propelled gun thus took on new importance, with its ability to 'shoot and scoot' before incoming fire reached its position.

ASU-57

The ASU-57 was designed for use by Soviet airborne divisions, to be parachuted with the troops, using pallets fitted with retro-rocket systems that softened impact on landing. For lightly armed, airborne troops such vehicles offer invaluable mobile artillery support.

SPECIFICATIONS	
Country of origin:	Soviet Union
Crew:	3
Weight:	3300kg (7260lb)
Dimensions:	Length: 4.995m (16ft 4.7in); width: 2.086m (6ft 10in); height: 1.18m (3ft 10.5in)
Range:	250km (155 miles)
Armour:	6mm (.23in)
Armament:	1 x 57mm (2.24in) CH-51M gun; 1 x 7.62mm (.3in) anti-aircraft MG
Powerplant:	1 x M-20E four-cylinder petrol engine developing 41kW (55hp)
Performance:	Maximum road speed: 45km/h (28mph); vertical obstacle: .5m (20in); trench: 1.4m (4ft 7in)

M50 Ontos

The M50 Ontos was an air-portable tank destroyer. It mounted six RCL 106mm (4.17in) recoilless rifles, three mounted either side of a small central turret. Attached to the top four guns were 12.7mm (.5in) spotting rifles, firing tracer rounds to help targeting.

SPECIFICATIONS	
Country of origin:	USA
Crew:	3
Weight:	8640kg (19,051lb)
Dimensions:	Length: 3.82m (12ft 6in); width: 2.6m (8ft 6in); height: 2.13m (6ft 11in)
Range:	240km (150 miles)
Armour:	13mm (.51in) maximum
Armament:	6 x RCL 106mm (4.17in) recoilless rifles; 4 x 12.7mm (.5in) M8C spotting rifles
Powerplant:	1 x General Motors 302 petrol engine developing 108kW (145hp)
Performance:	Maximum road speed: 48km/h (30mph)

TIMELINE

1951  1952  1955

POSTWAR SELF-PROPELLED GUNS

MK 61

The MK 61 105mm self-propelled gun was the first indigenous postwar artillery for the French army, entering service in 1952. The 105mm (4.1in) gun was mounted in a fixed, non-traversing casemate positioned towards the rear of the tracked chassis.

SPECIFICATIONS	
Country of origin:	France
Crew:	5
Weight:	Approx 16,500kg (36,382lb)
Dimensions:	Length: 6.4m (30ft); width: 2.65m (8ft 3.3in); height: 2.7m (8ft 10.3in)
Range:	350km (217.5 miles)
Armour:	Up to 20mm (.78in)
Armament:	1 x 105mm (4.13in) gun; 2 x 7.5mm (2.95in) MGs
Powerplant:	1 x SOFAM 8Gxb, 8-cylinder petrol engine developing 186.4kW (250hp)
Performance:	Maximum road speed: 60km/h (37.3mph)

Jagdpanzer Kanone (JPK)

The Jagdpanzer Kanone (JPK) was a tank destroyer armed with a 90mm (3.54in) gun. A variant was the Jagdpanzer Rakete, armed with French SS.12 anti-tank missiles, which were later replaced by HOT (Haute subsonique Optiquement Téleguidé) missiles.

SPECIFICATIONS	
Country of origin:	Germany
Crew:	4
Weight:	Approx 25,700kg (55,669lb)
Dimensions:	Length: 6.238m (20ft 5.6in); width: 2.98m (9ft 9.3in); height: 2.085m (6ft 10in)
Range:	Road 400km (248.5 miles)
Armour:	Up to 50mm (1.97in)
Armament:	1 x 90mm (3.54in) gun; 2 x 7.62mm (.3in) MGs
Powerplant:	1 x Daimler-Benz MB837, 8-cylinder diesel engine developing 372.9kW (500hp)
Performance:	Maximum speed: road, 70km/h (43.5mph)

SPz lang HS30

The HS30 carried five soldiers and three crew inside an all-welded steel hull. It was not amphibious, and lacked an NBC option. But it carried a turret-mounted 20mm (.78in) Hispano HS820, and the option of ATGWs and M40A1 106mm (4.17in) recoilless rifles.

SPECIFICATIONS	
Country of origin:	West Germany
Crew:	3 + 5
Weight:	14,600kg (32,200lb)
Dimensions:	Length: 5.56m (18ft 3in); width: 2.25m (7ft 4in); height: 1.85m (6ft 1in)
Range:	270km (170 miles)
Armour:	Up to 30mm (1.18in)
Armament:	1 x 20mm (.78in) Hispano HS820 gun and other options
Powerplant:	1 x Rolls-Royce 8-cylinder petrol engine developing 175kW (235hp) at 3800rpm
Performance:	Maximum road speed: 51km/h (32mph); fording .7m (2ft 4in); gradient: 60 per cent; vertical obstacle: .6m (2ft); trench: 1.6m (5ft 4in)

 1958  1959

COLD WAR TANKS AND AFVS

Postwar Tracked APCs

During the Cold War era, armoured personnel carriers were increasingly employed as infantry fighting vehicles rather than just troop carriers. The conflicts of the late twentieth century, including Vietnam, proved that their usefulness was undiminished. Some tracked Soviet APCs resembled small tanks, which indeed they were.

M75

The design of the International Harvester M75 was flawed, relying on expensive tank components, especially the running gear, engine and transmission. The M75 featured a steel armoured box with a sloped glacis front big enough for two crew and 10 soldiers.

SPECIFICATIONS	
Country of origin:	USA
Crew:	2 + 10
Weight:	18,828kg (41,516lb)
Dimensions:	Length: 5.19m (17ft .36in); width 2.84m (9ft 3in); height 2.77m (9ft 1in)
Range:	185km (115 miles)
Armour:	15.9mm (.63in)
Armament:	1 x 12.7mm (.5in) Browning M2 HB HMG
Powerplant:	1 x Continental AO-895-4 6-cylinder petrol, developing 220kW (295hp) at 2660rpm
Performance:	Maximum road speed: 71km/h (44mph); fording: 1.22m (4ft); gradient: 60 per cent; vertical obstacle: .46m (1.5ft); trench: 1.68m (5.51ft)

M59

From 1954 to 1960, the M59 armoured personnel carrier replaced the M75 in US Army service. It was amphibious, had a lower profile and was cheaper to produce, using two smaller engines instead of one powerful unit. Some 6300 were built.

SPECIFICATIONS	
Country of origin:	United States
Crew:	2 + 10
Weight:	19,323kg (38,197lb)
Dimensions:	Length: 5.61m (18ft 5in); width: 3.26m (10ft 8in); height: 2.27m (7ft 5in)
Range:	164km (102 miles)
Armour:	16mm (0.63in)
Armament:	1 x 12.7mm (0.5in) Browning M2 HB HMG
Powerplant:	2 x General Motors Model 302 6-cylinder petrol engines, developing 95kW (127hp) at 3350rpm
Performance:	Maximum road speed: 51km/h (32mph); fording: amphibious; gradient: 60 percent; vertical obstacle: 0.46m (1ft 6in); trench: 1.68m (5ft 7in)

TIMELINE

1951

1954

SU 60

The SU-60 was Japan's first postwar tracked APC. One crew member sat behind the commander, operating the 12.7mm (.5in) roof-mounted machine gun. Variants include two mortar carriers, an NBC detection vehicle, an anti-tank vehicle and a bulldozer.

SPECIFICATIONS
Country of origin:	Japan
Crew:	4 + 6
Weight:	11,800kg (26,000lb)
Dimensions:	Length: 4.85m (15ft 11in); width: 2.4m (7ft 10in); height: 1.7m (5ft 6in)
Range:	300km (190 miles)
Armour:	Not available
Armament:	1 x 12.7mm (0.5in) Browning M2 HB HMG
Powerplant:	1 x Mitsubishi 8 HA 21 WT 8-cylinder diesel, developing 164kW (220hp) at 2400rpm
Performance:	Maximum road speed: 45km/h (28mph); fording: 1m (3ft 4in); gradient: 60 percent; vertical obstacle: 0.6m (2ft); trench: 1.82m (5ft 11in)

M113

With production exceeding 80,000 units, the M113 is the most widely used armoured fighting vehicle of all time. Introduced in 1960, it was developed to transport airborne troops in transport aircraft like the Lockheed C-130 Hercules.

SPECIFICATIONS
Country of origin:	Soviet Union
Crew:	2 + 11
Weight:	12,329 kg (27,180lb)
Dimensions:	Length: 2.686m (8ft 9in); width: 2.54m (8ft 4in); height: 2.52m (8ft 3in)
Range:	483km (300 miles)
Armour:	Not available
Armament:	1 x 12.7mm (.5in) MG
Powerplant:	1 x Detroit Diesel 6V53T developing 205kW (275hp)
Performance:	Maximum road speed: 66km/h (41 mph)

M113A1

The M113A1 with a GM 160kw (215hp) diesel engine replaced the M113 in production from 1964. With additional gun shields and roof-mounted weapons, the M113A1s in service with cavalry units in Vietnam were known as armoured cavalry assault vehicles (ACAVs).

SPECIFICATIONS
Country of origin:	USA
Crew:	2 + 11
Weight:	11,341kg (24,950lb)
Dimensions:	Length: 2.686m (8ft 9in); width: 2.54m (8ft 4in); height: 2.52m (8ft 3in)
Range:	483km (300 miles)
Armour:	Up to 44mm (1.73in)
Armament:	1 x 12.7mm (.5in) HMG; 2 x 7.62mm (.3in) MGs
Powerplant:	1 x 6-cylinder water-cooled diesel engine developing 160kW (215bhp)
Performance:	Maximum road speed: 67.59km/h (42mph); maximum water speed: 5.8km/h (3.6mph); fording: amphibious; vertical obstacle: 0.61m (2ft); trench: 1.68m (5ft 6in)

Postwar Rocket Systems

Rocket artillery, a concept with its roots in ancient China, came into its own during World War II and continued to be a vital component of an army's arsenal in the following years. Its accuracy continued to develop as new tracking and targeting systems came into being, and the Soviets continued to deploy it in large numbers.

BM-24

The BM-24 was mounted on a ZIL-151 6x6 truck, but later transferred to a ZIL-157. The system consisted of two rows of six tubular frame rails. Rocket types included high-explosive, smoke and chemical, a typical range being 11km (6.84 miles).

SPECIFICATIONS	
Country of origin:	Soviet Union
Crew:	6
Weight:	9200kg (20,240lb)
Dimensions:	Length: 6.7m (21ft 11in); width: 2.3m (7ft 6in); height: 2.91m (9ft 6in)
Range:	430km (269 miles)
Armour:	None
Armament:	12 x 240mm (9.4in) rocket-launcher tubes
Powerplant:	1 x six-cylinder water-cooled petrol engine developing 81kW (109hp)
Performance:	Maximum road speed: 65km/h (40.6mph); fording: .85m (2ft 9in);

Type 70

The Type 63 consisted of three rows of four barrels. To fire, the wheels were removed and the launcher supported by two legs at the front, a spade at the rear absorbing recoil. The type 70 (pictured) uses components of the YW 531C armoured personnel carrier.

SPECIFICATIONS	
Country of origin:	China
Crew:	2
Weight:	12,600kg (27,720lb)
Dimensions:	Length: 5.4m (17ft 9in); width: 3m (9ft 9in); height: 2.58m (8ft 6in)
Range:	500km (312 miles)
Armour:	10mm (.39in)
Armament:	12 x 107mm (4.2in) rocket-launcher tubes
Powerplant:	1 x V-8 diesel engine developing 238kW (320hp)
Performance:	Maximum road speed: 65km/h (38mph); fording: amphibious; vertical obstacle: .6m (1ft 11in); trench: 2m (6ft 6in)

TIMELINE

1953  1958  1963

BM-21

The BM-21 entered service in the early 1960s and became the standard multiple rocket-launcher of Warsaw Pact armies, as well as most Soviet client states. Variants were produced in China, India, Egypt and Romania. Most variants have been used in action.

SPECIFICATIONS
Country of origin:	Soviet Union
Crew:	6
Weight:	11,500kg (25,300lb)
Dimensions:	Length: 7.35m (42ft 1in); width: 2.69m (8t 9in); height: 2.85m (9ft 4in)
Range:	405km (253miles)
Armour:	None
Armament:	40 x 122mm (4.8in) rocket-launcher tubes
Powerplant:	1 x V-B water-cooled petrol engine developing 134kW (180hp)
Performance:	Maximum road speed: 75km/h (46.8mph); fording: 1.5m (4ft 11in); vertical obstacle: .65m (2ft 1in)

Walid

Based on the Walid 4x4 armoured personnel carrier, a full salvo of D-3000 rockets is able to produce a smoke screen 1000m (3280ft) long, lasting for up to 15 minutes, sufficient to cover most activities on the battlefield.

SPECIFICATIONS
Country of origin:	Egypt
Crew:	2
Weight:	Not available
Dimensions:	Length: 6.12m (20ft); width: 2.57m (8ft 5in); height: 2.3m (7ft 6in)
Range:	800km (500 miles)
Armour:	8mm (.31in)
Armament:	12 x 80mm (3.15in) rocket-launcher tubes
Powerplant:	1 x diesel engine developing 125kW (168hp)
Performance:	Maximum road speed: 86km/h (54mph); fording: .8m (2ft 7in); vertical obstacle: .5m (1ft 7in)

LARS II

The LARS II was mounted on the MAN 4x4 chassis. With 36 launcher tubes, the full complement could be launched in only 17.5 seconds and reloaded in around 15 minutes with a variety of different rounds, including parachute-retarded anti-tank mines.

SPECIFICATIONS
Country of origin:	Germany
Crew:	3
Weight:	17,480kg (38,537lb)
Dimensions:	Length: 8.28m (27ft 2in); width: 2.5m (8ft 2in); height: 2.99m (9ft 10in)
Range:	480km (300 miles)
Armour:	None
Armament:	36 x 110mm (4.3in) rocket tubes
Powerplant:	1 x V-8 liquid-cooled diesel engine developing 238kW (320hp)
Performance:	Maximum road speed: 100km/h (62.5mph); fording: 1.4m (94ft 7in); vertical obstacle: .6m (1ft 11in); trench: 1.6m (5ft 3in)

1967

1969

M113

With production exceeding 80,000 units, the M113 is the most widely used armoured fighting vehicle of all time. Introduced in 1960, it was developed to transport airborne troops in transport aircraft such as the Lockheed C-130 Hercules.

M113

TRACKS
The M113 uses its tracks to propel itself through water at about 5km/h (3.1mph). The rubber track shroud controls the flow of water over the tracks when swimming

ALUMINIUM ARMOUR
The M113's hull is made of aluminium, much lighter than steel but much less effective as armour. The hull is proof against small-arms fire and shell splinters

M113

Extra armour protection was added to the M113 to produce an ACAV (Armoured Cavalry) version, which was transformed into a true armoured fighting vehicle by the addition of shields for its machine guns. The M113 serves or has served with 50 countries.

Machine gun
The M125 retains the standard 12.7mm (.5in) machine gun armament of the M113 APC. This was essential in Vietnam, where convoys were frequently ambushed at very short range

SPECIFICATIONS	
Country of origin:	USA
Crew:	2 + 11
Weight:	11,343kg (25,007lb)
Dimensions:	Length: 2.52m (8ft 3in); width: 2.69m (8ft 10in); height (to hull top): 1.85m (6ft 1in)
Range:	480km (298 miles)
Armour:	Up to 45mm (1.77in)
Armament:	Various, but minimum usually 1 x 12.7mm (.5in) HMG
Powerplant:	1 x General Motors 6V53 6-cylinder diesel engine developing 158kW (212hp)
Performance:	Maximum road speed: 61km/h (38mph)

Splash plate
Lowered for amphibious operation, the splash plate was often dropped and piled full of sandbags to provide a high degree of protection against RPG-7 anti-tank rockets

Engine compartment
The first M113s had a petrol engine, which was a significant fire risk. From the M113A1 onwards, all models have been fitted with a GMC Detroit Diesel 6-cylinder water-cooled engine

SALADIN

The long-serving Saladin armoured car was popular with the armed forces of several countries, and proved to be an excellent vehicle for patrol duty on behalf of the United Nations. Its few variants included an amphibious model propelled by its wheels.

SALADIN

BROWNING .30 CAL MACHINE GUN
This was mounted as an anti-aircraft weapon only. To fire it, the commander had to stand up in the turret, making himself visible and vulnerable to small arms fire

COMMANDER
The commander saw through four forward-mounted periscopes and one swivelling periscope at the rear of the hatch

MAIN GUN
Up to 42 rounds of ammunition were carried for the main armament. It could fire HE, HESH, smoke, canister and illuminating shells and was able to knock out all armoured cars and many types of medium tank

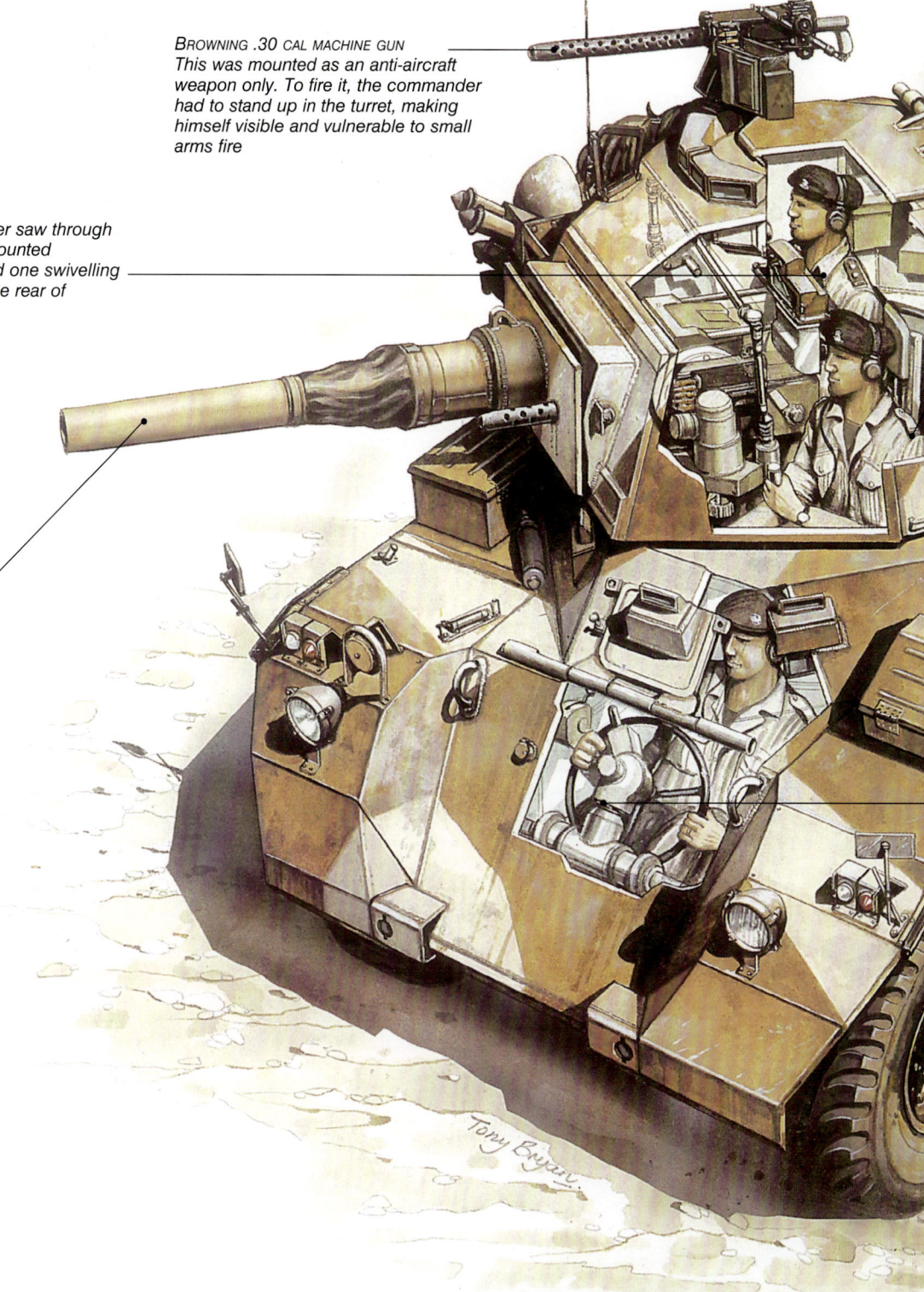

SALADIN

Production of the Saladin began in 1959 and ended in 1972, totalling 1177 vehicles. In addition to the British Army, the Saladin was used by Indonesia, Jordan, the Federal German Police, the Yemen and Lebanon. All six wheels of the Saladin were powered, with steering on the front four.

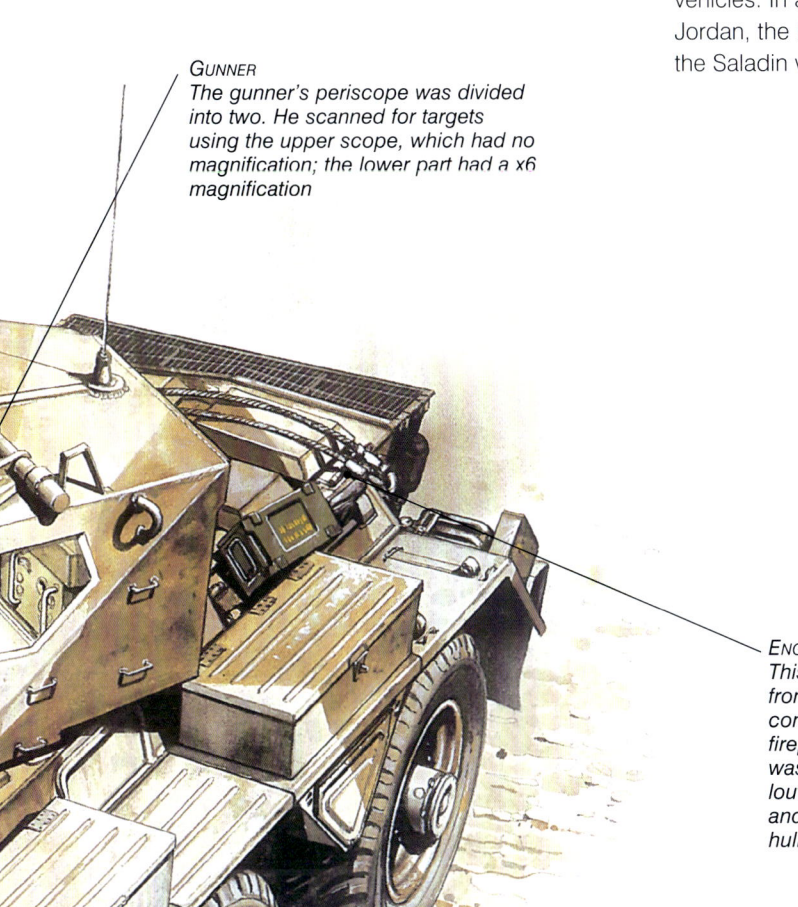

Gunner
The gunner's periscope was divided into two. He scanned for targets using the upper scope, which had no magnification; the lower part had a x6 magnification

Engine compartment
This was separated from the fighting compartment by a fireproof bulkhead. Air was drawn down via six louvred engine covers and blown out from the hull rear

Driver
With the hatch cover folded back, the driver could see directly out of the Saladin, as well as through his three periscopes

SPECIFICATIONS	
Country of origin:	United Kingdom
Crew:	3
Weight:	11,500kg (25,300lb)
Dimensions:	Length: 5.284m (17ft 4in); width: 2.54m (8ft 4in); height: 2.93m (9ft 7.3in)
Range:	400km (250 miles)
Armour:	8–32mm (.31–1.25in)
Armament:	1 x 76mm (3in) gun; 1 x 7.62mm (.3in) coaxial MG; 1 x 7.62mm (.3in) AA MG
Powerplant:	1 x 8-cylinder petrol engine developing 127kW (170hp)
Performance:	Maximum road speed: 72km/h (45mph); fording: 1.07m (3ft 6in); vertical obstacle: .46m (1ft 6in); trench: 1.52m (5ft)

Postwar AA Vehicles

With air power dominating the battlefield, it was logical that armies would continue to place great emphasis on the development of mobile anti-aircraft systems. In the early postwar years, the Soviet Union produced a formidable array of mobile AA weapons, forcing NATO air forces to develop new tactics.

M42

The M42 anti-aircraft system, or 'Duster', was based on the M41 Bulldog tank. Its petrol engine restricted its operating range, and it lacked a radar fire-control system, forcing the gunner to rely on optic sights. The open-topped turret afforded the crew little protection.

SPECIFICATIONS	
Country of origin:	USA
Crew:	6
Weight:	22,452kg (49,394lb)
Dimensions:	Length: 6.356m (20ft 10in); width: 3.225m (10ft 7in); height: 2.847m (9ft 4in)
Range:	161km (100 miles)
Armour:	12–38mm (.47–1.5in)
Armament:	2 x 40mm (1.57in) anti-aircraft guns; 1 x 7.62mm (.3in) MG
Powerplant:	1 x Continental AOS-895-3 six-cylinder air-cooled petrol engine developing 373kW (500hp)
Performance:	Maximum road speed: 72.4km/h (45mph); fording: 1.3m (4ft 3in); vertical obstacle: 1.711m (2ft 4in); trench: 1.829m (6ft)

M53/59

The M53/59 self-propelled anti-aircraft gun was a clear-weather system, lacking infrared night vision equipment and a nuclear, biological and chemical (NBC) defence system. Poor cross-country mobility prevented effective operation with tracked vehicles.

SPECIFICATIONS	
Country of origin:	Czechoslovakia
Crew:	6
Weight:	10,300kg (22,660lb)
Dimensions:	Length: 6.92m (22ft 8in); width: 2.35m (7ft 9in); height: 2.585m (8ft 6in)
Range:	500km (311 miles)
Armour:	None (vehicle as a whole)
Armament:	2 x 30mm (1.2IN) guns
Powerplant:	1 x Tatra T912-2 six-cylinder diesel engine developing 82kW (110hp)
Performance:	Maximum road speed: 60km/h (37mph); vertical obstacle: .46m (1ft 6in); trench: .69m (2ft 3in)

TIMELINE

1951 　　1953 　　1955

POSTWAR AA VEHICLES

ZSU-57-2

With a large, open-topped turret and a chassis that was a lightened version of the T-54 main battle tank, the ZSU-57-2 had a greater power-to-weight ratio than the T-54. Coupled with extra fuel tanks, this gave the gun good mobility and an effective operating range.

SPECIFICATIONS
Country of origin:	Soviet Union
Crew:	6
Weight:	28,100kg (61,820lb)
Dimensions:	Length: 8.48m (27ft 10in); width: 3.27m (10ft 9in); height: 2.75m (9ft)
Range:	420km (260 miles)
Armour:	15mm (.59in)
Armament:	2 x 57mm (2.24in) anti-aircraft guns
Powerplant:	1 x Model V-54 V-12 diesel engine developing 388kW (520hp)
Performance:	Maximum road speed: 50km/h (31mph); fording: 1.4m (4ft 7in); vertical obstacle: .8m (2ft 7in); trench: 2.7m (8ft 10in)

BRDM-1 SA-9 Gaskin

The BRDM-1 was the first of the BRDM series of amphibious armoured scout cars and entered service in 1959. The vehicle was fully amphibious and was propelled by a single rear-mounted water jet. The twin SA-9 missile launchers replaced the usual turret.

SPECIFICATIONS
Country of origin:	Soviet Union
Crew:	5
Weight:	5600kg (5.5 tons)
Dimensions:	Length: 5.7m (18ft 8in); width: 2.25m (7ft 3in); height: 1.9m (6ft 3in)
Range:	500km (311 miles)
Armour:	10mm (.394in)
Armament:	1 x 7.62mm (.3in) SGMB MG; twin SA-9 AA missile launchers
Powerplant:	1 x GAZ-40P 6-cylinder petrol engine 67.2kW (90bhp) at 3400rpm
Performance:	Maximum speed: 80km/h (50mph); 9km/h (5.6mph) in water

M727 HAWK

The HAWK ('Homing All the Way to the Kill') entered service in 1960, the launcher unit first being towed by a 2.5-ton 6x6 truck. A self-propelled version, the M727 SP HAWK, used a modified M548 tracked cargo carrier to support the three-missile launcher.

SPECIFICATIONS
Country of origin:	USA
Crew:	4
Weight:	12,925kg (28,494lb)
Dimensions:	Length: 5.87m (19ft 3in); width: 2.69m (8ft 9in); height: 2.5m (8ft 2in)
Range:	489km (304 miles)
Armour:	Not available
Armament:	3 x HAWK SAM
Powerplant:	1 x Detroit 6V53 6-cylinder diesel engine developing 160kW (214hp)
Performance:	Maximum road speed: 61km/h (38mph); fording: 1m (3.3ft); gradient: 60 per cent; vertical obstacle: .61m (2ft); trench: 1.68m (5.51ft)

 1959    1960

COLD WAR TANKS AND AFVS

Postwar Amphibious Vehicles

Many modern military vehicles, from light wheeled command and reconnaissance vehicles to armoured personnel carriers and tanks, are manufactured with amphibious capabilities. During the Cold War, the Soviet bloc states developed a wide range of amphibious APCs, fighting vehicles and tanks, both wheeled and tracked.

BAV-485

The chassis of the BAV-485 was based on the ZIL-151 6x6 truck. With a drop-down tailgate at the rear of the cargo compartment, it could carry 25 fully equipped troops or 2500kg (5500lb) of cargo. A single propeller at the rear of the hull offered amphibious propulsion.

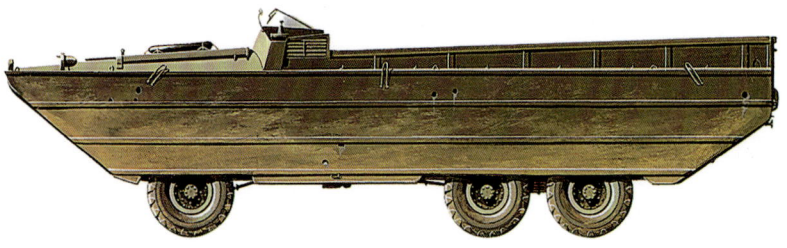

SPECIFICATIONS	
Country of origin:	Soviet Union
Crew:	2
Weight:	9650kg (21,278lb)
Dimensions:	Length: 9.54m (31ft 3in); width: 2.5m (8ft 2in); height: 2.66m (8ft 8in)
Range:	530km (330 miles)
Armour:	Not available
Armament:	1 x 12.7mm (.5in) DShKM MG (optional)
Powerplant:	1 x ZIL-123 6-cylinder petrol engine, developing 82kW (110hp)
Performance:	Maximum road speed: 60km/h (37mph); maximum water speed: 10km/h (6mph); gradient 60 per cent; vertical obstacle: .4m (1.3ft)

GAZ-46 MAV

Used almost entirely for reconnaissance, the GAZ-46 mounted no armament. It carried three troops in the rear, with the driver and commander in front. A fold-down windscreen protected against water spray. A trim vane had to be fitted before entering the water.

SPECIFICATIONS	
Country of origin:	Soviet Union
Crew:	1 + 4
Weight:	2480kg (5470lb)
Dimensions:	Length: 5.06m (16ft 7in); width: 1.74m (5ft 8in); height: 2.04m (8ft 6in)
Range:	500km (310 miles)
Armour:	Not available
Armament:	None
Powerplant:	1 x M-20 4-cylinder petrol engine developing 41kW (55hp)
Performance:	Maximum road speed: 90km/h (56mph); maximum water speed: 9km/h (5.5mph); gradient 60 per cent

TIMELINE

 1949    1952

POSTWAR AMPHIBIOUS VEHICLES

K-61

The K-61 was a popular amphibious transporter, able to carry 60 fully armed troops or 5000kg (11,000lb) of cargo on water. Two propellers at the rear of the hull propelled it at a speed of 10km/h (6mph). The K-61 was also used as an amphibious weapons platform.

SPECIFICATIONS	
Country of origin:	Soviet Union
Crew:	2
Weight:	14,550kg (32,083lb)
Dimensions:	Length: 9.15m (30ft); width: 3.15m (10ft 4in); height: 2.15m (7ft 6in)
Range:	260km (160 miles)
Armour:	Not available
Armament:	Various
Powerplant:	1 x YaAZ M204VKr 4-cylinder diesel engine developing 101kW (135hp)
Performance:	Maximum road speed: 36km/h (22mph); maximum water speed: 10km/h (6mph); gradient: 40 per cent; vertical obstacle: .65m (2.13ft)

Gillois PA

The Gillois PA ('Pont Amphibian') had an inflatable flotation chamber on each side. Each chamber had nine watertight sections to prevent flooding in case of partial damage. For bridging operations, the PA could transport an 8 x 4m (26.25 x 13.12ft) bridge slab.

SPECIFICATIONS	
Country of origin:	France
Crew:	4
Weight:	26,915kg (59,425lb)
Dimensions:	Length: 11.86m (38ft 11in); width: 3.2m (10ft 6in); height: 3.99m (13ft 1in)
Range:	780km (480 miles)
Armour:	Not applicable
Armament:	None
Powerplant:	1 x Deutz V12 diesel engine developing 164kW (220hp)
Performance:	Maximum road speed: 64km/h (40mph); maximum water speed 12km/h (7mph)

GT-S

The GT-S was the first vehicle able to move from deep snow to water without any preparation. Its tracks were wide – 300mm (11.81in) – and ground pressure was only .24kg/sq cm (3.4lb/sq in). Tough and reliable, it could carry 11 soldiers or 1000kg (2200lb) of cargo.

SPECIFICATIONS	
Country of origin:	Soviet Union
Crew:	1 + 11
Weight:	4600kg (10,100lb)
Dimensions:	Length: 4.93m (16ft 2in); width: 2.4m (7ft 10in); height: 1.96m (6ft 5in)
Range:	400km (250miles)
Armour:	Not available
Armament:	None
Powerplant:	1 x GAZ-61 6-cylinder petrol engine developing 63kW (85hp)
Performance:	Maximum road speed: 35km/h (22mph); maximum water speed: 4km/h (2.5mph); gradient: 60 per cent; vertical obstacle: .6m (2ft)

 1955

COLD WAR TANKS AND AFVS

T-62

A straightforward development of the Soviet T-55, the T-62 was innovative in that it mounted the world's first smoothbore tank gun, but it proved something of a disaster in combat, being poorly armoured and prone to catching fire.

T-62

U-5TS 115MM (4.53IN) SMOOTHBORE GUN
This has a shorter effective range than the NATO 105mm (4.1in) gun, but it is highly accurate up to 1500m (1640 yards)

GLACIS
This is protected by 102mm (4.01in) of armour sloped at 54° on the lower half and 60° on the upper. A shell on a flat trajectory thus has about 200mm (7.87in) of armour to punch its way through

T-62

Entering production in 1961, the T-62 was the leading Soviet battle tank by the late 1960s and formed the backbone of the Warsaw Pact armies deployed against Western Europe during the 1970s. During the Yom Kippur War, Israel captured several hundred T-62s from the Egyptians and Syrians and pressed them into service, adding thermal imaging equipment and laser rangefinders.

SPECIFICATIONS

Country of origin:	Soviet Union
Crew:	4
Weight:	39,912kg (87,808lb)
Dimensions:	Length: 9.34m (28ft 6in); width: 3.3m (10ft 1in); height: 2.4m (7ft 5in)
Range:	650km (40 miles)
Armour:	15–242mm (.59–9.52in)
Armament:	1 x 115mm (4.53in) U-5TS gun; 1 x 7.62mm (.3in) coaxial MG
Powerplant:	1 x V-55-5 V-12 liquid-cooled diesel engine developing 432kW (580hp)
Performance:	Maximum road speed: 60km/h (37.5mph); fording: 1.4m (4ft 7in); vertical obstacle: .8m (2ft 7.5in); trench: 2.7m (8ft 10.25in))

12.7mm (.5in) DshKM machine gun
The DshKM can be operated only manually, necessitating opening the hatch and venturing outside

External fuel tank
This jettisonable drum contains 400l of fuel. Extra tanks increase the T-62's range to 650km (404 miles)

Tracks
Made of very tough maganese steel, the track is hard wearing, but T-62s tend to shed their track if turned too quickly or if slammed suddenly into reverse

M60 Main Battle Tank

The M60 main battle tank was developed to counter the threat from the Soviet Union's new T-62 medium tank. The M60 underwent various upgrades during its operational life, the first in 1963, when the M60A1 appeared with a larger and better-designed turret, improved armour and more efficient shock-absorbers.

M60A2

To counter the Soviet T-62 medium tank, the M48 received a more powerful engine and the British 105mm (4.1in) L7. Designated the M60, it underwent various upgrades. The M60A2 featured a redesigned low-profile turret with a commander's machine-gun cupola on top, giving the commander a good view and field of fire while remaining protected. It was also armed with a 152mm (5.9in) calibre main gun. The M60A3 incorporated a new rangefinder and ballistic computer and a turret stabilization system.

SPECIFICATIONS	
Country of origin:	USA
Crew:	4
Weight:	52,617kg (51.8 tons)
Dimensions:	Length (over gun): 9.44m (31ft); width: 3.63m (11ft 11in); height: 3.27m (10ft 8in)
Range:	500km (311 miles)
Armour:	Up to 143mm (5.63in)
Armament:	1 x 105mm (4.13in) M68 gun; 1 x 7.62mm (.3in) MG; 1 x 12.7mm (.5in) HMG
Powerplant:	1 x Continental AVDS-1790-2A V-12 turbo-charged diesel developing 559.7kW (750hp)
Performance:	Maximum speed: 48km/h (30mph)

Turret
The M60A2, featured redesigned low-profile turret with a commander's machine-gun cupola on top, giving the commander a good view and field of fire while remaining protected

Engine
The M60 incorporated a Continental V-12 559.7kW (750hp) air-cooled, twin-turbocharged diesel engine. Power was transmitted to a final drive through a cross drive transmission, a combined transmission, differential, steering, and braking unit

M60

Armament
The M60A2 was also armed with a 152mm (5.9in) calibre main gun similar to that of the M551 Sheridan, which was able to fire the Shillelagh gun-launched anti-tank missile as well as normal rounds

Equipment
The M60A3 version incorporated a number of technological advances such as a new rangefinder and ballistic computer and a turret stabilization system. All American M60s were upgraded to this standard

Armour
The M60 was the first and last US main battle tank to utilize homogeneous steel armour for protection. It was also the last to feature either the M60 machine gun or an escape hatch under the hull

Main Battle Tanks

Although the medium tank dominated the armoured scene, the main battle tank had an important role to play throughout the Cold War, and technological advancements enhanced the survivability of these AFVs. For the first time, main battle tanks could fire with great accuracy while moving at high speed.

T-62

The T-62 had an unusual integral shell case ejection system. The recoil of the gun ejected the case out of a trapdoor in the turret, saving space but reducing the overall rate of fire. The tank could create a smoke screen by injecting diesel into the exhaust.

SPECIFICATIONS	
Country of origin:	Soviet Union
Crew:	4
Weight:	39,912kg (87,808lb)
Dimensions:	Length: 9.34m (28ft 6in); width: 3.3m (10ft 1in); height: 2.4m (7ft 5in)
Range:	650km (406 miles)
Armour:	15–242mm (.59–9.52in)
Armament:	1 x 115mm (4.5in) U-5TS gun; 1 x 7.62mm (.3in) coaxial machine gun
Powerplant:	1 x V-55-5 V-12 liquid-cooled diesel, developing 432kW (580hp)
Performance:	Maximum road speed: 60km/h (37.5mph); fording: 1.4m (4ft 7in); vertical obstacle: .8m (2ft 7.5in); trench: 2.7m (8ft 10.25in)

Pz 61 and Pz 68 Main Battle Tanks

The Pz 61 was armed with the British 105mm (4.1in) L7 gun. Further development resulted in the Pz 68, which had a stabilization system for the gun and wider tracks with replaceable rubber pads and a greater length of track in contact with the ground.

SPECIFICATIONS	
Country of origin:	Germany
Crew:	4
Weight:	39,700kg (87,538lb)
Dimensions:	Length: 6.88m (22ft 7in); width: 3.14m (10ft 4in); height: 2.75m (9ft)
Range:	350km (215 miles)
Armour:	120mm (4.72in)
Armament:	1 x 105mm (4.13in) gun; 2 x MGs
Powerplant:	1 x MTU MB-837 V8 diesel engine developing 492.5kW (660bhp) at 2200rpm
Performance:	Maximum speed: 55km/h (34mph)

TIMELINE

1960  1961

MAIN BATTLE TANKS

M60A1

In 1963, the M60 was upgraded to the M60A1. This new variant featured a larger, better-shaped turret and improvements to the armour protection and shock absorbers. The M60A1 was also equipped with a stabilization system for the main gun.

SPECIFICATIONS	
Country of Origin:	U.S.A
Crew:	4
Weight:	52,617kg (116,020lb)
Dimensions:	Length: 9.44m (31ft); width: 3.63m (11ft 11in); height: 3.27m (10ft 8in)
Range:	500km (311 miles)
Armour:	143mm (5.63in)
Armament:	1 X M68 105mm (4.13in) gun; 1 X 7.62mm (0.3in) MG; 1 X 12.7mm (0.50in) MG
Powerplant:	1 x Continental AVDS-1790-2A V12 turbo-charged diesel engine developing 559.7kW (750hp)
Performance:	48k/h (30mph)

Vijayanta (Vickers Mark 1 and Mark 3)

The Vickers main battle tank was developed for export. It used the standard 105mm (4.1in) L7 rifled tank gun and the automotive components of the Chieftain. The tank was also manufactured in India under the name Vijayanta (Victorious).

SPECIFICATIONS	
Country of origin:	India/ United Kingdom
Crew:	4
Weight:	38,600kg (83,775lb)
Dimensions:	Length: 7.92m (26ft); width: 3.168m (10ft 6in); height: 2.44m (8ft)
Range:	480km (300 miles)
Armour:	80mm (3.15in)
Armament:	1 x 105mm (4.1in) gun; 1 x 12.7mm (0.5in) MG; 2 x 7.62mm (0.3in) MGs
Powerplant:	1 x Leyland L60 6-cylinder multi-fuel developing 484.7kW (650bhp) at 2670rpm
Performance:	Maximum speed: 48km/h (30mph)

T-64 Main Battle Tank

The T-64 Main Battle Tank was used only by Soviet forces. The only variant of the T-64A was the T-64AK command vehicle, with more radio equipment and a telescopic mast. The T-64B fired the Kobra (AT-8 Songster) gun-launched anti-tank missile.

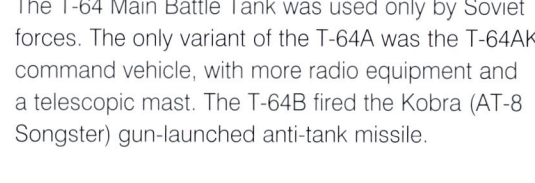

SPECIFICATIONS	
Country of origin:	Soviet Union
Crew:	3
Weight:	42,000kg (92,610lb)
Dimensions:	Length (hull): 7.4m (24ft 3in); width: 3.64m (11ft 11in); height: 2.2m (7ft 3in)
Range:	400km (248 miles)
Armour:	Up to 200mm (7.87in)
Armament:	1 x 125mm (4.92in) D-81TM (2A46 Rapira 3) smoothbore gun; 1 x co-axial 7.62mm (.3in) PKT MG; 1 x 12.7mm (.5in) NSVT AA MG
Powerplant:	1 x 5DTF 5-cylinder opposed diesel engine developing 560kW (750bhp)
Performance:	Maximum speed: 75km/h (47mph)

1964

1966

AMX-30

Until the mid-1950s, both France and Germany relied on American M47s for their armour, though France also had a number of the excellent German Panther tanks. A requirement was drawn up for a new main battle tank, lighter and more powerfully armed than the M47, to supply both countries.

AMX-30

Both Germany and France adopted their own design. The French produced the AMX-30, the first production tanks appearing in 1966, half of which were destined for export. The AMX-30 chassis has been used for a number of other vehicles, including the Pluton tactical nuclear missile launcher, as well as a self-propelled anti-aircraft gun, a recovery vehicle, bridge-layer and engineer vehicles. The tank has also seen service with the Iraqi, Saudi Arabian and Spanish armies.

Armament
The coaxial 20mm (.78in) gun is unusual in that it can be elevated independently of the main armament, enabling it to be used against low-flying aircraft

Fire-control system
The AMX-30 has an integrated fire-control system incorporating a laser rangefinder and low-light TV system

SPECIFICATIONS	
Country of origin:	France
Crew:	4
Weight:	35,941kg (79,072lb)
Dimensions:	Length: 9.48m (31ft 1in); width: 3.1m (10ft 2in); height: 2.86m (9ft 4in)
Range:	600km (373 miles)
Armour:	15–80mm (.6–3.1in)
Armament:	1 x 105mm (4.1in) gun; 1 x 20mm (.78in) gun; 1 x 7.62mm (.3in) MG
Powerplant:	1 x Hispano-Suiza 12-cylinder diesel, developing 537kW (720hp)
Performance:	Maximum road speed: 65km/h (40mph); vertical obstacle: .93m (3ft 0.6in); trench: 2.9m (9ft 6in)

AMX-30

INACCURATE FIRE
The AMX-30 is one of the few Western MBTs that is not fitted with a two-axis stabilization system for its main armament, and therefore cannot fire with any accuracy while on the move

TURRET
The AMX-30 has a welded hull of rolled steel plate, while the three-man turret is of cast construction

ANTI-TANK ROUND
The primary anti-tank round is the OCC (HEAT) type, which has a muzzle velocity of 1000m (3280ft) per second

AMMUNITION
The AMX-30 carries 47 105mm (4.1in) rounds, as well as 1050 20mm (.78in) and 2050 7.62mm (.3in) rounds

Leopard 1

Germany's design for the intended 1960s Franco-German joint venture was the Leopard 1 and this was eventually adopted independently of the French. Built by Krauss-Maffei, the first vehicles appeared in 1965 and production continued until 1979, a total of 2237 being built for the German Army and many more for export.

Leopard 1

There were four basic versions of the Leopard, differing in armour, turret type and fire-control systems. The tank formed the basis of a complete family of vehicles designed to support the vehicle on the battlefield. Optional equipment for the tank included a snorkel for deep-wading and a hydraulic blade to be attached to the front. The Leopard 1 is one of the best tank designs to have come out of Europe.

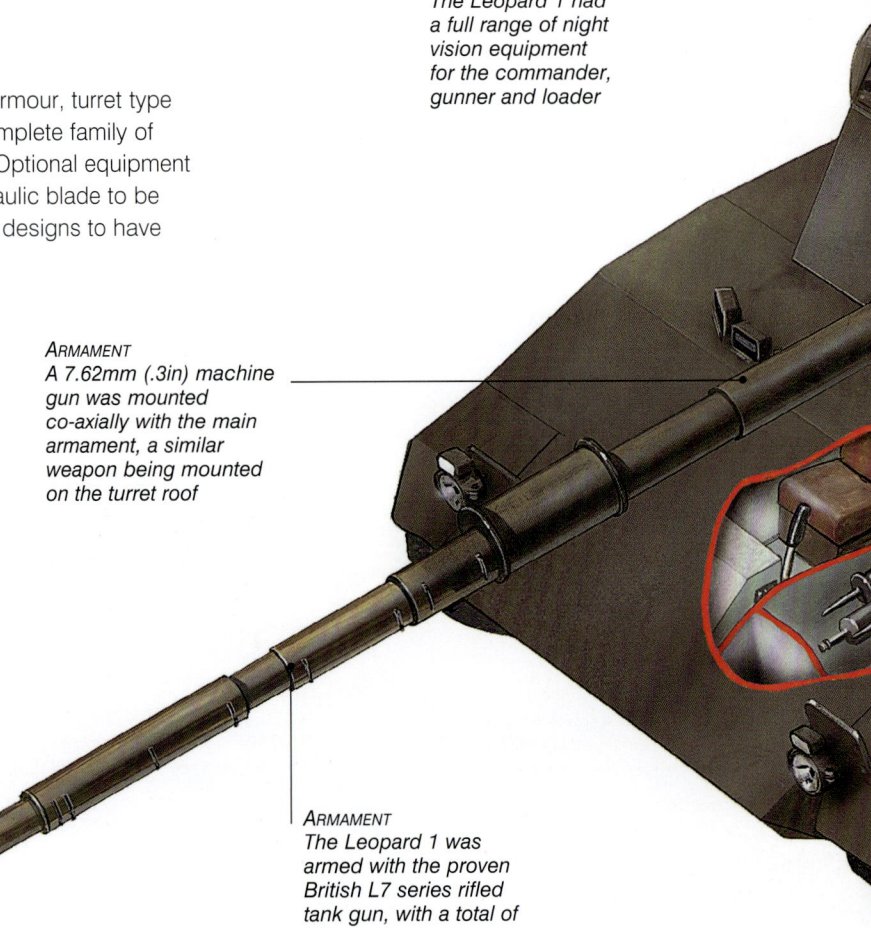

NIGHT VISION
The Leopard 1 had a full range of night vision equipment for the commander, gunner and loader

ARMAMENT
A 7.62mm (.3in) machine gun was mounted co-axially with the main armament, a similar weapon being mounted on the turret roof

ARMAMENT
The Leopard 1 was armed with the proven British L7 series rifled tank gun, with a total of 60 rounds

SPECIFICATIONS	
Country of origin:	West Germany
Crew:	4
Weight:	39,912kg (87,808lb)
Dimensions:	Length: 9.543m (31ft 4in); width: 3.25m (10ft 8in); height: 2.613m (8ft 7in)
Range:	600km (373 miles)
Armour:	Classified
Armament:	1 x 105mm (4.1in) gun; 1 x 7.62mm (.3in) coaxial MG; 7.62mm (.3in) anti-aircraft MG; 4 x smoke dischargers
Powerplant:	1 x MTU 10-cylinder diesel engine developing 619kW (830hp)
Performance:	Maximum road speed: 65km/h (40.4mph); fording: 2.25m (7ft 4in); vertical obstacle: 1.15m (3ft 9.25in); trench: 3m (9ft 10in)

Leopard 1 A3

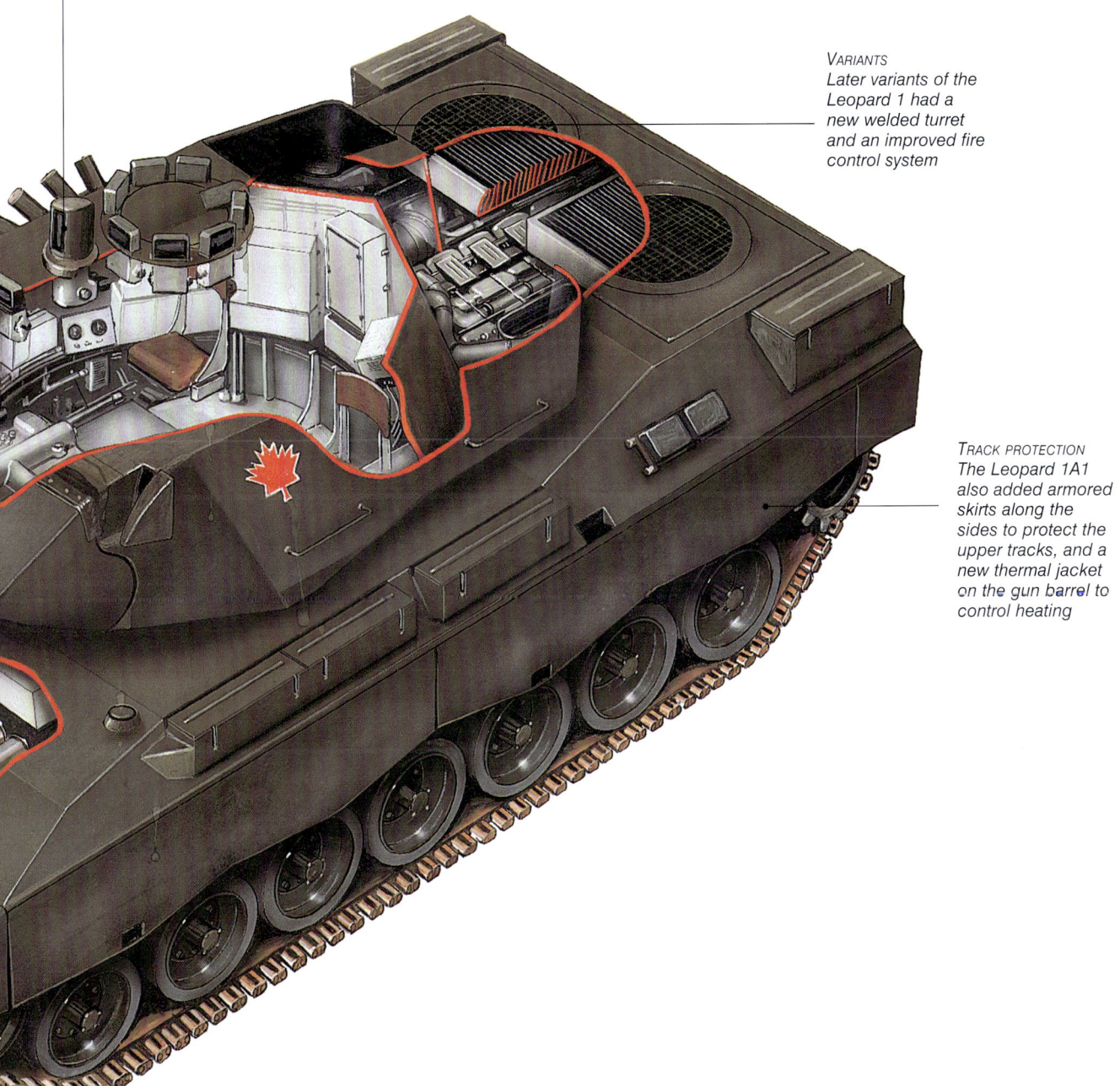

TURRET
The production Leopard 1 featured a new cast turret and several hull changes to raise the rear deck in order to make a roomier engine compartment. An optical range-finder system was also added

VARIANTS
Later variants of the Leopard 1 had a new welded turret and an improved fire control system

TRACK PROTECTION
The Leopard 1A1 also added armored skirts along the sides to protect the upper tracks, and a new thermal jacket on the gun barrel to control heating

Chieftain

The Chieftain was designed in the late 1950s to succeed the Centurion. Over 900 entered service with the British Army, with many being sold to Kuwait and Iran (they saw service in the Iran-Iraq war). Until the Leopard 2 entered German service in 1980, the Chieftain was the best armed and armoured main battle tank in the world.

Chieftain Mk 5

Until the late 1980s, the Chieftain remained the mainstay of British armoured forces on NATO's frontline in Germany, with frequent technological additions such as laser rangefinders and thermal-imaging devices. It included a bridge-layer, engineer tank and recovery vehicle amongst its variants. It was gradually replaced by the Challenger, and the Chieftains in British Army service are now in reserve.

NBC PACK
The NBC pack was mounted on the turret bustle, and a fire detection and extinguishing system was installed in the engine compartment

ARMAMENT
Until the introduction of the West German Leopard 2, the Chieftain with its 120mm (4.7in) rifled gun was the most powerful and well armoured of all NATO's main battle tanks

FIRE CONTROL SYSTEM
Chieftains in British service were retrofitted with the Improved Fire Control System, used in conjunction with a laser rangefinder

SPECIFICATIONS	
Country of origin:	United Kingdom
Crew:	4
Weight:	54,880kg (120,736lb)
Dimensions:	Length: 10.795m (35ft 5in); width: 3.657m (11ft 8.5in); height: 2.895m (9ft 6in)
Range:	500km (310 miles)
Armour:	Classified
Armament:	1 x 120mm (4.7in) rifled gun; 1 x 7.62mm (.3in) coaxial machine gun; 12 x smoke dischargers
Powerplant:	1 x Leyland 6-cylinder multi-fuel engine developing 560kW (750hp)
Performance:	Maximum road speed: 48km/h (30mph); fording: 1.066m (3ft 6in); vertical obstacle: .914m (3ft); trench: 3.149m (10ft 4in)

Chieftain Mk 5

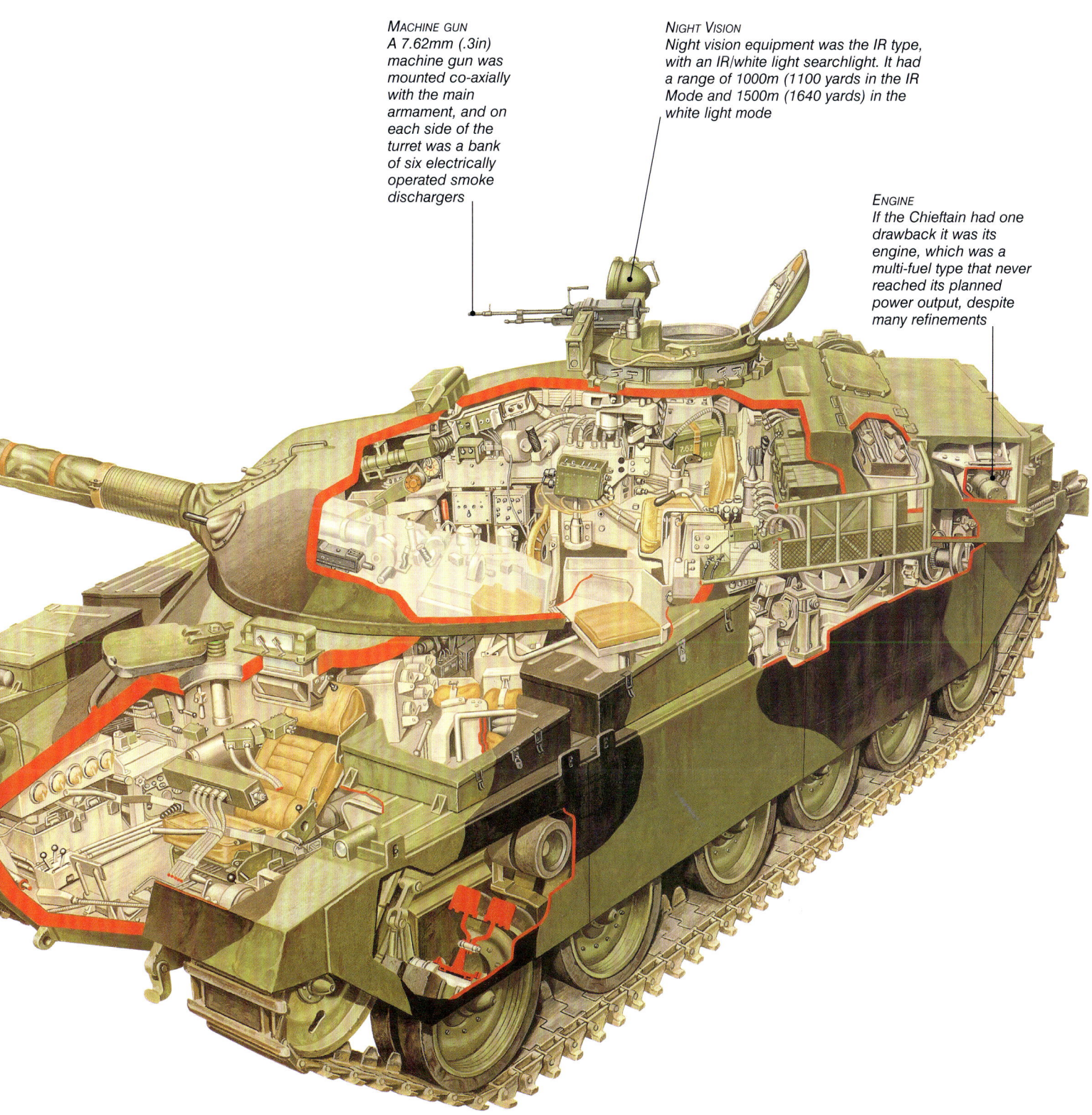

MACHINE GUN
A 7.62mm (.3in) machine gun was mounted co-axially with the main armament, and on each side of the turret was a bank of six electrically operated smoke dischargers

NIGHT VISION
Night vision equipment was the IR type, with an IR/white light searchlight. It had a range of 1000m (1100 yards in the IR Mode and 1500m (1640 yards) in the white light mode

ENGINE
If the Chieftain had one drawback it was its engine, which was a multi-fuel type that never reached its planned power output, despite many refinements

COLD WAR TANKS AND AFVS

1970s Main Battle Tanks

Several new main battle tanks made their appearance in the 1970s, some of which were modifications of existing designs. Israel, in particular, made extensive use of main battle tanks during this period, principally in the Yom Kippur War of October 1973, as did India and Pakistan in their conflict of 1971.

Sho't (Israeli Centurion)

So good a design was the Centurion that it continued to serve in some numbers into the 1990s. Armed with the 105mm (4.1in) L7 gun, it entered service with the Israeli Army in 1970.

SPECIFICATIONS	
Country of origin:	United Kingdom
Crew:	4
Weight:	51,723kg (113,792lb)
Dimensions:	Length: 9.854m (32ft 4in); width: 3.39m (11ft 1.5in); height: 3.009m (9ft 10.5in)
Range:	205km (127 miles)
Armour:	51–152mm (2–6in)
Armament:	1 x 105mm (4.1in) gun; 2 x 7.62mm (.3in) MGs; 1 x 12.7mm (.5in) HMG
Powerplant:	1 x Rolls-Royce Meteor Mk IVB V-12 petrol engine developing 485kW (650hp)
Performance:	Maximum road speed: 43km/h (27mph); fording: 1.45m (4ft 9in); vertical obstacle: 0.91m (3ft); trench: 3.352m (11ft)

Type 74

The Type 74 main battle tank has an unusual cross-linked hydro-pneumatic suspension system. This allows it to raise or lower parts of the chassis to cross diffcult terrain or engage targets outside the gun's normal elevation/depression range.

SPECIFICATIONS	
Country of origin:	Japan
Crew:	3
Weight:	38,000kg (83,600lb)
Dimensions:	Length: 9.42m (28ft 8in); width: 3.2m (9ft 10in); height: 2.48m (7ft 7in)
Range:	300km (188 miles)
Armour:	Classified
Armament:	1 x 105mm (4.1in) gun; 1 x 7.62mm (.3in) coaxial MG; 1 x 12.7mm (.5in) anti-aircraft MG; 2 x smoke dischargers
Powerplant:	1 x 10ZF V-10 liquid-cooled diesel engine developing 536kW (720hp)
Performance:	Maximum road speed: 55km/h (34.4mph); fording: 1m (3ft 4in); vertical obstacle: 1m (3ft 4in); trench: 2.7m (8ft 10in)

TIMELINE

1970    1975    1978

OF 40

1970S MAIN BATTLE TANKS

Designed exclusively for export, the OF 40 benefitted from experience building the German Leopard 1 tank. The commander has eight periscopes for all-round observation and the gunner has one periscope and a roof-mounted optical sight.

SPECIFICATIONS	
Country of origin:	Italy
Crew:	4
Weight:	45,500kg (100,327lb)
Dimensions:	Length: 6.893m (22ft 7in); width: 3.35m (11ft); height: 2.76m (9ft 1in)
Range:	600km (373 miles)
Armour:	Not available
Armament:	1 x 105mm (4.1in) gun, two 7.62mm (.3in) MGs, 4 x smoke-dischargers
Powerplant:	1 x MTU 90° diesel engine developing 620kW (831.5bhp)
Performance:	Maximum speed 60km/h (37.3mph)

Olifant Mk 1A

The South Africans upgraded the Centurion tank to suit their needs, improving firepower and mobility. The tank was fitted with a hand-held laser rangefinder for the commander and image-intensifier for the gunner. Variants include an armoured recovery vehicle.

SPECIFICATIONS	
Country of origin:	South Africa
Crew:	4
Weight:	56,000kg (123,200lb)
Dimensions:	Length: 9.83m (30ft); width: 3.38m (10ft 4in); height: 2.94m (8ft 11in)
Range:	500km (313 miles)
Armour:	17–118mm (.66–4.6in)
Armament:	1 x 105mm (4.1in) gun; 1 x 7.62mm (.3in) coaxial MG; 1 x 7.62mm (.3in) anti-aircraft MG; 2 x 4 smoke dischargers
Powerplant:	1 x V-12 air-cooled turbocharged diesel engine developing 559kW (750hp)
Performance:	Maximum road speed: 45km/h (28.1mph); fording: 1.45m (4ft 9in); vertical obstacle: .9m (2ft 11in); trench: 3.352m (11ft)

Khalid

The Khalid was a late-production Chieftain, with a modified fire control system and powerpack. It was such an important advance over the Chieftain that the British government ordered further development, and it later emerged as the Challenger 1.

SPECIFICATIONS	
Country of origin:	United Kingdom
Crew:	4
Weight:	58,000kg (127,890lb)
Dimensions:	Length: 6.39m (21ft 10in); width: 3.42m (11ft 7in); height: 2.435m (9ft 11in)
Range:	(estimated) 400km (248.5 miles)
Armour:	Not available
Armament:	1 x 120mm (4.72in) L11A5 gun; 2 x 7.62mm (.3in) MGs
Powerplant:	1 x Perkins Engines (Shrewsbury) Condor V-12 1200 12-cylinder water-cooled diesel engine developing 894.8kW (1200bhp) at 2300rpm
Performance:	Maximum speed: 48km/h (30mph)

1980  1981

T-72

The T-72 came into production in 1971. Smaller and faster than such tanks as the Chieftain, the T-72 was poorly armoured with less versatility and effective firepower than its competitors. This became brutally clear in 1982, when Syrian T-72s proved no match for Israeli Merkava tanks and were knocked out in droves.

T-72

The T-72 was designed for a conscript army and thus is easy to operate and maintain. This accounts for its export success, being transferred to 14 other countries. It is quite versatile, though, and within a matter of minutes it can be equipped for deep-fording, unlike most other tanks. It is also nuclear, biological and chemical (NBC) protected. Variants include a command vehicle, an anti-tank Cobra missile launcher and an armoured recovery vehicle.

SPECIFICATIONS	
Country of origin:	Soviet Union
Crew:	3
Weight:	38,894kg (85,568lb)
Dimensions:	Length: 9.24m (30ft 4in); width: 4.75m (15ft 7in); height: 2.37m (7ft 9in)
Range:	550km (434 miles)
Armour:	Classified
Armament:	1 x 125mm (4.9in) gun; 1 x 7.62mm (.3in) coaxial MG; 1 x 12.7mm (.5in) anti-aircraft HMG
Powerplant:	1 x V-46 V-12 diesel engine developing 626kW (840hp)
Performance:	Maximum road speed: 80km/h (50mph); fording: 1.4m (4ft 7in); vertical obstacle: .85m (2ft 9in); trench: 2.8m (9ft 2in)

ARMAMENT
The main armament is a 125mm (4.9in) D-81 smoothbore gun with an automatic loader. The tank carries 45 rounds of main gun ammunition

WEIGHT
At 41 tonnes, the T-72 is extremely lightweight, and very small compared to Western main battle tanks. Its light weight enables it to traverse bridges that would be denied to other MBTs

T-72 'Ural'

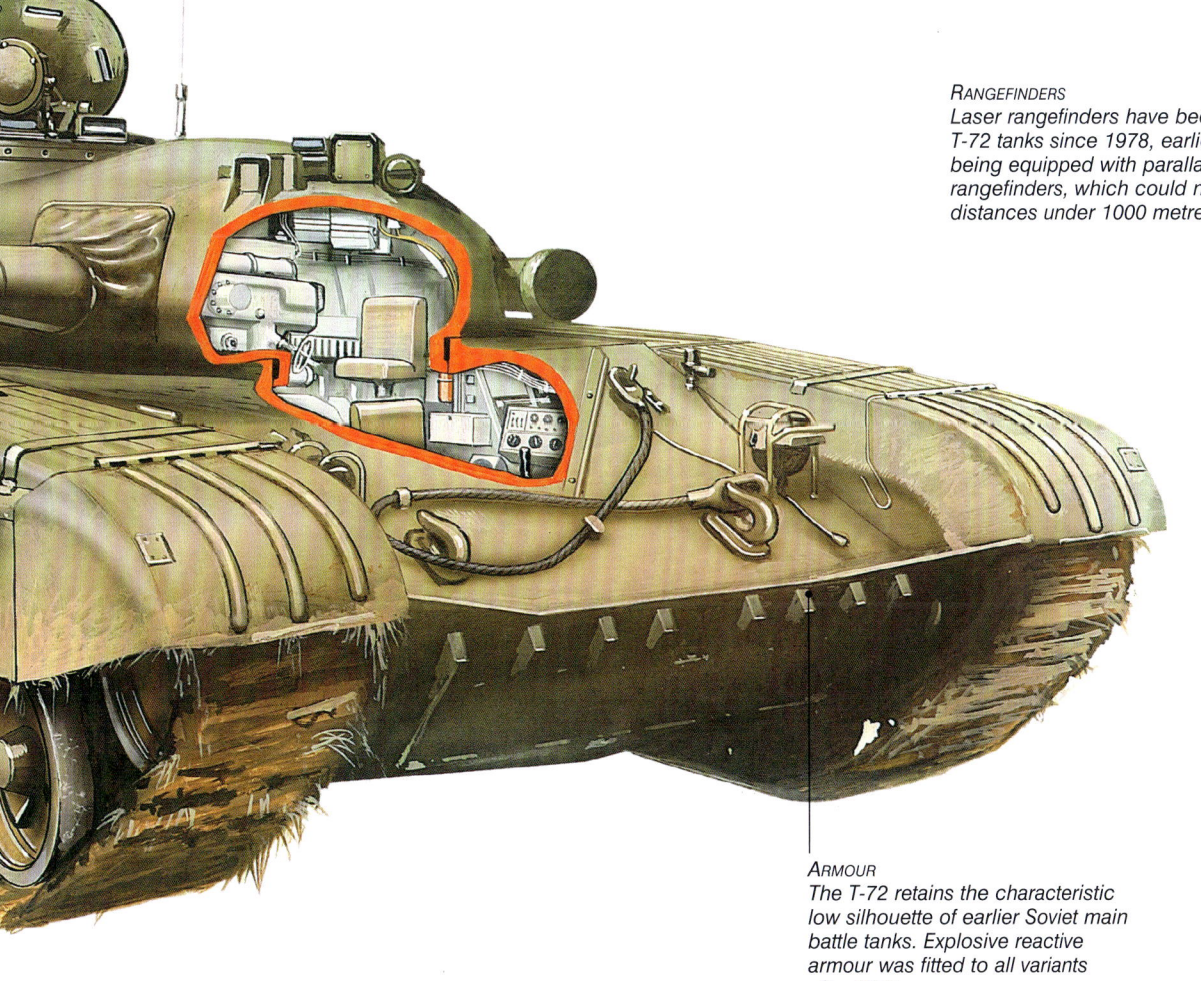

NBC PROTECTION
The T-72 has a comprehensive nuclear, biological and chemical (NBC) protection system. The inside of both hull and turret is lined with a synthetic fabric made of a boron compound, meant to reduce the penetrating radiation from neutron bomb explosions

VARIANTS
Later variants can fire the 9K119 Refleks (NATO codename AT-11 Sniper) anti-armour missile, which can also engage low-flying aircraft

RANGEFINDERS
Laser rangefinders have been fitted in T-72 tanks since 1978, earlier examples being equipped with parallax optical rangefinders, which could not be used for distances under 1000 metres (1100 yd)

ARMOUR
The T-72 retains the characteristic low silhouette of earlier Soviet main battle tanks. Explosive reactive armour was fitted to all variants after 1988

T-72 Variants

The T-72 gave rise to numerous variants. The original version was the T-72A, which had a laser rangefinder and improved armour, while the T-72B had additional front turret armour. The T-72BM was the first to incorporate Kontakt-5 explosive reactive armour; the T-72S is an export version, and the T-72BK is a command vehicle.

T-72G

The T-72G is the Polish-manufactured version of the T-72, with thinner armour than its Russian equivalent. Many parts and tools are not interchangeable between the Russian, Polish and Czechoslovakian versions, which caused logistical problems.

SPECIFICATIONS	
Country of origin:	Poland
Crew:	3
Weight:	46,500kg (102,515lb)
Dimensions:	Length: 9.53m (31ft 4in); width: 3.59m (11ft 10in); height: 2.29m (7ft 4in)
Range:	550km (342 miles)
Armour:	Classified
Armament:	1 x 125mm (4.9in) 2A46M smoothbore gun; 1 x AT-11 'Sniper' ATGW; 1 x 12.7mm (.5in) NSVT MG; 1 x 7.62 (.3in) PKT coaxial MG
Powerplant:	1 x V12 multifuel (V-84) diesel developing 626kW (840hp)
Performance:	Maximum road speed: 60km/h (37mph)

T-72M

The T-72M is an export version of the T-72, intended for the armies of the former Warsaw Pact. It has thinner armour and downgraded weapon systems, and replaces the right-side coincident rangefinder with a TPDK-1 LRF mounted on the centre line.

SPECIFICATIONS	
Country of origin:	Soviet Union
Crew:	3
Weight:	46,500kg (102,515lb)
Dimensions:	Length: 9.53m (31ft 4in); width: 3.59m (11ft 10in); height: 2.29m (7ft 4in)
Range:	550km (342 miles)
Armour:	Classified
Armament:	1 x 125mm (4.9in) 2A46M smoothbore gun; 1 x AT-11 'Sniper' ATGW;1 x 12.7mm (.5in) NSVT MG; 1 x 7.62 (.3in) PKT coaxial MG
Powerplant:	1 x V12 multifuel (V-84) diesel developing 626kW (840hp)
Performance:	Maximum road speed: 60km/h (37mph)

TIMELINE

 1979    1985    1986

T-72 VARIANTS

T-72S

The T-72S is an export version of the T-72BM, which has explosive reactive armour similar to that fitted to the later T-90. A number of upgrade packages are available for this vehicle.

SPECIFICATIONS	
Country of origin:	Soviet Union/Russia
Crew:	3
Weight:	46,500kg (102,515lb)
Dimensions:	Length: 9.53m (31ft 4in); width: 3.59m (11ft 10in); height: 2.29m (7ft 4in)
Range:	550km (342 miles)
Armour:	Classified
Armament:	1 x 125mm (4.9in) 2A46M smoothbore gun; 1 x AT-11 'Sniper' ATGW
Powerplant:	1 x V12 multifuel (V-84) diesel engine developing 626kW (840hp)
Performance:	Maximum road speed: 60km/h (37mph)

T-72CZ M4

The T-72CZ is an upgraded version of the Czech Republic's T-72M1, featuring Western fire control systems and an Israeli powerpack incorporating a Perkins diesel engine and Allison transmission.

SPECIFICATIONS	
Country of origin:	Czech Republic
Crew:	3
Weight:	46,500kg (102,515lb)
Dimensions:	Length: 9.53m (31ft 4in); width: 3.59m (11ft 10in); height: 2.29m (7ft 4in)
Range:	550km (342 miles)
Armour:	Classified
Armament:	1 x 125mm (4.9in) 2A46M smoothbore gun; 1 x AT-11 'Sniper' ATGW; 1 x 12.7mm (.5in) NSVT MG; 1 x 7.62 (.3in) PKT coaxial MG
Powerplant:	Perkins Condor V-12 diesel engine developing 746kW (1000hp)
Performance:	Maximum road speed: 60km/h (37mph)

ZTS T-72

In 1979–80, the government of what was then Czechoslovakia began licence production of the T-72, with ZTS Martin undertaking production of the complete chassis and ZTS Dubnica' nad Vahom responsible for the complete T-72 turret.

SPECIFICATIONS	
Country of origin:	Slovakia
Crew:	3
Weight:	46,500kg (102,515lb)
Dimensions:	Length: 9.53m (31ft 4in); width: 3.59m (11ft 10in); height: 2.27m (7ft 4in)
Range:	550km (342 miles)
Armour:	Classified
Armament:	1 x 125mm (4.9in) 2A46M smoothbore gun; 1 x AT-11 'Sniper' ATGW; 1 x 12.7mm (.5in) NSVT MG; 1 x 7.62 (.3in) PKT coaxial MG
Powerplant:	1 x Model V-84MS 4-stroke, 12-cylinder multifuel diesel engine developing 626kW (840hp)
Performance:	Maximum road speed: 60km/h (37mph)

1996  2004

Light Tanks

Although armies had generally moved away from the light tank concept during World War II, there was a resurgence of interest in these AFVs during the postwar years. This was because they were small enough to be air-deployed as part of the rapid reaction forces that evolved to counter trouble spots around the world.

M41 Walker Bulldog

The Light Tank M41 Little Bulldog was renamed the Walker Bulldog in honour of General Walton Walker (killed in an accident in Korea in 1950). It was well armed and agile, but it was noisy and heavy, which caused problems when transported by air.

SPECIFICATIONS	
Country of origin:	USA
Crew:	5
Weight:	46,500kg (102,533lb)
Dimensions:	Length: 12.3m (40ft 6.5in); width: 3.2m (10ft 8in); height: 2.4m (8ft 2in)
Range:	144km (90 miles)
Armour:	Up to 38mm (1.5 in)
Armament:	1 x 76mm (3in) gun; 1 x 7.62mm (.3in) MG; 1 x 12.7mm (.5in) MG
Powerplant:	1 x 261.1kW (350hp) Bedford 12-cylinder petrol engine
Performance:	Maximum speed: 21km/h (13.5mph)

M551 Sheridan

The M551 was an air-portable tank for the US Army's airborne divisions. In Vietnam, it was susceptible to landmines, but could fire canister ammunition to beat off mass guerrilla attacks. Its main armament was a 152mm (6in) weapon firing a Shilelagh missile.

SPECIFICATIONS	
Country of origin:	USA
Crew:	4
Weight:	15,830kg (34,826lb)
Dimensions:	Length: 6.299m (20ft 8in); width: 2.819m (9ft 3in); height (overall): 2.946m (9ft 8in)
Range:	600km (310 miles)
Armour:	40–50mm (1.57–2in)
Armament:	1 x 152mm (6in) gun/missile launcher; 1 x coaxial 7.62mm (.3in) machine MG; 1 x 12.7mm (.5in) anti-aircraft HMG
Powerplant:	1 x 6-cylinder Detroit 6V-53T diesel developing 224kW (300hp)
Performance:	Maximum road speed: 70km/h (43mph); fording: amphibious; vertical obstacle: .838m (2ft 9in); trench: 2.54m (8ft 4in)

TIMELINE

 1951    1963   1965

LIGHT TANKS

Stridsvagn 103 (S-tank)

This Swedish tank carried a gun that was fixed to the chassis rather than a turret. To aim, the vehicle was turned and the suspension raised or lowered – an entirely new concept in tank design. Its only real drawback was that it was unable to fire on the move.

SPECIFICATIONS

Country of origin:	Sweden
Crew:	3
Weight:	38,894kg (85,568lb)
Dimensions:	Length (with gun): 8.99m (29ft 6in); length (hull): 7.04m (23ft 1in); width: 3.26m (10ft 8.3in); height (overall): 2.5m (8ft 2.5in)
Range:	390km (242 miles)
Armour:	90–100mm (3.54–3.94in)
Armament:	1 x 105mm (4.1in) gun; 3 x 7.62mm (.3in) MGs
Powerplant:	1 x diesel engine developing 119kW (240hp) and a Boeing 553 gas turbine, developing 366kW (490hp)
Performance:	Maximum road speed: 50km/h (31mph); fording: 1.5m (4ft 11in); vertical obstacle: .9m (2ft 11.5in); trench: 2.3m (7ft 6.5in)

Norinco Type 63 Amphibious Light Tank

The Type 63 carries an 85mm (3.4in) gun, and is fitted with a powerful engine derived from the Type 77 series armoured personnel carrier. Fully amphibious, it is propelled in water by two water jets mounted at the rear of the hull.

SPECIFICATIONS

Country of origin:	China
Crew:	4
Weight:	18,400kg (40,572lb)
Dimensions:	Length 8.44m (27ft 8in); width 3.2m (10ft 6in); height 2.52m (8ft 4in)
Range:	370km (230 miles)
Armour:	14mm (.55in) steel
Armament:	1 x 85mm (3.4in) gun; 1 x 7.62mm (.3in) MG; 1 x 12.7mm (.5in) HMG
Powerplant:	1 x Model 12150-L V-12 diesel, 298kW (400bhp)
Performance:	Maximum speed: land 64km/h (40mph); water: 12km/h (7mph)

Steyr SK 105 Light Tank

This light tank, known as the Kürassier, was developed as a fast anti-armour vehicle. Variants include the Greif (Griffin) armoured recovery vehicle, the Pionier combat engineer vehicle, and the Fahrschulpanzer driver training vehicle.

SPECIFICATIONS

Country of origin:	Austria
Crew:	3
Weight:	17,500kg (38,580lb)
Dimensions:	Length: 5.582m (18.31ft); width: 2.5m (8.2ft); height: 2.53m (8.3ft)
Range:	Not available
Armour:	Not available
Armament:	1 x 105mm (4.13in) GIAT-Cn-105-57 rifled gun; 1 x 7.62mm (.3in) MG74 co-axial MG
Powerplant:	1 x Steyr 7FA diesel engine developing 238kW (320bhp)
Performance:	Maximum speed: 65.3km/h (40.6mph)

1966

Self-Propelled Guns

Long-range artillery was largely a thing of the past by the 1960s, having been replaced by tactical missiles, but there was still a requirement for modern self-propelled artillery. This was equipped now with fire control systems which ensured that its operators could place their rounds on target with unparalleled accuracy.

M107

The M107 was designed along with the M110 203mm (8in) self-propelled howitzer, with the same chassis and gun mount. The M107 had a crew of 13, but carried only the commander, driver and three gunners. The rest travelled in a truck, with the rounds, charges and fuses.

SPECIFICATIONS	
Country of origin:	USA
Crew:	5
Weight:	28,168kg (61,970lb)
Dimensions:	Length: 11.256m (36ft 7.75in); width: 3.149m (10ft 4in); height: 3.679m (12ft)
Range:	725km (450 miles)
Armour:	Classified
Armament:	1 x 175mm (6.9in) gun
Powerplant:	1 x Detroit Diesel Model 8V-71T diesel engine developing 302kW (405hp)
Performance:	Maximum road speed: 56km/h (35mph); fording: 1.06m (3ft 6in); vertical obstacle: 1.016m (3ft 4in); trench: 2.326m (7ft 9in)

Abbot

The 105mm (4.1in) Abbot self-propelled gun was developed for the British Army of the Rhine. Reliable, it was a potent force. The Value Engineered Abbot was produced for the Indian Army without night vision and nuclear, biological and chemical (NBC) protection.

SPECIFICATIONS	
Country of origin:	United Kingdom
Crew:	4
Weight:	16,494kg (36,288lb)
Dimensions:	Length: 5.84m (19ft 2in); width: 2.641m (8ft 8in); height: 2.489m (8ft 2in)
Range:	390km (240 miles)
Armour:	6–12mm (.23–.47in)
Armament:	1 x 105mm (4.1in) gun; 1 x 7.62mm (.3in) anti-aircraft MG; 3 x smoke dischargers
Powerplant:	1 x Rolls-Royce six-cylinder diesel engine developing 179kW (240hp)
Performance:	Maximum road speed: 47.5km/h (30mph); fording: 1.2m (3ft 11in); vertical obstacle: .609m (2ft); trench: 2.057m (6ft 9in)

TIMELINE

1958  1962  1963

SELF-PROPELLED GUNS

Bandkanon

The Bandkanon 1A was the first fully automatic self-propelled gun to serve with any army. Ammunition was kept in a 14-round clip carried externally at the rear of the hull. Once the first round was loaded manually, remaining rounds were loaded automatically.

SPECIFICATIONS	
Country of origin:	Sweden
Crew:	5
Weight:	53,000kg (116,600lb)
Dimensions:	Length: 11m (36ft 1in); width: 3.37m (11ft 0.7in); height: 3.85m (12ft 7.5in)
Range:	230km (143 miles)
Armour:	10–20mm (.4–.8in)
Armament:	1 x 155mm (6.1in) gun; 1 x 7.62mm (.3in) anti-aircraft MG
Powerplant:	1 x Rolls-Royce diesel engine developing 179kW (240hp) and Boeing gas turbine, developing 224kW (300hp)
Performance:	Maximum road speed: 28km/h (17.4mph); fording: 1m (3ft 3in); vertical obstacle: .95m (3ft 1.5in); trench: 2m (6ft 6.75in)

Type 54-1

The Type 54-1 consists of a 122mm (4.8in) howitzer mounted on the chassis of the YW 531 armoured personnel carrier. The gun is outclassed by most NATO and Russian artillery, but the simplicity of the Type 54-1 vehicle makes it a dependable weapon.

SPECIFICATIONS	
Country of origin:	China
Crew:	7 (max)
Weight:	15,400kg (34,000lb)
Dimensions:	Length: 5.65m (18ft 6in); width: 3.06m (10ft); height: 2.68m (8ft 9in)
Range:	500km (310 miles)
Armour:	Not available
Armament:	1 x 122mm (4.8in) howitzer
Powerplant:	1 x Deutz 6150L 6-cylinder diesel engine developing 192kW (257hp)
Performance:	Maximum road speed: 56km/h (35mph)

Mk F3 155mm

Based on the AMX-13 tank chassis, the Mk F3 was equipped with two rear spades, which were reversed into the ground for stability. It fired a standard 155mm (6.1in) high-explosive projectile, other types of ammunition being rocket-assisted, smoke and illumination.

SPECIFICATIONS	
Country of origin:	France
Crew:	2
Weight:	17,410kg (38,304lb)
Dimensions:	Length: 6.22m (20ft 5in); width: 2.72m (8ft 11in); height: 2.085m (6ft 10in)
Range:	300km (185 miles)
Armour:	20mm (.78in)
Armament:	1 x 155mm (6.1in) gun
Powerplant:	1 x SOFAM 8Gxb 8-cylinder petrol engine developing 186kW (250hp)
Performance:	Maximum road speed: 60km/h (37mph); vertical obstacle: .6m (2ft); trench: 1.5m (4ft 11in)

1965  1966

M109

The M109 was developed for a 1952 requirement for a self-propelled howitzer to replace the M44. The first production vehicles were completed in 1962 and survived numerous adaptations and upgrades to become the most widely used howitzer in the world, seeing action in Vietnam, in the Arab-Israeli Wars and the Iran-Iraq War.

M109

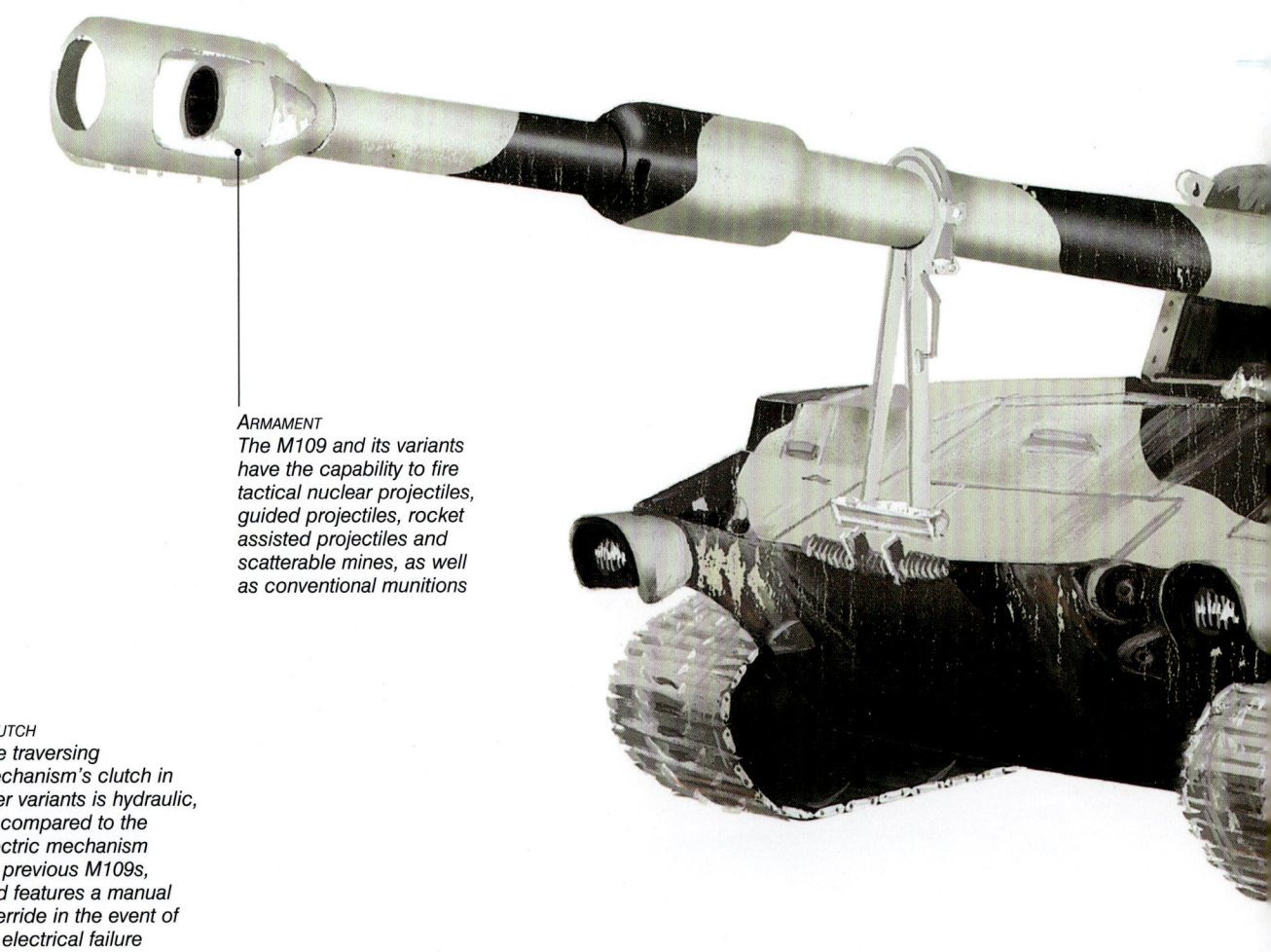

CREW
The six-man crew of the M109 comprises a section chief, driver and gunners to prepare and load the ammunition and aim and fire the cannon

ARMAMENT
The M109 and its variants have the capability to fire tactical nuclear projectiles, guided projectiles, rocket assisted projectiles and scatterable mines, as well as conventional munitions

CLUTCH
The traversing mechanism's clutch in later variants is hydraulic, as compared to the electric mechanism on previous M109s, and features a manual override in the event of an electrical failure

M109

Secondary armament
Later versions of the M109 have an inertial navigation system, sensors detecting the weapons' lay, automation, and an encrypted digital communication system that allows the howitzer to send grid location and altitude to the battery fire direction centre

The M109 has an amphibious capability and fires a variety of projectiles, including tactical nuclear shells. It has undergone numerous upgrades, including a new gun mount, new turret with longer barrel ordnance, automatic fire control, upgraded armour and improved armour.

SPECIFICATIONS	
Country of origin:	USA
Crew:	6
Weight:	23,723kg (52,192lb)
Dimensions:	Length: 6.612m (21ft 8.25in); width: 3.295m (10ft 9.75in); height: 3.289m (10ft 9.5in)
Range:	390km (240 miles)
Armour:	Classified
Armament:	1 x 155mm (6.1in) howitzer; 1 x 12.7mm (.5in) AA HMG
Powerplant:	1 x Detroit Diesel Model 8V-71T diesel engine developing 302kW (405hp)
Performance:	Maximum road speed: 56km/h (35mph); fording: 1.07m (3ft 6in); vertical obstacle: .533m (1ft 9in); trench: 1.828m (6ft)

Survivability
The M109 is equipped with many refinements to improve its survivability on the nuclear, chemical and biological battlefield, including air purifiers, heaters, and Mission Oriented Protective Posture (MOPP) gear

Gunner
The gunner aims the cannon left or right (deflection), while the the assistant gunner aims the cannon up and down (quadrant)

COLD WAR TANKS AND AFVS

Anti-tank Vehicles

By the 1960s, vehicle-mounted anti-tank missiles were an important component of the world's armies. The originals were wire-guided, but more sophisticated systems were developed, allowing greater range and accuracy. All the world's anti-tank missiles of today have their origin in German research during World War II.

Hornet Malkara

Developed to offer a long-range anti-tank capability, the Hornet Malkara carried a Malkara missile launcher. The operator fired the missile from the cab and controlled it by a joystick attached to a wire unreeling from the missile sights.

SPECIFICATIONS	
Country of origin:	United Kingdom
Crew:	3
Weight:	5700kg (12,540lb)
Dimensions:	Length: 5.05m (16ft 7in); width: 2.22m (7ft 3.5in); height: 2.34m (7ft 8in)
Range:	402km (250 miles)
Armour:	8–16mm (.31–.62in)
Armament:	2 x Malkara anti-tank missiles
Powerplant:	1 x Rolls-Royce B60 Mk 5A 6-cylinder petrol engine developing 89kW (120hp)
Performance:	Maximum road speed: 64km/h (40mph); trench: not applicable

ASU-85 Tank Destroyer

The ASU-85 was the logical follow up to the ASU-57 in terms of both armament and tactical docrine. Design work began in the late 1950s, and this culminated in a new air-transportable SP gun/tank destroyer, which entered Red Army service in 1962.

SPECIFICATIONS	
Country of origin:	Soviet Union
Crew:	4
Weight:	15,500kg (15.25 tons)
Dimensions:	Length: 6m (19ft 8in); width: 2.8m (9ft 2in); height: 2.1m (6ft 10in)
Range:	260km (161 miles)
Armour:	8–40mm (.31–1.57in)
Armament:	1 x 85mm (3.4in) D-70 gun; 1 x 7.62mm (.3in) SGMT MG
Powerplant:	1 x V-6 Diesel engine developing 179kW (240bhp) at 1800rpm
Performance:	Maximum road speed: 45km/h (28mph)

TIMELINE

1958

1960

1961

ANTI-TANK VEHICLES

Raketenjadgpanzer 1

The Raketenjadgpanzer 1 was a ground-breaking anti-tank vehicle, mounting two French SS-11 ATGW launchers. When one missile was ready to fire, the other was withdrawn inside the hull for rearming. The missile had a range of 3000m (3266.6 yards).

SPECIFICATIONS	
Country of origin:	West Germany
Crew:	4
Weight:	13,000kg (28,665lb)
Dimensions:	Length: 5.56m (18ft 3in); width: 2.25m (7ft 4.5in); height: 1.7m (5ft 7in)
Range:	270km (170 miles)
Armour:	30mm (1.18in)
Armament:	10 x SS-11 ATGWs
Powerplant:	1 x Rolls-Royce B81 Mk80F 8-cylinder petrol engine developing 175kW (235hp) at 3800rpm
Performance:	Maximum road speed: 51km/h (32mph); fording: .7m (2.3ft); gradient: 60 per cent; vertical obstacle: .6m (2ft); trench: 1.6m (5.2ft)

Type 60

The Komatsu Type 60 is still in service, its armament being two RCL 106mm (4.17in) recoilless rifles. The guns are only able to traverse 10° either side, forcing the driver to align the vehicle with the target. It is suitable only for engaging light armoured vehicles.

SPECIFICATIONS	
Country of origin:	Japan
Crew:	3
Weight:	8000kg (17,600lb)
Dimensions:	Length: 4.3m (14ft 1in); width: 2.23m (7ft 4in); height: 1.59m (5ft 3in)
Range:	130km (80 miles)
Armour:	12mm (.47in)
Armament:	2 x RCL 106mm (4.17in) recoilless rifles, 1 x 12.7mm (.5in) HMG
Powerplant:	1 x Komatsu 6T 120-2 6-cylinder diesel, developing 89kW (120hp) at 2400rpm
Performance:	Maximum road speed: 55km/h (34mph); fording: .7m (2.3ft); gradient 60 per cent; vertical obstacle: .6m (2ft); trench: 1.8m (5.9ft)

FV102 Striker

The FV102 Striker is a Scorpion reconnaissance vehicle, its turret replaced by a hydraulically raised launcher box with five BAC Swingfire anti-tank guided weapons. Five more missiles are stored internally. The Swingfire has a range of 4000m (13,100ft).

SPECIFICATIONS	
Country of origin:	United Kingdom
Crew:	3
Weight:	8221kg (18,127lb)
Dimensions:	Length: 4.76m (15ft 7in); width: 2.18m (7ft 2in); height: 2.21m (7ft 3in)
Range:	483km (300 miles)
Armour:	Classified
Armament:	5 + 5 x Swingfire anti-tank missiles; 1 x 7.62mm (.3in) MG
Powerplant:	1 x Jaguar J60 No 1 Mk100B 6-cylinder petrol, developing 145kW (195hp)
Performance:	Maximum road speed: 100km/h (60mph); fording: 1m (3ft 4in); gradient: 60 per cent; vertical obstacle: .6m (2ft)

1962

1975

Early Tactical Missile Launchers

By the early 1960s, the world's leading armies possessed large arsenals of tactical missiles, most of which could be configured to deliver either conventional or tactical nuclear warheads. Their launchers were mostly purpose-built vehicles, highly mobile and capable of operating in all types of terrain.

Honest John

The Douglas MGR-1A Honest John was the first US tactical battlefield nuclear weapon. It and the MGR-1B were unguided rockets used like conventional tube artillery. As well as 2-, 20-, or 40-kiloton W31 nuclear warheads, conventional warheads could be fitted.

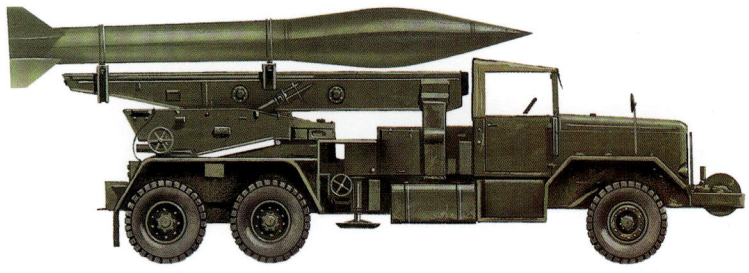

SPECIFICATIONS	
Country of origin:	USA
Crew:	3
Weight:	16,400kg (36,200lb)
Dimensions:	Length: 9.89m (32ft 5in); width: 2.9m (9ft 6in); height: 2.67m (8ft 9in)
Range:	480km (300 miles)
Armour:	Not applicable
Armament:	4 x tactical battlefield missiles supplied from independent transport truck
Powerplant:	1 x AM General 6-cylinder petrol, developing 104kW (139hp)
Performance:	Maximum road speed: 90km/h (56mph)

FROG-7

FROG stands for 'Free Rocket Over Ground'. The FROG-7 was the result of experience gained with earlier models. It was a large single-stage solid fuel rocket provided with a variety of warheads and carried on an 8x8 transporter-erector-launcher vehicle.

SPECIFICATIONS	
Country of origin:	Soviet Union
Crew:	4
Weight:	20,300kg (44,900lb)
Dimensions:	Length: 10.69m (35ft .8in); width: 2.8m (9ft 2in); height (with missile): 3.35m (11ft)
Range:	650km (400 miles)
Armour:	Not applicable
Armament:	1 x FROG-7 rocket
Powerplant:	2 x ZIL-375 8-cylinder petrol engine developing 132kW (177hp)
Performance:	Maximum road speed: 40km/h (25mph)

TIMELINE

1953    1961    1963

EARLY TACTICAL MISSILE LAUNCHERS

Pershing

The Martin Marietta MGM-31A Pershing tactical nuclear missile was conceived to deliver 60- to 400-kiloton W50 airburst nuclear warheads at ranges of between 160 and 740km (100 and 460 miles). Under nuclear weapon treaties, the last Pershings were destroyed in 1991.

SPECIFICATIONS
Country of origin:	USA
Crew:	1
Weight:	8100kg (17,900lb)
Dimensions:	Length (with rocket): 10.6m (34ft 9in); width: 2.5m (8ft 2in); height: 3.79m (12ft 5in)
Range:	320km (200 miles)
Armour:	None
Armament:	1 x Pershing tactical nuclear missile
Powerplant:	1 x Chrysler A-710-B 8-cylinder petrol engine developing 160kW (215hp)
Performance:	Maximum road speed: 65km/h (40mph)

M548 Carrier with Lance Missile

The M548 has been used for many specialized vehicles including the Vought Lance missile launcher (as shown); a radar and electronic warfare vehicle; a mine layer; a 35mm (1.37in) self-propelled anti-aircraft gun; and a mine clearance and recovery vehicle.

SPECIFICATIONS
Country of origin:	USA
Crew:	1 + 3
Weight:	12,882kg (28,340lb)
Dimensions:	Length: 5.893m (19ft 4in); width: 2.692m (8ft 10in); height: 2.82m (9ft 3in)
Range:	483km (300 miles)
Armour:	44mm (1.73in)
Armament:	Varied. Model pictured: 1 x Vought Lance missile
Powerplant:	1 x Detroit-Diesel Model 6V-53 six-cylinder liquid-cooled diesel engine developing 160kW (215hp)
Performance:	Maximum road speed: 64km/h (40mph); fording: amphibious; vertical obstacle: .61m (2ft); trench: 1.68m (5ft 6in)

SS-1c Scud B Launcher

The Scud B is a medium-range tactical missile with nuclear or conventional high-explosive warheads. Its guidance system works for only the first 80 seconds of flight. After this, the warhead detaches and flies unguided to the target – hence its inaccuracy.

SPECIFICATIONS
Country of origin:	Russia/Soviet Union
Crew:	4
Weight:	37,400kg (82,500lb)
Dimensions:	Length: 13.36m (43ft 10in); width: 3.02m (9ft 11in); height: 3.33m (10ft 11in)
Range:	450km (280 miles)
Armour:	Not applicable
Armament:	1 x Scud B tactical ballistic missile
Powerplant:	1 x D12A-525A 12-cylinder diesel engine developing 391kW (525hp) at 2100rpm
Performance:	Maximum road speed: 45km/h (28mph)

1969

1972

Anti-aircraft Vehicles

During the Cold War, vehicle-mounted anti-aircraft weaponry fell into two distinct classes: guns and missiles. The missiles were designed to intercept incoming aircraft at maximum range, while the guns acted as close range anti-aircraft artillery to deal with any that managed to break through to the target area.

Type 63

The Type 63 was based on the Soviet T-34 tank, and had an open-topped turret with twin anti-aircraft guns. It had no provision for radar control and had to be sighted and elevated manually, a major drawback when faced with fast, low-flying aircraft.

SPECIFICATIONS	
Country of origin:	China
Crew:	6
Weight:	32,000kg (70,400lb)
Dimensions:	Length: 6.432m (21ft 1in); width: 2.99m (9ft 10in); height: 2.995m (9ft 10in)
Range:	300km (186 miles)
Armour:	18–45mm (.7–1.8in)
Armament:	2 x 37mm anti-aircraft guns
Powerplant:	1 x V-12 water-cooled diesel engine developing 373kW (500hp)
Performance:	Maximum road speed: 55km/h (34mph); fording: 1.32m (4ft 4in); vertical obstacle: .73m (2ft 5in); trench: 2.5m (8ft 2in)

76mm Otomatic Air Defence Tank

The 76mm (3in) Otomatic Air Defence Tank defended ground troops from helicopters and ground attack aircraft. The Otomatic System was a turret assembly with a 76mm (3in) gun, surveillance and tracking radars, electro-optical sights and fire control computer.

SPECIFICATIONS	
Country of origin:	Italy
Crew:	4
Weight:	47,000kg (46.26 tons)
Dimensions:	Length: 7.08m (23ft 3in); width: 3.25m (10ft 8in); height: 3.07m (10ft 1in) (radar stowed)
Range:	500km (310 miles)
Armour:	Not disclosed
Armament:	1 x 76mm (3in) gun
Powerplant:	1 x MTU V-10 multi-fuel engine developing 238.6kW (830bhp) at 2200rpm
Performance:	Maximum road speed: 60km/h (37mph); fording: 1.2m (3ft 11in); vertical obstacle: 1.15m (3ft 9in); trench: 3m (9ft 10in)

TIMELINE

 1962    1963    1964

SA-4 Ganef

ANTI-AIRCRAFT VEHICLES

The SA-4 Ganef SAM system entered service in 1964. Each Ganef missile weighs 1800kg (4000lb). The missiles sit on a turntable mounting capable of 360° traverse and 45° elevation. NBC, air filtration and infrared night-vision devices are fitted as standard.

SPECIFICATIONS
Country of origin:	Russian Federation/Soviet Union
Crew:	3–5
Weight:	30,000kg (66,150lb)
Dimensions:	Length (with missiles): 9.46m (31ft .3in); width: 3.2m (10ft 6in); height (with missiles): 4.47m (14ft 8in)
Range:	450km (280 miles)
Armour:	15–20mm (.59–.78in)
Armament:	2 x SA-4 Ganef SAM missiles
Powerplant:	1 x V-59 12-cylinder diesel engine developing 388kW (520hp)
Performance:	Maximum road speed: 45km/h (28mph)

Roland

The Roland SAM system is a Franco-German production. Known as the Euromissile, it has two versions: clear-weather (Roland 1) and all-weather (Roland 2). The radar system has a computer to offer an Identification Friend or Foe (IFF) assessment.

SPECIFICATIONS
Country of origin:	France/West Germany
Crew:	3
Weight:	34,800kg (76,7004lb)
Dimensions:	Length: 6.92m (22ft 8in); width: 3.24m (10ft 7in); height: 2.92m (9ft 7in)
Range:	600km (370 miles)
Armour:	Steel (details classified)
Armament:	2 + 8 Roland SAM missiles
Powerplant:	1 x MTU mB 833Ea-500 6-cylinder diesel, developing 440kW (590hp)
Performance:	Maximum road speed: 60km/h (37mph); fording: 1.5m (4ft 11in); gradient: 60 per cent; vertical obstacle: 1m (3ft 4in); trench: 2.5m (8ft 2in)

M163 Vulcan

In the early 1960s, Rock Island Arsenal developed a self-propelled air-defence system based on the M113 armoured personnel carrier chassis. This had adjustable rates of fire, depending on whether it was being used for ground attack or anti-aircraft defence.

SPECIFICATIONS
Country of origin:	USA
Crew:	4
Weight:	12,310kg (27,082lb)
Dimensions:	Length: 4.86m (15ft 11in); width: 2.85m (9ft 4in); height: 2.736m (9ft 11in)
Range:	83km (300 miles)
Armour:	38mm (1.5in)
Armament:	1 x 20mm (.76in) six-barrelled M61 series gun
Powerplant:	1 x Detroit 6V-53 six-cylinder diesel engine developing 160kW (215hp)
Performance:	Maximum road speed: 67km/h (42mph); fording: amphibious; vertical obstacle: 1.61m (2ft); trench: 1.68m (5ft 6in)

1965

ZSU-23-4 Shilka

The ZSU-23-4 Shilka was developed in the 1960s as the replacement for the ZSU-57-2. Although having a shorter range, the radar fire-control and an increased firing rate made the weapon much more effective.

ZSU-23-4

'GUN DISH' RADAR
This is the NATO codename for the B-76 radar fitted to the ZSU-2304, which allows it to operate in all weathers and at night. It is a good tracking radar, difficult to evade or detect

TURRET ARMOUR
The turret is protected by a meagre 9mm (.35in) of armour, which can be penetrated by 12.7mm (.5in) calibre machine guns

ZSU-23-4

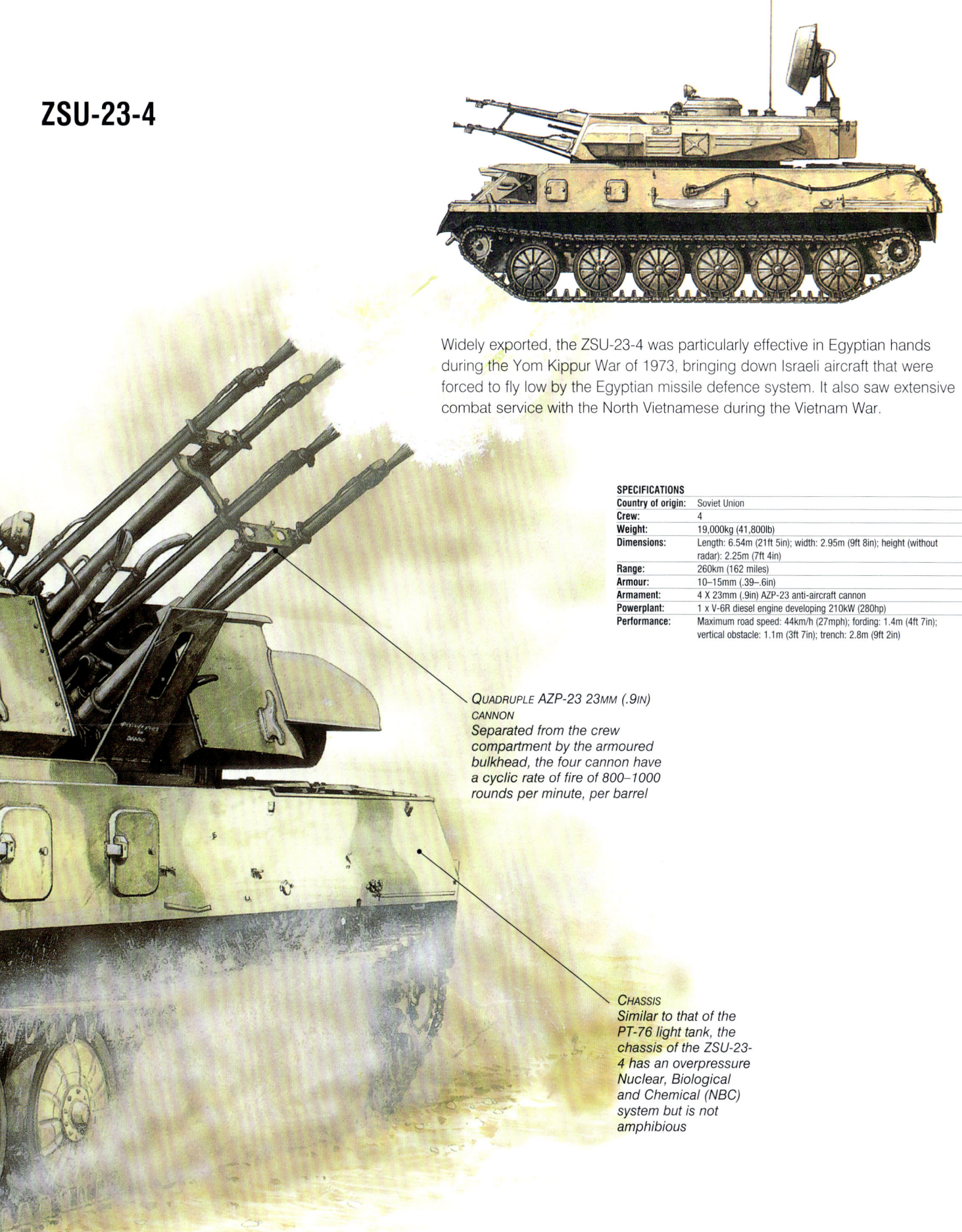

Widely exported, the ZSU-23-4 was particularly effective in Egyptian hands during the Yom Kippur War of 1973, bringing down Israeli aircraft that were forced to fly low by the Egyptian missile defence system. It also saw extensive combat service with the North Vietnamese during the Vietnam War.

SPECIFICATIONS

Country of origin:	Soviet Union
Crew:	4
Weight:	19,000kg (41,800lb)
Dimensions:	Length: 6.54m (21ft 5in); width: 2.95m (9ft 8in); height (without radar): 2.25m (7ft 4in)
Range:	260km (162 miles)
Armour:	10–15mm (.39–.6in)
Armament:	4 X 23mm (.9in) AZP-23 anti-aircraft cannon
Powerplant:	1 x V-6R diesel engine developing 210kW (280hp)
Performance:	Maximum road speed: 44km/h (27mph); fording: 1.4m (4ft 7in); vertical obstacle: 1.1m (3ft 7in); trench: 2.8m (9ft 2in)

QUADRUPLE AZP-23 23MM (.9IN) CANNON
Separated from the crew compartment by the armoured bulkhead, the four cannon have a cyclic rate of fire of 800–1000 rounds per minute, per barrel

CHASSIS
Similar to that of the PT-76 light tank, the chassis of the ZSU-23-4 has an overpressure Nuclear, Biological and Chemical (NBC) system but is not amphibious

1960s Anti-aircraft Vehicles

During the 1960s, when the Cold War was at its height, the world's leading armies fielded a formidable arsenal of mobile anti-aircraft weaponry, a mixture of surface-to-air missiles and anti-aircraft artillery, most designed to counter attacks by low-flying fast jets. Some of the closer-range weapons were based on air-to-air missiles.

SA-6 Gainful

The SA-6 Gainful was one of the most successful anti-aircraft weapons. Three SA-6 missiles were transported on the modified chassis of a ZSU-23-4. The vehicle had radiation warning systems, NBC protection and fire-control equipment.

SPECIFICATIONS	
Country of origin:	Soviet Union
Crew:	3
Weight:	14,000kg (30,900lb)
Dimensions:	Length: 7.39m (24ft 3in); width: 3.18m (10ft 5in); height (with missiles): 3.45m (11ft 4in)
Range:	260km (160 miles)
Armour:	15mm (.59in)
Armament:	3 x SA-6 Gainful SAMs
Powerplant:	1 x Model V-6R 6-cylinder diesel engine developing 179kW (240hp)
Performance:	Maximum road speed: 44km/h (27mph); fording: 1m (3ft 4in); gradient: 60 per cent; vertical obstacle: 1.1m (3ft 7in); trench: 2.8m (9ft 2in)

M48 Chaparral

The M48 Chaparral used the MIM-72C SAM for short-range low-altitude interception missions. Guidance to target was via a fire-and-forget infrared system. An enemy aircraft warning system for the Chaparral was provided by a Forward Area Alerting Radar (FAAR).

SPECIFICATIONS	
Country of origin:	USA
Crew:	5
Weight:	11,500kg (25,360lb)
Dimensions:	Length: 6.06m (19ft 10in); width: 2.69m (8ft 9in); height: 2.68m (8ft 9in)
Range:	489km (304 miles)
Armour:	Not applicable
Armament:	12 x MIM-72C SAM
Powerplant:	1 x Detroit 6V53 6-cylinder diesel engine developing 160kW (214hp)
Performance:	Maximum road speed: 61km/h (38mph); fording: 1m (3.3ft); gradient: 60 per cent; vertical obstacle: .61m (2ft); trench: 1.68m (5.51ft)

TIMELINE

 1967  1969

1960S ANTI-AIRCRAFT VEHICLES

AMX-13 DCA

The AMX-13 DCA was an AMX-13 main battle tank chassis fitted with a cast-steel turret. Up to the 1980s, it was the only self-propelled anti-aircraft gun in use with French forces. To aid fire control, an Oeil Noir 1 radar scanner was fitted to the back of the turret.

SPECIFICATIONS

Country of origin:	France
Crew:	3
Weight:	17,200kg (37,840lb)
Dimensions:	Length: 5.4m (17ft 11in); width: 2.5m (8ft 2in); height (radar up): 3.8m (12ft 6in); height (radar down): 3m (9ft 10in)
Range:	300km (186 miles)
Armour:	25mm (.98in)
Armament:	2 x 30mm (1.2in) Hispano (now Oerlikon) guns
Powerplant:	1 x SOFAM Model 8Gxb 8-cylinder water-cooled petrol engine developing 186kW (250hp)
Performance:	Maximum road speed: 60km/h (37mph); fording: .6m (1ft 11in); vertical obstacle: .65m (2ft 2in); trench: 1.7m (5ft 7in)

GDF-CO3

The GDF series was designed as a highly mobile anti-aircraft defence system to protect rear-area targets such as factories and air bases. The GDF-CO3 has a day/night fire-control system with laser rangefinder, in addition to a Contraves search radar.

SPECIFICATIONS

Country of origin:	Switzerland
Crew:	3
Weight:	18,000kg (39,600lb)
Dimensions:	Length: 6.7m (22ft); width: 2.813m (9ft 3in); height: 4m (13ft 2in)
Range:	480km (297 miles)
Armour:	8mm (.31in)
Armament:	2 x 35mm (1.38in) (KDF cannon
Powerplant:	1 x GMC 6V-53T 6-cylinder diesel engine developing 160kW (215hp)
Performance:	Maximum road speed: 45km/h (28mph); fording: .6m (1ft 11in); vertical obstacle: .609m (2ft); trench: 1.8m (5ft 11in)

SA-8 Gecko

The SA-8 Gecko was the first Soviet air-defence system to combine surveillance, target acquisition and missile launcher in one vehicle. It is carried by the chassis of the ZIL-167 6x6 vehicle. Amphibious, it has an NBC-protected crew compartment.

SPECIFICATIONS

Country of origin:	Russian Federation/Soviet Union
Crew:	5
Weight:	17,499kg (38,587lb)
Dimensions:	Length: 9.14m (30ft); width: 2.8m (9ft 2in); height (with radar lowered): 4.2m (13ft 11in)
Range:	250km (155 miles)
Armour:	Not available
Armament:	6 x SA-8 Type 9M33 SAMs
Powerplant:	1 x 5D20 B-300 diesel with gas-turbine auxiliary drive, developing 223kW (299hp)
Performance:	Maximum road speed: 80km/h (50mph); fording: amphibious

1970  1972

1970s Anti-aircraft Vehicles

The sophistication of anti-aircraft vehicles continued to grow throughout the 1970s. The Israelis learned this to their cost in the Yom Kippur War of 1973, when Soviet-made anti-aircraft systems, many of them tracked and highly mobile, cost the Israeli Air Force some 40 per cent of its first-line strength in the initial stages of the conflict.

Flakpanzer 1 Gepard

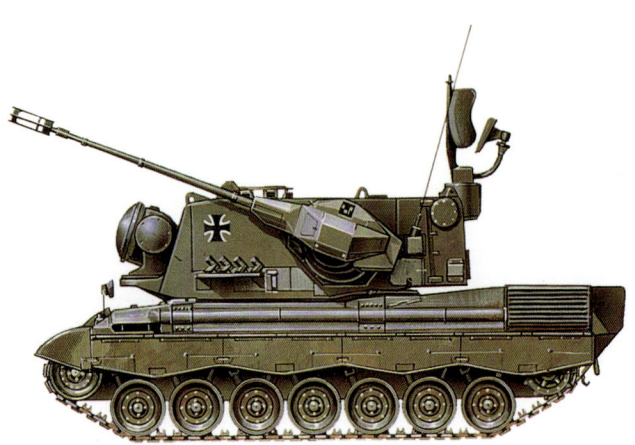

The Gepard was a self-propelled anti-aircraft gun (SPAAG) for the West German Army. Its chassis was based on the Leopard 1 MBT with downgraded armour requirements. Fire control was computerized with ground- or aerial-engagement options.

SPECIFICATIONS	
Country of origin:	Germany
Crew:	4
Weight:	47,300kg (104,000lb)
Dimensions:	Length: 7.68m (25ft 2in); width: 3.27m (10ft 8in); height 3.01m (9ft 10in)
Range:	550km (340 miles)
Armour:	Up to 40mm (1.57in)
Armament:	2 x 35mm (1.38in) guns; 8 smoke dischargers
Powerplant:	1 x MTU MB 838 Ca M500 10-cylinder multi-fuel engine developing 619kW (830hp)
Performance:	Maximum road speed: 64km/h (40.5mph); fording: 2.5m (8ft 2in)

Spartan

The Alvis Spartan has specialized roles, such as carrying Javelin surface-to-air missiles (SAMs) or Royal Engineer assault teams. It has a three-man crew and room in the back for four fully equipped troops, though the vehicle has no firing ports.

SPECIFICATIONS	
Country of origin:	United Kingdom
Crew:	3 + 4
Weight:	8172kg (17,978lb)
Dimensions:	Length: 5.125m (16ft 9in); width: 2.24m (7ft 4in); height: 2.26m (7ft 5in)
Range:	483km (301 miles)
Armour:	Classified
Armament:	1 x 7.62mm (.3in) MG; 1 x Milan missile launcher
Powerplant:	1 x Jaguar six-cylinder petrol engine developing 142kW (190hp)
Performance:	Maximum road speed: 80km/h (50mph); fording: 1.067m (3ft 6in); vertical obstacle: .5m (1ft 7in); trench: 2.057m (6ft 9in)

TIMELINE

1976    1978

1970S ANTI-AIRCRAFT VEHICLES

Crotale

The Crotale was an advanced all-weather SAM system. The firing vehicle mounted four R.440 missiles, each with a range of 8500m (28,000ft) and a blast radius of 8m (26ft). The target-acquisition vehicle carried a Doppler pulse search and surveillance radar.

SPECIFICATIONS	
Country of origin:	France
Crew:	3
Weight:	(Launcher vehicle) 27,300kg (60,200lb)
Dimensions:	Length: 6.22m (20ft 3in); width: 2.65m (8ft 8in); height (vehicle): 2.04m (6ft 8in)
Range:	500km (310 miles)
Armour:	3–5mm (.12–.2in)
Armament:	4 x Matra R.440 SAM missiles
Powerplant:	1 x diesel generator and 4 x electric motors, developing 176kW (236hp)
Performance:	Maximum road speed: 70km/h (43mph); fording: .68m (2ft 2.5in); gradient: 40 per cent; vertical obstacle: .3m (1ft)

SA-13 Gopher

The SA-13 Gopher consists of four SA-13 infrared-guided missiles transported by a MT-LB Multipurpose Armoured Vehicle acting as a TELAR (Transporter Erector Launcher and Radar). The missiles' range is 5000m (16,400ft) to an altitude of 3500m (11,500ft).

SPECIFICATIONS	
Country of origin:	Russian Federation/Soviet Union
Crew:	3
Weight:	12,080kg (26,636lb)
Dimensions:	Length: 6.93m (22ft 9in); width: 2.85m (9ft 4in); height 3.96m (13ft)
Range:	500km (310 miles)
Armour:	7–14mm (.28–.55in)
Armament:	4 + 4 9M37 SAM missiles
Powerplant:	1 x YaMZ-239V 8-cylinder diesel, developing 529kW (709hp)
Performance:	Maximum road speed: 61.5km/h (38mph); fording: amphibious; gradient: 60 per cent; vertical obstacle: .7m (2ft 4in); trench: 2.7m (8ft 10in)

Shahine

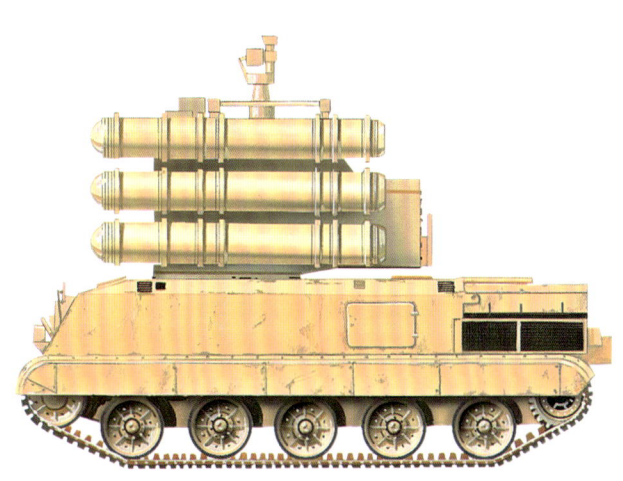

Unlike many mobile SAM systems, the Shahine is heavily armoured. Its firing and acquisition units are mounted on an AMX-30 MBT chassis with armour 15–80mm (.59–3.15in) thick. The Matra R.460 SAM missiles are guided by the command-control centre.

SPECIFICATIONS	
Country of origin:	France
Crew:	3
Weight:	(Launcher vehicle) 38,799kg (85,554lb)
Dimensions:	Length: 6.59m (21ft 7.4in); width: 3.1m (10ft 2in); height: 5.5m (18ft)
Range:	600km (370 miles)
Armour:	15–80mm (.59–3.15in)
Armament:	6 x Matra R.460 SAM missiles
Powerplant:	1 x Hispano-Suiza HS110 12-cylinder multi-fuel, developing 515kW (690hp)
Performance:	Maximum road speed: 65km/h (40mph); fording: 1.3m (4ft 4in); gradient: 60 per cent; vertical obstacle: .93m (3ft .6in)

 1979

Tracked APCs

In the 1950s, the Soviet Union, whose war plan was based on a fast surprise attack on Western Europe by massive concentrations of armour, produced huge numbers of tracked personnel carriers, the logical argument being that an armoured thrust will fail if the breakthrough is not supported by sufficient troops.

OT-810

The OT-810 was a modification of the German Sdfz.251 half-track. The engine was replaced by a Tatra six-cylinder air-cooled diesel. Armoured roof hatches were added to the troop compartment. An anti-tank variant was produced.

SPECIFICATIONS

Country of origin:	Czechoslovakia
Crew:	2 + 10
Weight:	9000kg (19,800lb)
Dimensions:	Length: 5.71m (18ft 9in); width: 2.1m (6ft 10in); height: 1.88m (6ft 2in)
Range:	600km (370 miles)
Armour:	Up to 12mm (.47in)
Armament:	1 x 7.62mm (.3in) MG
Powerplant:	1 x Tatra 928-3 6-cylinder diesel engine developing 89kW (120hp)
Performance:	Maximum road speed: 55km/h (34mph); fording: .5m (1ft 7in); gradient: 24 per cent; vertical obstacle: .23m (9in); trench: 1.98m (6ft 4in)

AMX VCI

The AMX VCI (Véhicule de Combat d'Infanterie) was constructed around the modified chassis of the AMX-13 light tank. It carried three crew and 10 soldiers. Its standard armament was a 12.7mm (.5in) M2 HB machine gun pintle-mounted on the roof.

SPECIFICATIONS

Country of origin:	France
Crew:	3 + 10
Weight:	15,000kg (33,100lb)
Dimensions:	Length: 5.7m (18ft 8in); width: 2.67m (8ft 9in); height: 2.41m (7ft 11in)
Range:	350km (220 miles)
Armour:	Up to 30mm (1.18in)
Armament:	1 x 12.7mm (.5in) M2 HB MG
Powerplant:	1 x SOFAM 8Gxb 8-cylinder petrol engine developing 186kW (250hp) at 3200rpm
Performance:	Maximum road speed: 60km/h (37mph); fording: 1m (3ft 4in); gradient: 60 per cent; vertical obstacle: 1m (3ft 4in); trench 1.6m (5ft 3in)

TIMELINE

1957

TRACKED APCS

BTR-50

Between 1957 and the early 1970s, the BTR-50 was the standard Soviet APC, able to carry 20 personnel. The low-profile hull meant that the occupants entered and exited over the vehicle's side. Water jets powered it when in amphibious mode.

SPECIFICATIONS	
Country of origin:	Soviet Union
Crew:	2 + 20
Weight:	14,200kg (31,300lb)
Dimensions:	Length: 7.03m (23ft 1in); width: 3.14m (10ft 4in); height: 2.07m (6ft 9in)
Range:	400km (250 miles)
Armour:	10mm (.39in)
Armament:	1 x 7.62mm (.3in) MG
Powerplant:	1 x Model V6 6-cylinder diesel engine developing 179kW (240hp) at 1800rpm
Performance:	Maximum road speed: 44km/h (27mph); fording: amphibious; gradient: 70 per cent; vertical obstacle: 1.1m (3ft 7in); trench: 2.8m (9ft 2in)

M-60P

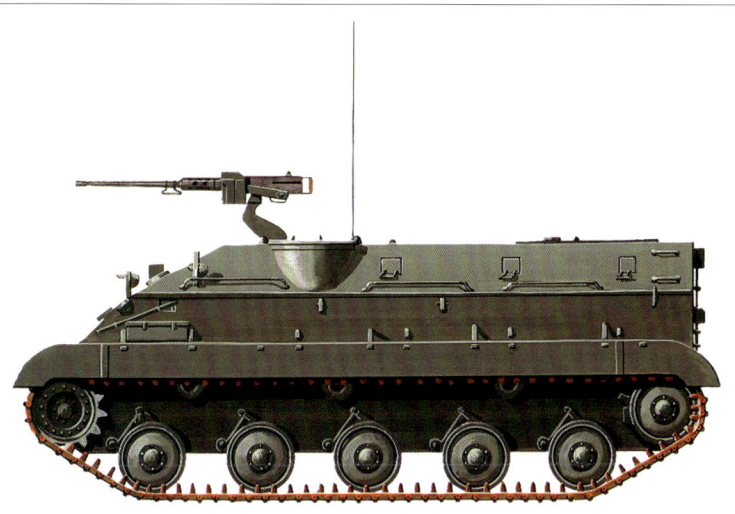

The M60P is a medley of foreign design elements. The Soviet SU-76 self-propelled gun chassis provides the basis for the suspension, an Austrian Steyr-type engine gives the power, and Western APCs, such as the US M59, contribute to the overall design.

SPECIFICATIONS	
Country of origin:	Yugoslavia
Crew:	3 + 10
Weight:	11,000kg (24,300lb)
Dimensions:	Length: 5.02m (16ft 5in); width: 2.77m (9ft 1in); height: 2.77m (9ft 1in)
Range:	400km (250 miles)
Armour:	25mm (.98in)
Armament:	1 x 12.7mm (.5in) HMG; 1 x 7.62mm (.3in) MG
Powerplant:	1 x FAMOS 6-cylinder diesel engine developing 104kW (140hp)
Performance:	Maximum road speed: 45km/h (28mph); fording: 1.25m (4ft 1in); gradient: 60 per cent; vertical obstacle: .6m (2ft); trench: 2m (6ft 7in)

Saurer 4K 4FA-G1

The basic Saurer 4K 4FA armoured personnel carrier began a line of variants. In its standard form, it was a steel-armoured APC with a two-plus-eight personnel capacity and a single 12.7mm (.5in) Browning M2 HB machine gun mounted on a forward cupola.

SPECIFICATIONS	
Country of origin:	Austria
Crew:	2 + 8
Weight:	12,200kg (26,900lb)
Dimensions:	Length: 5.35m (17ft 7in); width: 2.5m (8ft 2in); height: 1.65m (5ft 5in)
Range:	370km (230 miles)
Armour:	Up to 20mm (.78in)
Armament:	1 x 12.7mm (.5in) HMG
Powerplant:	1 x Saurer 4FA 6-cylinder turbo diese engine developing 186kW (250hp) at 2400rpm
Performance:	Maximum road speed: 65km/h (40mph); fording: 1m (3ft 4in); gradient: 75 per cent; vertical obstacle: .8m (2ft 7in); trench: 2.2m (7ft 2in)

1960  1961

1960s Tracked APCs

The development of tracked APCs continued in the 1960s, and these vehicles were widely exported throughout the world. In areas where trucks might find it difficult to operate, the APC continued to be the lifeblood of a mechanized army, forming an integral part of a modern battle group.

FV432

The FV432, known at one time as the Trojan, was used to transport up to 10 men to the battlefield. It was fitted with a nuclear, biological and chemical (NBC) defence system. Variants included a command vehicle, ambulance, mortar carrier and mine-layer.

SPECIFICATIONS	
Country of origin:	United Kingdom
Crew:	2 + 10
Weight:	15,280kg (33,616lb)
Dimensions:	Length: 5.251m (17ft 7in); width: 2.8m (9ft 2in); height (with machine gun): 2.286m (7ft 6in)
Range:	483km (300 miles)
Armour:	12mm (.47in)
Armament:	1 x 7.62mm (.3in) MG
Powerplant:	1 x Rolls-Royce K60 six-cylinder multi-fuel engine developing 170kW (240hp)
Performance:	Maximum road speed: 52.2km/h (32mph); fording: 1.066m (3ft 6in); vertical obstacle: .609m (2ft); trench: 2.05m (6ft 9in)

OT-62

The OT-62 was the Czech version of the Russian BTR-50PK. It had an all-welded armoured hull with slightly less carrying capacity (18 rather than 20 passengers) but more powerful engines. Two waterjets at the rear of the vehicle provided propulsion for amphibious use.

SPECIFICATIONS	
Country of origin:	Czechoslovakia
Crew:	3 + 18
Weight:	15,100kg (33,300lb)
Dimensions:	Length: 7m (22ft 11in); width: 3.22m (10ft 7in); height: 2.72m (8ft 11in)
Range:	460km (290 miles)
Armour:	Up to 14mm (.55in)
Armament:	1 x 7.62mm (.3in) PKY MG and various other configurations
Powerplant:	1 x PV6 6-cylinder turbo diesel engine developing 224kW (300hp) at 1200rpm
Performance:	Maximum road speed: 60km/h (37mph); fording: amphibious; gradient: 70 per cent; vertical obstacle: 1.1m (3ft 7in); trench 2.8m (9ft 2in)

TIMELINE

1963  1964  1965

Pbv

The Pbv was one of the first of its type to have a fully enclosed weapon station. Its troop compartment at the rear held 10 fully equipped soldiers, firing through the hatches on the top. Variants included a command vehicle, observation vehicle and ambulance.

SPECIFICATIONS	
Country of origin:	Sweden
Crew:	2 + 10
Weight:	13,500kg (29,700lb)
Dimensions:	Length: 5.35m (17ft 7in); width: 2.86m (9ft 5in); height: 2.5m (8ft 2in)
Range:	300km (186 miles)
Armour:	Classified
Armament:	1 x 20mm (.78in) Hispano gun
Powerplant:	1 x Volvo-Penta Model THD 100B 6-cylinder inline diesel engine developing 209kW (280hp)
Performance:	Maximum road speed: 66km/h (41mph); fording: amphibious; vertical obstacle: .61m (2ft); trench: 1.8m (5ft 11in)

BMP-1

The primary armament of the BMP-1 is a turret-mounted 73mm (2.87in) short-recoil gun, fed with fin-stabilized rocket-assisted ammunition from a 40-round magazine. The weapon features a low-pressure system to negate excessive backblast into the cabin.

SPECIFICATIONS	
Country of origin:	Russia/Soviet Union
Crew:	3 + 8
Weight:	13,900kg (30,650lb)
Dimensions:	Length: 6.74m (22ft 1in); width 2.94m (9ft 8in); height 1.9m (6ft 3in)
Range:	600km (370 miles)
Armour:	(steel) 33mm (1.29in)
Armament:	1 x 73mm (2.87in) gun; 1 x Sagger ATGW missile; 1 x 7.62mm (.3in) coaxial MG
Powerplant:	1 x UTD-20 6-cylinder diesel engine developing 223kW (300hp)
Performance:	Maximum road speed: 80km/h (50mph); fording: amphibious; gradient: 60 per cent; vertical obstacle: .8m (2ft 7in); trench: 2.2m (7ft 2in)

YW 531

The YW 531 armoured personnel carrier held 13 troops in the rear of the vehicle, with a two-man crew of driver and commander. The engine was located behind the commander, who viewed the battlefield through a hatch or a 360° rotatable periscope integral to the hatch.

SPECIFICATIONS	
Country of origin:	China
Crew:	2 + 13
Weight:	12,500kg (27,600lb)
Dimensions:	Length: 5.74m (18ft 10in); width: 2.99m (9ft 9in); height: 2.11m (6ft 11in)
Range:	425km (260 miles)
Armour:	Not available
Armament:	1 x 12.7mm (.5in) MG
Powerplant:	1 x Deutz Type 6150L 6-cylinder diesel engine developing 192kW (257hp)
Performance:	Maximum road speed: 50km/h (31mph); fording: amphibious; vertical obstacle: .6m (2ft); trench: 2m (6ft 7in)

1966

1970s Personnel Carriers

During the 1970s, the armies of the world continued to develop new personnel carriers. The designs were varied in order to meet the individual requirements of the operators worldwide, and consequently the appearance of these AFVs also differed considerably.

Marder

With excellent armour and high cross-country speed, the Marder Schützenpanzer was able to operate with Leopard main battle tanks. The troops fired from inside by means of a periscope and firing ports. A remote-controlled machine gun was also provided.

SPECIFICATIONS	
Country of origin:	West Germany
Crew:	4 + 6
Weight:	28,200kg (62,040lb)
Dimensions:	Length: 6.79m (22ft 3in); width: 3.24m (10ft 8in); height: 2.95m (9ft 8in)
Range:	520km (323 miles)
Armour:	Classified
Armament:	1 x 20mm (.78in) Rh 202 gun; 1 x 7.62mm (.3in) coaxial MG
Powerplant:	1 x MTU MB 833 6-cylinder diesel engine developing 447kW (600hp)
Performance:	Maximum road speed: 75km/h (46.6mph); fording: 1.5m (4ft 11in); vertical obstacle: 1m (3ft 3in); trench: 2.5m (8ft 2in)

BMD

The BMD was designed for airborne forces, increasing firepower and mobility behind enemy lines. Based on the BMP-1, it had a hydraulic suspension-adjustment system to alter the level of ground clearance. In 1979, it spearheaded the Soviet invasion of Afghanistan.

SPECIFICATIONS	
Country of origin:	Soviet Union
Crew:	3 + 4
Weight:	6700kg (14,740lb)
Dimensions:	Length: 5.4m (17ft 9in); width: 2.63m (9ft 8in); height: 1.97m (6ft 6in)
Range:	320km (200 miles)
Armour:	15–23mm (.59–.9in)
Armament:	1 x 73mm (2.87in) gun; 1 x 7.62mm (.3in) coaxial MG; 2 x 7.62mm (.3in) front-mounted MGs; 1 x AT-3 'Sagger' ATGW
Powerplant:	1 x V-6 liquid-cooled diesel engine developing 179kW (240hp)
Performance:	Maximum road speed: 70km/h (43mph); fording: amphibious; vertical obstacle: .8m (2ft 8in); trench: 1.6m (5ft 3in)

TIMELINE

1970

AMX 10P

1970S PERSONNEL CARRIERS

With an all-aluminium hull, the French AMX-10 was fully amphibious, propelled by two waterjets. It carried a nuclear, biological and chemical (NBC) defence system. Variants include an ambulance, a repair vehicle and a mortar tractor for towing a Brandt 120mm (4.7in) mortar.

SPECIFICATIONS	
Country of origin:	France
Crew:	3 + 8
Weight:	14,200kg (31,240lb)
Dimensions:	Length: 5.778m (18ft 11in); width: 2.78m (9ft 1in); height: 2.57m (8ft 5in)
Range:	600km (373 miles)
Armour:	Classified
Armament:	1 x 20mm (.78in) gun; 1 x 7.62mm (.3in) coaxial MG
Powerplant:	one HS-115 V-8 water-cooled diesel, developing 280hp (209kW)
Performance:	Maximum road speed: 65km/h (40mph); fording: amphibious; vertical obstacle: .7m (2ft 4in); trench: 1.6m (5ft 3in)

AIFV

The shortcomings of the M113 prompted the AIFV. Better armed and armoured (with steel appliqué armour layers), it is fully amphibious. The main gun is enclosed for better protection and the seven troops in the back all have firing ports.

SPECIFICATIONS	
Country of origin:	USA
Crew:	3 + 7
Weight:	13,687kg (30,111lb)
Dimensions:	Length: 5.258m (17ft 3in); width: 2.819m (9ft 3in); height (overall): 2.794m (9ft 2in)
Range:	490km (305 miles)
Armour:	Classified
Armament:	1 x 25mm (.98in) Oerlikon gun; 1 x 7.62mm (.3in) coaxial MG
Powerplant:	1 x Detroit-Diesel 6V-53T V-6 diesel engine developing 197kW (264hp)
Performance:	Maximum road speed: 61.2km/h (38mph); fording: amphibious; vertical obstacle: .635m (2ft 1in); trench: 1.625m (5ft 4in)

Type 73 (command variant)

The Type 73 has an aluminium-armoured hull. NBC and night-vision equipment is standard, but the vehicle is not amphibious unless an optional swim-kit is fitted. To date, the Type 73 has been produced in only one variant, a command post vehicle.

SPECIFICATIONS	
Country of origin:	Japan
Crew:	3 + 9
Weight:	13,300kg (29,300lb)
Dimensions:	Length: 5.8m (19ft); width: 2.8m (9ft 2in); height: 2.2m (7ft 2in)
Range:	300km (190 miles)
Armour:	Aluminium (details classified)
Armament:	1 x 12.7mm (.5in) HMG; 1 x 7.62mm (.3in) MG
Powerplant:	1 x Mitsubishi 4ZF V4 diesel, developing 202kW (300hp) at 2200rpm
Performance:	Maximum road speed: 70km/h (43mph); fording: amphibious with swim-kit; gradient: 60 per cent; vertical obstacle: .7m (2ft 4in); trench 2m (5ft 7in)

1973

Reconnaissance Vehicles

Despite the increasing use of aircraft and helicopters for battlefield reconnaissance, the armoured reconnaissance vehicle still had a role. Such vehicles had the ability to operate in all types of terrain. Most armoured regiments of the world's leading armies incorporated an armoured reconnaissance squadron.

Schützenpanzer SPz 11-2 kurz

The SPz kurz was a German version of the French Hotchkiss SP 1A. It featured an all-welded hull of APC type. It was a reconnaissance vehicle variant of the 11-2, with the small turret carrying a Hispano-Suiza 20mm (.78in) gun.

SPECIFICATIONS	
Country of origin:	Germany
Crew:	5
Weight:	8200kg (18,100lb)
Dimensions:	Length: 4.51m (14ft 9in); width: 2.28m (7ft 3in); height: 1.97m (6ft 5.5in)
Range:	400km (250 miles)
Armour:	15mm (.59in)
Armament:	1 x 20mm (.78in) Hispano-Suiza 820/L35 gun; 3 x smoke grenade launchers
Powerplant:	1 x Hotchkiss 6-cylinder petrol developing engine 122kW (164hp) at 3900rpm
Performance:	Maximum speed: 58km/h (36mph); fording: 1m (3ft 3in); gradient: 60 per cent; vertical obstacle: .6m (2ft); trench: 1.5m (4ft 10in)

Lynx CR

The Lynx Command and Reconnaissance Vehicle was based on the M113A1 APC, sharing the all-welded aluminium hull and tracked configuration and being amphibious. Its armament was a 12.7mm (.5in) Browning M2 HB HMG and 7.62mm (.3in) Browning M1919 MG.

SPECIFICATIONS	
Country of origin:	USA
Crew:	3
Weight:	8775kg (19,300lb)
Dimensions:	Length: 4.6m (15ft 1in); width: 2.41m (7ft 11in); height: 1.65m (5ft 5in)
Range:	525km (325 miles)
Armour:	Aluminium (details not available)
Armament:	1 x 12.7mm (.5in) MG; 1 x 7.62mm (.3in) MG
Powerplant:	1 x Detroit Diesel GMC 6V53 6-cylinder diesel, developing 160kW (215hp) at 2800rpm
Performance:	Maximum road speed: 70km/h (43mph); fording: amphibious; gradient: 60 per cent; vertical obstacle: .61m (2ft); trench: 1.47m (4ft 9.8in)

TIMELINE
1958 1960 1961

YP-104

RECONNAISSANCE VEHICLES

The YP-104 was a Dutch scout car that relied mainly on its top speed of 98km/h (61mph) for defence, though a single 7.62mm (0.3in) machine gun was an optional fitting. The armour was sufficient to protect against light small-arms fire only.

SPECIFICATIONS	
Country of origin:	Netherlands
Crew:	2
Weight:	5400kg (11,900lb)
Dimensions:	Length: 4.33m (14ft 2.5in); width: 2.08m (6ft 9.8in); height: 2.03m (6ft 8in)
Range:	500km (310 miles)
Armour:	Up to 16mm (.63in)
Armament:	1 x 7.62mm (.3in) MG (optional)
Powerplant:	1 x Herkules JXLD 6-cylinder petrol, developing 98kW (131hp)
Performance:	Maximum road speed: 98km/h (61mph); fording .91m (2ft 11.8in); gradient 46 per cent; vertical obstacle .41m (1ft 4.2in); trench 1.22m (4ft)

FUG

The FUG was an amphibious scout car based upon the Soviet BRDM-1, and was propelled by two waterjets set in the rear of the hull. Three more modern features were central tyre-pressure regulation, infrared headlights and NBC options.

SPECIFICATIONS	
Country of origin:	Hungary
Crew:	2 + 4
Weight:	7000kg (15,400lb)
Dimensions:	Length: 5.79m (18ft 11in); width: 2.5m (8ft 2in); height: 1.91m (6ft 3in)
Range:	600km (370 miles)
Armour:	10mm (.39in)
Armament:	1 x 7.62mm (.3in) SGMB MG
Powerplant:	1 x Csepel D.414.44 4-cylinder diesel engine developing 75kW (100hp)
Performance:	Maximum road speed: 87km/h (54mph); fording: amphibious; gradient: 32 per cent; vertical obstacle: .4m (1ft 4in)

EE-3 Jararaca

The EE-3 Jararaca is a 4x4 scout car, only 4.12m (13ft 6.2in) long and 1.56m (5ft 1.4in) high. Its small size and light weight give it excellent manoeuvrability and speed. It takes various armaments, from the 12.7mm (.5in) Browning machine gun to a MILAN ATGW.

SPECIFICATIONS	
Country of origin:	Brazil
Crew:	3
Weight:	5500kg (12,100lb)
Dimensions:	Length: 4.12m (13ft 6.2in); width: 2.13m (6ft 11in); height: 1.56m (5ft 1.4in)
Range:	750km (470 miles)
Armour:	Double-layer steel (details classified)
Armament:	1 x 12.7mm (.5in) Browning M2 HB HMG as standard
Powerplant:	1 x Mercedes-Benz OM 314A 4-cylinder turbo diesel, developing 89kW (120hp) at 2800rpm
Performance:	Maximum road speed: 100km/h (60mph); fording: .6m (2ft); gradient: 60 per cent; vertical obstacle: .4m (1.3ft); trench: .4m (1.3ft)

 1964    1980

Scorpion

The first prototype of the Alvis Scorpion, officially named Combat Vehicle Reconnaissance (Tracked), appeared in 1969, following a British Army requirement for a tracked reconnaissance vehicle to replace the Saladin armoured car. It entered service in 1972 and was exported to countries all over the world, particularly Belgium.

Scorpion

TURRET
Seated on the left of the turret, the commander has a roof-mounted sight and seven periscopes. The commander also loads the main armament

POWERPLANT
The Scorpion is powered by a 4.2l Jaguar engine, de-rated and with a reduced compression ratio to allow the use of military fuel

CREW
Located in a semi-reclining position in the front of the vehicle, the driver is provided with overhead cover by a hatch that lifts and rotates

Scorpion

The Scorpion saw action in the Falklands in 1982, where its flotation screens giving amphibious capability were particularly useful during the landings. The Scorpion proved its worth in both reconnaissance and in high-speed advances. The Scorpion chassis has been used for a complete range of tracked vehicles, such as the Sultan, Spartan and Scimitar.

SPECIFICATIONS

Country of origin:	United Kingdom
Crew:	3
Weight:	8073kg (17,760lb)
Dimensions:	Length: 4.794m (15ft 8.75in); width: 2.235m (7ft 4in); height: 2.102m (6ft 10.75in)
Range:	644km (400 miles)
Armour:	Up to 12.7mm (.5in)
Armament:	1 x 76mm (2.99in) gun; 1 x 7.62mm (.3in) coaxial MG
Powerplant:	1 x Jaguar 4.2-litre petrol engine developing 142kW (190hp)
Performance:	Maximum road speed: 80km/h (50mph); fording: 1.067m (3ft 6in); vertical obstacle: .5m (1ft 8in); trench: 2.057m (6ft 9in)

Smoke dischargers
A bank of three or four smoke dischargers is located on each side of the turret

Armour
The Scorpion's all-welded aluminium front armour is proof against machine gun rounds up to 14.5mm (.57in) calibre, as fitted to the Russian BRDM-2 amphibious scout car

Suspension
The torsion bar suspension consists of five aluminium road wheels with rubber tyres; the first and last wheels have hydraulic shock absorbers

COLD WAR TANKS AND AFVS

Combat Engineer Vehicles

Just as had been the case with their forebears of World War II, the main battle tanks of the Cold War provided the chassis for many types of combat engineer vehicles, all of which had a vital part to play on the modern battlefield, from blasting and bulldozing gaps in fortifications to vehicle recovery.

Centurion AVRE

In the 1960s, the Centurion Mk V Armoured Vehicle Royal Engineers (AVRE) carried a heavy demolition gun for blasting enemy fortifications and a dozer blade for removing obstacles. It was also capable of laying down tracks for wheeled vehicles.

SPECIFICATIONS	
Country of origin:	United Kingdom
Crew:	5
Weight:	51,809kg (113,979lb)
Dimensions:	Length: 8.69m (28ft 6in); width: 3.96m (13ft); height: 3m (9ft 10in)
Range:	177km (110 miles)
Armour:	17–118mm (.66–4.6in)
Armament:	1 x 165mm (6.5in) demolition gun; 1 x 7.62mm (.3in) coaxial MG; 1 x 7.62mm (.3in) anti-aircraft MG
Powerplant:	1 x Rolls-Royce Meteor Mk IVB 12-cylinder petrol engine developing 484.7kW (650hp)
Performance:	Maximum road speed: 34.6km/h (21.5mph); fording: 1.45m (4ft 9in); vertical obstacle: .94m (3ft 1in); trench: 3.35m (11ft)

M728

The M728 was a combat engineer vehicle based on the chassis of the M60 main battle tank. The M60's gun was replaced by a demolition-charge launcher and a hydraulically operated dozer blade was added. The vehicle could be equipped for deep-fording.

SPECIFICATIONS	
Country of origin:	USA
Crew:	4
Weight:	53,200kg (117,040lb)
Dimensions:	Length (travelling): 8.92m (29ft 3in); width (overall): 3.71m (12ft 2in); height (travelling): 3.2m (10ft 6in)
Range:	451km (280 miles)
Armour:	Classified
Armament:	1 x 165mm (6.5in) demolition gun; 1 x 7.62mm (.3in) MG
Powerplant:	1 x Teledyne Continental AVDS-1790-2A 12-cylinder diesel engine developing 559.3kW (750bhp)
Performance:	Maximum road speed: 48.3km/h (30mph); fording: 1.22m (4ft); vertical obstacle: .76m (2ft 6in); trench: 2.51m (8ft 3in)

TIMELINE

1960  1963 1968

Leopard 1 Armoured Engineer Vehicle

The Leopard 1 Armoured Engineer Vehicle was based on the chassis of the Leopard 1 main battle tank. A dozer blade, fitted to the front, could be widened and fitted with 'scarifiers' for digging up road surfaces (to defend West Germany from an attack from the east).

SPECIFICATIONS	
Country of origin:	West Germany
Crew:	4
Weight:	40,800kg (89,760lb)
Dimensions:	Length: 8.98m (26ft 2.2in); width: 3.75m (12ft 3.6in); height: 2.69m (8ft 9.9in)
Range:	850km (528 miles)
Armour:	10–70mm (.39–2.75in)
Armament:	1 x 7.62mm (.3in) MG
Powerplant:	1 x MTU MB 838 Ca.M500 10-cylinder diesel engine developing 618.9kW (830hp)
Performance:	Maximum road speed: 65km/h (40.4mph); fording: 2.1m (6ft 11in); vertical obstacle: 1.15m (3ft 9.3in); trench: 3m (9ft 10in)

Chieftain AVRE

The Chieftain AVRE had a similar chassis and engine to the Chieftain MK 5 main battle tank. The main difference is that the gun turret was removed and replaced with an armoured 'penthouse'. The vehicle carried three large hampers for storing equipment.

SPECIFICATIONS	
Country of origin:	United Kingdom
Crew:	3
Weight:	51,809kg (113,979lb)
Dimensions:	Length: 7.52m (24ft 9in); width: 3.663m (11ft 9in); height: 2.89m (9ft 4.8in)
Range:	500km (300 miles)
Armour:	Classified
Armament:	1 x 7.62mm (.3in) MG; 2 x 6 smoke dischargers (front); 2 x 4 smoke dischargers (rear)
Powerplant:	1 x Leyland L60 (No.4 Mk 8) 12-cylinder diesel engine developing 559kW (750hp)
Performance:	Maximum road speed: 48km/h (28.8mph); fording: 1.067m (3ft 6in); vertical obstacle: .9m (3ft); trench: 3.15m (10ft 4in)

AMX-30 Tractor

The AMX-30 EBG was based on the AMX-30 main battle tank chassis and designed to carry out a number of roles, including laying mines and clearing battlefield obstacles. At the rear was a winch for recovering damaged vehicles.

SPECIFICATIONS	
Country of origin:	France
Crew:	3
Weight:	38,000kg (83,600lb)
Dimensions:	Length: 7.9m (25ft 11in); width: 3.5m (11ft 5.8in); height: 2.94m (9ft 7.7in)
Range:	500km (311 miles)
Armour:	80mm (3.14in)
Armament:	1 x 142mm (5.59in) demolition-charge launcher; 1 x 7.62mm (.3in) MG
Powerplant:	1 x Hispano-Suiza HS 110-2 12-cylinder multi-fuel engine developing 522.0kW (700hp)
Performance:	Maximum road speed: 65km/h (40.4mph); fording: 2.50m (8ft 2.4in); vertical obstacle: .9m (2ft 11.4in); trench: 2.9m (9ft 6.2in)

1970

LATE COLD WAR TANKS AND AFVS

Merkava

Prior to the Six-Day War in 1967, Israel had relied on Sherman and Centurion tanks for its armoured forces. However, doubts as to future supplies, as well as concerns that these tanks did not fully meet Israeli requirements, prompted development of the Merkava. It saw action for the first time against Syrian forces in Lebanon in 1982.

Merkava

Compared to other modern main battle tanks, the Merkava is slow and has a poor power-to-weight ratio. However, it is designed for specific tactical requirements, which differ from those of most other tank producers. The emphasis is on crew survivability. Hence the Merkava's small cross-section, which makes it less of a target, and well-sloped armour for greatest protection.

SPECIFICATIONS	
Country of origin:	Israel
Crew:	4
Weight:	55,898kg (122,976lb)
Dimensions:	Length: 8.36m (27ft 5.25in); width: 3.72m (12ft 2.5in); height: 2.64m (8ft 8in)
Range:	500km (310 miles)
Armour:	Classified
Armament:	1 x 105mm (4.1in) rifled gun; 1 x 7.62mm (.3in) MG
Powerplant:	1 x Teledyne Continental AVDS-1790-6A V-12 diesel engine developing 671kW (900hp)
Performance:	Maximum road speed: 46km/h (28.6mph); vertical obstacle: 1m (3ft 3.3in); trench: 3m (9ft 10in)

UPDATES
A 1995 version, the Mk 3B (also known as Merkava Baz), has an improved fire-control system and a built-in NBC protection and air conditioning system

LAYOUT
The Merkava is optimized for operations in the rough terrain of northern Israel and the Golan Heights and is unusual in layout, with the engine at the front and the fighting compartment to the rear

ARMOUR
A modular armour package (called 'Kasag') was retrofitted to the tank, making the Merkava one of the most protected tanks in the world

MERKAVA

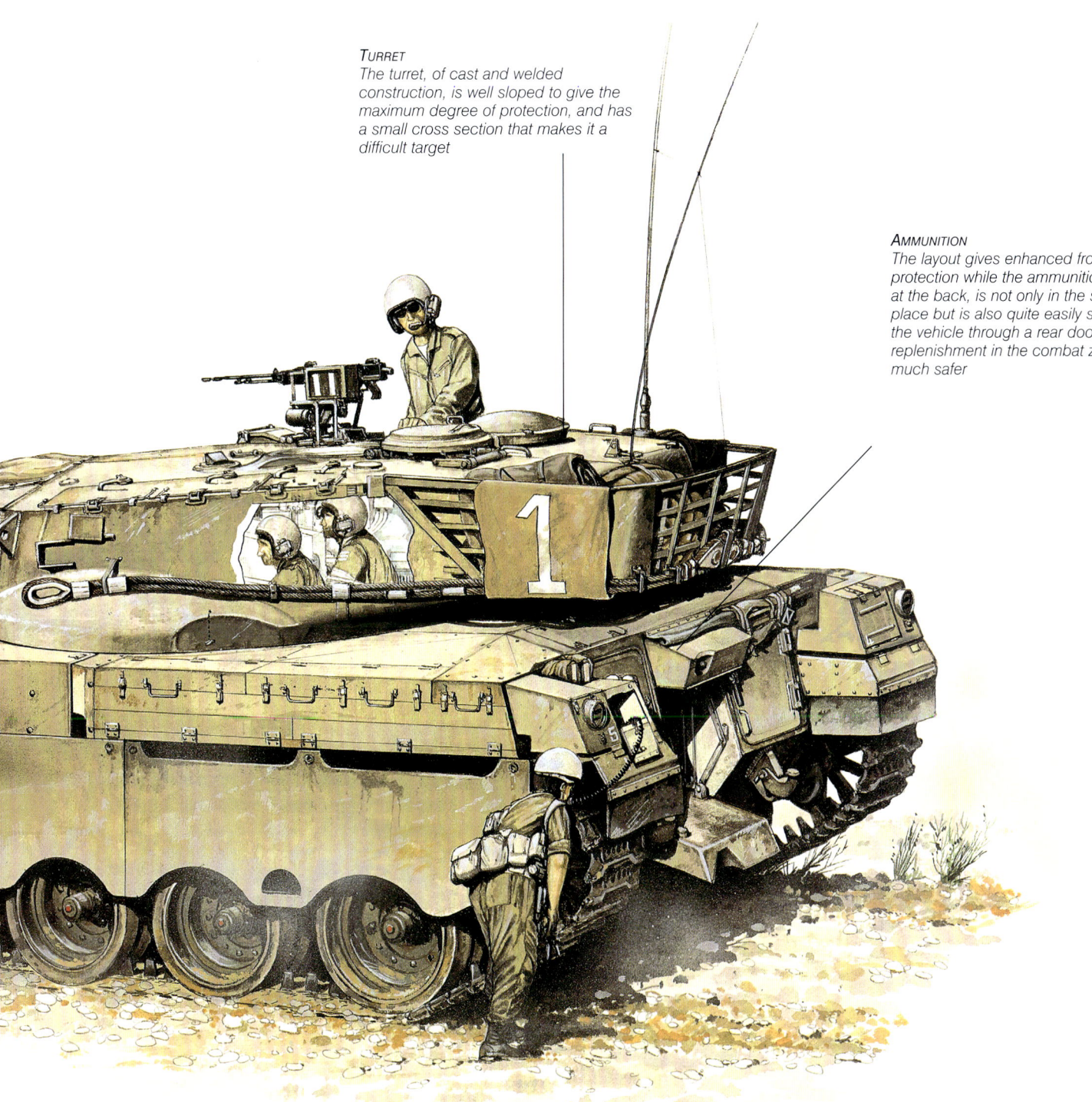

TURRET
The turret, of cast and welded construction, is well sloped to give the maximum degree of protection, and has a small cross section that makes it a difficult target

AMMUNITION
The layout gives enhanced frontal protection while the ammunition, kept at the back, is not only in the safest place but is also quite easily stowed in the vehicle through a rear door, making replenishment in the combat zone much safer

LATE COLD WAR TANKS AND AFVS

1980s Main Battle Tanks

It takes up to 15 years from the initial design of a new tank to its entry into service, so the tanks that began to be used in the early 1980s were the fruits of 1960s technology. One of the most radical improvements was the introduction of composite armour. Tanks began to take on a different appearance.

Type 69

The Type 69 Mk 1 was fitted with a 100mm (3.9in) gun, which was probably based on that of the Soviet T-62, examples of which were captured during border clashes with the Soviet Union. Variants include an armoured bridgelayer and armoured recovery vehicle.

SPECIFICATIONS	
Country of origin:	China
Crew:	4
Weight:	36,500kg (80,300lb)
Dimensions:	Length: 8.68m (26ft 6in); width: 3.3m (10ft 1in); height: 2.87m (8ft 10in)
Range:	375km (250 miles)
Armour:	100mm (3.94in)
Armament:	1 x 100mm (3.9in) gun; 2 x 7.62mm (.3in) MGs; 1 x 12.7mm (.5in) HMG; 2 x smoke rocket dischargers
Powerplant:	1 x V-12 liquid-cooled diesel engine developing 432kW (580hp)
Performance:	Maximum road speed: 50km/h (31.3mph); fording: 1.4m (4ft 7in); vertical obstacle: .8m (2ft 7in); trench: 2.7m (8ft 10in)

T-80

The T-80 was a development of the T-72 main battle tank. Its main gun was a fully stabilized 125mm (4.9in) weapon as fitted in the T-72, but was capable of firing a greater range of ammunition, including depleted uranium rounds for extra armour-piercing capability.

SPECIFICATIONS	
Country of origin:	Soviet Union
Crew:	3
Weight:	48,363kg (106,400lb)
Dimensions:	Length: 9.9m (32ft 6in); width: 3.4m (11ft 2in); height: 2.2m (7ft 3in)
Range:	450km (281 miles)
Armour:	Classified
Armament:	1 x 125mm (4.9in) gun; 1 x 7.62mm (.3in) coaxial MG; 1 x 12.7mm (.5in) anti-aircraft HMG
Powerplant:	1 x multi-fuel gas turbine engine developing 745kW (1000hp)
Performance:	Maximum road speed: 70km/h (43.75mph); fording: 5m (16ft 5in); vertical obstacle: 1m (3ft 4in); trench: 2.85m (9ft 4in)

TIMELINE

1982  1984 1986

Type 88

The Type 88 is also known as the K1. The main smoothbore armament has a thermal sleeve and fume extractor and uses a computerized fire-control system based on the M1 ballistic computer and an environmental sensor package.

SPECIFICATIONS	
Country of origin:	South Korea
Crew:	4
Weight:	52,000kg (114,400lb)
Dimensions:	Length: 9.67m (29ft 6in); width: 3.59m (10ft 11in); height: 2.25m (6ft 10in)
Range:	500km (313 miles)
Armour:	classified
Armament:	1 x 105mm (4.1in) gun; 1 x 7.62mm (.3in) coaxial MG; 1 x 12.7mm (.5in) anti-aircraft HMG; 2 x 6 smoke dischargers
Powerplant:	1 x liquid-cooled turbocharged diesel engine developing 895kW (1200hp)
Performance:	Maximum road speed: 65km/h (40.6mph); fording: 1.2m (3ft 11in); vertical obstacle: 1m (3ft 4in)

T-80BV

The T-80 BV is a T-80B with Kontakt-1 explosive reactive armour. The smoke grenade launchers were moved from either side of the main armament back to either side of the turret, positioned between the turret side and the ERA panels.

SPECIFICATIONS	
Country of origin:	Soviet Union
Crew:	3
Weight:	46,000kg (101,413lb)
Dimensions:	Length: 9.66m (31ft 8in); width: 3.59m (11ft 10in); height: 2.20m (7ft 2in)
Range:	440 km (273 miles)
Armour:	Classified
Armament:	1 x 125mm (4.9in) smoothbore gun; 1 x 12.7mm (.5in) MG; 1 x 7.62mm (.3in) MG; 1 x AT-8 songster ATGW
Powerplant:	1 x GTD-1250 multi-fuel gas turbine engine developing 932kW (1250hp)
Performance:	Maximum road speed: 70km/h (44mph)

T-80U

This variant of the T-80A has a new 820kW (1100hp) GTD-1000F multi-fuel gas turbine engine, a 1A46 fire control system and a new turret. The T-80U can fire the 9M119 Svir (AT-11 Sniper) and 9M119M Refleks (AT-11B Sniper) anti-tank guided weapons.

SPECIFICATIONS	
Country of origin:	Soviet Union
Crew:	3
Weight:	46,000kg (101,413lb)
Dimensions:	Length: 9.66m (31ft 8in); width: 3.59m (11ft 10in); height: 2.20m (7ft 2in)
Range:	440 km (273 miles)
Armour:	Classified
Armament:	1 x 125mm (4.9in) smoothbore gun; 1 x 12.7mm (.5in) MG; 1 x 7.62mm (.3in) MG; 1 x 9K119 Refleks missile system
Powerplant:	1 x GTD-1250 multi-fuel gas turbine engine developing 932kW (1250hp)
Performance:	Maximum road speed: 70km/h (44mph)

 1988

LATE COLD WAR TANKS AND AFVS

Challenger 1

The Challenger I was introduced in 1982 as a replacement for the Chieftain. It was slower than contemporary Warsaw Pact vehicles, but made up for this with its composite Chobham armour, which was virtually impenetrable to enemy rounds, and the greater accuracy of its armament.

Challenger 1

POWERPLANT
Powerplant was a Rolls-Royce (Perkins) Condor 12V-1200 engine, which used high-efficiency turbochargers and had a low specific fuel consumption

Challenger 1 tanks were fitted with the TOGS (Thermal Observation and Gunnery Sight) in an armoured barbette on the right of the turret. This provided separate outputs for the commander and gunner.

ARMOUR
The Challenger 1 was fitted with the highly classified Chobham armour, which provided an effective defence against anti-tank rockets and anti-tank guided missiles

TRACK
The Challenger's track was a single-pin type with removable rubber pads, and consideration was given to replacing it with a new track offering a longer life and less rolling resistance

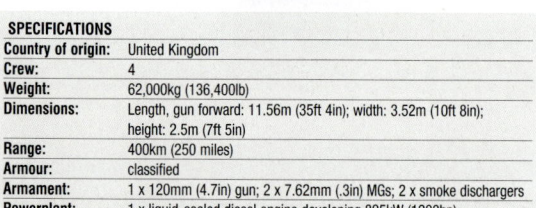

SPECIFICATIONS	
Country of origin:	United Kingdom
Crew:	4
Weight:	62,000kg (136,400lb)
Dimensions:	Length, gun forward: 11.56m (35ft 4in); width: 3.52m (10ft 8in); height: 2.5m (7ft 5in)
Range:	400km (250 miles)
Armour:	classified
Armament:	1 x 120mm (4.7in) gun; 2 x 7.62mm (.3in) MGs; 2 x smoke dischargers
Powerplant:	1 x liquid-cooled diesel engine developing 895kW (1200hp)
Performance:	Maximum road speed: 55km/h (35mph); fording: 1m (3ft 4in); vertical obstacle: 0.9m (2ft 10in); trench: 2.8m (9ft 2in)

Challenger 1

CUPOLA
The commander had a modified No 15 cupola fitted with either a day sight or image intensifier sight for night combat. Nine periscopes provided the commander with all-round vision

PERISCOPE
The driver had a wide-angle periscope for day driving, and this could be replaced by a Pilkington passive night sight for driving in conditions of darkness

LATE COLD WAR TANKS AND AFVS

M1 Abrams

First entering service in 1980, the M1 Abrams is the main battle tank of the United States' Army today. It is one of the best-armed, most heavily armoured, and highly mobile tanks. Notable features include a powerful gas turbine engine, sophisticated composite armour, and separate ammunition storage in a blow-out compartment.

M1A1 Abrams

The M1 Abrams main battle tank will forever be remembered for its epic dash through the Iraqi desert during the Gulf War of 1991, the aim being to cut off and destroy Saddam Hussein's Republican Guard. The Abrams produced an excellent performance, dispelling concerns by its critics that it might not show up well in combat against the latest Russian equipment used by the Iraqis.

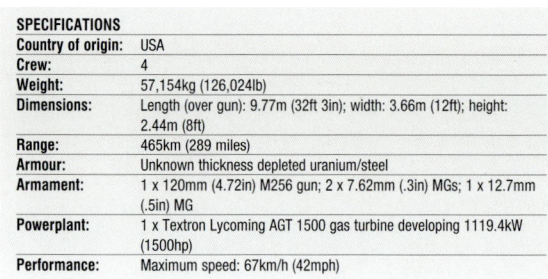

SPECIFICATIONS	
Country of origin:	USA
Crew:	4
Weight:	57,154kg (126,024lb)
Dimensions:	Length (over gun): 9.77m (32ft 3in); width: 3.66m (12ft); height: 2.44m (8ft)
Range:	465km (289 miles)
Armour:	Unknown thickness depleted uranium/steel
Armament:	1 x 120mm (4.72in) M256 gun; 2 x 7.62mm (.3in) MGs; 1 x 12.7mm (.5in) MG
Powerplant:	1 x Textron Lycoming AGT 1500 gas turbine developing 1119.4kW (1500hp)
Performance:	Maximum speed: 67km/h (42mph)

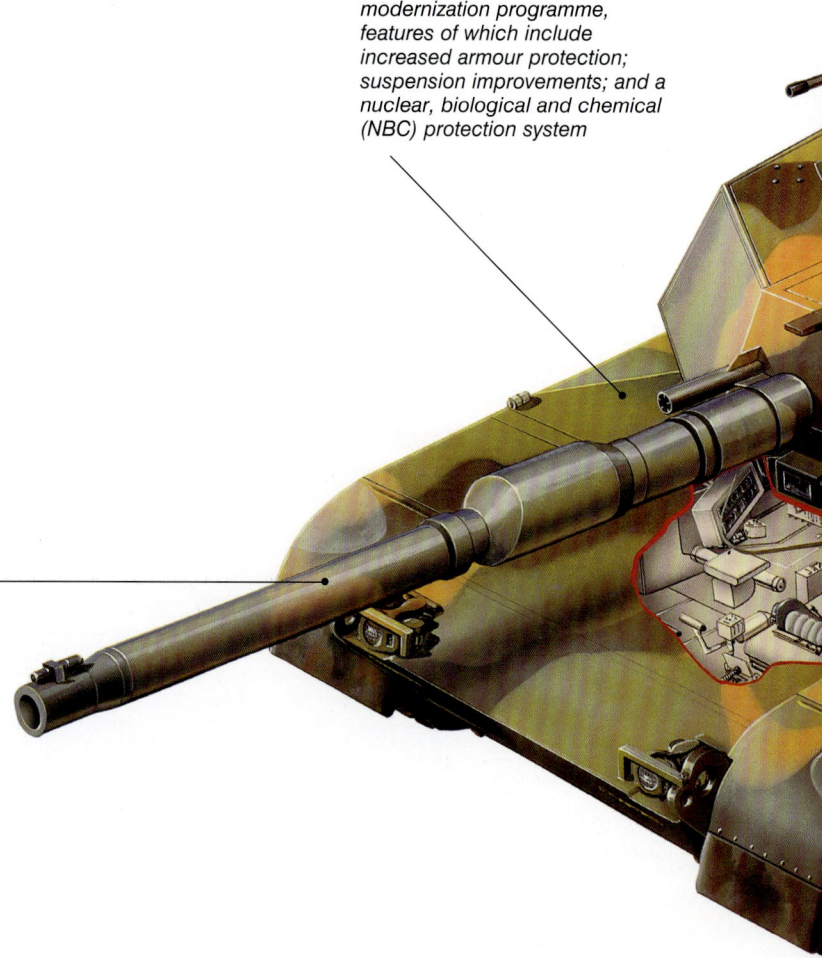

MODERNIZATION
The M1A1 has undergone a modernization programme, features of which include increased armour protection; suspension improvements; and a nuclear, biological and chemical (NBC) protection system

ARMAMENT
The Abrams' 120mm (4.7in) main gun, combined with the powerful 1119.4kW (1500hp) turbine engine and special armour, make the tank particularly suitable for attacking or defending against large concentrations of heavy armoured forces on a highly lethal battlefield

M1A2 Abrams

EQUIPMENT
The Abrams' equipment includes a commander's independent thermal viewer, an improved commander's weapon station, position navigation equipment, a distributed data and power architecture, an embedded diagnostic system and improved fire control systems

PERISCOPES
The commander has six periscopes to provide 360° vision and a sight for the 12.7mm (.5in) Browning anti-aircraft gun. He is also linked to the gunner's primary sight

TECHNOLOGY
A System Enhancement Programme includes upgrades to processors/memory that enable the M1A2 to use the Army's common command and control software, enabling the rapid transfer of digital situational data and overlays

ENGINE
The gas turbine engine in the M1 was unique in the MBT world until the introduction of the Russian T-80. Weight for weight, it is twice as efficient as a conventional engine

LATE COLD WAR TANKS AND AFVS

1980s Main Battle Tanks Pt 2

In the 1980s, a number of armies that were not aligned to the principal power blocs endeavoured to modernize their forces. Main battle tanks were on their arms shopping list, and the arms manufacturers of both West and East took advantage of this requirement to boost their MBT exports.

AMX-40

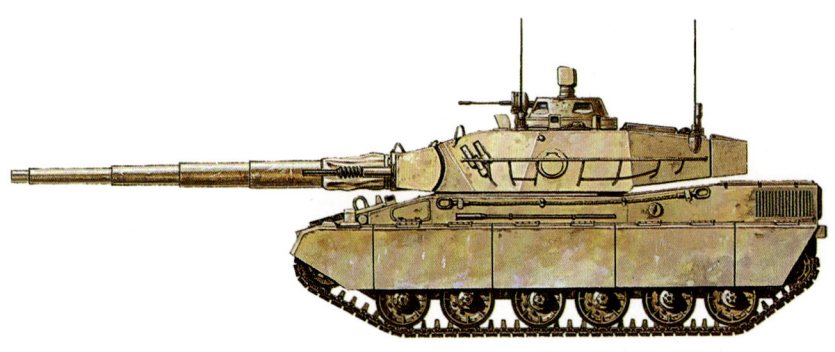

The AMX-40 has a laser rangefinder, gun stabilization equipment and a low-light television for night-fighting. The ammunition compartment is carried in the turret and surrounded by bulkheads. If hit, the ammunition explodes upwards, away from the crew below.

SPECIFICATIONS	
Country of origin:	France
Crew:	4
Weight:	43,000kg (94,600lb)
Dimensions:	Length: 10.04m (32ft 11.3in); width: 3.36m (11ft 0.3in); height: 3.08m (10ft 1.3in)
Range:	600km (373 miles)
Armour:	Classified
Armament:	1 x 120mm (4.7in) gun; 1 x 20mm (.78in) gun in cupola; 1 x 7.62mm (.3in) MG
Powerplant:	1 x Poyaud 12-cylinder diesel engine developing 820kW (1100hp)
Performance:	Maximum road speed: 70km/h (44mph); fording: 1.3m (4ft 3in); vertical obstacle: 1m (6ft 7in); trench: 3.2m (10ft 6in)

C1 Ariete

The C1 Ariete main battle tank uses the latest Galileo computerized fire-control system. The thermal vision night sight and laser rangefinder give high single-shot kill probability, whether moving or stationary. Variants include the XA-181 air defence variant.

SPECIFICATIONS	
Country of origin:	Italy
Crew:	4
Weight:	54,000kg (118,800lb)
Dimensions:	Length: 9.67m (29ft 6in); width: 3.6m (11ft); height: 2.5m (7ft 7in)
Range:	600km (375 miles)
Armour:	Classified
Armament:	1 x 120mm (4.7in) gun; 1 x 7.62mm (.3in) coaxial MG; 1 x 7.62mm (.3in) anti-aircraft MG; 2 x 4 smoke dischargers
Powerplant:	1 x IVECO FIAT MTCA V-12 turbocharged diesel engine developing 932kW (1250hp)
Performance:	Maximum road speed: 66km/h (41.3mph); fording: 1.2m (3ft 11in); vertical obstacle: 2.1m (6ft 11in); trench: 3m (9ft 10in)

TIMELINE

1985

1986

1980S MAIN BATTLE TANKS PT2

Type 80

The Chinese Type 80 main battle tank has a computerized fire-control system, including a laser rangefinder. The vehicle carries a snorkel, which can be fitted for deep fording. It also has an in-built fire-detection/suppression system.

SPECIFICATIONS
Country of origin:	China
Crew:	4
Weight:	38,000kg (83,600lb)
Dimensions:	Length: 9.33m (28ft 6in); width: 3.37m (10ft 4in); height: 2.3m (7ft)
Range:	570km (356 miles)
Armour:	Classified
Armament:	1 x 105mm (4.1in) gun; 1 x 7.62mm (.3in) coaxial MG; 1 x 12.7mm (.5in) coaxial HMG
Powerplant:	1 x V-12 diesel engine developing 544kW (730hp); manual transmission
Performance:	Maximum road speed: 60km/h (37.5mph); fording: 1.4m (4ft 7in); vertical obstacle: .8m (2ft 7in); trench: 2.7m (8ft 10in)

ENGESA EE-T1 Osorio

The Osorio features a laser rangefinder, stabilizers for firing on the move and thermal-imaging cameras, as well as a full nuclear, biological and chemical (NBC) defence system. Variants include an anti-aircraft vehicle, armoured recovery vehicle and bridgelayer.

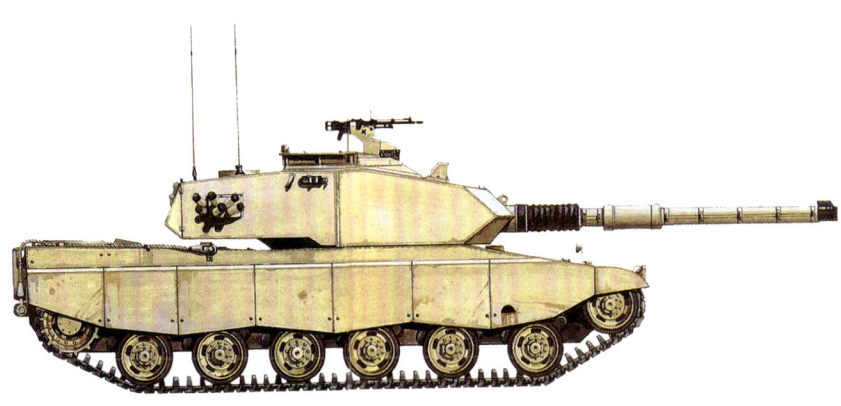

SPECIFICATIONS
Country of origin:	Brazil
Crew:	4
Weight:	39,000kg (85,800lb)
Dimensions:	Length: 9.995m (32ft 9.5in); width: 3.26m (10ft 8.3in); height: 2.371m (7ft 9.3in)
Range:	550km (342 miles)
Armour:	Classified
Armament:	1 x British 105mm (4.1in)/French 120mm (4.7in) gun; 1 x 7.62mm (.3in) MG
Powerplant:	1 x 12-cylinder diesel engine developing 745kW (1000hp)
Performance:	Maximum road speed: 70km/h (43.5mph); fording: 1.2m (3ft 11in); vertical obstacle: 1.15m (3ft 4in); trench: 3.0m (9ft 10in)

Type 85

The armament of the Type 85 main battle tank is fed by an automatic loader, which allows the crew to be reduced to a minimum. Even so, space inside is limited for the ammunition itself is made up of a separate projectile and charge.

SPECIFICATIONS
Country of origin:	China
Crew:	3
Weight:	41,000kg (90,200lb)
Dimensions:	Length: 10.28m (31ft 5in); width: 3.45m (10ft 6in); height: 2.3m (7ft)
Range:	500km (312 miles)
Armour:	Classified
Armament:	1 x 125mm (4.9in) gun; 1 x 7.62mm (.3in) coaxial MG; 1 x 12.7mm (.5in) anti-aircraft HMG; 2 x smoke grenade launchers
Powerplant:	1 x V-12 supercharged diesel engine developing 544kW (730hp)
Performance:	Maximum road speed: 57.25km/h (35.8mph); fording: 1.4m (4ft 7in); vertical obstacle: .8m (2ft 7in); trench: 2.7m (8ft 10in)

 1988

LATE COLD WAR TANKS AND AFVS

Light Tanks

The Falklands War of 1982, where the British Scorpion light tank played a prominent part, showed that there was an ongoing requirement for vehicles of this type. Apart from providing mobile firepower, they were also light enough to negotiate terrain such as marshy ground without fear of becoming bogged down.

Scorpion 90

The Scorpion 90 is an export version of the Scorpion light tank, armed with the long-barrelled Cockerill Mk3 M-A1 90mm (3.54in) gun, fitted with a prominent muzzle brake. It was purchased by the Indonesian, Malaysian and Venezuelan armed forces.

SPECIFICATIONS	
Country of Origin:	United Kingdom
Crew:	3
Weight:	8573kg (17,798lb)
Dimensions:	Length: 5.28m (17ft 4in); width: 2.24m (7ft 4in); height: 2.10m (6ft 11in)
Range:	644km (400 miles)
Armour:	12.7mm (.5in)
Armament:	1 x 90mm (3.54in) gun, 1 x 7.62mm (0.30in) MG
Powerplant:	1 x Jaguar J60 No 1 Mk 100B 4.2 litre, 6-cylinder petrol engine, developing 142kW (190hp)
Performance:	Maximum road speed: 73km/h (45mph)

AMX-10 PAC 90

The PAC-90 is a fire support and anti-tank variant of the AMX-10, designed for fire support and anti-tank work. It is armed with a 90mm (3.54in) GIAT gun and is one of seven main AMX-10 versions in service.

SPECIFICATIONS	
Country of origin:	France
Crew:	3 plus 8 troops
Weight:	14,500kg (31967lb)
Dimensions:	Length: 5.9m (19ft 4.3in); width: 2.83m (9ft 3.4in); height: 2.83m (9ft 3.4in)
Range:	500km (310.7 miles)
Armour:	Classified
Armament:	1 x 20mm (0.79in) gun; 1 x coaxial 7.62mm (0.3in) MG
Powerplant:	Hispano-Suiza HS 115, V-8 diesel, 193.9kW (260hp)
Performance:	Maximum road speed: 65km/h (40.4mph); water: 7km/h (4.35mph)

TIMELINE

1975       1978

LIGHT TANKS

TAM

The hull of the TAM was based on that of the MICV used by the West German Army. The armour is comparatively poor but is well sloped to give as much protection as possible. It was not produced in time to have any impact on the 1982 Falklands conflict.

SPECIFICATIONS	
Country of origin:	Argentina
Crew:	4
Weight:	30,500kg (67,100lb)
Dimensions:	Length (gun forward): 8.23m (25ft 2in); width: 3.12m (9ft 6in); height: 2.42m (7ft 5in)
Range:	900km (560 miles)
Armour:	Classified
Armament:	1 x 105mm (4.1in) gun; 1 x 7.62mm (.3in) coaxial MG; 1 x 7.62mm (.3in) anti-aircraft MG
Powerplant:	1 x V-6 turbo-charged diesel engine developing 537kW (720hp)
Performance:	Maximum road speed: 75km/h (46.9mph); fording: 1.5m (4ft 11in); vertical obstacle: 1m (3ft 3in); trench: 2.5m (8ft 2in)

Stingray

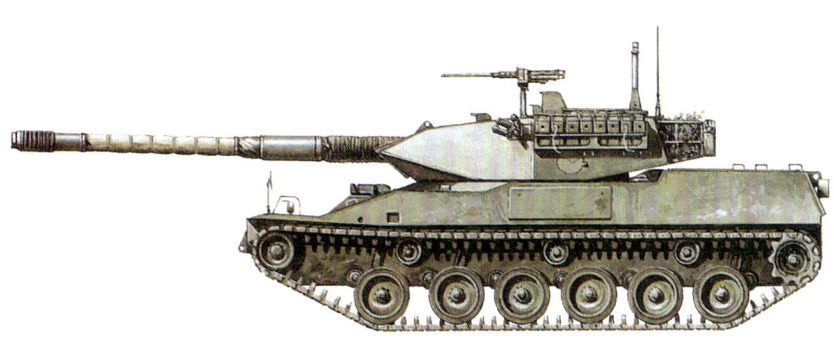

To save development costs, the Stingray adapted parts proven by other vehicles. It is equipped with a laser rangefinder, stabilization devices for the gun, and nuclear, biological and chemical (NBC) protection. Its drawback is its light armour.

SPECIFICATIONS	
Country of origin:	USA
Crew:	4
Weight:	19,051kg (41,912lb)
Dimensions:	Length: 9.35m (30ft 8in); width: 2.71m (8ft 11in); height: 2.54m (8ft 4in)
Range:	483km (300 miles)
Armour:	Classified
Armament:	1 x 105mm (4.1in) gun; 1 x 7.62mm (.3in) coaxial MG; 1 x 7.62mm (.3in) anti-aircraft MG
Powerplant:	1 x Diesel Model 8V-92 TA diesel engine developing 399kW (535hp)
Performance:	Maximum road speed: 69km/h (43mph); fording: 1.22m (4ft); vertical obstacle: .76m (2ft 6in); trench: 1.69m (5ft 7in)

Close Combat Vehicle – Light

The prototype of the CCV-L created interest required only a three-man crew, thanks to an automatic loading system for the main armament. This gave a rate of fire of 12 round per minute. To save costs, the CCV-L borrowed parts from various other vehicles.

SPECIFICATIONS	
Country of origin:	USA
Crew:	3
Weight:	19,414kg (42,710lb)
Dimensions:	Length: 9.37m (30ft 9in); width: 2.69m (8ft 10in); height: 2.36m (7ft 9in)
Range:	483km (300 miles)
Armour:	Classified
Armament:	1 x 105mm (4.1in) gun
Powerplant:	1 x Detroit Diesel Model 6V-92 TA 6-cylinder diesel engine developing 412kW (552hp)
Performance:	Maximum road speed: 70km/h (43.5mph); fording: 1.32m (4ft 4in); vertical obstacle: .76m (2ft 6in); trench: 2.13m (7ft)

1984

1985

LATE COLD WAR TANKS AND AFVS

Self-Propelled Guns

Although self-propelled guns were by no means as prolific in the world's armies by 1980 as they had been previously, they provided a form of long-range artillery that was used in limited conflicts around the world, notably the Iran-Iraq War of that decade, Israeli operations in the Lebanon and Soviet operations in Afghanistan.

L33

The L33 is based on the Sherman M4A3E8 tank, upon which has been built a superstructure for the crew of eight. Doors on both sides of the hull allow access, while two doors at the rear allow for ammunition re-supply. The gun's systems are manually operated.

SPECIFICATIONS	
Country of origin:	Israel
Crew:	8
Weight:	42,250kg (93,161lb)
Dimensions:	Length: 6.5m (21ft 4in); width: 3.27m (10ft 9in); height: 2.46m (8ft 1in)
Range:	260km (160 miles)
Armour:	Up to 60mm (2.36in)
Armament:	1 x 155mm (6.1in) howitzer
Powerplant:	1 x Cummins VTA-903 V8 diesel engine 343kW (460bhp) at 2600rpm
Performance:	Maximum speed: 36km/h (22mph)

M1973

This was known as the 2S3 Akatsiya in the former Soviet Union, and 18 vehicles supported each tank division and motorized rifle division. It was fitted with nuclear, biological and chemical (NBC) protection and a tactical nuclear capability.

SPECIFICATIONS	
Country of origin:	Soviet Union
Crew:	6
Weight:	24,945kg (54,880lb)
Dimensions:	Length: 8.4m (27ft 6.7in); width: 3.2m (10ft 6in); height: 2.8m (9ft 2.25in)
Range:	300km (186 miles)
Armour:	15–20mm (.59–.78in)
Armament:	1 x 152mm (5.9in) gun; 1 x 7.62mm (.3in) anti-aircraft MG
Powerplant:	1 x V-12 diesel engine developing 388kW (520hp)
Performance:	Maximum road speed: 55km/h (34mph); fording: 1.5m (4ft 11in); vertical obstacle: 1.1m (3ft 7in); trenchC: 2.5m (8ft 2.5in)

TIMELINE

1963  1970   1973

SELF-PROPELLED GUNS

M110A2

At first, the 203mm (8in) howitzer was put on the same tracked carriage as the M107 175mm (6.88in) gun. This did not live up to expectations, so a new, longer howitzer barrel was developed. The M110A2 had the addition of a muzzle brake, increasing range.

SPECIFICATIONS	
Country of origin:	USA
Crew:	5
Weight:	28,350kg (62,512lb)
Dimensions:	Length: 5.72m (18ft 9in); width: 3.14m (1ft 4in); height: 2.93m (9ft 8in)
Range:	520km (325 miles)
Armour:	Not disclosed
Armament:	1 x 203mm (8in) howitzer
Powerplant:	1 x Detroit Diesel V-8 turbocharged engine developing 335.5kW (405hp) at 2300 rpm
Performance:	Maximum road speed: 56km/h (34mph); fording: 1.06m (3ft 6in); vertical obstacle: 1.01m (3ft 4in); trench: 2.32m (7ft 9in)

M1974

Known as the Gvozdika in the Soviet Union, the M1974 was deployed in large numbers (36 per tank division, 72 per motorized rifle division). Fully amphibious, the chassis has been used for armoured command and chemical warfare reconnaissance vehicles.

SPECIFICATIONS	
Country of origin:	Soviet Union
Crew:	4
Weight:	15,700kg (34,540lb)
Dimensions:	Length: 7.3m (23ft 11.5in); width: 2.85m (9ft 4in); height: 2.4m (7ft 10.5in)
Range:	500km (310 miles)
Armour:	15–20mm (.59–.78in)
Armament:	1 x 122mm (4.8in) howitzer; 1 x 7.62mm (.3in) anti-aircraft MG
Powerplant:	1 x YaMZ-238V V-8 water-cooled diesel engine developing 179kW (240hp)
Performance:	maximum road speed 60km/h (37mph); fording amphibious; vertical obstacle 1.1m (3ft 7in); trench 3m (9ft 10in)

SM-240

The SM-240 self-propelled mortar was unique. The chassis of an SA-4 SAM launcher created a moving platform for a massive 2B8 240mm (9.45in) mortar, requiring a crew of nine. Ammunition types included HE, fragmentation, smoke and nuclear warheads.

SPECIFICATIONS	
Country of origin:	USSR
Crew:	9
Weight:	27,500kg (60,600lb)
Dimensions:	Length: 7.94m (26ft .6in); width: 3.25m (10ft 8in); height 3.22m (10ft 6.7in)
Range:	500km (310 miles)
Armour:	15–20mm (.59–.78in)
Armament:	1 x 2B8 240mm (9.45in) mortar
Powerplant:	1 x V-59 12-cylinder diesel engine 388kW (520hp)
Performance:	Maximum road speed: 45km/h (28mph)

1974   1978

LATE COLD WAR TANKS AND AFVS

Heavy SPGs

There is still a place for the heavy self-propelled gun in the modern world. During the Cold War, these AFVs would have played an important part in forestalling a Warsaw Pact attack on the West, and their ongoing importance was confirmed in the Gulf War of 1991 and subsequent conflicts in Iraq and Afghanistan.

GCT 155mm

The GCT 155mm (6.1in) succeeded the Mk F3 in the French Army. Its improvements included an automatic loading system, giving a rate of fire of eight rounds a minute, and the ability to fire a range of projectiles, including a round carrying multiple anti-tank mines.

SPECIFICATIONS	
Country of origin:	France
Crew:	4
Weight:	41,949kg (92,288lb)
Dimensions:	Length: 10.25m (33ft 7.5in); width: 3.15m (10ft 4in); height: 3.25m (10ft 8in)
Range:	450km (280 miles)
Armour:	Up to 20mm (.78in)
Armament:	1 x 155mm (6.1in) gun; 1 x 7.62mm/12.7mm (.3in/.5in) anti-aircraft MG
Powerplant:	1 x Hispano-Suiza HS 110 12-cylinder water-cooled multi-fuel engine developing 537kW (720hp)
Performance:	Maximum road speed: 60km/h (37mph); vertical obstacle: .93m (3ft 0.7in); trench: 1.90m (6ft 3in)

DANA

The DANA was the first wheeled self-propelled howitzer to enter service in modern times. Three hydraulic stabilizers are lowered into the ground before firing. The rate of fire is three rounds per minute for a period of 30 minutes.

SPECIFICATIONS	
Country of origin:	Czechoslovakia
Crew:	4 to 5
Weight:	23,000kg (50,600lb)
Dimensions:	Length: 10.5m (34ft 5in); width: 2.8m (9ft 2in); height: 2.6m (8ft 6in)
Range:	600km (375 miles)
Armour:	Up to 12.7mm (.5in)
Armament:	1 x 152mm (5.9in) gun; 1 x 12.7mm (.5in) HMG
Powerplant:	1 x V-12 diesel engine developing 257kW (345hp)
Performance:	Maximum road speed: 80km/h (49.71mph); fording: 1.4m (4ft 7in); vertical obstacle: 1.5m (4ft 11in); trench: 1.4m (4ft 7in)

TIMELINE			
	1975	1977	1980

HEAVY SPGS

2S7

The 2S7 mounts a long-barrelled 2A44 203mm (8in) gun, which fires conventional ammunition to a range of over 37km (23 miles). Coping with the recoil of the 43kg (95lb) shells requires three recoil pistons with maximum travel of 140cm (55.11in).

SPECIFICATIONS	
Country of origin:	Soviet Union
Crew:	7
Weight:	46,500kg (102,500lb)
Dimensions:	Length (with gun): 13.12m (43ft .4in); width: 3.38m (11ft 1in); height: 3m (9ft 9in)
Range:	650km (400 miles)
Armour:	Not available
Armament:	1 x 203mm (8in) 2A44 gun
Powerplant:	1 x V-46l 12-cylinder diesel engine developing 626kW (839hp)
Performance:	Maximum road speed: 50km/h (31mph); fording: 1.2m (3ft 10in); gradient: 22 per cent; vertical obstacle: 1m (3ft 4in)

Palmaria

Unusually, the Palmaria has an auxiliary power unit for the turret, to conserve fuel for the main engine. It comes with a range of munitions, including rocket-assisted projectiles. An automatic loader gives a rate of fire of one round every 15 seconds.

SPECIFICATIONS	
Country of origin:	Italy
Crew:	5
Weight:	46,632kg (102,590lb)
Dimensions:	Length: 11.474m (37ft 7.75in); width: 2.35m (7ft 8.5in); height: 2.874m (9ft 5.25in)
Range:	400km (250 miles)
Armour:	Classified
Armament:	1 x 155mm (6.1in) howitzer; 1 x 7.62mm (.3in) MG
Powerplant:	1 x 8-cylinder diesel engine developing 559kW (750hp)
Performance:	Maximum road speed: 60km/h (37mph); fording: 1.2m (3ft 11in); vertical obstacle: 1m (3ft 3in); trench: 3m (9ft 10in)

G6 Rhino

The G6 is a highly mobile, heavily armoured and technologically advanced 155mm (6.1in) self-propelled gun system. The gun can be used in both direct- and indirect-fire roles. Direct-fire range is 3km (1.9 miles); indirect-fire range is 30km (19 miles).

SPECIFICATIONS	
Country of origin:	South Africa
Crew:	6
Weight:	47,000kg (103,600lb)
Dimensions:	Length (chassis): 10.2m (33ft 4.3in); width: 3.4m (11ft 2in); height: 3.5m (11ft 6in)
Range:	700km (430 miles)
Armour:	Not available
Armament:	1 x 155mm (6.1in) gun
Powerplant:	1 x diesel engine developing 391kW (525hp)
Performance:	Maximum road speed: 90km/h (56mph); fording: 1m (3ft 4in)

1982    1987

Anti-tank Vehicles

The modern tank faces an array of anti-tank weapons, which are available in greater numbers than ever before. As well as direct-fire weapons, there are artillery-launched scattered mines, smart 'search and destroy' mines and projectiles that loiter over the battlefield on parachutes, their sensors waiting for a target to appear. below.

Raketenjadgpanzer 2

The Raketenjadgpanzer 2 (RJPZ 2) replaced the RJPZ 1, using updated missile technology and boasting improved vehicle performance. The same SS-11 ATGW missile launchers were fitted, and TOW and HOT ATGWs could also be fitted.

SPECIFICATIONS	
Country of origin:	West Germany
Crew:	4
Weight:	23,000kg (50,700lb)
Dimensions:	Length: 6.43m (21ft 1in); width: 2.98m (9ft 9in); height: 2.15m (7ft .6in)
Range:	400km (250 miles)
Armour:	Up to 50mm (1.97in)
Armament:	14 x SS-11 ATGWs; 2 x 7.62mm (.3in) MG3 MGs
Powerplant:	1 x Daimler-Benz MB 837A 8-cylinder diesel engine developing 373kW (500hp) at 2000rpm
Performance:	Maximum road speed: 70km/h (43mph); fording: 1.4m (4ft 7in); gradient: 60 per cent; vertical obstacle: .75m (2ft 5in); trench 2m (6ft 7in)

Wiesel 1 TOW

Designed to be air-portable, the Wiesel weighs only 2750kg (6063lb) and is only 3.26m (10ft 8in) long. It carries the Hughes TOW 1 ATGW weapon, to give air-landed units immediate anti-armour support. There are seven missiles, two set for immediate use.

SPECIFICATIONS	
Country of origin:	Germany
Crew:	3
Weight:	2750kg (6063lb)
Dimensions:	Length: 3.26m (10ft 8in); width: 1.82m (5ft 11in); height: 1.89m (6ft 2in)
Range:	200km (125 miles)
Armour:	Not available
Armament:	1 x TOW 1 ATGW system
Powerplant:	1 x Volkswagen Type 069 5-cylinder turbo diesel engine developing 73kW (98hp)
Performance:	Maximum road speed: 80km/h (50mph); gradient: 60 per cent; vertical obstacle: .4m (1ft 4in); trench: 1.5m (4ft 10in)

TIMELINE

1962  1965  1978

ANTI-TANK VEHICLES

Jagdpanzer Jaguar

The Jadgpanzer Jaguar 1 self-propelled anti-tank vehicle upgraded the Raketenjadgpanzer 2. It mounted the Euromissile K3S HOT ATGW, a command-to-line-of-sight system with a range of 4000m (4374 yards), able to penetrate modern explosive-reactive armour.

SPECIFICATIONS
Country of origin:	Germany
Crew:	4
Weight:	25,500kg (56,200lb)
Dimensions:	Length: 6.61m (21ft 8in); width: 3.12m (10ft 3in); height: 2.55m (8ft 4in)
Range:	400km (250 miles)
Armour:	Up to 50mm (1.97in)
Armament:	1 x HOT ATGW system; 1 x 7.62mm (.3in) MG3 MG
Powerplant:	1 x Daimler-Benz MB837A 8-cylinder diesel engine developing 373kW (500hp) at 2000rpm
Performance:	Maximum road speed: 70km/h (43mph); fording: 1.4m (4.6ft); gradient: 60 per cent; vertical obstacle: .75m (2.46ft); trench 2m (6.6ft)

Panzerjäger 90

The Panzerjäger 90 is an ATGW launcher version of the MOWAG Piranha APC. The first versions were armed with a single Tube-Launched Optically Tracked Wire-Guided (TOW) 2 ATGW, but later versions have received a turret-mounted twin TOW launcher.

SPECIFICATIONS
Country of origin:	Switzerland
Crew:	5
Weight:	11,000kg (24,300lb)
Dimensions:	Length: 6.23m (20ft 5in); width: 2.5m (8ft 2in); height: 2.97m (9ft 9in)
Range:	600km (370 miles)
Armour:	10mm (.39in)
Armament:	1 x turret-mounted twin TOW launcher; 1 x 7.62mm(.3in) coaxial MG
Powerplant:	1 x Detroit Diesel 6V-53T 6-cylinder diesel engine developing 160kW (215hp)
Performance:	Maximum road speed: 102km/h (63mph); fording: amphibious; gradient: 60 per cent; vertical obstacle: .5m (1ft 7in)

VBL

The Panhard Véhicule Blindé Léger (Lightly Armoured Vehicle) is an armoured scout car with tank-hunting capabilities. A single Anti-Tank Guided Weapon (ATGW) is mounted on the roof and operated by the crew member using a roof hatch.

SPECIFICATIONS
Country of origin:	France
Crew:	3
Weight:	3550kg (7828lb)
Dimensions:	Length: 3.87m (12ft 8in); width: 2.02m (6ft 7in); height: 1.7m (5ft 7in)
Range:	600km (370 miles)
Armour:	Up to 11.5mm (.45in)
Armament:	1 x MILAN ATGW; 1 x 7.62mm (.3in) GPMG
Powerplant:	1 x Peugeot XD3T 4-cylinder turbo diesel engine developing 78kW (105hp) at 4150rpm
Performance:	Maximum road speed: 95km/h (59mph); fording: .9m (2ft 10in); gradient: 50 per cent; trench: .5m (1ft 7in)

1988    1990

LATE COLD WAR TANKS AND AFVS

MLRS

The Vought Multiple Launch Rocket System (MLRS) had its origins in a 1976 feasibility study into what was known as a General Support Rocket System. Following trials, the Vought system was chosen and entered service with the US Army in 1982.

MLRS

Cab
The danger from rocket fumes is a serious problem for rocket-launcher crews. MLRS has an overpressure system in the cab to keep the fumes out, as well as a full NBC system

The Vought MLRS was licensed to the United Kingdom, France, Italy, West Germany and the Netherlands for production. It saw action during the 1991 Gulf War, when Allied MLRS batteries tore large holes in Iraqi defence lines prior to the ground offensive to free Kuwait.

SPECIFICATIONS	
Country of origin:	USA
Crew:	3
Weight:	25,191kg (55,420lb)
Dimensions:	Length: 6.8m (22ft 4in); width: 2.92m (9ft 7in); height: 2.6m (8ft 6in)
Range:	483km (302 miles)
Armour:	Classified
Armament:	2 x rocket pod containers, each holding 6 rockets
Powerplant:	1 x Cummings VTA-903 turbo-charged 8-cylinder diesel engine developing 373kW (500hp)
Performance:	Maximum road speed: 64km/h (40mph); fording: 1.1m (3ft 7in); vertical obstacle: 1m (3ft 4in); trench: 2.29m (7ft 6in)

Aluminium armour
This protects the crew from small arms fire and shell splinters, but cannot save the vehicle from a direct hit by shells or cannon

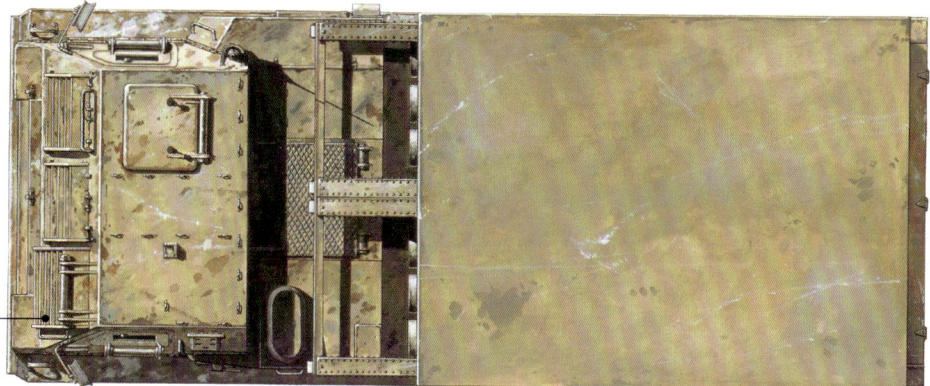

FIRE CONTROL
The fire control system has to relay the MLRS after it has fired a rocket since the massive blast moves the vehicle's position

ROCKETS
MLRS fires three basic types of rocket: one dispensing anti-personnel mines, one dropping anti-tank mines, and one chemical warhead (used only by the US Army)

1980s Anti-aircraft Vehicles

In the last decade of the Cold War, many believed that the sophistication of anti-aircraft systems meant that an attacking aircraft stood little chance of reaching its target once it was detected, such was the effectiveness of mobile anti-aircraft systems – hence the American emphasis on developing 'stealth' technology.

SA-10 Grumble

The S-300PMU1 is a cutting-edge air-defence system. The key advantage of the SA-10 is that it can acquire and engage multiple targets simultaneously across a very broad spectrum of altitude – 25m (82ft) to 30,000m (98,400ft).

SPECIFICATIONS	
Country of origin:	Russian Federation/Soviet Union
Crew:	Unknown
Weight:	43,300kg (95,500lb)
Dimensions:	Length: 11.47m (37ft 7.5in); width: 10.17m (33ft 4.3in); height: 3.7m (12ft 1in)
Range:	650km (400 miles)
Armour:	Classified
Armament:	4 x 5V55K SA-10 SAMs
Powerplant:	1 x D12A-525A 12-cylinder diesel engine developing 386kW (517hp)
Performance:	Maximum road speed: 60km/h (37mph)

SA-11 Gadfly

The 9K37M1 BUK-1 SAM system is a medium-range radar-guided missile with a 90 per cent kill probability against aircraft, 40 per cent against cruise missiles. The target is acquired by a SNOW DRIFT warning and acquisition radar at a range of 70km (43 miles).

SPECIFICATIONS	
Country of origin:	Russian Federation/Soviet Union
Crew:	4
Weight:	(Launch vehicle) 32,340kg (71,309lb)
Dimensions:	Length: 9.3m (30ft 6in); width: 3.25m (10ft 8in); height: 3.8m (12ft 6in)
Range:	500km (310 miles)
Armour:	Not applicable
Armament:	4 x Type 9M38M1 (SA-11) SAM missiles
Powerplant:	1 x V-64-4 12-cylinder diesel, developing 529kW (709hp)
Performance:	Maximum road speed: 65km/h (40mph)

TIMELINE

1980   1982

1980S ANTI-AIRCRAFT VEHICLES

Tracked Rapier

The Tracked Rapier carried eight missiles on its launcher, and the optical tracker was located in the cab, exiting through the roof. The tracking antenna elevated above the launcher itself. Protection for the Tracked Rapier crew was provided by a fully armoured cab.

SPECIFICATIONS	
Country of origin:	United Kingdom
Crew:	3
Weight:	14,010kg (30,892lb)
Dimensions:	Length: 6.4m (21ft); width: 9.19m (30ft 2in); height: 2.5m (8ft 2in)
Range:	300km (190 miles)
Armour:	Aluminium (details classified)
Armament:	8 x Rapier SAMs
Powerplant:	1 x GMC 6-cylinder turbocharged diesel engine developing 186kW (250hp) at 2600rpm
Performance:	Maximum road speed: 80km/h (50mph); fording: amphibious; gradient: 60 per cent; vertical obstacle: .6m (2ft); trench 1.75m (5ft 9in)

Sergeant York

The M247 has a comprehensive fire-control system, including both surveillance and tracking radar. It is capable of engaging both aircraft and helicopters, as well as tactical missiles. The Sergeant York, however, never entered service.

SPECIFICATIONS	
Country of origin:	USA
Crew:	3
Weight:	54,430kg (119,746lb)
Dimensions:	Length: 7.674m (25ft 2in); width: 3.632m (11ft 11in); height (radar up): 4.611m (15ft 2in)
Range:	500km (311 miles)
Armour:	Up to 120mm (4.72in)
Armament:	2 x 40mm (1.57in) L/70 Bofors guns
Powerplant:	1 x Teledyne Continental AVDS-1790-2D diesel engine developing 559kW (750hp)
Performance:	Maximum road speed: 48km/h (30mph); fording: 1.219m (4ft); vertical obstacle: 1.914m (3ft); trench: 2.591m (8ft 6in)

M1 Tunguska

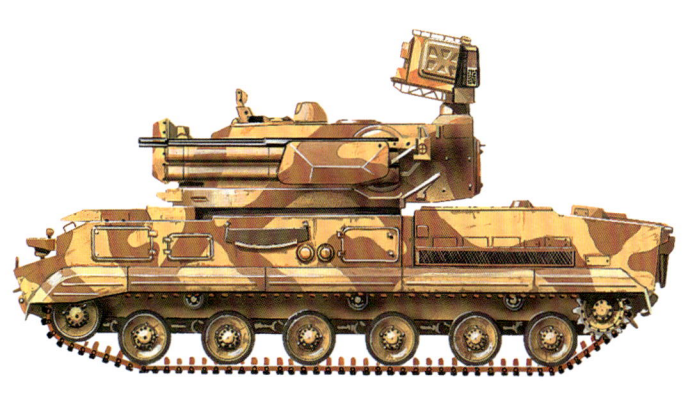

The Tunguska mixes SAM and gun technology to make a versatile aerial-interception platform. Eight SA-19 Grison missiles are mounted for medium-range interception, and two twin-barrel 2A38M 30mm (1.2in) guns for close-range targets.

SPECIFICATIONS	
Country of origin:	Russia
Crew:	4
Weight:	34,000kg (75,000lb)
Dimensions:	Length: 7.93m (26ft); width: 3.24m (10ft 7.5in); height: 4.02m (13ft 1in)
Range:	500km (310 miles)
Armour:	Classified
Armament:	2 x 30mm (1.18in) twin-barrel 2A38M guns; 8 x SA-19 Grison SAMs
Powerplant:	1 x V-64-4 12-cylinder diesel engine developing 529kW (709hp)
Performance:	Maximum road speed: 65km/h (40mph)

1983    1986

M113 Variants

LATE COLD WAR TANKS AND AFVS

The original M113 model gave rise to several variants. For example, more armour protection was added to produce an ACAV (Armoured Cavalry) version, which was transformed into a true armoured fighting vehicle by the addition of shields for its machine guns. Other versions included a command vehicle.

M114A1E1

The M114 was a command and reconnaissance derivative of the M113 APC. It was intially armed with manually operated machine guns, but later a remote-controlled 12.7mm (.5in) or even a 20mm (.78in) Hispano-Suiza cannon was fitted to the cupola.

SPECIFICATIONS	
Country of origin:	USA
Crew:	3 or 4
Weight:	6930kg (15,280lb)
Dimensions:	Length: 4.46m (14ft 7.5in); width: 2.33m (7ft 7.6in); height: 2.16m (7ft 1in)
Range:	440km (270 miles)
Armour:	37mm (1.46in)
Armament:	1 x 12.7mm (.5in) HMG or 1 x 20mm (.78in) Hispano-Suiza gun; 1 x 7.62mm (.3in) MG
Powerplant:	1 x Chevrolet 283-V8 8-cylinder petrol engine developing 119kW (160hp)
Performance:	Maximum road speed: 58km/h (36mph); fording: amphibious; gradient: 60 per cent; vertical obstacle: .5m (1ft 7in); trench: 1.5m (4ft 10in)

Arisgator

The Arisgator is a M113A2 armoured personnel carrier fitted with a flotation kit. At the front, a bow-shaped bolt-on section improves handling in turbulent waters. Two tail sections contain the propeller units. Other modifications include redirected air intakes.

SPECIFICATIONS	
Country of origin:	Italy
Crew:	2 + 11
Weight:	c.12,000kg (26,500lb)
Dimensions:	Length: 6.87m (22ft 6.4in); width: 2.95m (9ft 8in); height: 2.05m (6ft 8.7in)
Range:	550km (340 miles)
Armour:	38mm (1.49in)
Armament:	1 x 12.7mm (.5in) Browning M2 HB HMG
Powerplant:	1 x Detroit 6V-53N 6-cylinder diesel, developing 160kW (215hp) at 2800rpm
Performance:	Maximum road speed: 68km/h (42mph)

TIMELINE

1964  1967  1978

M113 VARIANTS

Green Archer

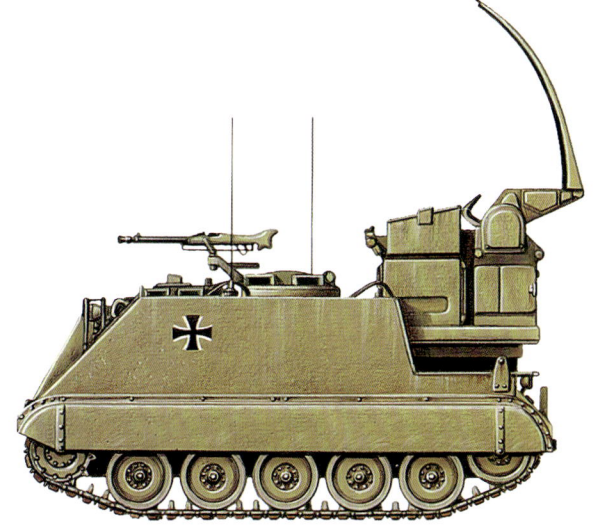

The Green Archer was a mortar-detecting radar system, which the British mounted on the FV432 armoured personnel carrier. By detecting the firing and detonation of a shell, the radar could compute the location of an enemy mortar at ranges of up to 30km (18.6 miles).

SPECIFICATIONS	
Country of origin:	Germany/UK/Netherlands
Crew:	4
Weight:	11,900kg (26,200lb)
Dimensions:	Length: 4.86m (154ft 11in); width: 2.7m (8ft 10in); height: 4.32m (14ft 2in)
Range:	480km (300 miles)
Armour:	12–38mm (.47–1.5in)
Armament:	1 x 7.62mm (.3in) MG
Powerplant:	1 x Detroit Diesel 6V-53N 6-cylinder diesel engine developing 160kW (215hp) at 2800rpm
Performance:	Maximum road speed: 68km/h (42mph); fording: amphibious; gradient: 60 per cent; vertical obstacle: .61m (2ft); trench: 1.68m (5ft 6in)

Zelda

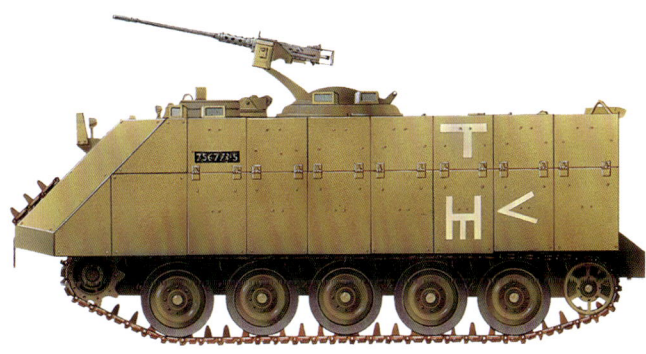

The M113 was an air-transportable armoured vehicle, using aluminium to keep weight low. The Israeli Zelda is a standard M113 but with additional side and floor armour to protect against rocket-propelled grenades and mine detonations respectively.

SPECIFICATIONS	
Country of origin:	Israel
Crew:	2 + 11
Weight:	12,500kg (27,600lb)
Dimensions:	Length: 5.23m (17ft 2in); width: 3.08m (10ft 1in); height: 1.85m (6ft .8in)
Range:	480km (300 miles)
Armour:	38mm (1.49in)
Armament:	various MG configurations
Powerplant:	1 x Detroit Diesel 6V-53T 6-cylinder diesel, developing 158kW (212hp) at 2800rpm
Performance:	Maximum road speed: 61km/h (38mph); fording: amphibious; gradient: 60 per cent; vertical obstacle: .6m (2ft); trench: 1.68m (5ft 6in)

M901

The M901 is an upgrade of the M113A1 APC, mounted with the two-tube M27 TOW 2 cupola on the roof. It must come to a stop before it can fire, though it takes only 20 seconds for the TOW system to target and fire. Reloading takes around 40 seconds.

SPECIFICATIONS	
Country of origin:	USA
Crew:	4 or 5
Weight:	11,794kg (26,005lb)
Dimensions:	Length: 4.88m (16ft .1in); width: 2.68m (8ft 9in); height: 3.35m (10ft 11.8in)
Range:	483km (300 miles)
Armour:	Up to 44mm (1.73in)
Armament:	1 x TOW 2 ATGW system
Powerplant:	1 x Detroit Diesel 6V-53N 6-cylinder diesel, developing 160kW (215hp)
Performance:	Maximum road speed: 68km/h (42mph); fording: amphibious; gradient: 60 per cent; vertical obstacle: .61m (2ft); trench: 1.68m (5ft 6in)

1986  2006

LATE COLD WAR TANKS AND AFVS

MARDER

Entering service in 1971, the Marder I was NATO's first infantry fighting vehicle, combining the qualities of an armoured personnel carrier with those of a light tank. Upgraded over the years, it is still in service today but is to be replaced by a lighter and cheaper AFV in due course.

MARDER

RHEINMETALL 20MM (.78IN) RH202 CANNON
Mounted above the turret, this avoids filling the vehicle with fumes when firing the cannon and it also allows the gun to depress by 17°, which is useful in hull-down positions

GUNNER
The 20mm (.78in) cannon is served by three separate belts so that the gunner can rapidly alter his choice of ammunition as different targets appear

TROOP COMPARTMENT
On most Marders, this accommodates six infantrymen, but the A1 version has four crew and only five infantrymen. All Marders have a Nuclear, Biological, Chemical (NBC) system fitted as standard

MARDER

When the Marder entered service, it meant that for the first time in NATO countries, infantry travelling in an armoured personnel carrier carried with them their own effective fire support. Versions have been sold to Chile, and the Argentinian TAM is based on the Marder but with different armaments.

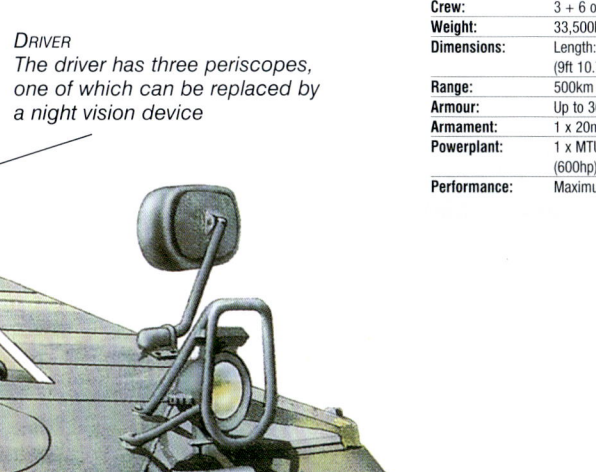

STEEL TURRET
The front of the turret is armoured to withstand 20mm (.78in) cannon shells

DRIVER
The driver has three periscopes, one of which can be replaced by a night vision device

SPECIFICATIONS	
Country of origin:	Germany
Crew:	3 + 6 or 7 troops
Weight:	33,500kg (73,855lb)
Dimensions:	Length: 6.88m (22ft 6.8in); width: 3.38m: (11ft 1in); height: 3.02m (9ft 10.7in)
Range:	500km (310 miles)
Armour:	Up to 30mm (1.18in)
Armament:	1 x 20mm (.7in) gun; 1 x 7.62mm (.3in) MG
Powerplant:	1 x MTU MB 833 Ea-500 6-cylinder diesel engine developing 447kW (600hp)
Performance:	Maximum road speed: 65km/h (40.4mph)

LATE COLD WAR TANKS AND AFVS

Tracked APCs

The production of tracked armoured personnel carriers continued virtually unabated into the last decade of the Cold War. As well as new systems designed to enhance crew comfort and survivability, vehicles were mostly equipped for NBC (Nuclear, Chemical and Biological) warfare conditions and were very well armed.

BMS-1 Alacran

The BMS-1 Alacran is unusual in being a modern armoured personnel carrier in half-track configuration. It can carry 12 fully equipped soldiers, and has seven firing ports and eight vision blocks around the troop compartment. A machine gun is fitted as standard.

SPECIFICATIONS	
Country of origin:	Chile
Crew:	2 + 12
Weight:	10,500kg (23,150lb)
Dimensions:	Length: 6.37m (20ft 10in); width: 2.38m (7ft 9in); height: 2.03m (6ft 8in)
Range:	900km (560 miles)
Armour:	Not available
Armament:	1 x 12.7mm (.5in) or 1 x 7.62mm (.3in) MG
Powerplant:	1 x Cummins V-555 turbo diesel engine developing 167kW (225hp) at 3000rpm
Performance:	Maximum road speed: 70km/h (43mph); fording: 1.6m (5ft 4in); gradient: 70 per cent

MT-LB

The MT-LB was a multi-purpose vehicle that fulfilled a range of roles: armoured personnel carrier; artillery tractor; mobile command and control centre; repair vehicle; engineer vehicle; ambulance; and Gopher SAM system. It was fully amphibious.

SPECIFICATIONS	
Country of origin:	Russia/Soviet Union
Crew:	2 + 11
Weight:	14,900kg (32,900lb)
Dimensions:	Length: 7.47m (24ft 6in); width: 2.85m (9ft 4in); height: 2.42m (7ft 11in)
Range:	525km (330 miles)
Armour:	3–10mm (.11–.39in)
Armament:	1 x 12.7mm (.5in) HMG; or 1 x 7.62mm (.3in) MG
Powerplant:	1 x YaMZ-238N 8-cylinder diesel engine developing 164kW (220hp) at 2400rpm
Performance:	Maximum road speed: 62km/h (39mph); fording: amphibious; gradient: 60 per cent; vertical obstacle: .6m (2ft); trench: 2.41m (7.9ft)

TIMELINE			
	1974 	1977 	1981 

TRACKED APCS

FV433 Stormer

The Stormer is valued for its versatility. It can be fitted with a variety of equipment and weaponry, including NBC systems, night-vision instruments, and guns ranging from 12.7mm (.5in) HMGs to 90mm (3.54in) guns and Starstreak missiles.

SPECIFICATIONS	
Country of origin:	United Kingdom
Crew:	3 + 8
Weight:	12,700kg (28,000lb)
Dimensions:	Length: 5.33m (17ft 5in); width: 2.4m (7ft 10in); height 2.27m (7ft 5in)
Range:	650km (400 miles)
Armour:	Aluminium (details classified)
Armament:	various
Powerplant:	1 x Perkins T6/3544 6-cylinder turbo diesel, developing 186kW (250hp) at 2600rpm
Performance:	Maximum road speed: 80km/h (50mph); fording: amphibious; gradient: 60 per cent; vertical obstacle: .6m (2ft); trench 1.75m (5ft 9in)

BMP-2

The low silhouette of the BMP-2 and long sloping front presents a small target, which is vital because its armour is extremely poor. One remarkable feature is that the rear doors are hollow and serve as fuel tanks, with obvious dangers for the troops inside.

SPECIFICATIONS	
Country of origin:	Soviet Union
Crew:	3 + 7
Weight:	14,600kg (32,120lb)
Dimensions:	Length: 6.71m (22ft); width: 3.15m (10ft 4in); height: 2m (6ft 7in)
Range:	600km (375 miles)
Armour:	Classified
Armament:	1 x 30mm (1.2in) gun; 1 x 7.62mm (.3in) coaxial MG; 1 x At-5 anti-missile launcher
Powerplant:	1 x Model UTD-20 6-cylinder diesel engine developing 223kW (300hp)
Performance:	Maximum road speed: 65km/h (40.6mph); fording: amphibious; vertical obstacle: .7m (2ft 4in); trench: 2.4m (8ft 2in)

Type 77

First observed in 1974, the Type 77 is a low-profile but capacious APC which is fully amphibious and has been produced in a variety of specialized versions. The Type 77 is based on the layout of the Russian BTR-50K with six large road wheels and no return rollers.

SPECIFICATIONS	
Country of origin:	Soviet Union
Crew:	2 + 16
Weight:	15,500kg (34,177lb)
Dimensions:	Length: 7.4m (24ft 4in); width: 3.2m (10ft 6in); height: 2.44m (8ft)
Range:	370km (240 miles)
Armour:	Up to 8mm (0.31in)
Armament:	1 x 12.7mm (0.5in) MG
Powerplant:	1 x Type 12150L-2A V-12 fuel-injected diesel engine developing 298kW (400bhp)
Performance:	Maximum speed: 60km/h (37mph)

1982  1982

1980s Tracked APCs

In some parts of the world, the terrain means that tracked APCs are preferred over their wheeled counterparts. In South Korea, for example, military operations in the event of an attack from the north would unfold in the same mountainous terrain that had caused United Nations forces so many problems in the Korean War of 1950–3.

M-80 MICV

The notable feature of the M-80 Mechanized Infantry Combat Vehicle (MICV) is the configuration of turret armament – a 30mm (1.12in) cannon and a 7.62mm (.3in) machine gun. Both are set in turret slits to allow anti-aircraft engagement at an elevation of 75˚.

SPECIFICATIONS	
Country of origin:	Yugoslavia
Crew:	3 + 7
Weight:	13,700kg (30,200lb)
Dimensions:	Length: 6.4m (20ft 11in); width: 2.59m (8ft 5in); height: 2.3m (7ft 6in)
Range:	500km (310 miles)
Armour:	30mm (1.18in)
Armament:	1 x 30mm (1.12in) cannon; 1 x 7.62mm (.3in) MG; 2 x Yugoslav Sagger ATGWs
Powerplant:	1 x HS-115-2 8-cylinder turbo diesel engine developing 194kW (260hp)
Performance:	Maximum road speed: 60km/h (37mph); fording: amphibious; gradient: 60 per cent; vertical obstacle: .8m (2ft 7in); trench 2.2m (7ft 2in)

KIFV K-200

The Korean Infantry Fighting Vehicle K-200 can carry nine infantry and its three-man crew at speeds of 74km/h (46mph). Two machine guns are standard armament, but optional configurations include 20mm (.78in) Vulcan cannon and two mortar carriers.

SPECIFICATIONS	
Country of origin:	South Korea
Crew:	3 + 9
Weight:	12,900kg (28,400lb)
Dimensions:	Length: 5.48m (19ft 2in); width: 2.84m (9ft 3in); height 2.51m (8ft 3in)
Range:	480km (300 miles)
Armour:	Aluminium and steel (details classified)
Armament:	1 x 12.7mm (.5in) HMG; 1 x 7.62mm (.3in) MG
Powerplant:	1 x MAN D-284T V8 diesel engine developing 208kW (280hp) at 2300rpm
Performance:	Maximum road speed: 74km/h (46mph); fording: amphibious; gradient: 60 per cent; vertical obstacle: .64m (2ft 1in); trench: 1.68m (5ft 6in)

TIMELINE 1980  1982 1985

1980S TRACKED APCS

Cobra

Fully amphibious, the Cobra armoured personnel carrier had a turbo engine that powered an electric generator driving both wheels and waterjets. A roof-mounted Browning M2 HB 12.7mm (.5in) machine gun and a 7.62mm (.3in) GPMG were its weapons.

SPECIFICATIONS	
Country of origin:	Belgium
Crew:	3 + 9
Weight:	8500kg (18,700lb)
Dimensions:	Length: 4.52m (14ft 9in); width: 2.75m (9ft .2in); height: 2.32m (7ft 7in)
Range:	600km (370 miles)
Armour:	Steel (details classified)
Armament:	1 x 12.7mm (.5in) Browning M2 HB HMG; 1 x 7.62mm (.3in) GPMG
Powerplant:	1 x Cummins VT-190 6-cylinder turbo diesel engine developing 141kW (190hp) at 3300rpm
Performance:	Maximum road speed: 75km/h (46mph); fording: amphibious

VCC-1 Armoured Personnel Carrier

The VCC-1 Camillino is based on the M113 tracked armoured personnel carrier. Modifications included introducing firing ports (enabling the infantry to fire from inside the vehicle) and armour protection for the machine gunner on top of the hull.

SPECIFICATIONS	
Country of origin:	Italy
Crew:	2 + 7
Weight:	11,600kg (25,578lb)
Dimensions:	Length: 5.04m (16ft 6in); width: 2.69m (8ft 10in); height: 2.03m (6ft 8in)
Range:	550km (340 miles)
Armour:	Not revealed
Armament:	2 x MGs
Powerplant:	1 x GMC V6 diesel engine developing 156kW (210bhp) at 2800rpm
Performance:	Maximum speed: 65km/h (40mph)

Type 89 Infantry Combat Vehicle

The Type 89 has a conventional layout, its driver seated at front right with the powerpack on his left and the troop compartment at the rear. Both crew members have a single-piece roof hatch, periscopes for observation and a main armament aiming sight.

SPECIFICATIONS	
Country of origin:	Japan
Crew:	3 + 7
Weight:	27,000kg (59,535lb)
Dimensions:	Length: 6.7m (22ft); width: 3.2m (10ft 6in); height: 2.5m (8ft 3in)
Range:	400km (250 miles)
Armour:	Not available
Armament:	1 x 35mm (1.38in) Oerlikion gun; 1 x 7.62mm (.3in) MG; 2 x Jyu-MAT wire-guided anti-tank missile launchers
Powerplant:	1 x Mitsubishi diesel engine developing 447kW (600bhp)
Performance:	Maximum speed: 70km/h (44mph)

 1990

Amphibious Vehicles

LATE COLD WAR TANKS AND AFVS

Nowhere are the advances made in the development of armoured fighting vehicles more apparent than in amphibious vehicles. These reached new heights of refinement during the Cold War, culminating in amphibious assault vehicles whose speed and off-road manoeuvrability matches that of any main battle tank.

LVTP7

The LVTP7 was designed specifically for the US Marine Corps. With an aluminium hull and propelled either by twin waterjets or its tracks, it was capable of carrying up to 25 fully equipped troops. Variants included a command vehicle and recovery vehicle.

SPECIFICATIONS	
Country of origin:	USA
Crew:	3 + 25
Weight:	22,837kg (50,241lb)
Dimensions:	Length: 7.943m (26ft 0.7in); width: 3.27m (10ft 8.7in); height: 3.263m (10ft 8.5in)
Range:	482km (300 miles)
Armour:	45mm (1.8in)
Armament:	1 x 12.7mm (.5in) HMG; optional 40mm (1.57in) grenade launcher
Powerplant:	1 x Detroit-Diesel Model 8V-53T engine developing 298kW (400hp)
Performance:	Maximum road speed: 64km/h (40mph); maximum water speed: 13.5km/h (8.5mph); fording: amphibious; vertical obstacle: .914m (3ft); trench: 2.438m (8ft)

CAMANF

The CAMANF is, in essence, a 6x6 F-7000 Ford chassis fitted with a watertight body. The payload is officially 5 tonnes (4.9 tons), but this is much reduced in rough waters. The CAMANF may be an unexceptional vehicle, but it is a reliable amphibian.

SPECIFICATIONS	
Country of origin:	Brazil
Crew:	3
Weight:	13,500kg (29,700lb)
Dimensions:	Length: 9.5m (31ft 2in); width: 2.5m (8ft 2.4in); height: 2.65m (8ft 3in)
Range:	430km (267 miles)
Armour:	6–10mm (.23–.39in)
Armament:	1 x 12.7mm (.5in) anti-aircraft HMG
Powerplant:	1 x Detroit-Diesel Model 40-54N diesel engine developing 142kW (190hp)
Performance:	Maximum road speed: 72km/h (45mph); maximum water speed: 14km/h (8.7mph); fording: amphibious; vertical obstacle: .4m (1ft 3.7in); trench: not applicable

TIMELINE

1971  1975

AMPHIBIOUS VEHICLES

Type 6640A

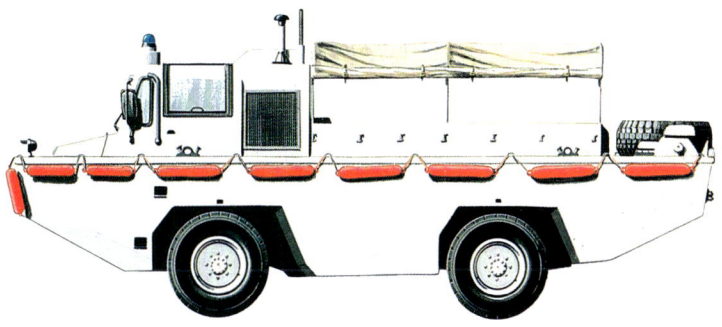

The hull of the the Type 6640A was constructed of aluminium for lightness and could carry a maximum payload of 2.14 tonnes (2 tons). Amphibious, it could be powered by its wheels or by a propeller, the rudder being coupled to the steering wheel.

SPECIFICATIONS	
Country of origin:	Italy
Crew:	2
Weight:	6950kg (15,290lb)
Dimensions:	Length: 7.3m (23ft 11.4in); width: 2.5m (8ft 2.4in); height: 2.715m (8ft 10.9in)
Range:	750km (466 miles)
Armour:	Up to 4mm (.16in)
Armament:	None
Powerplant:	1 x 6-cylinder diesel engine developing 87kW (117hp)
Performance:	Maximum road speed: 90km/h (56mph); maximum water speed with propeller: 11km/h (6.8mph) or with wheels: 5km/h (3.1mph); fording: amphibious; vertical obstacle: .43m (1ft 5in); trench: not applicable

Pegaso VAP 3550/1

The Pegaso VAP 3550/1 was designed for the Spanish marines, to be launched offshore from a Landing Ship Tank (LST). Eighteen troops or 3000kg (6600lb) of cargo are transported in the front of the vehicle. A hydraulic crane is fitted at the rear.

SPECIFICATIONS	
Country of origin:	Spain
Crew:	3 + 18
Weight:	12,500kg (27,550lb)
Dimensions:	Length: 8.85m (29ft .4in); width: 2.5m (8ft 2in); height: 2.5m (8ft 2in)
Range:	800km (500 miles)
Armour:	Up to 6mm (.24in)
Armament:	1 x 7.62mm (.3in) MG (export versions only)
Powerplant:	1 x Pegaso 9135/5 6-cylinder turbo diesel, developing 142kW (190hp)
Performance:	Maximum road speed: 87km/h (54mph); maximum water speed: 10km/h (6mph); gradient: 60 per cent

Bv 206

The Bv 206 has two units. The front unit contains the powerplant, transmission and operating crew. The rear unit is an 11-soldier transportation vehicle or a cargo-carrying unit. The two are linked by a steerable connector and are kept warm via an air heater.

SPECIFICATIONS	
Country of origin:	Sweden
Crew:	5 + 11
Weight:	Front unit: 2740kg (6042lb); rear unit: 1730kg (3815lb)
Dimensions:	Length: 6.9m (22ft 7in); width: 1.87m (6ft 1in); height: 2.4m (7ft 10in)
Range:	300km (190 miles)
Armour:	None
Armament:	None
Powerplant:	1 x Mercedes-Benz OM603.950 6-cylinder diesel, developing 101kW (136hp)
Performance:	Maximum road speed: 55km/h (34mph); fording: amphibious; gradient: 60 per cent; vertical obstacle: .5m (1ft 7in)

 1981

LATE COLD WAR TANKS AND AFVS

AAV7

Originally known as the LVTP7, the AAVP7 entered production in 1972 for the US Marine Corps. It is a fully tracked armoured amphibious assault landing vehicle, its function being to transport Marines from landing ship to shore and then inland. Its designation was changed to AAV7 in 1985.

AAV7

MACHINE GUN
The AAV7's M58 12.7mm (.5in) machine gun is mounted in a small turret behind the engine. Updated versions have a larger turret with a grenade launcher

ROOF HATCHES
The AAV7 has three torsion-spring-balanced roof hatches, which are used to load and unload the vehicle when it is in water

AAV7

The AAV7 Amtrak provides protected transport of up to 25 combat-equipped Marines through all types of terrain. The engine compartment can be completely water-sealed, making it seaworthy in rough water conditions. The vehicle is fitted with sandwich-plated steel armour, with a layer of Kevlar underneath. Updated versions have a grenade launcher added to the turret.

SPECIFICATIONS

Country of origin:	USA
Crew:	3 + 25
Weight:	22,838kg (50,349lb)
Dimensions:	Length: 7.94m (26ft); width: 3.27m (10ft 8in); height: 3.26m (10ft 8in)
Range:	482km (300 miles)
Armour:	Up to 45mm (1.77in)
Armament:	(Update) 1 x 40mm (1.57in) grenade launcher; 1 x 12.7mm (.5in) HMG
Powerplant:	1 x Detroit Diesel 8V-53T turbo-charged engine developing 298kW (400hp)
Performance:	Maximum road speed: 64km/h (40mph); maximum water speed: 14km/h (9mph)

POWER RAMP
The power-operated ramp in the rear of the vehicle has a small door built-in to allow access in the event of a mechanical failure

SIDE ARMOUR
The hull top, sides, and bottom of the AAV7 are protected by 30mm (1.2in) of aluminium armour, giving protection against small-arms fire and shell splinters

LATE COLD WAR TANKS AND AFVS

Engineering Vehicles

The principal task of the combat engineer was originally to undermine or blow up enemy fortifications. The job of today's engineers is to open the route, improve roads, breach enemy minefields, destroy obstacles and build bridges, and they have the best state-of-the-art equipment to carry out the work.

Combat Engineer Tractor

The Combat Engineer Tractor (CET) was built for a variety of roles, such as vehicle recovery, clearing obstacles and preparing river banks for crossings. A rocket-propelled anchor can be fired into the earth to enable the vehicle to winch itself out if stuck.

SPECIFICATIONS	
Country of origin:	United Kingdom
Crew:	2
Weight:	18,000kg (39,600lb)
Dimensions:	Length: 7.54m (24ft 9in); width: 2.9m (9ft 6in); height: 2.67m (8ft 9in)
Range:	322km (200 miles)
Armour:	Classified
Armament:	1 x 7.62mm (.3in) MG
Powerplant:	1 x Rolls-Royce C6TFR 6-cylinder inline diesel engine developing 238.6kW (320hp)
Performance:	Maximum road speed: 56km/h (35mph); fording: 1.83m (6ft); vertical obstacle: .61m (2ft); trench: 2.06m (6ft 9in)

BAT-2

The BAT-2 is a general engineer vehicle, ideally for earth-shifting or obstacle-clearing operations, using its large V-shaped hydraulically powered articulating dozer blade. It is fitted with a ripping device, for tearing up frost-hardened ground.

SPECIFICATIONS	
Country of origin:	Ukraine
Crew:	2 + 8
Weight:	39,700kg (87,500lb)
Dimensions:	Length: 9.64m (31ft 7in); width: 4.2m (13ft 9in); height: 3.69m (12ft 1in)
Range:	500km (310 miles)
Armour:	Not applicable
Armament:	None
Powerplant:	1 x V-64-4 12-cylinder multi-fuel diesel engine developing 522kW (700hp)
Performance:	Maximum road speed: 60km/h (37mph); fording: 1.3m (4ft 4in); vertical obstacle: .8m (2ft 7in)

TIMELINE

1978

ENGINEERING VEHICLES

Pionierpanzer Dachs 2

The Pionierpanzer Dachs 2 has a digging bucket at the end of its telescopic arm. During transit, the arm is laid down along the right-hand side of the hull. A dozer blade acts as a minesweeping shield and there are onboard tools to split concrete and steel girders.

SPECIFICATIONS

Country of origin:	Germany
Crew:	3
Weight:	43,000kg (94,800lb)
Dimensions:	Length: 9.01m (29ft 6in); width: 3.25m (10ft 8in); height: 2.57m (8ft 5in)
Range:	650km (400 miles)
Armour:	Not available
Armament:	1 x 7.62mm (.3in) MG
Powerplant:	1 x MTU MB838 CaM5000 10-cylinder diesel engine developing 610kW (818hp)
Performance:	Maximum road speed: 62km/h (39mph)

M9 ACE

Unusually for an earthmover, the M9 is amphibious and propels itself through water at speeds of 4.8km/h (3mph) on its tracks. At the front, a large apron/dozer blade is deployed by using hydraulic rotary actuators to lower the front of the vehicle.

SPECIFICATIONS

Country of origin:	USA
Crew:	3
Weight:	16,327kg (36,001lb)
Dimensions:	Length: 6.25m (20ft 6in); width: 3.2m (10ft 6in); height: 2.7m (8ft 10in)
Range:	322km (200 miles)
Armour:	Not available
Armament:	None
Powerplant:	1 x Cummins V903C 8-cylinder diesel engine developing 164kW (220hp)
Performance:	Maximum road speed: 48km/h (30mph)

AAAVR7A1

The Armoured Amphibian Assault Vehicle Recovery 7A1 (AAAVR7A1) recovers damaged vehicles or makes engineering repairs on or from amphibious landing zones. On board are a Miller Maxtron 300 welder, a portable generator and an air compressor.

SPECIFICATIONS

Country of origin:	USA
Crew:	5
Weight:	23,601kg (52,040lb)
Dimensions:	Length: 7.94m (26ft); width: 3.27m (10ft 9in); height: 3.26m (10ft 8in)
Range:	480km (300 miles)
Armour:	Not available
Armament:	1 x 12.7mm (.5in) Browning M2 HB or 7.62mm (.3in) M60 MG
Powerplant:	1 x Cummins VT400 8-cylinder multi-fuel engine developing 298kW (399hp)
Performance:	Maximum road speed: 72km/h (45mph); maximum water speed: 13km/h (8mph)

1983

1990

MODERN ERA TANKS AND AFVS

The full inventory of 'special' armoured vehicles was brought into play during Operation Desert Storm in 1991, breaching the defensive perimeter the Iraqis had built around occupied Kuwait.

The ongoing development of main battle tanks culminated in two excellent designs that saw action in Iraq in the Gulf War of 1991 and again in 2003. Capable of taking on the latest Russian-designed equipment and destroying it with ease, America's M1 Abrams and Britain's Challenger 2 – along with Germany's Leopard 2 and France's Leclerc – are likely to remain in service for years to come.

Left: Thanks to its sophisticated fire control systems, the M1 Abrams has an unprecedented first-shot kill power, as Iraq's Republican Guard discovered.

MODERN ERA TANKS AND AFVS

LEOPARD 2

The Leopard 2 is an extremely potent main battle tank and is one of the finest of its generation, offering a unique blend of firepower, protection and mobility. First delivered in 1978, it has been radically upgraded over the years. It is equipped with laser rangefinder, thermal-imaging and amphibious capability.

LEOPARD 2

FIRE CONTROL
The fire control computer receives the range of the target from the rangefinder and guides the main armament to the line-of-sight of the gunner's periscope

The Leopard 2 was an offshoot of the cancelled joint MBT-70 project between the United States and West Germany in the late 1960s. Its weapon system is unusual in that the cartridge cases are combustible. When the shell is fired, all that remains is the base of the cartridge, which frees up space. Exports of Leopard 2 tanks were soon being made to other armies.

SPECIFICATIONS	
Country of origin:	West Germany
Crew:	4
Weight:	Approx 59,700kg (131,616lb)
Dimensions:	Length: 9.97m (32ft 8.4in); width: 3.74m (12ft 3.25in); height: 2.64m (8ft 7.9in)
Range:	500km (310 miles)
Armour:	Classified
Armament:	1 x 120mm (4.7in) smooth-bore gun; 2 x 7.62mm (.3in) MGs
Powerplant:	1 x MTU MB 873 Ka501 12-cylinder diesel engine developing 1119kW (1500hp)
Performance:	Maximum road speed: 72km/h (45mph)

MAIN ARMAMENT
The Leopard 2's 120mm (4.7in) Rheinmetall gun has a barrel life of more than 1000 rounds and is chromium-plated to cope with the immense strain of firing large projectiles at four times the speed of sound

LEOPARD 2

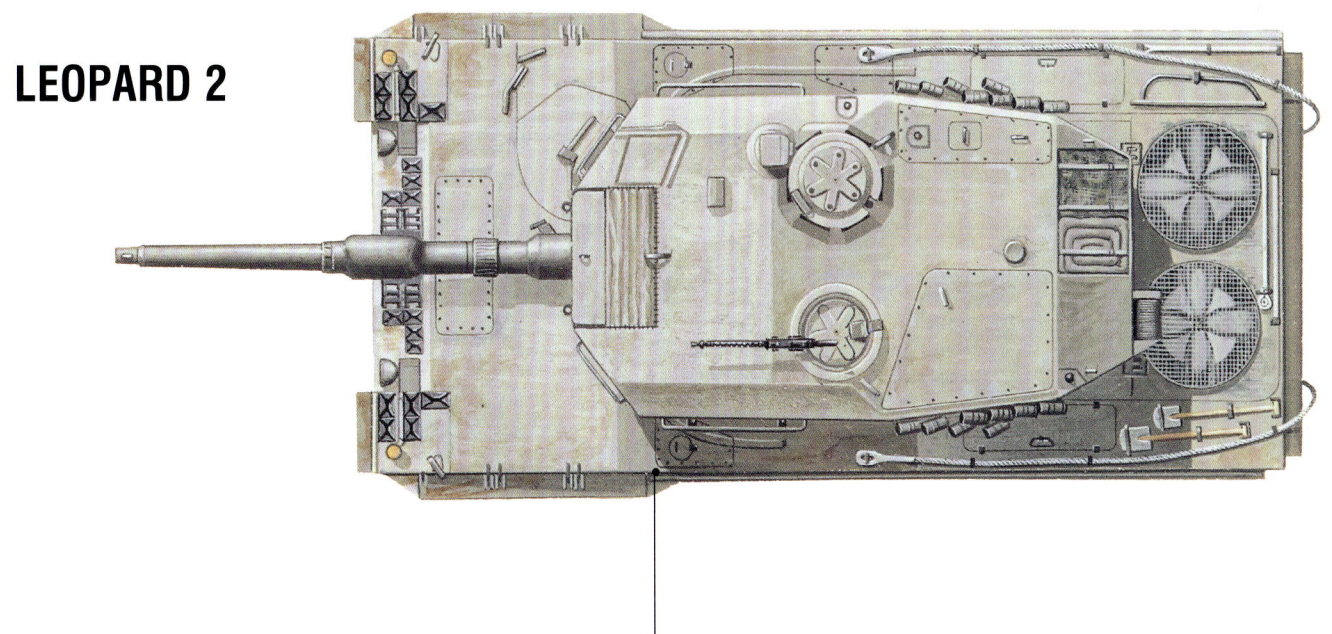

MOBILITY
On rough ground, the Leopard 2 is surprisingly mobile, throwing its great weight across the battlefield with the agility of an armoured personnel carrier

FIRE SUPPRESSION SYSTEM
To protect the crew against the consequences of a penetrative hit through the hull, the Leopard 2 is fitted with an inert gas fire-suppression system designed to put out a fire in one-fifth of a second

Main Battle Tanks

Far from being an anachronistic monster, the main battle tank is beginning a new lease of life. In the Yom Kippur War of 1973, most tanks were stopped dead in their tracks by lethal anti-tank weapons, but modern advances in armour mean that the tanks can now fight their way forward through serious opposition.

Ramses II

The Ramses II, developed for service in the Egyptian army, is essentially a T-54 modified and upgraded with current American systems. Testing of the modified vehicle went ahead in 1990, and production started in 2004.

SPECIFICATIONS	
Country of origin:	Egypt
Crew:	4
Weight:	45,800kg (101,972lb)
Dimensions:	Length: 9.9m (32ft 6in); width: 3.27m (10ft 8in); height: 2.4m (7ft 11in)
Range:	600km (373 miles)
Armour:	Not available
Armament:	1 x 105mm (4.73in) M68 rifled gun; 1 x 7.62mm (.3in) coaxial MG; 1 x 12.7mm (.5in) M2HB HMG
Powerplant:	1 x TCM AVDS-1790-5A turbocharged diesel engine developing 677kW (908hp)
Performance:	Maximum speed: 72km/h (45mph)

Al Khalid Main Battle Tank

The Al Khalid (meaning 'Sword') main battle tank is a joint venture between Pakistan, China and the Ukraine. Several prototypes underwent field trials and it was designed to be adaptable to different engines and transmissions. The current production variant uses a Ukrainian-designed diesel engine.

SPECIFICATIONS	
Country of origin:	Pakistan/China/Ukraine
Crew:	3
Weight:	46,000kg (101,430lb)
Dimensions:	Length: 6.9m (22ft 8in); width: 3.4m (11ft 2in); height: 2.3m (7ft 7in)
Range:	400km (250 miles)
Armour:	Not available
Armament:	1 x 125mm (4.92in) gun; 1 x 7.62mm (.3in) MG; 1 x 12.7mm (.5in) HMG
Powerplant:	1 x 8-cylinder 4-stroke water-cooled diesel engine developing 895kW (1200bhp)
Performance:	Maximum speed: 70km/h (44mph)

TIMELINE

1991 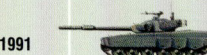 1992  1993

MAIN BATTLE TANKS

Type 90 Main Battle Tank

The Type 90 main battle tank mounts a 120mm (4.7in) as its main armament. This is essentially the German Rheinmetall gun built under licence, but it has its own recoil system and gun mount. The Type 90's crew is reduced to three as no loader is required.

SPECIFICATIONS	
Country of origin:	Japan
Crew:	3
Weight:	50,000kg (110,250lb)
Dimensions:	Length: 9.76m (32ft); width: 3.43m (11ft 4in); height: 2.34m (7ft 11in)
Range:	400km (250 miles)
Armour:	Not available
Armament:	1 x 120mm (4.7in) gun; 1 x 7.62mm (.3in) MG; 1 x 12.7mm (.5in) HMG
Powerplant:	1 x Mitsubishi 10ZG 10-cylinder diesel engine developing 1118kW (1500bhp)
Performance:	Maximum speed: 70km/h (43mph)

PT-91 Main Battle Tank

The Polish PT-91 main battle tank is a modification of the Soviet T-72M1, designed to modernize the later models of the Soviet tanks that were in service. Work proceeded slowly at first, but accelerated with the fragmentation of the Soviet Union.

SPECIFICATIONS	
Country of origin:	Poland
Crew:	3
Weight:	45,300kg (99,886lb)
Dimensions:	Length: 6.95m (22ft 10in); width: 3.59m (11ft 9in); height: 2.19m (7ft 2in)
Range:	650km (405 miles)
Armour:	Not available
Armament:	1 x 125mm (4.92in) gun; 1 x 7.62mm (.3in) MG; 1 x 12.7mm (.5in) HMG
Powerplant:	1 x S-12U V-12 supercharged diesel engine developing 634kW (850bhp) at 2300rpm
Performance:	Maximum speed: 60km/h (38mph)

T-90 Main Battle Tank

The T-90 main battle tank is a development of the T-72. It features the latest development of the Kontakt-5 explosive reactive armour, offering defence against chemical and kinetic energy warheads. The upgraded model from 1996 features a fully welded turret.

SPECIFICATIONS	
Country of origin:	Russia
Crew:	3
Weight:	46,500kg (102,532lb)
Dimensions:	Length (hull): 6.86m (22ft 6in); width: 3.37m (11ft 1in); height: 2.23m (7ft 4in)
Range:	650km (400 miles)
Armour:	Not available
Armament:	1 x 125mm (4.9in) 2A46M Rapira 3 smoothbore gun; 1 x 7.62mm (.3in) co-axial MG; one 12.7mm (.5in) anti-aircraft HMG
Powerplant:	1 x V-84MS 12-cylinder multi-fuel diesel engine developing 627kW (840bhp) at 2000rpm
Performance:	Maximum speed: 65km/h (40mph)

 2004

Current Main Battle Tanks

The Gulf War of 1991 demonstrated the awesome power of the main battle tank. The tanks used by the Coalition had the benefit of technology that had been under constant development and review for nearly 20 years, technology that gave them the ability to fight at night, and to hit their targets while still on the move.

Arjun

The Arjun is India's first indigenous main battle tank. It uses a German MTU diesel, but its main armament is a locally designed stabilized 120mm (4.7in) rifled gun able to fire a range of ammunition types: high explosive; high explosive anti-tank; high explosive squash head.

SPECIFICATIONS	
Country of origin:	India
Crew:	4
Weight:	58,000kg (127,600lb)
Dimensions:	Length: 9.8m (32ft 2in); width: 3.17m (10ft 5in); height: 2.44m (8ft)
Range:	400km (250 miles)
Armour:	Classified
Armament:	1 x 120mm (4.7in) gun; 1 x 7.62mm (.3in) MG
Powerplant:	1 x MTU MB 838 Ka 501 water-cooled diesel engine developing 1044kW (1400hp)
Performance:	Maximum road speed: 72km/h (45mph); fording: 1m (3ft 3in); vertical gradient: 1.1m (3ft 7in); trench: 3m (9ft 10in)

M1A2 with TUSK

The Tank Urban Survival Kit, or TUSK, can be installed in the field in tanks such as the Abrams M1A2, enabling them to be upgraded without being taken out of the line. Upgrades include slat armour on the rear to protect against rocket-propelled grenades.

SPECIFICATIONS	
Country of origin:	USA
Crew:	4
Weight:	(Without TUSK upgrade) 62,051kg (136,800lb)
Dimensions:	Length: 9.77m (32ft 1in); Width: 3.66m (12ft); Height: 2.44 (8ft)
Range:	391 km (243 miles)
Armour:	Classified; TUSK upgrade includes reactive side armour and slat armour on the rear
Armament:	1 x 120mm (4.7in) M256 smoothbore gun, 1 x 7.62mm (.3in) coaxial MG; 1 x 7.62mm (.3in) anti-aircraft MG, 1 x 12.7mm (.5in) Kongsberg Gruppen Remote Weapon Turret
Powerplant:	1 x Textron Lycoming AGT1500 gas turbine engine, developing 1119kW (1500hp)
Performance:	Maximum road speed: 72km/h (45mph)

TIMELINE

1990  1992  1994

CURRENT MAIN BATTLE TANKS

Challenger 2

The Challenger 2 boasts a carbon dioxide laser rangefinder, thermal-imaging and fully computerized fire-control systems, giving a high first-round hit probability. For greater combat capability, it can be fitted with the Battlefield Information Control System.

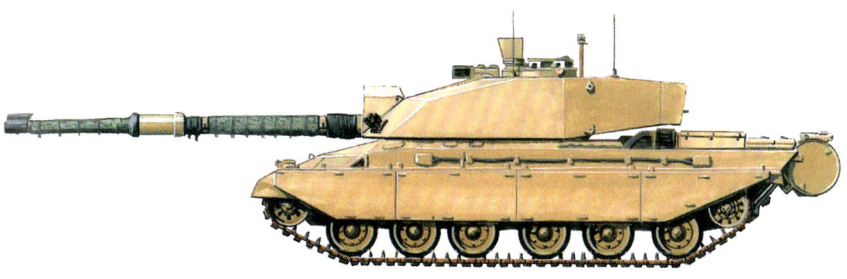

SPECIFICATIONS

Country of origin:	United Kingdom
Crew:	4
Weight:	62,500kg (137,500lb)
Dimensions:	Length: 11.55m (35ft 4in); width: 3.52m (10ft 8in); height: 2.49m (7ft 5in)
Range:	400km (250 miles)
Armour:	Classified
Armament:	1 x 120mm (4.7in) gun; 2 x 7.62mm (.3in) MGs; 2 x smoke rocket dischargers
Powerplant:	1 x liquid-cooled diesel engine developing 895kW (1200hp)
Performance:	Maximum road speed: 57km/h (35.6mph); fording: 1m (3ft 4in); vertical obstacle: .9m (2ft 10in); trench: 2.8m (9ft 2in)

Type 90-II

The Type 90-II was designed and developed in China for the export market. About 45 per cent of its components come from existing Chinese designs, such as the Type 59, Type 69 and Type 85/88C. Since 2001, it has been built under licence in Pakistan as the Al-Khalid.

SPECIFICATIONS

Country of origin:	China
Crew:	3
Weight:	48,000kg (105,822lb)
Dimensions:	Length: 10.1m (33ft 1in); width: 3.5m (11ft 6in); height: 2.20m (7ft 2in)
Range:	450km (249 miles)
Armour:	Not available
Armament:	1 x 125mm (4.9in) smoothbore gun; 1 x 12.7mm (.5in) external anti-aircraft MG; 1 x 7.62mm (.3in) coaxial MG
Powerplant:	1 x Perkins CV12-1200 TCA 12-Cylinder, water cooled electronically controlled diesel engine developing 895kW (1200hp)
Performance:	Maximum road speed: 62.3km/h (39mph)

Leopard 2A6

Armed with a Rheinmetall 120mm (4.7in) L55 smoothbore gun, the Leopard 2A6 is standard equipment of the Dutch Army's operational tank battalions and those of the German Army assigned to NATO's Crisis Intervention Force.

SPECIFICATIONS

Country of origin:	Germany
Crew:	4
Weight:	59,700kg (131,616lb)
Dimensions:	Length: 9.97m (32ft 8in); width: 3.5m (11ft 6in) height: 2.98m (9ft 10in)
Range:	500km (311 miles)
Armour:	Not available
Armament:	1 x 120mm (4.7in) LSS smoothbore gun; 2 x 7.62mm (.3in) MGs
Powerplant:	1 x MTU MB 873 four stroke 12-cylinder diesel engine, exhaust turbocharged, liquid cooled, developing 1119kW (1500hp)
Performance:	Maximum speed: 72km/h (45mph)

MODERN ERA TANKS AND AFVS

Leclerc

The Leclerc was designed to replace the French Army's fleet of AMX-30 tanks. Development began in 1983, and the first production Leclercs appeared in 1991. The Leclerc is an excellent vehicle. An automatic loading system for the main armament and remote-control machine guns allow the crew to be cut down to three.

Leclerc

The tank can be fitted with extra fuel tanks to increase operational range, and standard equipment includes a fire-detection/suppression system, thermal-imaging and laser rangefinder for the main gun, and a land navigation system. The on-board electronic systems are fully integrated to allow automatic reconfiguration in case of battlefield failure or damage.

MAIN ARMAMENT
Leclerc's 120mm (4.7in) smoothbore main armament is longer than the gun on the Leopard 2 and the M1 Abrams, but fires the same combustible cartridge ammunition

GUN
The gun is fully stabilized to allow firing on the move across country, and has an automatic loader that enables the Leclerc to sustain a rate of fire of 12 rounds per minute

AUTOLOADER
The autoloader can quickly switch between APFSDS and HEAT, the two principal types of projectile carried. Normal ammunition load is 22 ready-use rounds in the turret, with a further 18 in the hull

RANGEFINDER
The commander's panoramic sight incorporates a laser rangefinder and an image intensifier. The gunner's sight incorporates a thermal imager

Leclerc

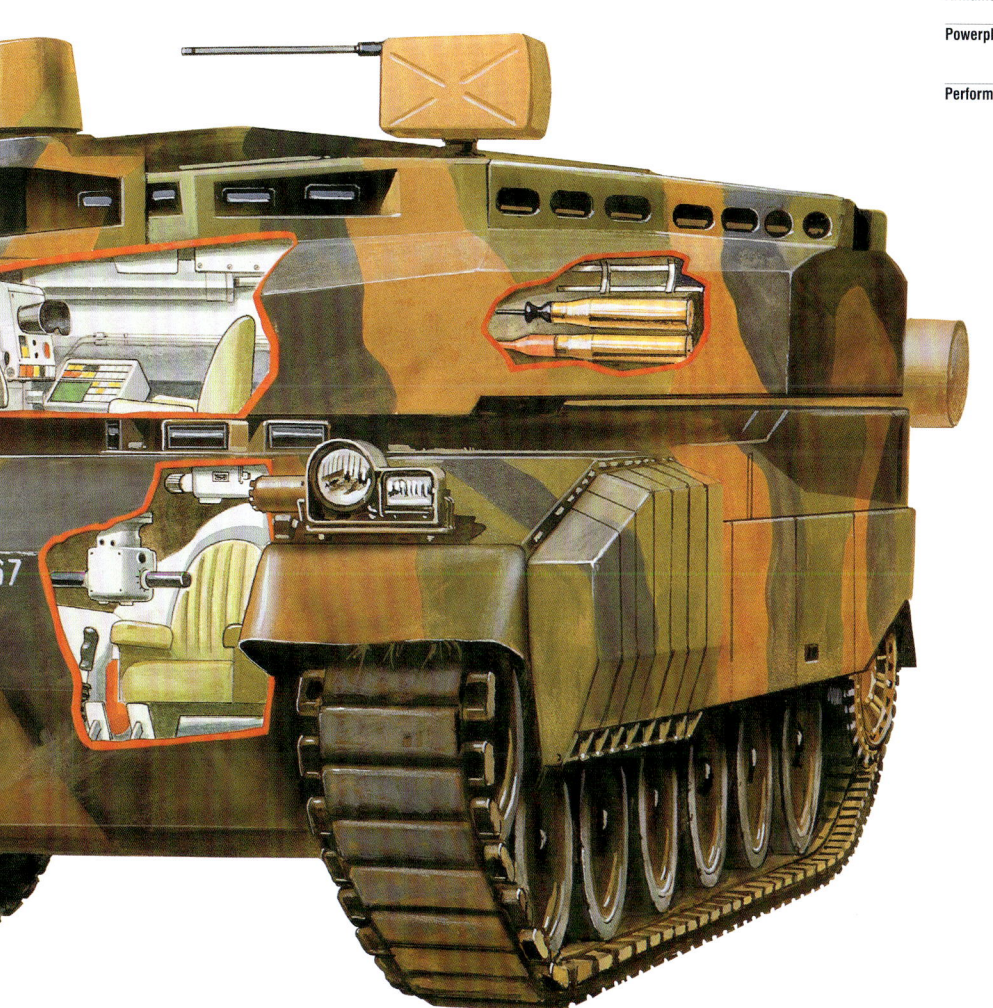

SPECIFICATIONS	
Country of origin:	France
Crew:	3
Weight:	53,500kg (117,700lb)
Dimensions:	Length: 9.87m (30ft); width: 3.71m (11ft 4in); height: 2.46m (7ft 6in)
Range:	550km (345 miles)
Armour:	Classified
Armament:	1 x 120mm (4.7in) gun, 1 x 12.7mm (.5in) MG, 1 x 7.62mm (.3in) MG, 3 x 9 smoke dischargers
Powerplant:	1 x SAEM UDU V8X 1500 T9 Hyperbar eight-cylinder diesel engine developing 1118.5kW (1500hp); SESM ESM500 automatic transmission
Performance:	Maximum road speed: 73km/h (45.6mph); fording: 1m (3ft 3in); vertical obstacle: 1.25m (4ft 1in); trench: 3m (9ft 10in)

POWERPLANT
Leclerc is powered by a SACM V8 high-pressure diesel engine, delivering 1119kW (1500hp) via an automatic hydrostatic transmission

ROUTE PLANNING
Leclerc is fitted with the FINDERS battle management system, which has a coloured map display on which the dispositions of allied and enemy forces can be projected. It can also be used for route and mission planning

MODERN ERA TANKS AND AFVS

Self-propelled Guns

Modern self-propelled guns have the latest in fire control technology, and use Precision Guided Munitions (PGMs) and Terminally-Guided Sub-Munitions (TGSMs), which burst over the general target area to release sub-munitions that float to earth under parachutes, their sensors homing in to attack the upper armour of tanks.

AS-90

Vickers Armstrong, while working on the failed SP70 project, saw the defects in the design and set about preparing an improved version. The AS-90 mounts a 39-calibre howitzer, but a number are being upgraded to 52-calibre weapons, called the AS90 Braveheart.

SPECIFICATIONS	
Country of origin:	United Kingdom
Crew:	5
Weight:	45,000kg (99,225lb)
Dimensions:	Length: 7.2m (23ft 8in); width: 3.4m (11ft 2in) height: 3m (9ft 10in)
Range:	240km (150 miles)
Armour:	17mm (.66in) maximum
Armament:	1 x 155mm (6.1in) howitzer; 1 x 12.7mm (.5in) MG
Powerplant:	1 x Cummins V-8 diesel engine developing 492kW (660hp) at 2800rpm
Performance:	Maximum road speed: 55km/h (34 mph); fording: 1.5m (5ft); vertical obstacle: .88m (2ft 11in); trench: 2.8m (9ft 2in)

155/45 Norinco SP Gun

The Chinese 155/45 Norinco SP Gun has a large turret carrying a 45-calibre 155mm (6.1in) gun. The gun has a muzzle brake and fume extractor, and is provided with mechanical assistance for loading and ramming at any angle of elevation.

SPECIFICATIONS	
Country of origin:	China
Crew:	5
Weight:	32,000kg (70,560lb)
Dimensions:	Length: 6.1m (20ft); width: 3.20m (10ft 6in); height: 2.59m (8ft 6in)
Range:	450km (20 miles)
Armour:	Not disclosed
Armament:	1 x 155mm (6.1in) WAC-21 gun
Powerplant:	1 x diesel engine developing 391.4kW (525hp)
Performance:	Maximum road speed: 56 km/h (35mph); fording: 1.2m (4ft); vertical obstacle: .7m (2ft 4in); trench: 2.7m (8ft 10in)

TIMELINE

 1993

 1994

SELF-PROPELLED GUNS

M109A6 Paladin

The 155mm (6.1in) M109A6 Paladin has the same hull and suspension as the M109, but everything else has been changed. The modifications have created a weapons system that can react more quickly to target opportunities and hit targets that are further away.

SPECIFICATIONS

Country of origin:	USA
Crew:	4
Weight:	28,738kg (63,367lb)
Dimensions:	Length: 6.19m (20ft 4in); width: 3.149m (10ft 4in); height: 3.236m (10ft 7in)
Range:	405km (252 miles)
Armour:	Not disclosed
Armament:	1 x 155mm (6.1in) howitzer M284
Powerplant:	1 x Detroit Diesel 8V-71T, V-8 turbocharged two-stroke diesel engine developing 302kW (405hp) at 2300 rpm
Performance:	Maximum road speed: 56km/h (35mph); fording: 1.95m (6ft 5in); vertical obstacle: .53m (1ft 9in); trench: 1.83m (6ft)

Rascal

The Rascal is a 20-ton 155mm (6.1in) self-propelled howitzer, light enough to be airlifted. Its low weight is combined with a powerful engine to provide fast speeds both on- and off-road. The howitzer has a range of 24km (15 miles).

SPECIFICATIONS

Country of origin:	Israel
Crew:	4
Weight:	19,500kg (43,000lb)
Dimensions:	Length (with gun): 7.5m (24ft 7in); width: 2.46m (8ft 1in); height: 2.3m (7ft 7in)
Range:	350km (220 miles)
Armour:	Not available
Armament:	1 x 155mm (6.1in) howitzer
Powerplant:	1 x diesel engine developing 261kW (350hp)
Performance:	Maximum road speed: 50km/h (31mph); fording: 1.2m (3ft 10in); gradient: 22 per cent; vertical obstacle: 1m (3ft 4in)

Panzerhaubitzer 2000

The hull and running gear of the Panzerhaubitzer are based on the Leopard II tank, but with the engine and transmission at the front of the hull. A large turret mounts the 52-calibre length gun, which has a sliding block breech and a large multi-baffle muzzle-brake.

SPECIFICATIONS

Country of origin:	Germany
Crew:	5
Weight:	55,000kg (121,275lb)
Dimensions:	Length: 7.87m (25ft 10in); width: 3.37m (11ft); height: 3.4m (11ft 2in)
Range:	420km (260 miles)
Armour:	Not disclosed
Armament:	1 x 155mm (6.1in) howitzer
Powerplant:	1 x MTU 881 V-12 diesel engine developing 745.7kW (1000hp)
Performance:	Maximum road speed: 60km/h (27mph); fording: 2.25m (7ft 5in); vertical obstacle: 1m (3ft 3in); trench: 3m (9ft 10in)

Anti-aircraft Vehicles

During the Cold War, both NATO and the Warsaw Pact developed air defence systems that provided their ground forces with an umbrella of interlocking sub-systems with a high degree of mobility. In the 1990s, however a new dimension appeared in the shape of mobile anti-missile systems like the Patriot.

Type 87 AWSP

The Type 87 Automatic Western Self-Propelled (AWSP) uses the turret of the Gepard air-defence system, mounted on the Type 74 MBT chassis. Like the Gepard, it has two 35mm (1.38in) Oerlikon KDA cannon but an improved fire-control system.

SPECIFICATIONS	
Country of origin:	Japan
Crew:	3
Weight:	36,000kg (79,400lb)
Dimensions:	Length: 7.99m (26ft 2in); width: 3.18m (10ft 5in); height: 4.4m (14ft 5in)
Range:	500km (310 miles)
Armour:	Steel (details classified)
Armament:	2 x 35mm (1.38in) Oerlikon KDA guns
Powerplant:	1 x 10F22WT 10-cylinder diesel engine developing 536kW (718hp)
Performance:	Maximum road speed: 60km/h (37mph); fording: 1m (3ft 4in); gradient: 60 per cent; vertical obstacle: 1m (3ft 4in); trench 2.7m (8ft 10in)

Sidam 25

In essence, the OTOBREDA Sidam 25 is four Oerlikon KBA 25mm (.98in) automatic cannon turret-mounted on the M113 APC. An optronic fire-control system makes the weapon highly accurate against low-flying aerial targets within its 2000m (6650ft) effective range.

SPECIFICATIONS	
Country of origin:	Italy
Crew:	3
Weight:	15,100kg (33,300lb)
Dimensions:	Length: 5.04m (16ft 6in); width: 2.67m (8ft 9in); height (without turret): 1.82m (5ft 11in)
Range:	550km (342 miles)
Armour:	Up to 38mm (1.49in)
Armament:	4 x 25mm (.98in) Oerlikon KBA guns
Powerplant:	1 x Detroit 6V-53T 6-cylinder diesel engine developing 198kW (266hp)
Performance:	Maximum road speed: 69km/h (40mph); fording: amphibious; gradient: 60 per cent; vertical obstacle: .61m (2ft); trench: 1.68m (5ft 6in)

TIMELINE

1987   1989

ANTI-AIRCRAFT VEHICLES

MIM-104 (GE) Patriot

The Raytheon MIM-104 (GE) Patriot is an advanced SAM system. Using a track-via-missile (TVM) onboard guidance system in tandem with a ground-based tracking unit, each missile can engage aircraft targets and incoming ballistic and cruise missiles.

SPECIFICATIONS	
Country of origin:	USA
Crew:	2
Weight:	(M109 launcher) 26,867kg (59,241lb)
Dimensions:	Length: 10.4m (34ft 1in); width: 2.49m (8ft 2in); height: 3.96m (13ft)
Range:	800km (500 miles)
Armour:	Not applicable
Armament:	4 x Patriot SAM missiles
Powerplant:	1 x MAN D2866 LGF 6-cylinder diesel engine developing 265kW (355hp)
Performance:	Maximum road speed: 80km/h (50mph)

ADATS

The Air Defense Anti-Tank System (ADATS) is the result of a troubled experiment to combine an anti-tank and anti-aircraft missile system. Eight ADATS missiles are mounted on the hull with a Doppler pulse radar and an electro-optical target-acquisition system.

SPECIFICATIONS	
Country of origin:	Switzerland/USA
Crew:	3
Weight:	15,800kg (34,800lb)
Dimensions:	Length: 4.86m (15ft 11in); width: 2.68m (8ft 9in); height (with radar antenna): 4.48m (14ft 8in)
Range:	400km (250 miles)
Armour:	12–38mm (.47–1.49in)
Armament:	8 x ADATS SAMs
Powerplant:	1 x Detroit 6V-53N 6-cylinder diesel engine developing 158kW (211hp)
Performance:	Maximum road speed: 58km/h (36mph); fording: amphibious; gradient: 60 per cent; vertical obstacle: .61m (2ft); trench 1.68m (5ft 6in)

Armoured Starstreak

The Starstreak SAM is an anti-aircraft missile for low-level defence against ground-attack aircraft. It can be launched from multiple platforms: shoulder; attack helicopter; and a vehicle, the Starstreak Self-Propelled High-Velocity Missile (SP HVM).

SPECIFICATIONS	
Country of origin:	United Kingdom
Crew:	3
Weight:	12,700kg (28,000lb)
Dimensions:	Length: 5.33 m (17ft 6in); width: 2.4m (7ft 10in); height 3.49m (11ft 5in)
Range:	650km (400 miles)
Armour:	Aluminium (details classified)
Armament:	8 +12 Starstreak SAMs
Powerplant:	1 x Perkins T6/3544 6-cylinder turbocharged diesel engine developing 186kW (250hp) at 2600rpm
Performance:	Maximum road speed: 80km/h (50mph); fording: amphibious; gradient: 60 per cent; vertical obstacle: .6m (2ft); trench 1.75m (5ft 9in)

 1997

M2 BRADLEY

The M2 Bradley had a protracted development history and by no means all US Army chiefs were in favour of it, but it made a big impression in the Gulf War of 1991, destroying more Iraqi armour than the M1 Abrams Main Battle Tank.

BRADLEY M2

CO-AXIAL MACHINE GUN
The 7.62mm (.3in) M2440C machine gun has 800 rounds at the ready and another 1540 in reserve. It is mainly for anti-personnel use

M257 SMOKE DISCHARGER
Electronically operated, the M257 fires a pattern of four smoke grenades in front of the Bradley as an emergency defence measure

ENGINE COMPARTMENT
The Cummins VTA-903T turbo-charged 8-cylinder diesel engine develops 372.8kW (500hp) at 2600rpm. It is equipped with a Halon (inert gas) fire suppression system

BRADLEY M2

TOW ANTI-TANK GUIDED MISSILE LAUNCHER
The sighting for the main gun and TOW missiles are fully integrated to make the gunner's life easier

The first production Bradley AFVs were completed in 1981 and production continued up to 1995, with about 6800 units being built. Of these, 400 were supplied to Saudi Arabia, the only overseas customer. The latest models of the Bradley are fitted with explosive reactive armour.

FIRING PORT
The infantrymen in the troop compartment are provided with the firing ports, each with a periscope above. These enable them to fire M231 Firing Port Weapons from within the vehicle, although accuracy is not high

SPECIFICATIONS	
Country of origin:	USA
Crew:	3 + 6
Weight:	22,940kg (50,574lb)
Dimensions:	Length: 6.55m (21ft 6in); width: 3.61m (11ft 9in); height (turret roof): 2.57m (8ft 5in)
Range:	483km (300 miles)
Armour:	Unknown thickness; aluminium/steel
Armament:	1 x 25mm (.98in) Bushmaster Chain Gun; 1 x 7.62mm (.3in) MG, 2 x TOW missile launchers
Powerplant:	1 x Cummins VTA-903T turbo-charged 8-cylinder diesel engine developing 372.8kW (500hp) at 2600rpm
Performance:	Maximum road speed: 64km/h (40mph); maximum water speed: 14km/h (9mph)

MODERN ERA TANKS AND AFVS

Warrior

The Warrior Mechanized Combat Vehicle began development in 1972 and entered service with the British Army in 1987. It was part of a movement to change the role of armoured personnel carriers from mere transporter of troops to a more capable infantry combat vehicle, this concept being inspired by the success of the Soviet BMP.

Warrior

Designed to supplement the FV432, the Warrior is heavier and much more heavily armoured. It is treated as a mobile fire base from which troops can fight, rather than a mere transport vehicle. Variants include a command vehicle, recovery vehicle, engineer and observation vehicle. It has been sold to Kuwait, whose Warriors have anti-tank missile launchers each side of the turret and air conditioning.

SPECIFICATIONS	
Country of origin:	United Kingdom
Crew:	3 + 7
Weight:	25,700kg (56,540lb)
Dimensions:	Length: 6.34m (20ft 10in); width: 3.034m (10ft); height: 2.79m (9ft 2in)
Range:	660km (412 miles)
Armour:	Classified
Armament:	1 x 30mm (1.2in) Rarden gun; 1 x 7.62mm co-axial MG; 4 x smoke dischargers
Powerplant:	1 x Perkins V-8 diesel engine developing 410kW (550hp)
Performance:	Maximum road speed: 75km/h (46.8mph); fording: 1.3m (4ft 3in); vertical obstacle: .75m (2ft 5in); trench: 2.5m (8ft 2in)

ARMAMENT
All variants are equipped with a 7.62mm (.3in) chain gun. The chain gun and the Rarden cannon have a low-level air defence capability against helicopters

ARMOUR
The Warrior's armour is designed to withstand an explosion from a 155mm (6.1in) shell at 10m and direct fire from machine guns up to a calibre of 14.5mm (.57in). During the first Gulf War and operations in the Balkans and Iraq, additional armoured protection was fitted for extra protection

Warrior

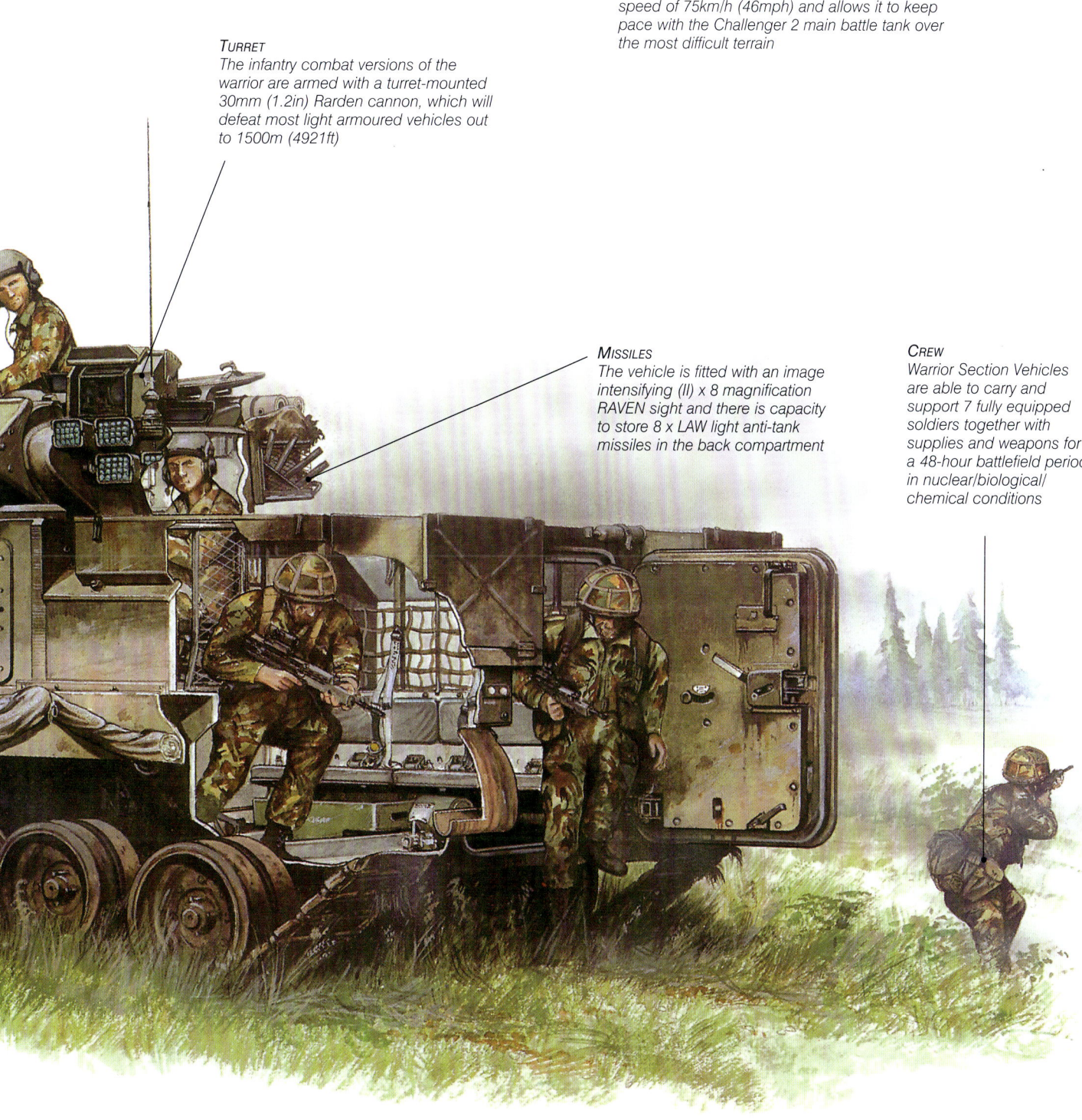

ENGINE
One of the most impressive features of the vehicle is its powerful diesel engine, which gives it a road speed of 75km/h (46mph) and allows it to keep pace with the Challenger 2 main battle tank over the most difficult terrain

TURRET
The infantry combat versions of the warrior are armed with a turret-mounted 30mm (1.2in) Rarden cannon, which will defeat most light armoured vehicles out to 1500m (4921ft)

MISSILES
The vehicle is fitted with an image intensifying (II) x 8 magnification RAVEN sight and there is capacity to store 8 x LAW light anti-tank missiles in the back compartment

CREW
Warrior Section Vehicles are able to carry and support 7 fully equipped soldiers together with supplies and weapons for a 48-hour battlefield period in nuclear/biological/chemical conditions

MODERN ERA TANKS AND AFVS

Tracked APCs

In modern warfare, tracked APCs are too vulnerable to too many different weapons for them to follow an armoured assault as closely as they did in the past. They must either be equipped with more armour and more countermeasures, or make the best use of ground cover some distance behind the tanks they are supporting.

M980

The layout of the M980 is standard for troop-carriers, with the troop compartment at the rear. The vehicle is fully amphibious, being propelled in water by its tracks. It has been employed by Serb, Bosnian and Croat forces in the wars in former Yugoslavia.

SPECIFICATIONS	
Country of origin:	Yugoslavia
Crew:	3 + 7
Weight:	11,700kg
Dimensions:	Length: 6.4m (21ft); width: 2.6m (8ft 6in); height: 1.8m (5ft 11in)
Range:	Not available
Armour:	8–30mm (0.315–1.18in)
Armament:	1 x 20mm (.79in) gun; 1x co-axial 7.62mm (.3in) MG; 2 x 'Sagger' ATGW missile launchers
Powerplant:	1 x HS 115-2 V-8 cylinder turbo-charged diesel engine developing 194kW (260bhp) at 3000rpm
Performance:	Not available

Bionix 25

Like the new generation of armoured combat vehicles, the Bionix 25 IFV is fast and manoeuvrable with enhanced survivability. Its front-drive system, capable of 70km/h (43mph), is stabilized for accuracy in rough terrain. A thermal sight enables night firing.

SPECIFICATIONS	
Country of origin:	Singapore
Crew:	3 + 7
Weight:	23,000kg (50,700lb)
Dimensions:	Length: 5.92m (19ft 5in); width: 2.7m (8ft 10in); height: 2.53m (8ft 4in)
Range:	415km (260 miles)
Armour:	Classified
Armament:	1 x 25mm (.98in) Boeing M242 gun; 1 x 7.62mm (.3in) coaxial MG; 1 x 7.62mm (.3in) turret-mounted MG; 2 x 3 smoke grenade launchers
Powerplant:	1 x Detroit Diesel Model 6V-92TA diesel engine developing 354kW (475hp)
Performance:	Maximum speed: 70km/h (43mph); fording: 1m (3ft 4in); gradient: 60 per cent; vertical obstacle: .6m (2ft); trench: 2m (6ft 7in)

TIMELINE

1985    1990    1997

TRACKED APCS

VCC-80/Dardo IFV

The Dardo IFV is derived from the VCC-80 MICV, a tracked infantry vehicle armed with a 25mm (.98in) Oerlikon Contraves KBA cannon. The Dardo is little different, though its turret is adapted to mount TOW anti-tank missiles.

SPECIFICATIONS	
Country of origin:	Italy
Crew:	2 + 7
Weight:	23,000kg (50,700lb)
Dimensions:	Length: 6.7m (21ft 11in); width: 3m (9ft 10in); height: 2.64m (8ft 8in)
Range:	600km (370 miles)
Armour:	Layered aluminium/steel (details classified)
Armament:	1 x 25mm (.98in) Oerlikon Contraves KBA gun; 1 x 7.62mm (.3in) coaxial machine gun; 2 x TOW launchers; 2 x 3 smoke grenade launchers
Powerplant:	1 x IVECO 8260 V-6 turbo diesel engine developing 388kW (520hp)
Performance:	Maximum speed: 70km/h (43mph); fording: 1.5m (4ft 10in); gradient: 60 per cent; vertical obstacle: .85m (2ft 9in); trench: 2.5m (8ft 2in)

BMP-3 Infantry Combat Vehicle

Developed by the Kurgan Machine Construction Plant, the BMP-3 infantry combat vehicle entered production in 1989. It was reportedly intended to replace the older BMP-2 on a one-for-one basis, but a lack of funds made this impossible.

SPECIFICATIONS	
Country of origin:	Russia
Crew:	3 + 7
Weight:	18,700kg (18.4 tons)
Dimensions:	Length: 7.2m (23ft 7in); width: 3.23m (10ft 7in); height: 2.3m (7ft 7in)
Range:	600km (373 km)
Armour:	Not available
Armament:	1 x 100mm (3.9in) 2A70 rifled gun; 1 x 30mm (1.2in) 2A72 automatic gun; 1 x 7.62mm (.3in) PKT MG
Powerplant:	1 x UTD-29 6-cylinder diesel engine developing 373kW (500bhp)
Performance:	Maximum speed: 70km/h (43mph)

AAAV

The Advanced Amphibious Assault Vehicle (AAAV) is the pinnacle of amphibious-vehicle technology. Entirely self-deploying, it carries combat-ready Marines across water at 47km/h (30mph). On land, it can keep up with main battle tanks, travelling at 72km/h (45mph).

SPECIFICATIONS	
Country of origin:	USA
Crew:	3 + 17
Weight:	33,525kg (73,922lb)
Dimensions:	Length: 9.01m (29ft 7in); width: 3.66m (12ft); height: 3.19m (10ft 6in)
Range:	480km (300 miles)
Armour:	Details classified
Armament:	1 x 30mm (1.18in) Bushmaster II gun; 1 x 7.62mm (.3in) MG
Powerplant:	1 x MTU MT883 12-cylinder multi-fuel engine developing 2015kW (2702hp)
Performance:	Maximum road speed: 72km/h (45mph); maximum water speed: 47km/h (30mph); gradient: 60 per cent; vertical obstacle: .9m (3ft); trench: 2.4m (8ft)

Other Tracked Vehicles

Tracked armoured vehicles come in a bewildering variety of shapes and sizes, designed to perform an equally bewildering variety of roles. In the armies of some smaller nations, local military engineers have shown considerable ingenuity in modifying existing tracked AFVs to suit their individual requirements.

PRAM-S

The Slovakian PRAM-S, also known as the Vzor 85, is a 120mm (4.7in) self-propelled mortar. A potent weapon, it offers sustained fire support. It can fire out to a range of 8000m (8749 yards) with a rate of fire of 18–20 rounds per minute via an automatic feeder.

SPECIFICATIONS

Country of origin:	Slovakia
Crew:	4
Weight:	16,970kg (37,419lb)
Dimensions:	Length: 7.47m (24ft 6in); width: 2.94m (9ft 8in); height: 2.25m (7ft 4in)
Range:	550km (340 miles)
Armour:	Up to 23mm (.9in)
Armament:	1 x 120mm (4.7in) Model 1982 mortar; 1 x 12.7mm (.5in) NSV HMG; RPG-75 anti-tank rocket launchers; 9K113 Konkurz ATGW
Powerplant:	1 x UTD-40 6-cylinder diesel engine developing 224kW (300hp)
Performance:	Maximum road speed: 63km/h (39mph); fording: amphibious; gradient: 60 per cent; vertical obstacle: .9m (2ft 10in); trench 2.7m (8ft 10in)

M577

The M577 has a troop compartment filled with communications and observation equipment. This lets it perform many command roles – directing fire, being a centre of mobile communications, and operating as a tactical liaison vehicle.

SPECIFICATIONS

Country of origin:	USA
Crew:	5
Weight:	11,513kg (25,386lb)
Dimensions:	Length: 4.86m (15ft 11in); width: 2.68m (8ft 9in); height: 2.68m (8ft 9in)
Range:	595km (370 miles)
Armour:	12–38mm (.47–1.5in)
Armament:	1 x 7.62mm (.3in) MG
Powerplant:	1 x GMC Detroit 6-cylinder diesel engine developing 160kW (215hp) at 2800rpm
Performance:	Maximum road speed: 68km/h (42mph); fording: amphibious; gradient: 60 per cent; vertical obstacle: .61m (2ft); trench 1.68m (5ft 6in)

TIMELINE

1985    1986    1990

OTHER TRACKED VEHICLES

M4 C2V

The M4 is a modern Command and Control Vehicle (C2V). An armoured module containing advanced C2V electronics is mounted on the chassis, including the Army Battle Command System (ABCS) Common Hardware and Software (CHS) communications suite.

SPECIFICATIONS
Country of origin:	USA
Crew:	1 + 8
Weight:	25,000–30,000kg (55,100–66,100lb)
Dimensions:	Length: 7.49m (24ft 7in); width: 2.97m (9ft 9in); height: 2.7m (8ft 10in)
Range:	400km (250 miles)
Armour:	Classified
Armament:	None
Powerplant:	1 x Cummins VTA-903T 8-cylinder turbo diesel engine developing 440kW (590hp)
Performance:	Maximum road speed: 65km/h (40mph)

ABRA/RATAC

The ABRA/RATAC vehicle is a M113 APC converted to use the RATAC artillery observation radar. This is a system that can detect or calculate the positions of active artillery from the flight or burst of shells or by simple radar detection of the artillery piece's location.

SPECIFICATIONS
Country of origin:	Germany
Crew:	3 or 4
Weight:	13,000kg (28,700lb)
Dimensions:	Length: 4.86m (15ft 11in); width: 2.7m (8ft 10in); height: 7.16m (23ft 6in)
Range:	300km (190 miles)
Armour:	12–38mm (.47–1.5in)
Armament:	None
Powerplant:	1 x GMC Detroit 6V-53N 6-cylinder diesel engine developing 160kW ((215hp) at 2800rpm
Performance:	Maximum road speed: 68km/h (42mph); fording: amphibious; gradient: 60 per cent; vertical obstacle: .61m (2ft); trench: 1.68m (5ft 6in)

M992 FAASV

The M992 Field Artillery Ammunition Support Vehicle (FAASV) is as fast and manoeuvrable as the self-propelled artillery piece it supports. It can move off-road at 56km/h (35mph) and can travel on all terrains, whether mud, snow, rock or sand.

SPECIFICATIONS
Country of origin:	USA
Crew:	2 + 6
Weight:	25,900kg (57,100lb) fully loaded
Dimensions:	Length: 6.27m (20ft 7in); width: 3.15m (10ft 4in); height: 3.24m (10ft 8in)
Range:	Not available
Armour:	Not available
Armament:	None
Powerplant:	1 x Detroit Diesel 8V-71T 8-cylinder diesel engine developing 297kW (398hp)
Performance:	Maximum road speed: 56km/h (35mph)

 1993

Warships 1945–Present

A significant development of this era was the digital computer and subsequent advances in electronic technology, which were seized on in ship design, construction and operation.

Huge merchant ships with automated navigational, mechanical and cargo-handling systems could be operated by a handful of crewmen. Satellite tracking and GPS locating were incorporated into navigation. In other developments, the passenger liner, made redundant by wide-bodied jet aircraft, evolved into the cruise ship. In naval operations, the nuclear reactor and the ballistic missile prompted a fundamental change of strategy, based on the nuclear submarine.

Left: *Nautilus* was the world's first nuclear-powered submarine. Early trials saw records broken including the first submerged transit across the North Pole.

Aircraft Carriers – USA

Among surface warships, the aircraft carrier took on the role of biggest and most dominant. The US Navy had many more and much larger carriers than any other fleet. Its *Kitty Hawk* and *Nimitz* classes, developed over decades, were vast ships with crew numbers running into thousands, and possessed of devastating firepower.

Forrestal

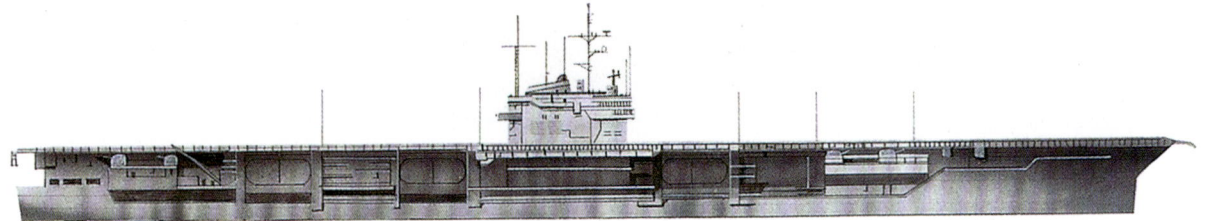

Forrestal and its three sisters were authorized in 1951. Large size was needed for combat jets, which needed more fuel than their piston-engined predecessors. *Forrestal* had space for 3.4 million litres (750,000 gallons) of aviation fuel. Decommissioned since 1993, it is to be sunk as a fishing reef.

SPECIFICATIONS	
Type:	US aircraft carrier
Displacement:	80,516 tonnes (79,248 tons)
Dimensions:	309.4m x 73.2m x 11.3m (1015ft x 240ft x 37ft)
Machinery:	Quadruple screws, turbines
Top speed:	33 knots
Main armament:	Eight 127mm (5in) guns
Aircraft:	90
Complement:	2764 plus 1912 air wing
Launched:	December 1954

Enterprise

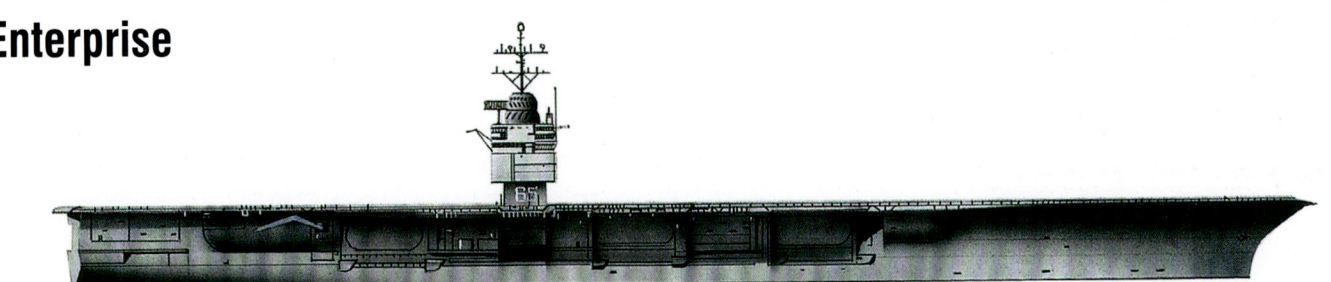

When completed in 1961, *Enterprise* was the largest ship in the world, and only the second nuclear-powered warship. Eight reactors gave it a range of 643,720km (400,000 miles) at 20 knots. It was re-fitted between 1979 and 1982, with a new island structure, and reconstructed in 1990–94 to serve until 2012–14.

SPECIFICATIONS	
Type:	US aircraft carrier
Displacement:	91,033 tonnes (89,600 tons) full load
Dimensions:	335.2m x 76.8m x 10.9m (1100ft x 252ft x 36ft)
Machinery:	Quadruple screws, turbines, steam supplied by eight nuclear reactors
Top speed:	35 knots
Aircraft:	99
Complement:	5500 including air wing
Launched:	September 1960

TIMELINE 1954 1960 1964

AIRCRAFT CARRIERS – USA

America

The *Kitty Hawk* class were the first carriers not to carry conventional guns. *America* was the first carrier equipped with an integrated Combat Information Centre (CIC). It took part in US engagements from Vietnam to Desert Storm (1991). Decommissioned in 1996, it was scuttled in 2005 after use as a target.

SPECIFICATIONS	
Type:	US aircraft carrier
Displacement:	81,090 tonnes (79,813 tons)
Dimensions:	324m x 77m x 10.7m (1063ft x 252ft 7in x 35ft)
Machinery:	Quadruple screws, geared turbines
Top speed:	33 knots
Main armament:	Three Mark 29 launchers for Sea Sparrow SAMs, three 20mm (0.79in) Phalanx CIWS
Aircraft:	82
Complement:	3306 excluding air wing
Launched:	1964

John F. Kennedy

Kennedy was the first to have an underwater protection system. Completed in May 1968, it was based in the North Atlantic and Mediterranean. In the 1980s, it was deployed off Lebanon and Libya, and off Iraq in 1991. It flew bombing missions against Al Qaeda targets in 2002. From 2007, it has been in the Reserve Fleet.

SPECIFICATIONS	
Type:	US carrier
Displacement:	82,240 tonnes (80,945 tons)
Dimensions:	320m x 76.7m x 11.4m (1052ft x 251ft 8in x 36ft)
Machinery:	Quadruple screws, geared turbines
Top speed:	33.6 knots
Main armament:	Three Sea Sparrow octuple launchers, three Mk15 Phalanx 20mm (0.79in) CIWS
Complement:	3306 plus 1379 air wing
Launched:	1967

George Washington

A *Nimitz*-class supercarrier, *George Washington* carries damage-control systems, including armour 63mm (2.5in) thick over parts of the hull, plus box protection over the magazines and machinery spaces. Aviation equipment includes four lifts and four steam catapults. In 2010, it was operating from Yokosuka, Japan.

SPECIFICATIONS	
Type:	US aircraft carrier
Displacement:	92,950 tonnes (91,487 tons)
Dimensions:	332.9m x 40.8m x 11.3m (1092ft 2in x 133ft 10in x 37ft)
Machinery:	Quadruple screws, two water-cooled nuclear reactors, turbines
Top speed:	30 knots+
Main armament:	Four Vulcan 20mm (0.79in) guns plus missiles
Aircraft:	70+
Launched:	September 1989

WARSHIPS 1945–PRESENT

Invincible

Commissioned in 1980, *Invincible*'s 7° (later 12°) 'ski-jump' let its Sea Harriers take off at low, fuel-saving speed. It was deployed with the Falkland Islands Task Force in April–June 1982; in the Adriatic Sea during the Yugoslav wars in 1993–95; and off Iraq in 1988–99. *Invincible* was decommissioned in August 2005.

COMBAT SYSTEM
BAE Systems ADIMP with communication links, multi-function consoles and flat-screen display. Astrium SCOT secure satellite communication system.

SONAR
Hull-mounted Type 2016 sonar active/passive system for search and attack was fitted. ASW helicopters could be rapidly deployed.

HULL
Below the hangar space Invincible *had crew quarters and messing and other facilities for crew and extra personnel required by the command/control function.*

Invincible

The Royal Navy maintains that *Invincible* could be deployed should the need arise and that navy policy assumes that it is still an active aircraft carrier. But *Invincible* was stripped of some parts for sister ships, so bringing the ship to a state of operational readiness would require 18 months.

SPECIFICATIONS	
Type:	British aircraft carrier
Displacement:	21,031 tonnes (20,700 tons)
Dimensions:	210m x 36m x 8.8m (689ft x 118ft 1in x 28ft 10in)
Machinery:	Twin screws, gas turbines
Top speed:	28 knots
Main armament:	Sea Dart anti-air and anti-missile missiles (removed c.1995), Goalkeeper CIWS
Aircraft:	Eight Harrier GR7/GR9, 11 Sea King helicopters
Complement:	726 plus 384 air wing
Launched:	1977

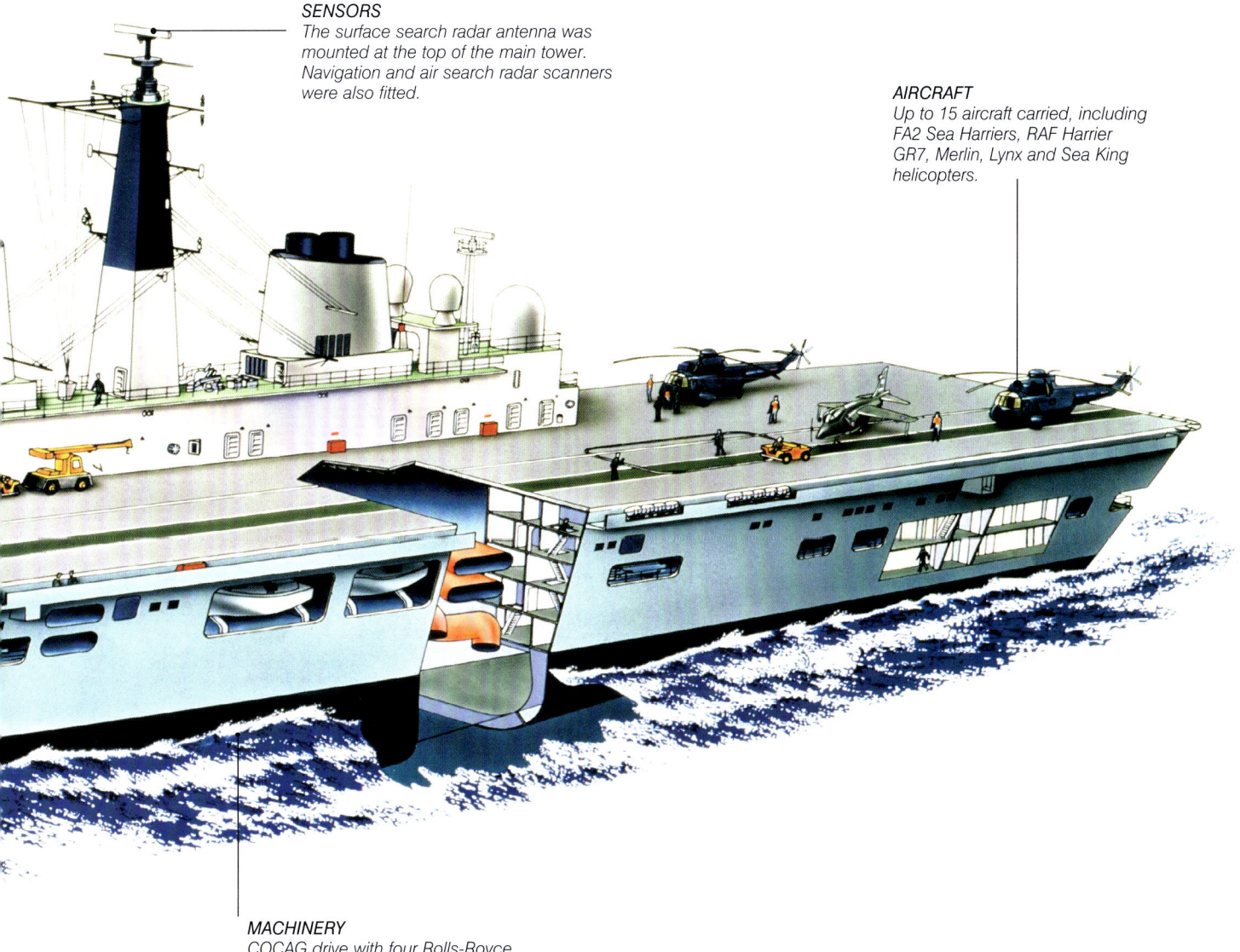

SENSORS
The surface search radar antenna was mounted at the top of the main tower. Navigation and air search radar scanners were also fitted.

AIRCRAFT
Up to 15 aircraft carried, including FA2 Sea Harriers, RAF Harrier GR7, Merlin, Lynx and Sea King helicopters.

MACHINERY
COCAG drive with four Rolls-Royce Olympus TM 3B marine gas turbines and eight Paxman Valenta diesel motors, producing 75MW (97,000hp).

Aircraft Carriers – Other Navies

Several carriers of World War II had unexpectedly long lives. Knowing the tactical importance of fixed-wing aircraft and helicopters, numerous countries sought to retain carriers or to acquire them second-hand. The introduction of angled flight decks and 'ski-jump' ramps enabled older ships to deploy modern aircraft.

Hermes

Plans for *Hermes* went back to 1943. After many design changes, the ship was finally completed in 1959. By 1979, it was handling the new Harrier vertical take-off jets. In 1982 it served as flagship in the Falklands War. In 1989, it was sold to India as *Viraat*, and was still operational in 2010.

SPECIFICATIONS	
Type:	British aircraft carrier
Displacement:	25,290 tonnes (24,892 tons)
Dimensions:	224.6m x 30.4m x 8.2m (737ft x 100ft x 27ft)
Machinery:	Twin screws, turbines
Top speed:	29.5 knots
Main armament:	Thirty-two 40mm (1.6in) guns
Aircraft:	42
Launched:	February 1953

Clemenceau

Clemenceau underwent modification during design and construction. It served in the Pacific, off Lebanon and in the 1991 Gulf War. Aircraft comprised 16 Super Etendards, three Etendard IVP, 10 F-8 Crusaders, seven Alize, plus helicopters. It was decommissioned in 2005 and sent for breaking in 2009.

SPECIFICATIONS	
Type:	French aircraft carrier
Displacement:	33,304 tonnes (32,780 tons)
Dimensions:	257m x 46m x 9m (843ft 2in x 150ft x 28ft 3in)
Machinery:	Twin screws, geared turbines
Main armament:	Eight 100mm (3.9in) guns
Aircraft:	40
Launched:	December 1957

TIMELINE — 1953 — 1957 — 1961

AIRCRAFT CARRIERS – OTHER NAVIES

Vikrant

Formerly the British light carrier *Hercules*, *Vikrant* was refitted in 1961 to carry Sea Hawk fighter-bombers. After the Indo-Pakistan war of 1971, it was refitted, including the provision of a ski-jump, in 1987–89, to carry Sea Harriers. It was withdrawn in 1996. India's new home-built carrier *Vikrant* is due for launch in 2010.

SPECIFICATIONS
Type:	Indian aircraft carrier
Displacement:	19,812 tonnes (19,500 tons)
Dimensions:	213.4m x 39m x 7.3m (700ft x 128ftx 24ft)
Machinery:	Twin screws, geared turbines
Top speed:	23 knots
Main armament:	Fifteen 40mm (1.57in) cannon
Aircraft:	16
Complement:	1250
Launched:	1945 (modernized 1961)

Veinticinco de Mayo

First HMS *Venerable,* then the Dutch *Karel Doorman*, *Veinticinco de Mayo* was bought by Argentina in 1968. The flight deck was extended in 1979, and flew A-4Q Skyhawks, Super Etendards, S-2A Trackers and Sea King helicopters. It was engaged in the Falklands War of 1982. In 1997, it was broken up.

SPECIFICATIONS
Type:	Argentinian aircraft carrier
Displacement:	20,214 tonnes (19,896 tons) full load
Dimensions:	211.3 x 36.9 x 7.6m (693ft 2in x 121ft x 25ft)
Machinery:	Twin screws, turbines
Top speed:	23 knots
Main armament:	Twelve 40mm (1.57in) cannon
Aircraft:	22
Complement:	1250
Launched:	1943 (modernized 1968)

Kiev

Completed in May 1975, *Kiev* was the first Russian aircraft carrier to be built with a full-length flight deck and a purpose-built hull. Apart from aircraft, it was armed with an array of missiles including the SS-N-12 Shaddock. Withdrawn in 1993, it has been an exhibit in a Chinese seaside theme park since 2004.

SPECIFICATIONS
Type:	Soviet aircraft carrier
Displacement:	38,608 tonnes (38,000 tons)
Dimensions:	273m x 47.2m x 8.2m (895ft 8in x 154ft 10in x 27ft)
Machinery:	Quadruple screws, turbines
Top speed:	32 knots
Main armament:	Four 76.2mm (3in) guns, plus missiles
Aircraft:	36
Launched:	December 1972

Giuseppe Garibaldi

Giuseppe Garibaldi has six decks with 13 watertight bulkheads. A 'ski-jump' launch ramp is mounted on the bows for vertical take-off and landing aircraft. This enables the aircraft to take off with a higher gross weight of fuel. It carries AV-8B Harrier jets or Agusta helicopters, or a combination of both, and has had several missile refits.

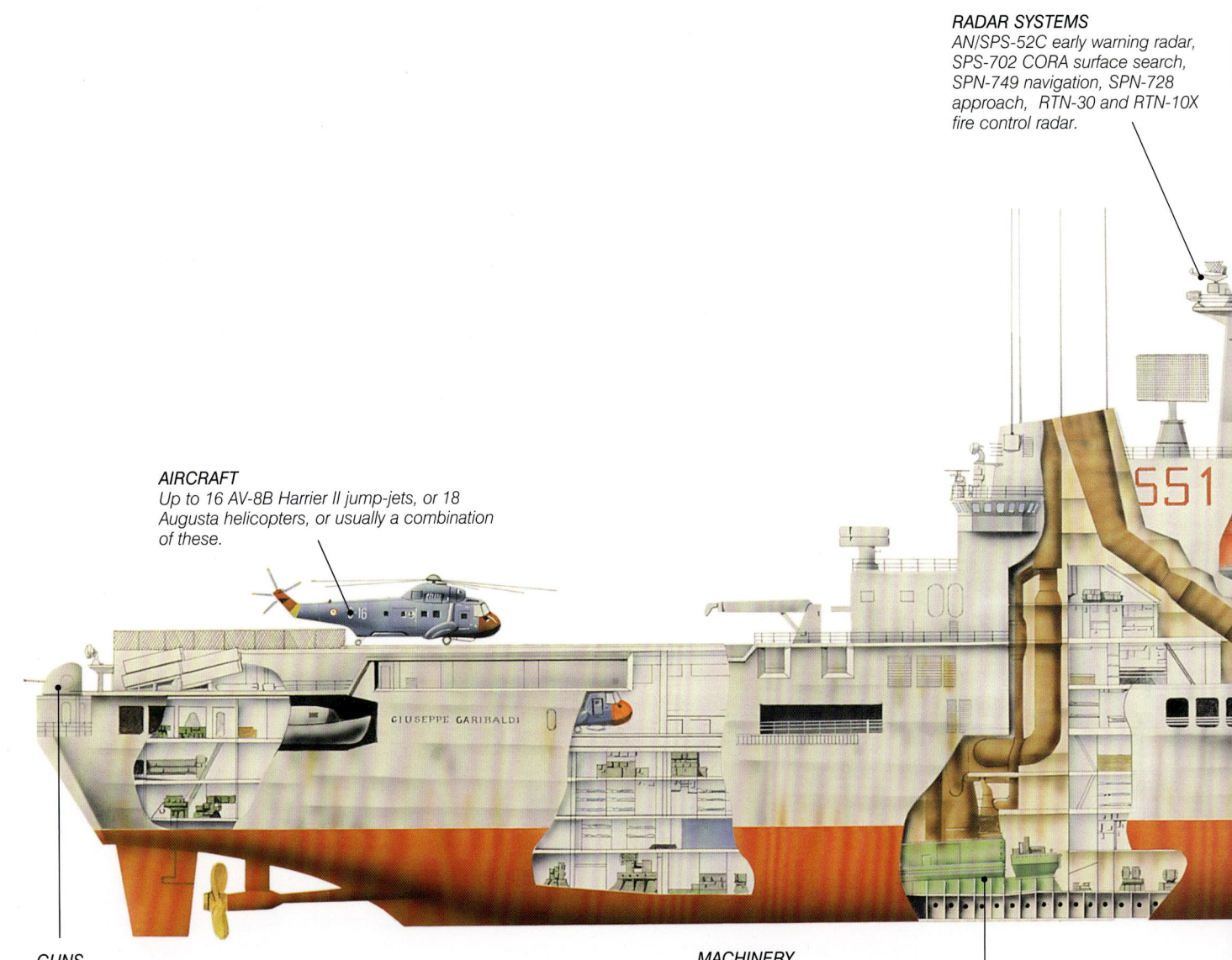

RADAR SYSTEMS
AN/SPS-52C early warning radar, SPS-702 CORA surface search, SPN-749 navigation, SPN-728 approach, RTN-30 and RTN-10X fire control radar.

AIRCRAFT
Up to 16 AV-8B Harrier II jump-jets, or 18 Augusta helicopters, or usually a combination of these.

GUNS
Three Selex NA 21 systems control three 40mm/70mm twin Oto Melara guns with an air target range of 4km (2.5 miles) and surface range of 12km (7.45 miles).

MACHINERY
COCAG drive with four Fiat-built General Electric LM2500 gas turbines and six diesel motors. Power output 60MW (81,000hp).

Giuseppe Garibaldi

SPECIFICATIONS	
Type:	Italian aircraft carrier
Displacement:	13,500 tonnes (13,370 tons)
Dimensions:	180m x 33.4m x 6.7m (590ft 6in x 109ft 6in x 22ft)
Machinery:	Quadruple screws, gas turbines
Top speed:	30 knots
Main armament:	Missile launchers, six torpedo tubes
Aircraft:	16 Harriers, or 18 helicopters
Complement:	550 plus 230 air wing
Launched:	1983

The WWII Peace Treaty banned Italy from having an aircraft carrier, which meant that at the time of launch *Giuseppe Garibaldi* did not receive its Harriers and was classed as an aircraft-carrying cruiser. The ban was eventually lifted and in 1989 the Italian Navy obtained fixed-wing aircraft to operate from the ship.

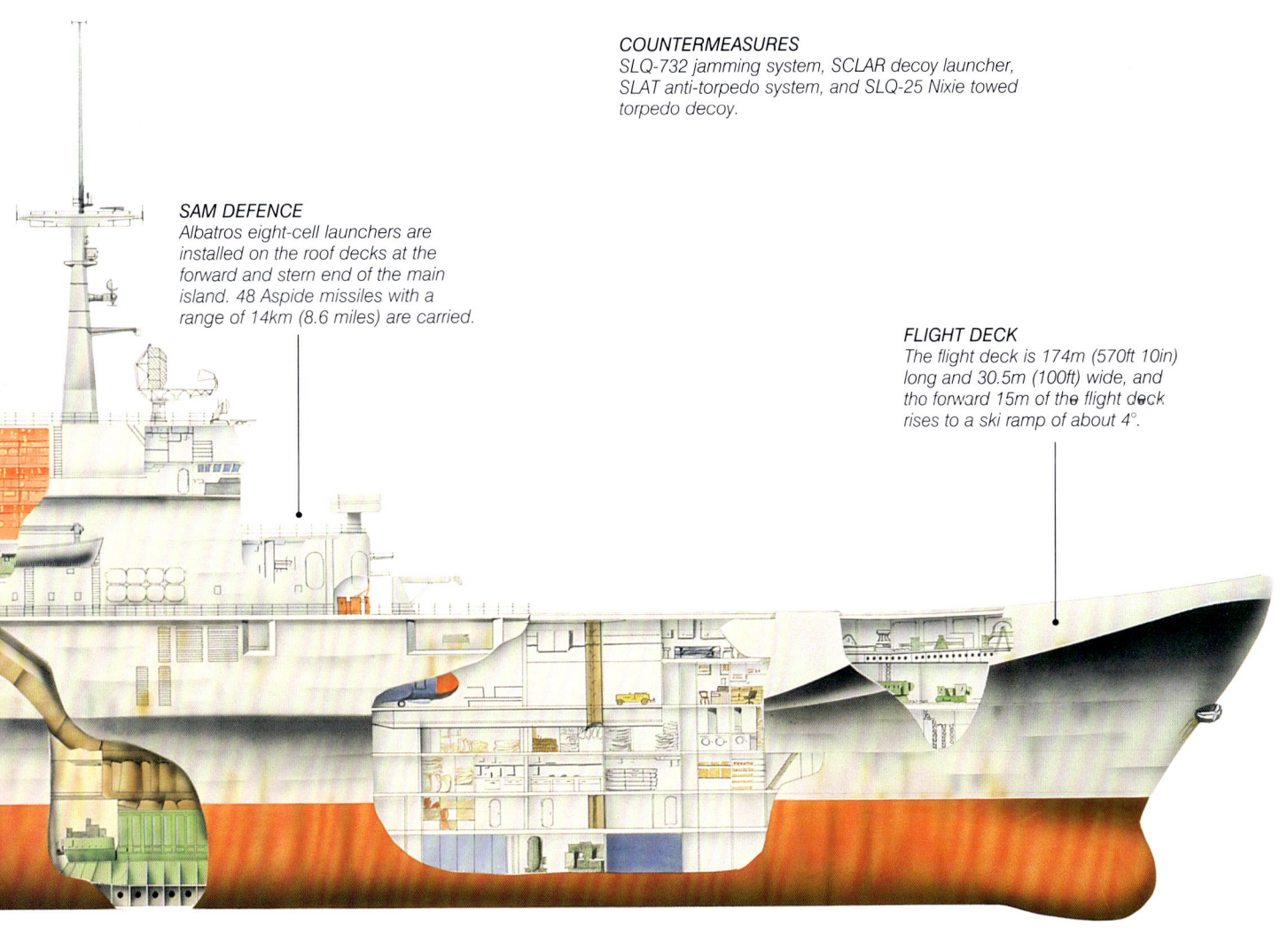

COUNTERMEASURES
SLQ-732 jamming system, SCLAR decoy launcher, SLAT anti-torpedo system, and SLQ-25 Nixie towed torpedo decoy.

SAM DEFENCE
Albatros eight-cell launchers are installed on the roof decks at the forward and stern end of the main island. 48 Aspide missiles with a range of 14km (8.6 miles) are carried.

FLIGHT DECK
The flight deck is 174m (570ft 10in) long and 30.5m (100ft) wide, and the forward 15m of the flight deck rises to a ski ramp of about 4°.

WARSHIPS 1945–PRESENT

Helicopter Carriers & Small Aircraft Carriers

The vastly increased importance of the helicopter as an element in anti-submarine action has made it, in effect, an extendable arm of the modern warship. In addition, the helicopter carrier can transport, service and fuel a larger number of helicopters to take part in a various situations, from military invasion to humanitarian aid.

Iwo Jima

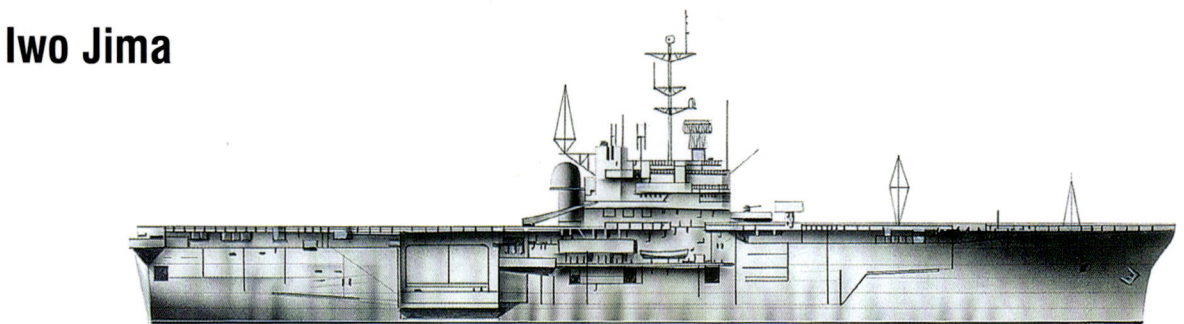

Iwo Jima was the first ship designed specifically to carry and operate helicopters, along with a Marine battalion of 2000 troops, plus artillery and support vehicles. In the 1970s, Sea Sparrow missile launchers were installed. A boiler explosion damaged the ship in 1990; it was stricken in 1993 and broken up in 1995.

SPECIFICATIONS	
Type:	US assault ship
Displacement:	18,330 tonnes (18,042 tons)
Dimensions:	183.6m x 25.7m x 8m (602ft 8in x 84ft x 26ft)
Machinery:	Single screw, turbines
Top speed:	23.5 knots
Main armament:	Four 76mm (3in) guns
Aircraft:	20 helicopters
Complement:	667, 2057 marines
Launched:	September 1960

Jeanne D'Arc

A multi-purpose cruiser, helicopter carrier and assault ship, Jeanne D'Arc could transport 700 men and eight large helicopters. In 1975, Exocet missiles were fitted, giving it a full anti-ship role. It also functioned as a training ship, providing facilities for up to 198 cadets at a time. It was struck from the list in 2009.

SPECIFICATIONS	
Type:	French helicopter carrier
Displacement:	13,208 tonnes (13,000 tons)
Dimensions:	180m x 25.9m x 6.2m (590ft 6in x 85ft x 20ft 4in)
Machinery:	Twin screws, turbines
Top speed:	26.5 knots
Main armament:	Four 100mm (3.9in) guns
Aircraft:	Eight
Complement:	627 including cadets
Launched:	September 1961

TIMELINE

1960   1961   1964

HELICOPTER CARRIERS & SMALL AIRCRAFT CARRIERS

Moskva

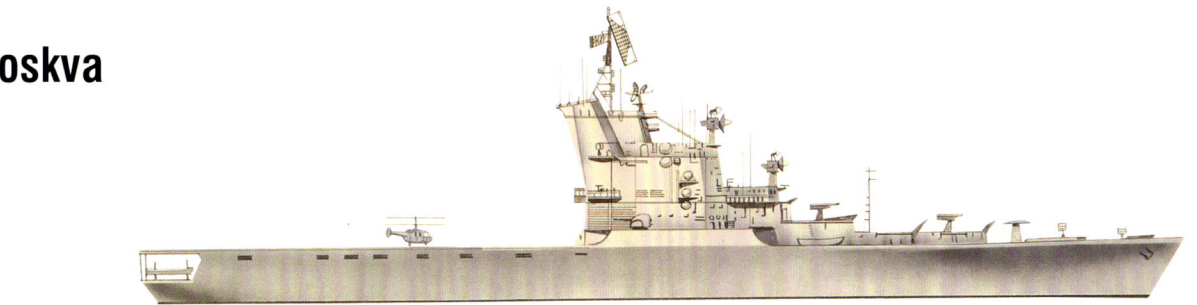

Moskva was the first helicopter carrier built for the Russian Navy, completed in 1967 to counteract the threat from the US nuclear-powered missile submarines that began to enter service in 1960. A central block dominated the vessel and housed the major weapons systems. *Moskva* was scrapped in the mid-1990s.

SPECIFICATIONS	
Type:	Soviet helicopter carrier
Displacement:	14,800 tonnes (14,567 tons)
Dimensions:	191m x 34m x 7.6m (626ft 8in x 111ft 6in x 25ft)
Machinery:	Twin screws, turbines
Top speed:	30 knots
Main armament:	One twin SUW-N-1 launcher, two twin SA-N-3 missile launchers
Aircraft:	18 helicopters
Complement:	850, including air wing
Launched:	1964

Vittorio Veneto

Vittorio Veneto was a purpose-built helicopter cruiser. A large central lift was set immediately aft of the superstructure, and two sets of fin stabilizers were fitted. Laid down in 1965, completed in 1969, the ship underwent a refit between 1981 and 1984. Withdrawn in 2003, it is intended to be a museum ship at Taranto.

SPECIFICATIONS	
Type:	Italian helicopter cruiser
Displacement:	8991 tonnes (8850 tons)
Dimensions:	179.5m x 19.4m x 6m (589ft x 63ft 8in x 19ft 8in)
Machinery:	Twin screws, turbines
Top speed:	32 knots
Main armament:	Twelve 40mm (1.6in) guns, eight 76mm (3in) guns, four Teseo SAM launchers, one ASROC launcher
Aircraft:	Nine helicopters
Complement:	550
Launched:	February 1967

Chakri Naruebet

Spanish-built, modelled on Spain's *Principe de Asturias*, Thailand's only carrier is also the world's smallest. *Chakri Naruebet* carries Harrier AV-8B VSTOL jump-jets and helicopters, the Harriers also being bought from Spain. The ship was in service in 2010, but its operational status is unclear and it rarely goes to sea.

SPECIFICATIONS	
Type:	Thai light carrier
Displacement:	11,480 tonnes (11,300 tons)
Dimensions:	182.5m x 30.5m x 6.15m (599ft 1in x 110ft 1in x 20ft 4in)
Machinery:	Twin screw, turbines and diesels
Top speed:	26 knots
Main armament:	Two launchers for Mistral SAM
Aircraft:	10
Complement:	455 plus 162 aircrew
Launched:	1996

1967  1996

WARSHIPS 1945–PRESENT

Tarawa

Commissioned in 1976, equipped for air-land assault, *Tarawa* had a floodable well-deck for landing craft, and command and control facilities to undertake a flagship role. It was the first of a class of five, all built within the space of a few years in response to perceived threats during the Cold War.

Tarawa

USS *Tarawa* was the lead ship of the Navy's first class of amphibious assault ships able to incorporate the best design features and capabilities of several amphibious assault ships then in service.

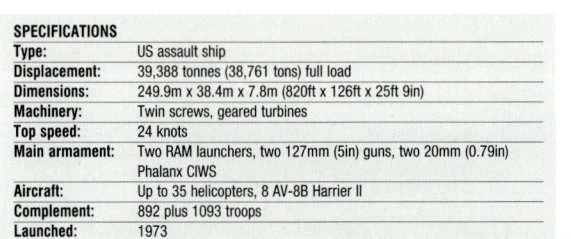

SPECIFICATIONS	
Type:	US assault ship
Displacement:	39,388 tonnes (38,761 tons) full load
Dimensions:	249.9m x 38.4m x 7.8m (820ft x 126ft x 25ft 9in)
Machinery:	Twin screws, geared turbines
Top speed:	24 knots
Main armament:	Two RAM launchers, two 127mm (5in) guns, two 20mm (0.79in) Phalanx CIWS
Aircraft:	Up to 35 helicopters, 8 AV-8B Harrier II
Complement:	892 plus 1093 troops
Launched:	1973

MEDICAL FACILITIES
Tarawa's facilities included 300-bed hospital, four med operating rooms, and three dental operating rooms.

ARMAMENT
Four Mk38 Mod 1 25mm (0.98in) Bushmaster cannon, five M2HB 12.7mm (0.5in) calibre machine guns, two Mk15 Phalanx CIWS, and two Mk49 RAM launchers.

WELL DECK
Internal roadways enabled vehicles to be driven from the garage space to the landing craft loading points.

TARAWA

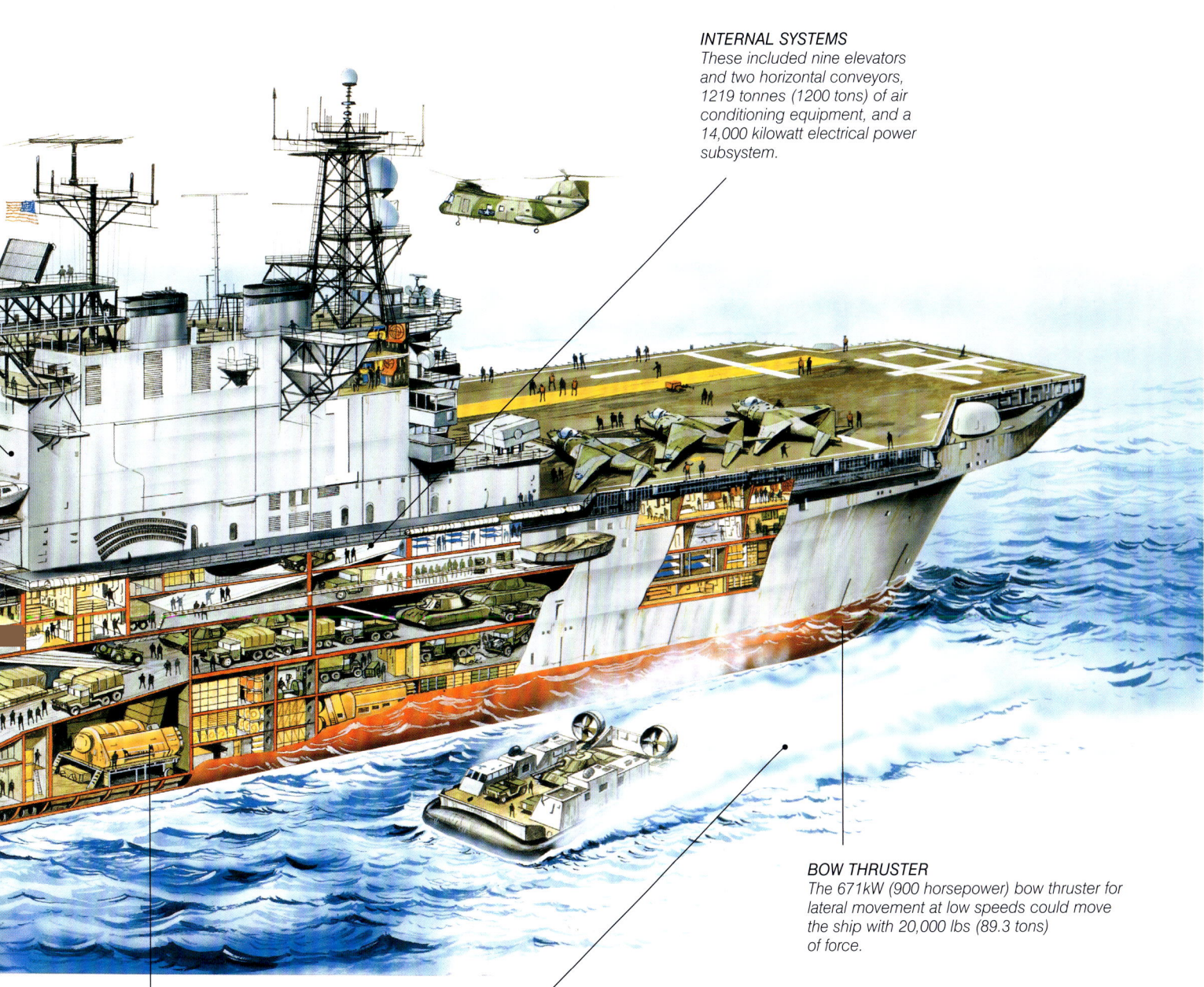

INTERNAL SYSTEMS
These included nine elevators and two horizontal conveyors, 1219 tonnes (1200 tons) of air conditioning equipment, and a 14,000 kilowatt electrical power subsystem.

BOW THRUSTER
The 671kW (900 horsepower) bow thruster for lateral movement at low speeds could move the ship with 20,000 lbs (89.3 tons) of force.

MACHINERY
The two boilers were the largest ever manufactured for the United States Navy, generating 406.4 tonnes (400 tons) of steam per hour, and developing 104,398kW (140,000hp).

BALLAST
Tarawa could ballast 12,192 tonnes (12,000 tons) of seawater for trimming the ship, while receiving and discharging landing craft from the well deck.

Assault Ships

The needs of amphibious warfare, mobilizing specialist equipment and specialized vehicles, and with sophisticated communications, have been answered by a new generation of landing ships and command ships. These have systems more developed than the LSTs and converted merchant ships of the 1940s and '50s.

Intrepid

A 'landing platform dock', *Intrepid* could operate its own set of landing craft, and carry up to 700 troops. Above the landing dock were hangar and flight deck for six helicopters. In the Falklands War, the Argentinian surrender was signed on board *Intrepid*. Decommissioned in 1999, it was scrapped shortly after.

SPECIFICATIONS	
Type:	British assault ship (LPD)
Displacement:	12,313 tonnes (12,120 tons)
Dimensions:	158m x 24m x 6.2m (520ft x 80ft x 20ft 6in)
Machinery:	Twin screws, turbines
Top speed:	21 knots
Main armament:	Two 40mm (1.57in) guns, four Seacat anti-aircraft missile launchers
Complement:	566
Launched:	June 1964

Denver

The 11 ships of this class, enlarged versions of the *Raleigh* group, have greater capacity for carrying troops and support vehicles. Assault and landing craft are held in the comprehensive docking facility forming the rear section of the vessel. In recent years, *Denver* has brought post-tsunami aid to Taiwan and Sumatra.

SPECIFICATIONS	
Type:	US command ship
Displacement:	9477 tonnes (9328 tons)
Dimensions:	174m x 30.5m x 7m (570ft 3in x 100ft x 23ft)
Machinery:	Twin screws, turbines
Top speed:	21 knots
Main armament:	Eight 76mm (3in) guns
Complement:	447 plus 840 marines
Launched:	January 1965

TIMELINE 1964  1965 1970

ASSAULT SHIPS

Mount Whitney

Modern warfare demands specialized command ships. *Mount Whitney*, flagship of the US Sixth Fleet from 2005, uses the same hull form and machinery as the Guam class, with flat open decks to allow maximum antenna placement. *Mount Whitney* has often been in 'hot-spot' situations, including the Black Sea in 2008.

SPECIFICATIONS

Type:	US command ship
Displacement:	19,598 tonnes (19,290 tons)
Dimensions:	189m x 25m x 8.2m (620ft 5in x 82ft x 27ft)
Machinery:	Single screw, turbines
Top speed:	23 knots
Main armament:	Four 76mm (3in) guns, two eight-tube Sea Sparrow missile launchers
Complement:	700
Launched:	January 1970

Ivan Rogov

A long-range assault ship, carrying 550 troops, plus 40 tanks and other support vehicles, *Ivan Rogov* was fitted with a bow ramp, and a docking area 76m (250ft) long. The aft superstructure housed a helicopter. In 1979, it was transferred from the Black Sea to the Pacific Fleet, and stricken in 1996.

SPECIFICATIONS

Type:	Soviet amphibious assault ship
Displacement:	13,208 tonnes (13,000 tons)
Dimensions:	158m x 24m x 8.2m (521ft 8in x 80ft 5in x 21ft 4in)
Machinery:	Twin screws, gas turbines
Top speed:	23 knots
Main armament:	Two 76mm (3in) guns, plus anti-aircraft missiles
Complement:	200
Launched:	1977

Whidbey Island

Whidbey Island's well deck accommodates four LCAC hovercraft or up to 21 smaller 61-tonne (60-ton) landing craft. The ship carries 450 troops, military vehicles and two assault and transport helicopters, and can fly Harrier jump jets. Assigned to Amphibious Group 2, it is currently deployed in the Persian Gulf.

SPECIFICATIONS

Type:	US dock landing ship
Displacement:	15,977 tonnes (15,726 tons)
Dimensions:	186m x 25.6m x 6.3m (609ft x 84ft x 20ft 8in)
Machinery:	Twin screws, diesel engines
Top speed:	20+ knots
Main armament:	Two 20mm (0.79in) Vulcan guns
Complement:	340
Launched:	June 1983

1977    1983

Corvettes/Patrol Ships: Part 1

Seaward extension of international boundaries and increasing levels of smuggling provide a role for the patrol ship. Lightweight but effective missile launchers and rapid-fire automatic guns endow small ships with heavy fire-power. Reconnaissance, surveillance and interception are their prime tasks.

Shanghai

The Chinese Navy has many coastal patrol craft. The *Shanghai* vessels carry a relatively powerful armament of light weapons, plus depth charges and mines. Many have been exported to Asia, the Middle East and Africa, and others built under licence by European navies. Type 1 was in service until the early 1990s.

SPECIFICATIONS	
Type:	Chinese fast attack/patrol boat
Displacement:	137 tonnes (135 tons)
Dimensions:	38.8m x 5.4m x 1.7m (127ft 4in x 17ft 8in x 5ft 7in)
Machinery:	Quadruple screws, diesel engines
Top speed:	28.5 knots
Main armament:	Four 37mm (1.45in) guns, four 25.4mm (1in) cannon
Launched:	1962

Dardo

Dardo was one of four vessels whose functions could switch from minelayer (with an anti-aircraft gun and eight mines) to torpedo boat (with one 40mm/1.6in gun and 21 x 533mm/21in torpedoes). Conversion could be achieved in under 24 hours. Hybrid designs are not always successful, but these proved effective.

SPECIFICATIONS	
Type:	Italian motor gunboat
Displacement:	218 tonnes (215 tons)
Dimensions:	46m x 7m x 1.7m (150ft x 23ft 9in x 5ft 6in)
Machinery:	Twin screws, diesels and gas turbines
Top speed:	40+ knots
Main armament:	One 40mm (1.6in) gun, four 533mm (21in) torpedo tubes
Launched:	1964

TIMELINE 1962 1964 1966

CORVETTES/PATROL SHIPS: PART 1

Spica

Spica was the first of a group of fast attack craft designed for Baltic waters. Bases are built into the rocky coastline, and can withstand most weapons except nuclear. The gas turbines develop 12,720hp, for rapid acceleration. The design was adopted by several other navies. Spica is now a museum ship in Stockholm.

SPECIFICATIONS	
Type:	Swedish fast attack torpedo craft
Displacement:	218 tonnes (215 tons)
Dimensions:	42.7m x 7m x2.6m (140ft x 23ft 4in x 8ft 6in)
Machinery:	Triple screws, gas turbines
Main armament:	One 57mm (2.24in) gun, six 533mm (21in) torpedo tubes
Launched:	1966

Nanuchka I

This class of light but heavily armed missile craft was known as the Nanuchka I. Some variants carried two 57mm guns and all had fire control radar and hull-mounted sonar systems. Nanuchka II types were built for India and III for Algeria and Libya. All Russian class members were decommissioned by the end of the 1990s.

SPECIFICATIONS	
Type:	Russian corvette
Displacement:	670.5 tonnes (660 tons)
Dimensions:	59.3m x 12.6m x 2.5m (194ft 7in x 41ft 4in x 7ft 11in)
Machinery:	Twin screws, diesels
Top speed:	32knots
Main armament:	Six SS-N-9 SSM, one SA-N-4 SAM launcher, one 76mm (3in) gun
Launched:	1969

D'Estienne d'Orves

D'Estienne d'Orves was one of a group of 20 frigates, small by contemporary standards, that followed on from the Commandant Rivière group, a class of larger ships. Designed for anti-submarine work in coastal waters, they can also operate at long range. This ship became the Turkish Navy's Beykoz in 1999.

SPECIFICATIONS	
Type:	French light frigate
Displacement:	1351 tonnes (1330 tons)
Dimensions:	80m x 10m x 3m (262ft 6in x 33ft 10in x 9ft 10in)
Machinery:	Twin screws, diesel engines
Top speed:	23.3 knots
Main armament:	Four Exocet launchers, one 100mm (3.9in) dual-purpose gun
Launched:	June 1973

1969  1975

Corvettes/Patrol Ships: Part 2

Longer-range patrol work needs a larger vessel than even a large motor boat, and several navies have responded with a modern version of the corvette. Once a slow light escort ship, it is now more likely to be a fast missile-bearing patrol craft with a full array of radar search and possibly sonar detection equipment.

Beskytteren

SPECIFICATIONS	
Type:	Danish patrol ship
Displacement:	2001.5 tonnes (1970 tons) full load
Dimensions:	74.4m x 12.5m x 4.5m (244ft x 41ft x 14ft 9in)
Machinery:	Single screw, three diesel motors
Top speed:	18 knots
Main armament:	One 76mm (3in) gun
Complement:	60
Launched:	1975

Designed for patrols in North Atlantic and Arctic waters, *Beskytteren* also has a fishery protection role. It is a smaller, modified version of the Danish *Hvidbjørnen* class frigate. Navigational radar and sonar equipment are fitted, and hangar and flight deck for a Lynx helicopter are incorporated, in a compact vessel.

Fremantle

SPECIFICATIONS	
Type:	Australian fast patrol boat
Displacement:	214 tonnes (211 tons)
Dimensions:	41.8m x 7.1m x 1.8m (137ft 2in x 23ft 4in x 6ft)
Machinery:	Triple screws, diesel engines
Top speed:	30 knots
Main armament:	One 40mm (1.6in) gun
Launched:	1979

Lead ship of a class of 15, *Fremantle* was built in Lowestoft, England; the others were built at Cairns, Australia. They were faster than the *Attack*-class vessels that previously fulfilled the role of long-range coastal patrols to prevent smuggling and the landing of illegal immigrants. *Fremantle* served until 2006.

TIMELINE

1975  1979  1980

Badr

One of six boats packing a lot of hardware into a small hull, *Badr* has four OTO Melara/Matra Otomat Mk 1 anti-ship missiles, fired from box launchers mounted behind the deckhouse, and its weapons are directed by fire-control and target-tracking systems. The air-surface search radar dome is a conspicuous feature.

SPECIFICATIONS

Type:	Egyptian fast patrol boat
Displacement:	355.6 tonnes (350 tons) full load
Dimensions:	52m x 7.6m x 2m (170ft 7in x 25ft x 6ft 7in)
Machinery:	Quadruple screws, four diesel engines
Top speed:	37 knots
Main armament:	Four SSM, one 76mm (3in) gun
Complement:	40
Launched:	1980

Kaszub

This Type 620 corvette, built in Poland, was to be the first of seven, but the collapse of the Warsaw Pact stopped development after *Kaszub*, which never received its intended Russian-made missile armament. Technical problems limit its usefulness, and it rarely puts to sea, though it is still on the active list in 2009.

SPECIFICATIONS

Type:	Polish corvette
Displacement:	1202 tonnes (1183 tons)
Dimensions:	82.3m x 10m x 2.8m (270ft 2in x 32ft 9in x 9ft 2in)
Machinery:	Quadruple screws, four diesel motors
Top speed:	28 knots
Main armament:	Two SA-N-5 SAM launchers, one 76mm (3in) gun, two 533mm (21in) torpedo tubes
Complement:	67
Launched:	1986

Eilat

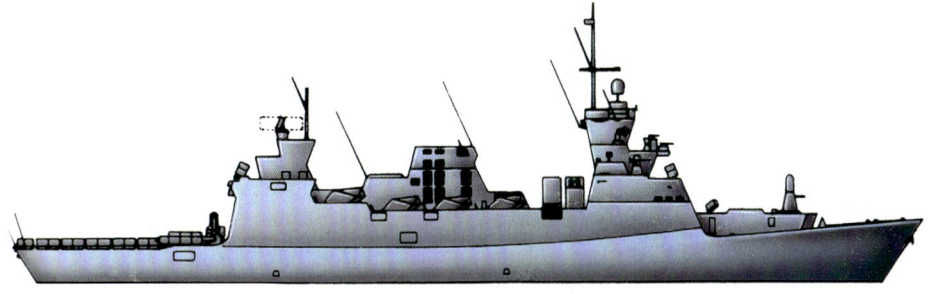

Designed on radar-dodging 'stealth' lines, *Eilat* is one of three well-armed small warships intended to lead Israel's large fleet of smaller attack craft. The forward gun position can be altered to mount different guns or a Phalanx CIWS system. A Dauphin helicopter is carried, and the vessel has sensor devices and radars.

SPECIFICATIONS

Type:	Israeli corvette
Displacement:	1295.4 tonnes (1275 tons) full load
Dimensions:	86.4m x 11.9m x 3.2m (283ft 6in x 39ft x 10ft 6in)
Machinery:	Twin screws, turbines plus turbo diesels
Top speed:	33 knots
Main armament:	Eight Harpoon SSM, eight Gabriel II SSM, one 76mm (3in) gun)
Complement:	74
Launched:	1994

1986  1994

WARSHIPS 1945–PRESENT
USS Long Beach

Completed in September 1961, *Long Beach* was the United States' largest non-carrier surface warship built since 1945. Off Vietnam in 1968, it shot down two MiG fighters in the first successful SAM naval action. In the 1980s, Harpoon missiles and Phalanx CIWS were installed. *Long Beach* was withdrawn from service in 1994.

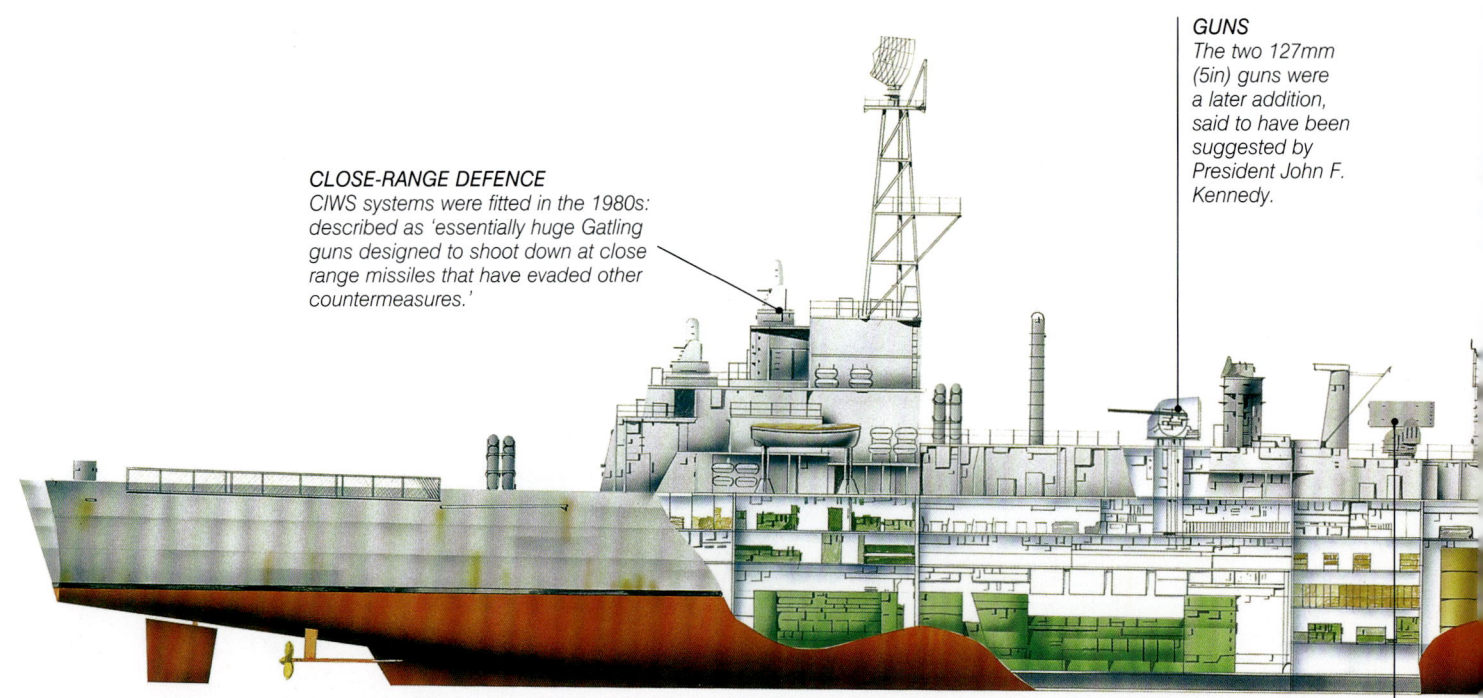

CLOSE-RANGE DEFENCE
CIWS systems were fitted in the 1980s: described as 'essentially huge Gatling guns designed to shoot down at close range missiles that have evaded other countermeasures.'

GUNS
The two 127mm (5in) guns were a later addition, said to have been suggested by President John F. Kennedy.

MACHINERY
Two C1W nuclear reactors powering two General Electric turbines. Total power output was 59,656kW (80,000shp). Range was effectively unlimited.

FINAL WEAPONS SUITE
Harpoon and BGM-109 Tomahawk missiles replaced the original Terrier and Talos systems. An 8-tube ASROC launcher was also fitted.

USS Long Beach

A deactivation ceremony took place on 2 July 1994 at Norfolk Naval Station. *Long Beach* was decommissioned on 1 May 1995, over 33 years after she had entered service. She is currently waiting to be recycled.

SPECIFICATIONS	
Type:	US missile cruiser
Displacement:	16,624 tonnes (16,602 tons)
Dimensions:	219.8m x 22.3m x 7.2m (721ft 3in x 73ft 4in x 23ft 9in)
Machinery:	Twin screws, two nuclear reactors driving geared turbines
Top speed:	30+ knots
Main armament:	Eight Harpoon SSM, two Terrier SSM systems, two 127mm (5in) guns, two 20mm (0.79in) Phalanx CIWS, one ASROC launcher, six 2324mm (12.75in) torpedo tubes
Complement:	1107
Launched:	1959

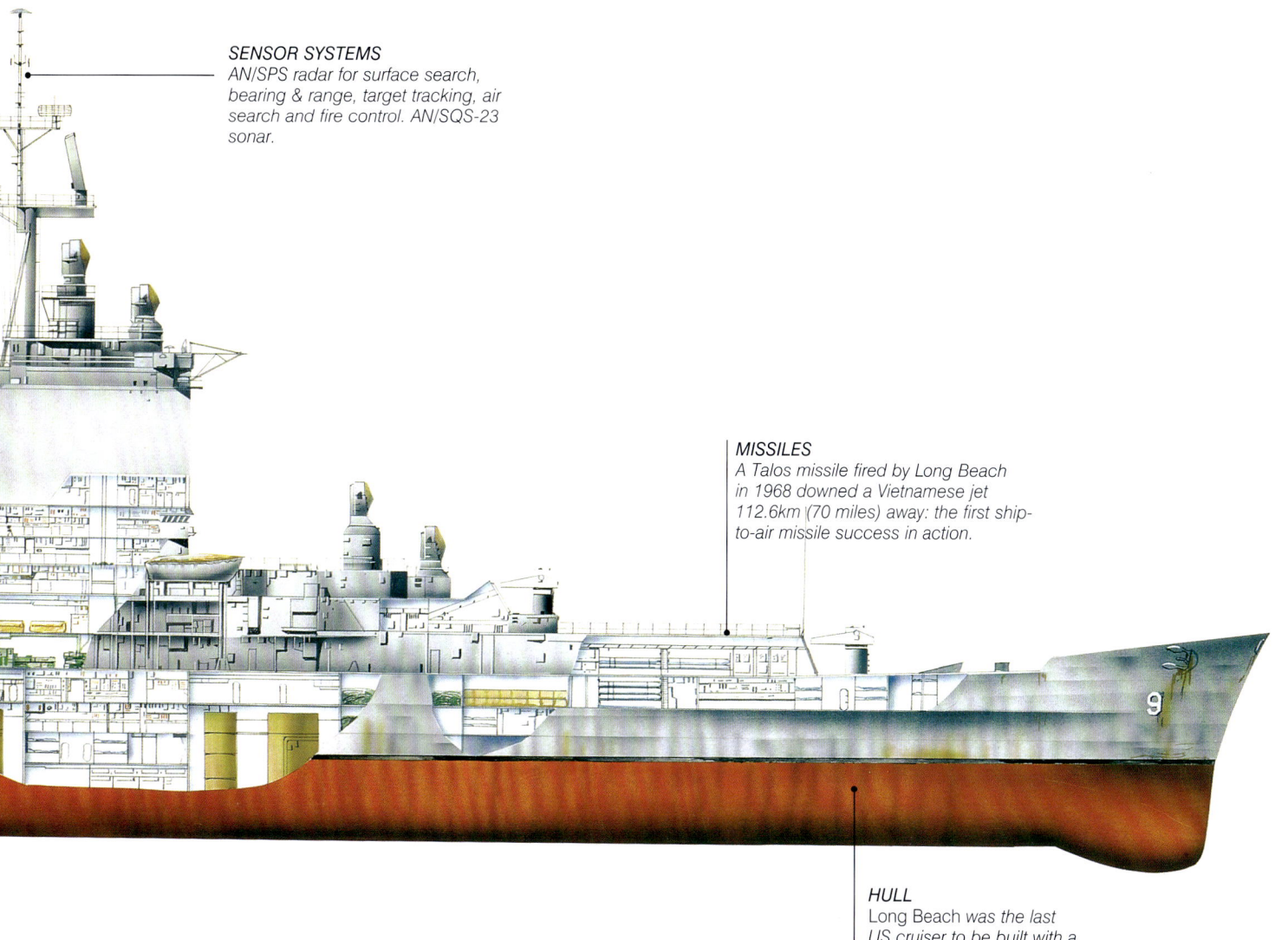

SENSOR SYSTEMS
AN/SPS radar for surface search, bearing & range, target tracking, air search and fire control. AN/SQS-23 sonar.

MISSILES
A Talos missile fired by Long Beach in 1968 downed a Vietnamese jet 112.6km (70 miles) away: the first ship-to-air missile success in action.

HULL
Long Beach was the last US cruiser to be built with a traditional cruiser-type hull, long and relatively narrow.

American Cruisers

While World War II marked the end of the battleship age, both the United States and Russia subsequently built many cruisers, though by the later 1950s the concept of the cruiser was obviously old-fashioned. New generation submarines, aircraft and missiles made them vulnerable. Smaller ships packed a more devastating punch.

Galveston

Galveston was one of six cruisers of World War II which were later refitted to carry Talos or Terrier missiles. The original guns were retained. The purpose was air defence and the ships were not expected to undertake deep sea missions. *Galveston* served until 1970, was stricken in 1973 and scrapped in 1975.

SPECIFICATIONS	
Type:	US cruiser
Displacement:	15,394 tonnes (15,152 tons)
Dimensions:	186m x 20m x 7.8m (610ft x 65ft 8in x 25ft 8in)
Machinery:	Quadruple screws, geared turbines
Top speed:	32 knots
Main armament:	Talos SAM system, six 152mm (6in) guns, six 127mm (5in) guns
Complement:	1382
Launched:	1945, converted 1958

Worden

Worden was one of nine vessels replacing cruisers from World War II. The layout incorporated masts and gantries for a complex radar system. The class was refitted in the late 1960s and again in the late 1980s. In 1991, *Worden* became an anti-aircraft command vessel. Stricken in 1993, it was sunk as a target in 2000.

SPECIFICATIONS	
Type:	US cruiser
Displacement:	8334 tonnes (8203 tons)
Dimensions:	162.5m x 16.6m x 7.6m (533ft 2in x 54ft 6in x 25ft)
Machinery:	Twin screws, turbines
Top speed:	32.7 knots
Main armament:	Two 20mm (0.8in) Vulcan guns, two quad Harpoon launchers, two twin launchers for Standard SM-2 ER missiles
Launched:	June 1962

TIMELINE

1958  1962  1976

Mississippi

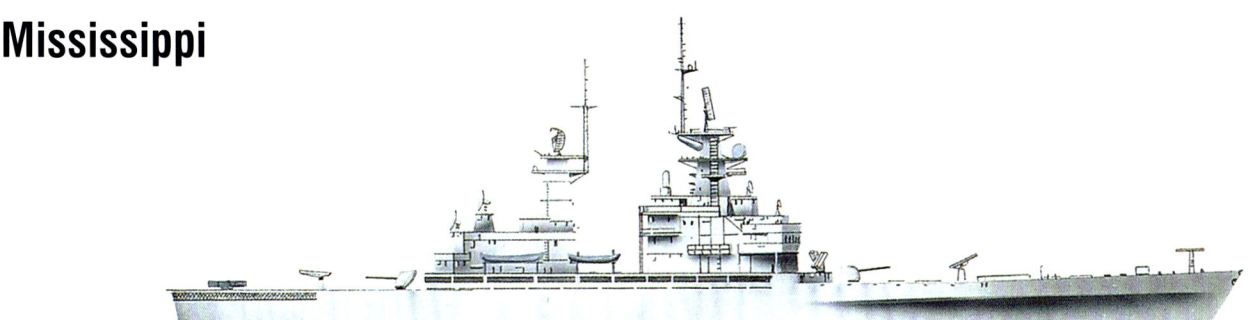

Mississippi was one of four ships with Mk 26 missile launchers and provision for a helicopter hangar and elevator in the stern. The hangars were replaced by three Tomahawk launchers in the 1980s. The 127mm (5in) Mk 45 guns fired 20 rounds per minute, with a range of over 14.6km (9.12 miles). It was broken up in 1997.

SPECIFICATIONS

Type:	US cruiser
Displacement:	11,176 tonnes (11,000 tons)
Dimensions:	178.5m x 19.2m x 9m (585ft 4in x 63ft x 29ft 6in)
Machinery:	Twin screws, nuclear-powered turbines
Main armament:	Two 127mm (5in) guns, two twin launchers for Tartar and Harpoon missiles, Asroc launcher
Launched:	July 1976

Ticonderoga

Ticonderoga was lead ship of a class of 27, originally frigates but reclassified due to the scale of their armament. Later vessels in the class were enlarged to carry greater missile stocks. A full array of sensor equipment was carried, including the Aegis air defence system, and one helicopter. It was decommissioned in 2004.

SPECIFICATIONS

Type:	US guided missile cruiser
Displacement:	9052.5 tonnes (8910 tons)
Dimensions:	171.6m x 19.81m x 9.45m (563ft x 65ft x 31ft)
Machinery:	Twin screws, four gas turbines
Top speed:	30 knots
Main armament:	Eight Harpoon SSM, two Mk 26 launchers for SAM and ASROC torpedoes, two 127mm (5in) guns, six 324mm (12.75in) torpedo tubes
Complement:	343
Launched:	1981

Bunker Hill

First ship of the large *Ticonderoga* class to be equipped with the Mk 41 Vertical Launching System, *Bunker Hill* was first deployed in the Persian Gulf in 1987, and later participated in the Desert Shield and Desert Storm operations. Weapons systems were upgraded in 2006. In 2010 the ship assisted with disaster relief after the Haiti earthquake.

SPECIFICATIONS

Type:	US guided missile cruiser
Displacement:	9,754 tonnes (9600 tons)
Dimensions:	173m x 16.8m x 10.2m (567ft x 55ft x 34ft)
Machinery:	Twin reversible-pitch screws, four gas turbines
Top speed:	32.5 knots
Main armament:	Two 61-cell Mk 41 VLS systems, 122 RIM-156 SM-2ER Bock IV, RIM-162 ESSM, BGM-109 Tomahawk or RUM-139 VL Asroc; 8 RGM-84 Harpoon missiles
Complement:	400
Launched:	1985

Kirov

Kirov's superstructure supports radars and early-warning antennae. Most missile-launching systems are forward, below deck, leaving the aft section for the helicopter hangar and machinery. The two nuclear reactors are coupled with oil-fired turbine superheaters to intensify the heat of the steam, increasing power for more speed.

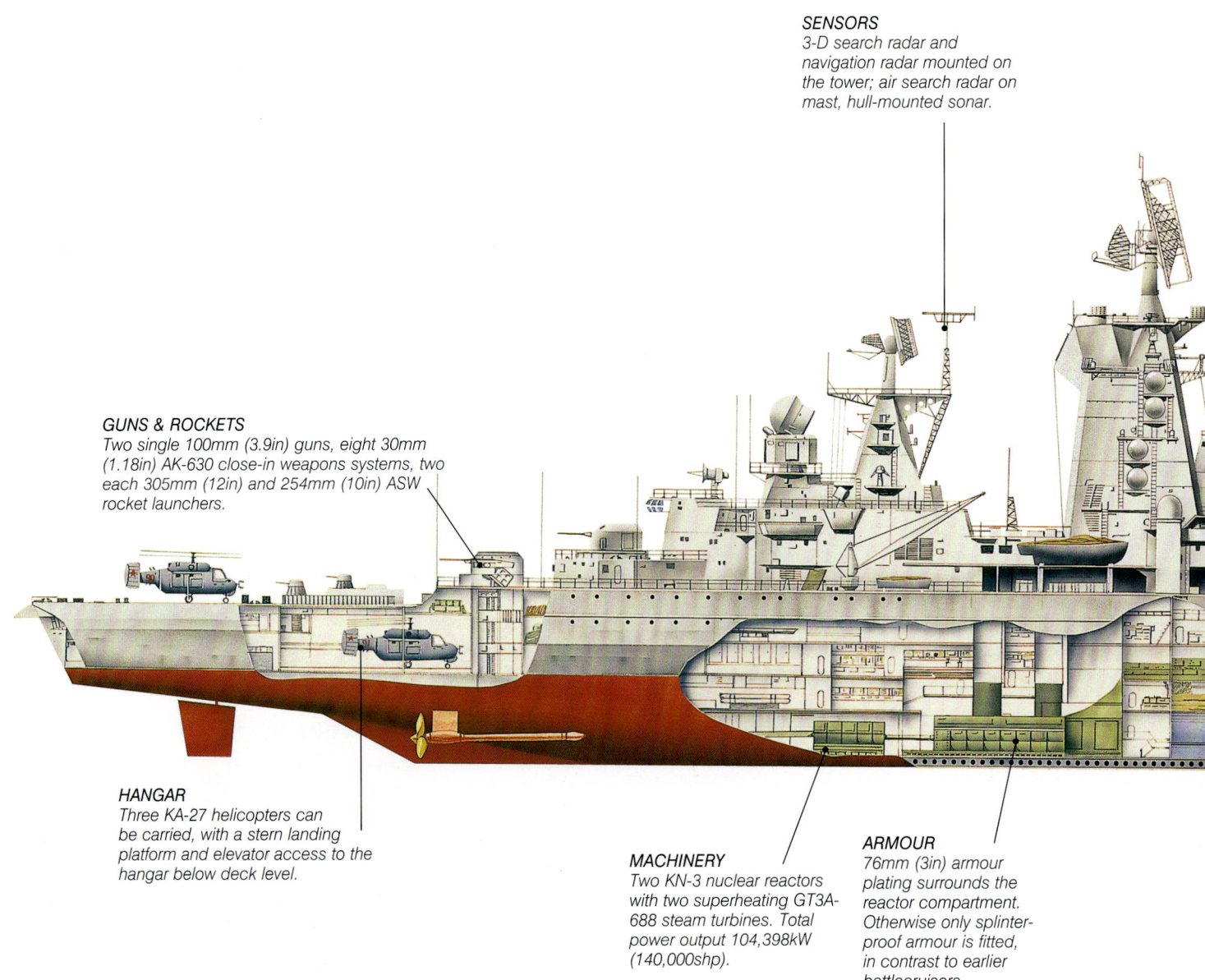

SENSORS
3-D search radar and navigation radar mounted on the tower; air search radar on mast, hull-mounted sonar.

GUNS & ROCKETS
Two single 100mm (3.9in) guns, eight 30mm (1.18in) AK-630 close-in weapons systems, two each 305mm (12in) and 254mm (10in) ASW rocket launchers.

HANGAR
Three KA-27 helicopters can be carried, with a stern landing platform and elevator access to the hangar below deck level.

MACHINERY
Two KN-3 nuclear reactors with two superheating GT3A-688 steam turbines. Total power output 104,398kW (140,000shp).

ARMOUR
76mm (3in) armour plating surrounds the reactor compartment. Otherwise only splinter-proof armour is fitted, in contrast to earlier battlecruisers.

Kirov

SPECIFICATIONS

Type:	Soviet guided-missile cruiser
Displacement:	28,448 tonnes (28,000 tons)
Dimensions:	248m x 28m x 8.8m (813ft 8in x 91ft 10in x 28ft 10in)
Machinery:	Twin screws, turbines, two pressurized water-type reactors
Top speed:	32 knots
Main armament:	Two 100mm (3.9in) guns, two twin SA-N-4 launchers, twelve SA-N-6 launchers plus 20 anti-ship missiles
Complement:	1600
Launch date:	December 1977

This ship had an impressive armament of missiles and guns as well as electronics. Its largest radar antenna was mounted on its foremast. *Kirov* suffered a reactor accident in 1990 while serving in the Mediterranean Sea. Repairs were never carried out, due to lack of funds and the changing political situation in the Soviet Union.

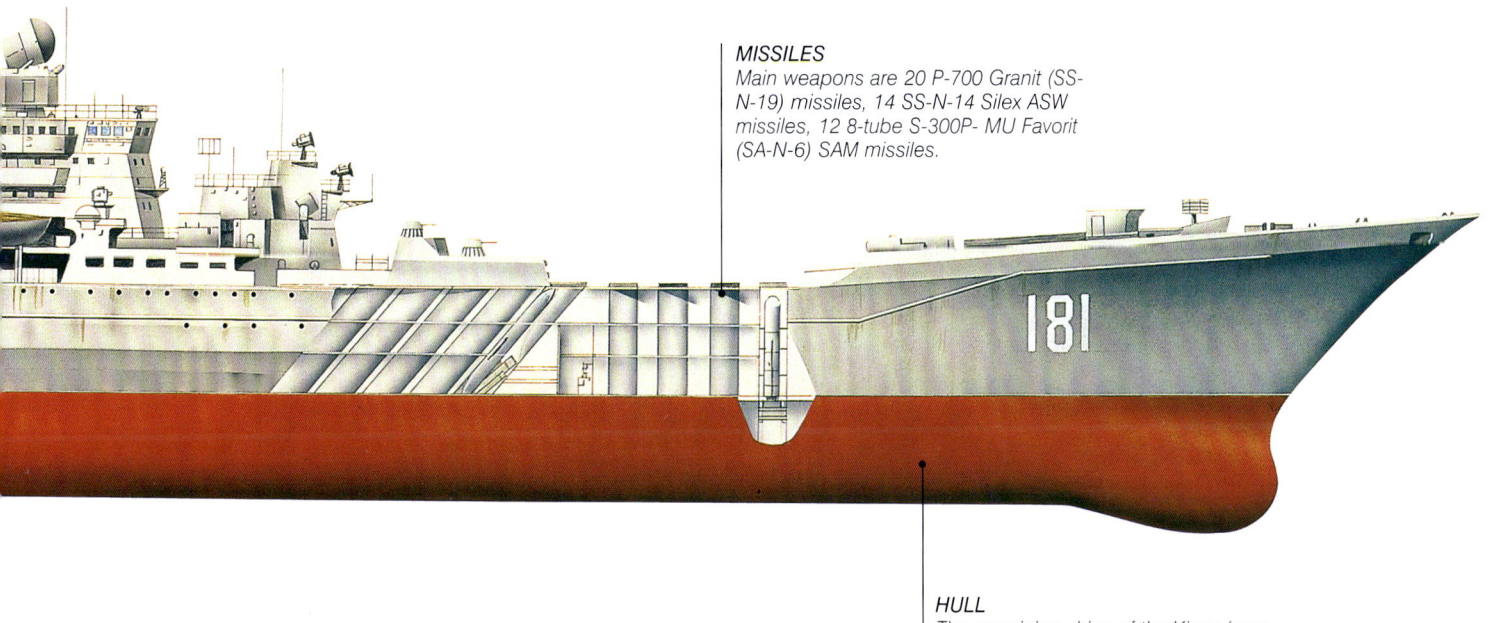

MISSILES
Main weapons are 20 P-700 Granit (SS-N-19) missiles, 14 SS-N-14 Silex ASW missiles, 12 8-tube S-300P- MU Favorit (SA-N-6) SAM missiles.

HULL
The remaining ships of the Kirov *(now renamed* Admiral Ushakov*) class are the world's largest non-carrier warships.*

Soviet Cruisers

The Soviet Union built up a substantial cruiser fleet in the 1950s, although building programmes were cut back from the original numbers. Despite efforts to modernize some of them, they were of diminishing strategic value. By later in the century, they were largely obsolescent and were retired or scrapped. Only a few remain active.

Admiral Senyavin

A *Sverdlov*-class heavy cruiser, this ship was extensively modified in 1971–72 to become a command ship in the event of nuclear war, and the rear guns were removed to install a helicopter deck and hangar. An SA-N-4 SAM launcher was fitted at the same time. It served with the Soviet fleet until stricken in 1991.

SPECIFICATIONS	
Type:	Soviet cruiser
Displacement:	16,723 tonnes (16,640 tons)
Dimensions:	210m x 22m x 6.9m (672ft 4in x 22ft 7in x 22ft)
Machinery:	Not known
Top speed:	32.5 knots
Main armament:	Twelve 152mm (6in) guns, ten 533mm (21in) torpedo tubes
Complement:	1250
Launched:	1952

Dmitry Pozharsky

Unlike *Admiral Senyavin*, this ship, sixth of the *Sverdlov* class to be built, retained its original (and increasingly out-of-date) armament. Sixteen ships of this class were built, out of a projected 30. Though obsolete, *Dmitry Pozharsky* remained on active service until 1987. Others were taken out of service 20 years earlier.

SPECIFICATIONS	
Type:	Soviet cruiser
Displacement:	16,906.3 tonnes (16,640 tons)
Dimensions:	210m x 22m x 6.9m (672ft 4in x 22ft 7in x 22ft)
Machinery:	Not known
Top speed:	32.5 knots
Main armament:	Twelve 152mm (6in) guns, twelve 100mm (3.9in) guns, ten 533mm (21in) torpedo tubes
Complement:	1250
Launched:	1953

TIMELINE
1952
1953

Dmitri Donskoi

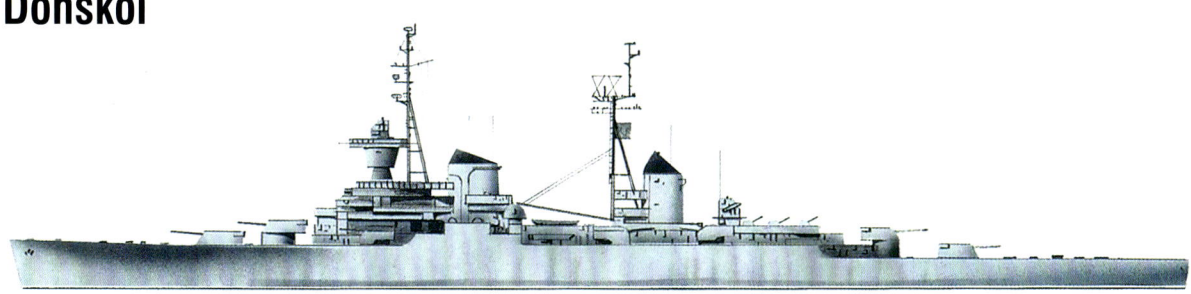

Of the 24 planned ships of this class, 17 were launched, but by the end of 1960 only 14 were operational. Nearly all were fitted for minelaying, mine stowage being on the main deck. The 152mm (6in) guns were mounted in triple turrets, two fore and two aft, each group having its own range-finders.

SPECIFICATIONS	
Type:	Soviet cruiser
Displacement:	19,507 tonnes (19,200 tons)
Dimensions:	210m x 21m x 7m (689ft x 70ft x 24ft 6in)
Machinery:	Twin screws, turbines
Main armament:	Twelve 152mm (6in) guns
Launched:	1953

Kerch

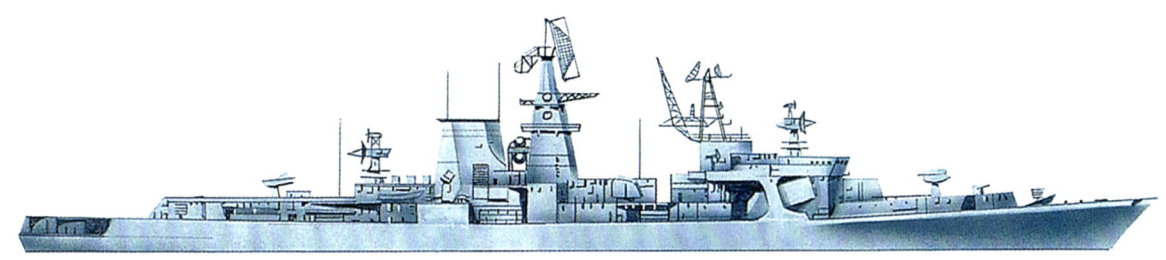

The largest vessels with all gas-turbine propulsion, the *Kerch* class developed 120,000hp, for a range of 5700km (3000 miles) at full speed, or 16,720km (8800 miles) at 15 knots. They had major AA and anti-submarine capabilities, a heavy gun armament, and a helicopter. *Kerch* remains in the Black Sea Fleet.

SPECIFICATIONS	
Type:	Soviet cruiser
Displacement:	9855 tonnes (9700 tons)
Dimensions:	173m x 18.6m x 6.7m (567ft 7in x 61ft x 22ft)
Machinery:	Twin screws, gas turbines
Main armament:	Four 76.2mm (3in) guns, two twin SA-N- 3 missile launchers
Complement:	525
Launched:	1973

Slava

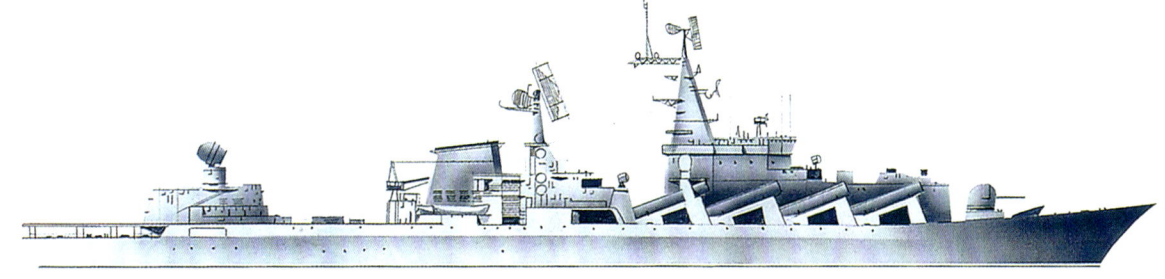

Slava is one of four ships, smaller versions of the *Kirov* class. The SS-N-12 missiles are in twin launchers along each side of the bridge. Complex radars are fitted on the massive foremast at the end of the bridge, with another mounted on top of the mainmast. *Slava* was in the South Ossetia war of 2008.

SPECIFICATIONS	
Type:	Soviet cruiser
Displacement:	11,700 tonnes (11,200 tons)
Dimensions:	186m x 20.8m x 7.6m (610ft 5in x 68ft 5in x 25ft)
Machinery:	Twin screws, gas turbines
Main armament:	Two 127mm (5in) guns, eight twin SS-N-12 launchers, eight launchers for SA-N-6 missiles plus two twin launchers for SAM missiles
Launched:	1979

1973  1979

Cruisers of Other Navies

The traditional 'big' surface warship lingered on through the second half of the twentieth century, encouraged by a sense of national prestige and by the desire of admirals to have flagships. Attitudes changed, particularly perhaps after the sinking of the Argentinian *General Belgrano* by the British submarine *Conqueror* in 1982.

Babur

The British cruiser *Diadem* was acquired by Pakistan in 1956 and comprehensively refitted in 1956–57. In 1961 it became a naval training ship, and rarely went to sea, later becoming harbour-bound. In 1982 it was renamed *Jahanqir* and a former British *Devonshire* class destroyer became *Babur*. It was broken up in 1985.

SPECIFICATIONS	
Type:	Pakistani cruiser
Displacement:	7638.3 tonnes (7518 tons)
Dimensions:	156m x 15m x 5.4m (512ft x 50ft 6in x 18ft)
Machinery:	Quadruple screws, geared tubines
Top speed:	32 knots
Main armament:	Eight 133mm (5.25in) guns
Armour:	76mm (3in) sides, 51–25mm (2–1in) deck
Complement:	530
Launched:	1942 (refitted 1956)

Coronel Bolognesi

One of two British *Fiji*-class cruisers sold to Peru, this was formerly HMS *Ceylon*. Modified before delivery, it received a new mast, increased anti-aircraft armament and improved bridge accommodation. Radar equipment was also updated, for long-range search, height-finding and fire-control. It remained active until 1982.

SPECIFICATIONS	
Type:	Peruvian cruiser
Displacement:	11,633 tonnes (11,450 tons)
Dimensions:	169.4m x 18.9m x 6.4m (555ft 6in x 62ft x 20ft 9in)
Machinery:	Quadruple screws, geared turbines
Top speed:	31.5 knots
Main armament:	Nine 152mm (6in) guns, eight 102mm (4in) guns
Complement:	920
Launched:	1942 (refitted 1960)

Prat

Formerly the US *Brooklyn* class cruiser *Nashville,* this World War II veteran was transferred to the Chilean Navy in 1951, as was *Brooklyn* itself, renamed *O'Higgins.* Both ships were refitted and modernized in the United States in 1957–58. *Prat* was decommissioned in 1984, while *O'Higgins* was stricken only in 1992.

SPECIFICATIONS	
Type:	Chilean cruiser
Displacement:	12,405 tonnes (12,210 tons)
Dimensions:	185.4m x 18.8m x 6.95m (608ft 4in x 61ft 9in x 22ft 9in)
Machinery:	Quadruple screws, geared turbines
Top speed:	30 knots
Main armament:	Fifteen 152mm (6in) guns, eight 127mm (5in) guns
Armour:	127mm (5in) belt, 51mm (2in) deck
Complement:	868
Launched:	1936 (modernized 1957)

Tiger

Tiger was originally laid down in 1941, but work stopped in 1946 and it was finally completed to new plans in 1959, as one of the last cruisers to enter British service. However, it proved unsuitable for the conditions and requirements of the period, and after little use, was withdrawn in the 1960s and scrapped in 1986.

SPECIFICATIONS	
Type:	British cruiser
Displacement:	12,273 tonnes (12,080 tons)
Dimensions:	170m x 20m x 6.4m (555ft 6in x 64ft x 21ft 3in)
Machinery:	Quadruple screws, turbines
Top speed:	31.5 knots
Main armament:	Four 152mm (6in) guns, six 76mm (3in) guns
Launched:	October 1945/refitted 1959

Caio Duilio

Caio Duilio and its sister-ship *Andrea Doria* were helicopter cruisers for anti-submarine and air defence, designed to a new plan, having a wide beam in relation to their length. They carried three AB 212SW armed helicopters. In 1980, *Caio Duilio* became a training ship, and was decommissioned in 1991.

SPECIFICATIONS	
Type:	Italian cruiser
Displacement:	6604 tonnes (6506 tons)
Dimensions:	144m x 17m x 4.7m (472ft 4in x 55ft 7in x 15ft 4in)
Machinery:	Twin screws, geared turbines
Top speed:	31 knots
Main armament:	Eight 76mm (3in) guns, plus Terrier surface-to-air missiles
Complement:	485
Launched:	December 1962

1959  1962

Destroyers (guided missile): Part 1

The introduction of the guided missile and the new generation of long-range guided or homing torpedoes, as well as radar-directed, rapid-fire guns, made the medium-sized warship a more formidable fighting craft than ever before. Destroyer design, construction and ship-management entered a phase of rapid, significant change.

Farragut

This class became the US Navy's first missile ships, and *Farragut* was among the first to carry the ASROC system, which fired rocket-boosted anti-submarine torpedoes. Later it was fitted with the Naval Tactical Data System for air defence command and control. Decommissioned in 1989, it was stricken in 1992.

SPECIFICATIONS	
Type:	US destroyer
Displacement:	5738.4 tonnes (5648 tons)
Dimensions:	156.3m x 15.9m x 5.3m (512ft 6in x 52ft 4in x 17ft 9in)
Machinery:	Twin screws, geared turbines
Top speed:	32 knots
Main armament:	One Terrier (later Standard) SAM missile system, one ASROC rocket-boosted ASW torpedo launcher, six 324mm (12.75in) torpedo tubes
Complement:	360
Launched:	1958

Gremyaschiy

The 'Krupny' class of nine destroyers were the first missile-armed ships in the Soviet fleet. The original SS-N-1 Scrubber missile system was soon removed and the class rearmed for anti-submarine warfare. With 16 57mm (2.24in) guns, the ship was lightly armed for self-defence, reducing its usefulness. It was stricken in 1995.

SPECIFICATIONS	
Type:	Soviet destroyer
Displacement:	4259 tonnes (4192 tons)
Dimensions:	138.9m x 14.84m x 4.2m (455ft 9in x 48ft 8in x 13ft 9in)
Machinery:	Twin screws, geared turbines
Top speed:	34.5 knots
Main armament:	Two SSN-N-1 SSM, two anti-submarine rocket launchers, six 533mm (21in) torpedo tubes
Complement:	310
Launched:	1959

TIMELINE 1958 1959 1960

DESTROYERS (GUIDED MISSILE): PART 1

Devonshire

Designed at the end of the 1950s, *Devonshire* and seven others were built to operate in the fall-out area of a nuclear explosion, with deck installations under cover. Later, Seacat missiles replaced the Seaslug and some of the class also carried the Exocet anti-ship missile. *Devonshire* was sunk as a target in 1984.

SPECIFICATIONS	
Type:	British destroyer
Displacement:	6299 tonnes (6200 tons)
Dimensions:	158m x 16m x 6m (520ft 6in x 54ft x 20ft)
Machinery:	Twin screws, turbines plus four gas turbines
Top speed:	32.5 knots
Main armament:	Four 114mm (4.5in) guns, twin launcher for Seaslug missile
Complement:	471
Launched:	June 1960

Boykiy

Boykiy was originally completed as a missile ship armed with SS-N-1 launchers. When these became obsolete in the mid-1960s, it was converted into an anti-submarine vessel. *Boykiy* served in the North Atlantic and the North Pacific. It was towed to Spain for breaking in 1988, grounding off Norway on the way.

SPECIFICATIONS	
Type:	Soviet destroyer
Displacement:	4826 tonnes (4750 tons)
Dimensions:	140m x 15m x 5m (458ft 9in x 49ft 5in x 16ft 6in)
Machinery:	Twin screws, geared turbines
Top speed:	35 knots
Main armament:	Eight 57mm (2.24in) guns, plus missiles
Complement:	310
Launched:	1960

Impavido

Impavido was one of Italy's first two missile-armed destroyers. Derived from the *Impetuoso*-class destroyers, it had a US Mk 13 launcher for Tartar surface-to-air missiles. The after funnel was heightened to clear the fire control tracker on top of the aft structure. Modernized in 1976–77, *Impavido* was decommissioned in 1992.

SPECIFICATIONS	
Type:	Italian destroyer
Displacement:	4054 tonnes (3990 tons)
Dimensions:	131.3m x 13.7m x 4.4m (430ft 9in x 45ft x 14ft 5in)
Machinery:	Twin screws, turbines
Top speed:	34 knots
Main armament:	Two 127mm (5in) guns, one Tartar missile launcher, six 533mm (21in) torpedo tubes
Complement:	340
Launched:	May 1962

Destroyers (guided missile): Part 2

As usual with warship types, destroyers tended to get larger, offering a more substantial launch platform for various missiles. The proposed US *Zumwalt* class, to replace the *Arleigh Burkes* in the years after 2010, has an anticipated displacement of 12,000 tonnes (11,808 tons), equivalent to a heavy cruiser of 60 years ago.

Ognevoy

Project 61 from the 1960s produced 20 guided missile destroyers, known as the 'Kashin' class. The third completed, *Ognevoy*, was given cruise missiles and sonar and new air defence in the mid-1970s. Sensor equipment included air-search, navigation and fire-control radar. *Ognevoy* was broken up in 1990.

SPECIFICATIONS	
Type:	Soviet destroyer
Displacement:	4460.3 tonnes (4390 tons)
Dimensions:	144m x 15.8m x 4.6m (472ft 5in x 51ft 10in x 15ft 1in)
Machinery:	Twin screws, four gas turbines
Top speed:	18 knots
Main armament:	Two SA-N-1 SSM launchers, two RBU-6000 and two RBU-1000 anti-submarine rocket launchers, four 76mm (3in) guns, five 533mm (21in) torpedo tubes
Complement:	266
Launched:	1963

Duquesne

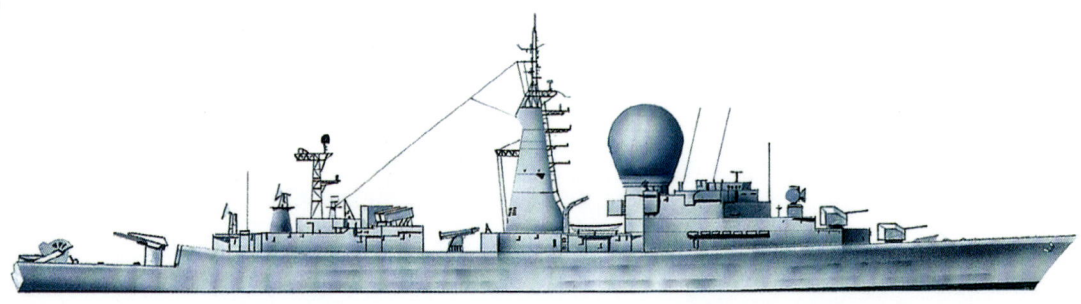

With its sister ship *Suffren*, *Duquesne* was the first French destroyer designed to carry surface-to-air missiles, serving as escort ships for the new generation of French aircraft carriers. *Duquesne* also received four Exocet missile launchers. Electronics were modernized in 1990–91, and it was decommissioned in 2007.

SPECIFICATIONS	
Type:	French destroyer
Displacement:	6187 tonnes (6090 tons)
Dimensions:	157.6m x 15.5m x 7m (517ft x 50ft 10in x 23ft 9in)
Machinery:	Twin screws, turbines
Top speed:	34 knots
Main armament:	Two 100mm (3.9in) guns, one Malafon anti-submarine missile launcher/four torpedo launchers
Launched:	February 1966

TIMELINE 1963 1966 1973

DESTROYERS (GUIDED MISSILE): PART 2

Duguay-Trouin

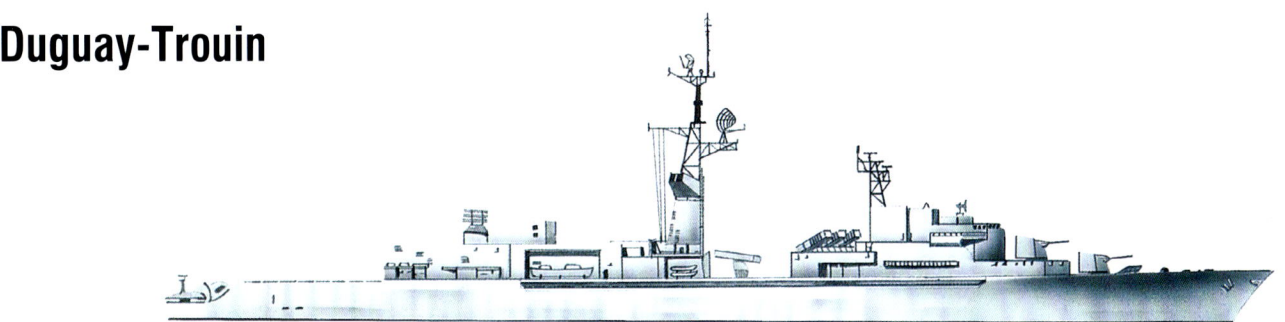

The three *Tourville*-class guided-missile destroyers were the first French warships of destroyer size purpose-built to operate two anti-submarine helicopters. In 1979, *Duguay-Trouin*'s Crotale missile launcher replaced a third gun turret, and new generation air-surveillance radar was fitted. It was decommissioned in 1999.

SPECIFICATIONS	
Type:	French destroyer
Displacement:	5892 tonnes (5800 tons)
Dimensions:	152.5m x 15.3m x 6.5m (500ft 4in x 50ft 2in x 21ft 4in)
Machinery:	Twin screws, turbines
Top speed:	32 knots
Main armament:	Two 100mm (3.9in) guns, one eight-cell Crotale launcher
Complement:	282
Launched:	June 1973

Spruance

Bigger than many former cruisers, the *Spruance* class was designed to offer a stable weapon-launcher, able to operate in difficult sea conditions. Its successful hull design was used, with modifications, on two other classes of US warship. After serving in the Atlantic fleet, *Spruance* was sunk as a target in 2006.

SPECIFICATIONS	
Type:	US destroyer
Displacement:	8168 tonnes (8040 tons)
Dimensions:	171.7m x 16.8m x 5.8m (563ft 4in x 55ft 2in x 19ft)
Machinery:	Twin screws, gas turbines
Top speed:	32.5 knots
Main armament:	Two 127mm (5in) guns, Tomahawk and Harpoon missiles
Complement:	296
Launched:	1973

Tachikaze

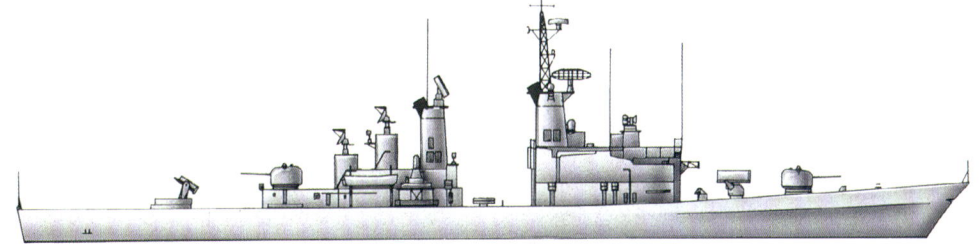

Like all Japanese post-war navy ships, *Tachikaze* was equipped with American weapons. Two 20mm (0.79in) Phalanx CIWS were added in 1983, when the Harpoon SSM were installed. It could also accommodate a SH-601 helicopter. From 1998 to decommissioning in January 2007, it was the Fleet Escort Force flagship.

SPECIFICATIONS	
Type:	Japanese destroyer
Displacement:	4877 tonnes (4800 tons)
Dimensions:	143m x 14.3m x 4.6m (469ft 2in x 46ft 10in x 15ft 1in)
Machinery:	Twin screws, geared turbines
Top speed:	32 knots
Main armament:	Eight Harpoon SSM, one Mk 13 Standard SAM launcher, one ASROC launcher, two 127mm (5in) guns
Complement:	277
Launched:	1974

Destroyers (guided missile): Part 3

Air and surface combat are usually seen as the prime role of the destroyer, though some navies have also fitted out destroyers for anti-submarine roles. CIWS (close-in weapons system) response systems are generally fitted, based on 20mm or 30mm (0.79in or 1.18in) guns, with a variety of missile and anti-missile systems, depending on the supplier.

Santisima Trinidad

Based on the Royal Navy's Type 42 destroyer, and carrying a Lynx helicopter, *Santisima Trinidad* was commissioned in 1981. In April 1982, it was the lead ship in Argentina's invasion of the Falkland Islands. In the late 1980s and 1990s, much of its equipment was cannibalized to keep its sister ship *Hercules* active.

SPECIFICATIONS	
Type:	Argentinian destroyer
Displacement:	4419.6 tonnes (4350 tons)
Dimensions:	125m x 14m x 5.8m (410ft x 46ft x 19ft)
Machinery:	Twin screws, four gas turbines
Top speed:	30 knots
Main armament:	One GWS30 Sea Dart SAM missile launcher, one 114mm (4.5in) gun, six 324mm (12.5in) torpedo tubes
Complement:	312
Launched:	1974

Dupleix

Dupleix was one of eight ships built at Brest as anti-submarine vessels. A major innovation was the use of gas turbine engines, developing 52,000hp for a speed of 30 knots. All in the class had a double hangar aft for helicopters. The ships are well adapted for surface work, as well as the standard ASW role.

SPECIFICATIONS	
Type:	French destroyer
Displacement:	4236 tonnes (4170 tons)
Dimensions:	139m x 14m x 5.7m (456ft x 46ft x 18ft 8in)
Machinery:	Twin screws, gas turbines, plus diesels
Top speed:	30 knots
Main armament:	One 100mm (3.9in) gun, four MM38 Exocet SSM launchers, octuple Crote Navale SAM launcher, two fixed torpedo launchers
Complement:	216
Launched:	December 1978

TIMELINE 1974 1978 1982

Euro

Euro entered service in 1983. The aft flight deck is 27m (88ft 6in) long and 12m (39ft 4in) wide, and a Variable Depth Sonar is streamed out on a 900m (984yd) long cable from a stern well. As the Italian Navy does not use the term 'destroyer', it is classed as a guided-missile frigate.

SPECIFICATIONS	
Type:	Italian destroyer
Displacement:	3088 tonnes (3040 tons)
Dimensions:	122.7m x 12.9m x 8.4m (402ft 6in x 42ft 4in x 27ft 6in)
Machinery:	Twin screws, diesels and gas turbines
Top speed:	29 knots (diesel), 32 knots (turbines)
Main armament:	One 127mm (5in) gun plus missiles
Launched:	December 1982

Mutenia

Built in Romania to Russian designs and using mostly Russian equipment, *Mutenia* was well equipped, having two Alouette III helicopters. It spent lengthy periods out of service, suffering from the problems of a one-off type as well as a lack of operating funds. In 1990–92, it was refitted with anti-submarine weapons.

SPECIFICATIONS	
Type:	Romanian destroyer
Displacement:	5882.6 tonnes (5790 tons)
Dimensions:	144.6m x 14.8m x 7m (474ft 4in x 48ft 6in x 23ft)
Machinery:	Twin screws, diesels
Top speed:	31 knots
Main armament:	Eight SSN-N-2C SSM, four 76mm (3in) guns, six 533mm (21in) torpedo tubes, two RBU-120 anti-submarine rocket launchers
Complement:	270
Launched:	1982

Hamayuki

Japan's Maritime Self Defence Force produced some versatile, well-armed ships in the 1980s. *Hamayuki* was fifth of 12 *Hatsuyuki* class ships and carried a Mitsubishi-built HSS 2B Sea King helicopter. Sensor equipment includes hull-mounted sonar, and air-search, sea-search and fire-control radar.

SPECIFICATIONS	
Type:	Japanese destroyer
Displacement:	3759.2 tonnes (3700 tons)
Dimensions:	131.7m x 13.7m x 4.3m (432ft 4in x 44ft 11in x 14ft 3in)
Machinery:	Twin screws, four gas turbines
Top speed:	30 knots
Main armament:	Eight Harpoon SSM, one Sea Sparrow SAM, one 76mm (3in) gun, two Mk15 Phalanx 20mm (0.79in) CIWS, one ASROC, six 324mm (12.7in) torpedo tubes
Complement:	190
Launched:	1983

Destroyers (guided missile): Part 4

Gas turbine propulsion, enabling rapid acceleration to maximum speed, is favoured for modern destroyers. The Soviet 'Kashin' class used gas turbines in the 1960s, as did the Canadian *Iroquois* class of the 1970s and the ships of the US *Spruance* class, launched between 1972 and 1980, and the *Arleigh Burke* class, still in service.

Edinburgh

Edinburgh was designed as an air-defence/ASW ship to work with a naval or amphibious task force. Its two helicopters carry air-to-surface weapons for use against lightly defended surface ships. Refitted in 1990, *Edinburgh* saw service off Iraq in the second Gulf War (2003) and had a further full refit in 2004–5.

SPECIFICATIONS	
Type:	British destroyer
Displacement:	4851 tonnes (4775 tons)
Dimensions:	141m x 14.9m x 5.8m (463ft x 48ft x 19ft)
Machinery:	Twin screws, gas turbines
Top speed:	30 knots
Main armament:	One 114mm (4.5in) gun, helicopter-launched Mk44 torpedoes, two triple mounts for Mk46 anti-submarine torpedoes, one Sea Dart launcher
Launched:	March 1983

Arleigh Burke

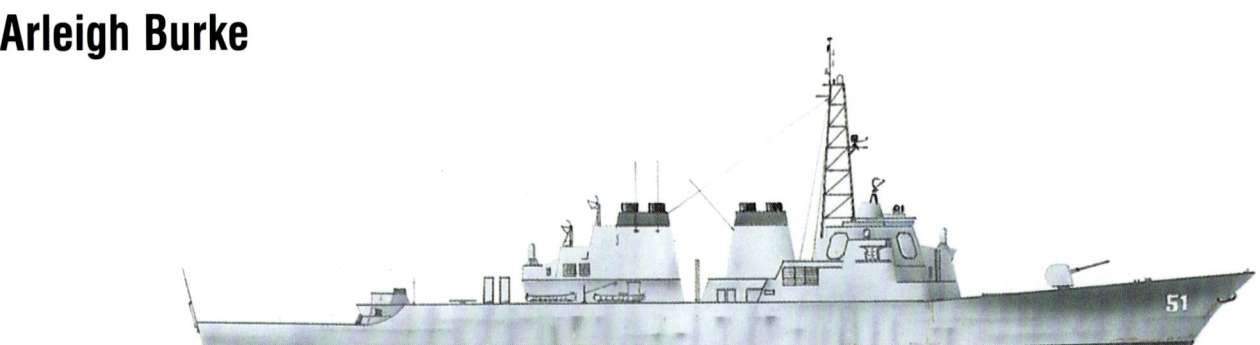

This class was designed to replace the *Adams* and *Coontz* class destroyers, which entered service in the early 1960s. *Arleigh Burke* was commissioned in 1991, to provide effective anti-aircraft cover, for which the SPY 1 D version of the Aegis system was fitted. It also has anti-surface and anti-submarine weapons.

SPECIFICATIONS	
Type:	US guided missile destroyer
Displacement:	8534 tonnes (8400 tons)
Dimensions:	142.1m x 18.3m x 9.1m (266ft 3in x 60ft x 30ft)
Machinery:	Twin screws, gas turbines
Top speed:	30+ knots
Main armament:	Harpoon and Tomahawk missiles, 127mm (5in) gun
Launched:	1989

TIMELINE

1983   1989   1991

DESTROYERS (GUIDED MISSILE): PART 4

Kongo

Kongo is based on the US *Arleigh Burke* class, its dimensions easily justifying cruiser designation. The flight deck has no hangar but can take a SH-60J Seahawk helicopter. Fitted with the Aegis air defence radar and missile system, the class has recently been modified to intercept North Korean ballistic missiles.

SPECIFICATIONS	
Type:	Japanese destroyer
Displacement:	9636.8 tonnes (9485 tons)
Dimensions:	160.9m x 20.9m x 6.2m (520ft 2in x 68ft 7in x 20ft 4in)
Machinery:	Twin screws, gas turbines
Top speed:	30 knots
Main armament:	Eight Harpoon SSM, two Mk41 VLS with Standard missiles and ASROC torpedoes, one 127mm (5in) gun, two 20mm (0.79in) Phalanx CIWS, six 324mm (12.75in) torpedo tubes
Complement:	300
Launched:	1991

Brandenburg

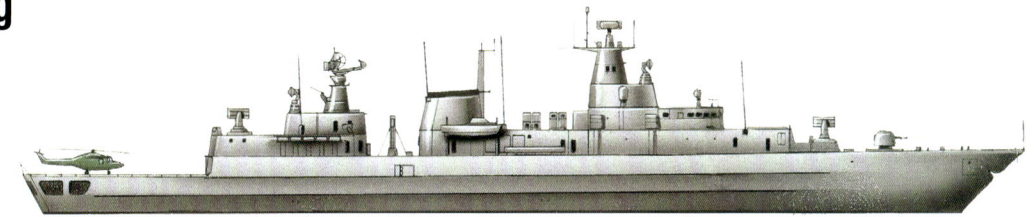

Brandenburg was lead ship of a class of four, identified as Type 123, and intended for air defence. They are equipped with air-search and air/surface search radar, two fire-control trackers and hull-mounted sonar. Two Lynx Mk88 helicopters are carried. Germany has built similar ships for Portugal and Turkey.

SPECIFICATIONS	
Type:	German destroyer
Displacement:	4343.4 tonnes (4275 tons)
Dimensions:	138.9m x 16.7m x 6.3m (455ft 8in x 57ft 1in x 20ft 8in)
Machinery:	Twin screws, gas turbines plus diesels
Top speed:	29 knots
Main armament:	Four MM38 Exocet SSM, one VLS for Sea Sparrow SAM, two 21-cell RAM launchers, one 76mm (3in) gun, six 324mm (12.75in) torpedo tubes
Complement:	219
Launched:	1992

Murasame

Commissioned in 1996, designed primarily for air defence but with anti-submarine potential, *Murasame* is a typically all-round member of the modern Japanese fleet. An SH-60J helicopter is carried, with hangar. The *Murasame* class has full radar search equipment, and both hull-mounted and towed sonar.

SPECIFICATIONS	
Type:	Japanese destroyer
Displacement:	5181.6 tonnes (5100 tons)
Dimensions:	151m x 16.9m x 5.2m (495ft 5in x 55ft 7in x 17ft 1in)
Machinery:	Twin screws, gas turbines
Top speed:	33 knots
Main armament:	Eight Harpoon SSM, two Mk41 VLS with Standard missiles and ASROC torpedoes, one 127mm (5in) gun, two 20mm (0.79in) Phalanx CIWS, six 324mm (12.75in) torpedo tubes
Complement:	170
Launched:	1994

Destroyers (anti-submarine): Part 1

From its inception as a warship-type, torpedoes were the main armament of a destroyer, enabling it to be a threat to much larger ships. While many destroyers still carry torpedo tubes, these no longer define the destroyer as such. Some ships classed as destroyers have no torpedoes, relying on missiles and and ASW mortars.

St Laurent

Seven ships formed the *St Laurent* class, the lead vessel being completed in 1955. In the early 1960s, the armament was updated and variable-depth sonar was mounted at the stern. A helicopter deck and hangar were built in. *St Laurent* was decommissioned in 1979 and sank under tow to the breakers in 1980.

SPECIFICATIONS	
Type:	Canadian destroyer
Displacement:	2641.6 tonnes (2600 tons)
Dimensions:	111.6m x 12.8m x 4m (366ft x 42ft x 13ft 2in)
Machinery:	Twin screws, turbines
Top speed:	28 knots
Main armament:	Four 76mm (3in) guns, two Limbo Mk10 anti-submarine mortars
Complement:	290
Launched:	1951

Neustrashimyy

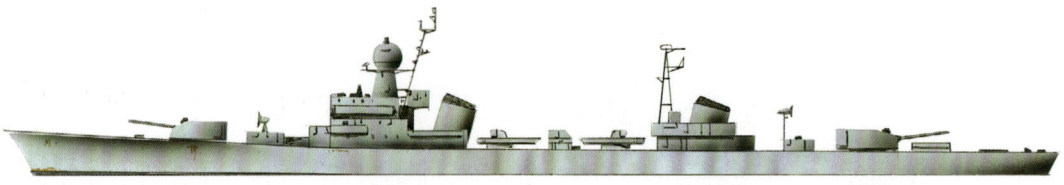

Planned as lead ship of a large class, completed in 1955, *Neustrashimyy* ended up as a one-off. But it displayed many features to reflect the nuclear era, including air-conditioning and sealable crew accommodation. Its pressure-fired boiler design was used in many subsequent ships. It was broken up in 1975.

SPECIFICATIONS	
Type:	Soviet destroyer
Displacement:	3434 tonnes (3830 tons)
Dimensions:	133.8m x 13.6m x 4.4m (439ft 1in x 44ft 6in x 14ft 6in)
Machinery:	Twin screws, geared turbines
Top speed:	36 knots
Main armament:	Four 130mm (5in) guns, 10 533mm (21in) torpedo tubes
Complement:	305
Launched:	1951

TIMELINE 1951   1952

DESTROYERS (ANTI-SUBMARINE): PART 1

Grom

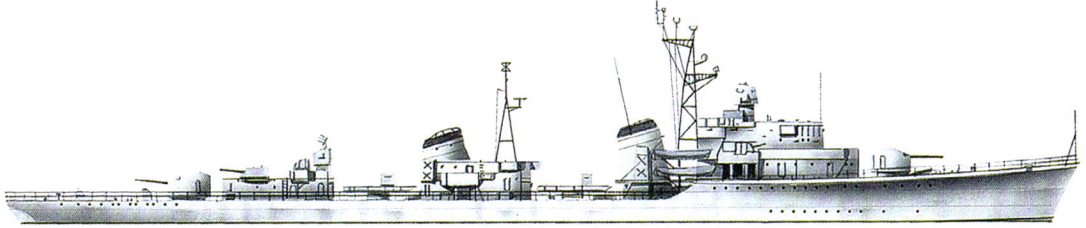

Grom was the former Soviet *Smetlivy*, one of two destroyers transferred from the USSR Baltic Fleet to Poland in 1957. It was a member of a class comprising the first Russian destroyers built after World War II, and incorporating features from German destroyers. In service until 1973, *Grom* was scrapped in 1977.

SPECIFICATIONS	
Type:	Polish destroyer
Displacement:	3150 tonnes (3100 tons)
Dimensions:	120.5m x 11.8m x 4.6m (395ft 4in x 38ft 9in x 15ft)
Machinery:	Twin screws, turbines
Main armament:	Four 130mm (5.1in) guns, two 76mm (3in) anti-aircraft guns
Launched:	1952

Groningen

Groningen's class of eight had some side armour as well as deck protection. They were also some of the first destroyers to have no torpedo capability. Two short-range ASW rocket launchers were fitted. It was one of seven of the class sold to Peru in the 1980s. As *Galvez*, it was deleted in 1991.

SPECIFICATIONS	
Type:	Dutch destroyer
Displacement:	3119 tonnes (3070 tons)
Dimensions:	116m x 11.7m x 3.9m (380ft 3in x 38ft 6in x 13ft)
Machinery:	Twin screws, turbines
Top speed:	36 knots
Main armament:	Four 120mm (4.7in) guns
Launched:	January 1954

Almirante Riveros

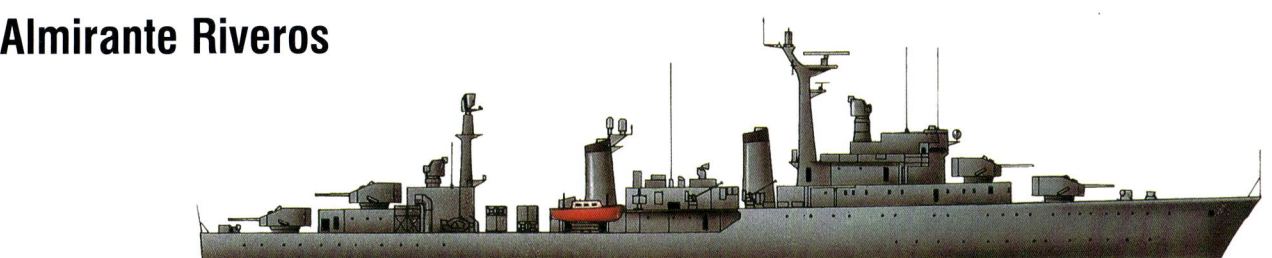

This heavily-armed destroyer was one of a pair built in England. It returned to the UK for modernization work in 1975, when missile systems were fitted, replacing its secondary 40mm (1.57in) armament, and Squid anti-submarine mortars. It served for another 20 years. Decommissioned in 1998, it was sunk as a target in that year.

SPECIFICATIONS	
Type:	Chilean destroyer
Displacement:	3650 tonnes (3300 tons)
Dimensions:	122.5m x 13.1m x 4m (402ft x 43ft x 13ft 4in)
Machinery:	Twin screws, turbines
Top speed:	34.5 knots
Main armament:	Four 102mm (4in) guns, four MM38 Exocet SSM, one Seacat SAM system, six 324mm (12.75in) torpedo tubes
Complement:	266
Launched:	December 1958

1954  1958

Destroyers (anti-submarine): Part 2

Destroyer-borne ASW helicopters are fitted with sonobuoys, dipping sonar and magnetic anomaly detectors to identify potential submerged targets, and armed with torpedoes or depth charges. Helicopters have become of such combat value that some ships are identified as 'helicopter destroyers' or 'helicopter cruisers'.

Coronel Bolognesi

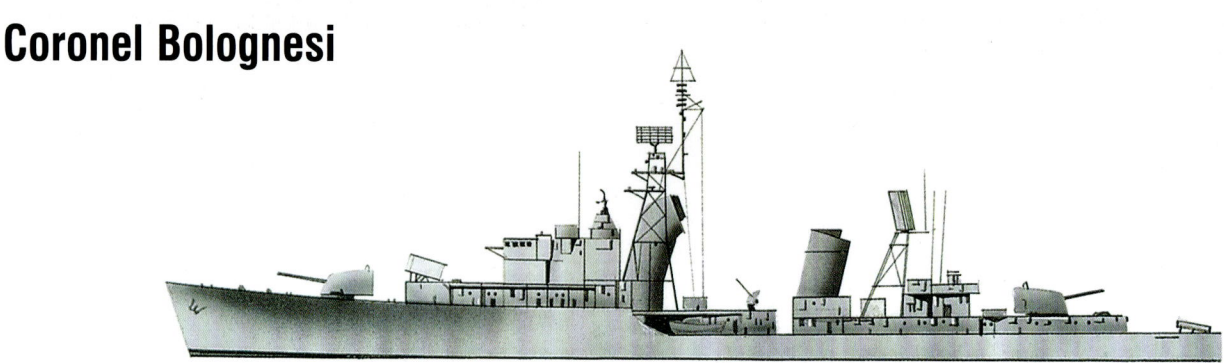

Coronel Bolognesi was formerly the Dutch *Overijssel*, of the *Friesland* class. Between 1980 and 1982, they were transferred to the Peruvian Navy, and fitted with Exocet missiles and other new weapons systems and sensors. *Coronel Bolognesi* arrived in Peru in July 1982 and was decommissioned in 1990.

SPECIFICATIONS	
Type:	Peruvian destroyer
Displacement:	3150 tonnes (3100 tons)
Dimensions:	116m x 12m x 5m (380ft 7in x 38ft 5in x 17ft)
Machinery:	Twin screws, turbines
Top speed:	36 knots
Main armament:	Four 120mm (4.7in) guns
Launched:	August 1955

Aragua

British-built, *Aragua* was one of the three *Nueva Esparta* class destroyers. Its two sister-ships were later given more up-to-date armament, including Seacat surface-to-air missiles, and their radar systems were also modernized. *Aragua* remained very much as delivered. It was withdrawn from service in 1975.

SPECIFICATIONS	
Type:	Venezuelan destroyer
Displacement:	3353 tonnes (3300 tons)
Dimensions:	122.5m x 13.1m x 3.9m (402ft x 43ft x 12ft 9in)
Machinery:	Twin screws, geared turbines
Top speed:	34.5 knots
Main armament:	Six 114mm (4.5in) guns, three 533mm (21in) torpedo tubes, two Squid anti-submarine mortars, two depth charge racks
Complement:	254
Launched:	1955

TIMELINE

1955　　1966

DESTROYERS (ANTI-SUBMARINE): PART 2

Asagumo

Asagumo and five sister-ships were typical of the mid period of Japanese destroyers after World War II. With 711 tonnes (700 tons) of oil fuel, Asagumo had a range of 11,400km (6000 miles) at 20 knots. The gunnery, radar and sensors were all supplied by the United States. It was decommissioned in 1998.

SPECIFICATIONS

Type:	Japanese destroyer
Displacement:	2083 tonnes (2050 tons)
Dimensions:	114m x 11.8m x 4m (374ft x 38ft 9in x 13ft)
Machinery:	Twin screws, diesel engines
Main armament:	Four 76mm (3in) guns, six torpedo tubes
Launched:	1966

Audace

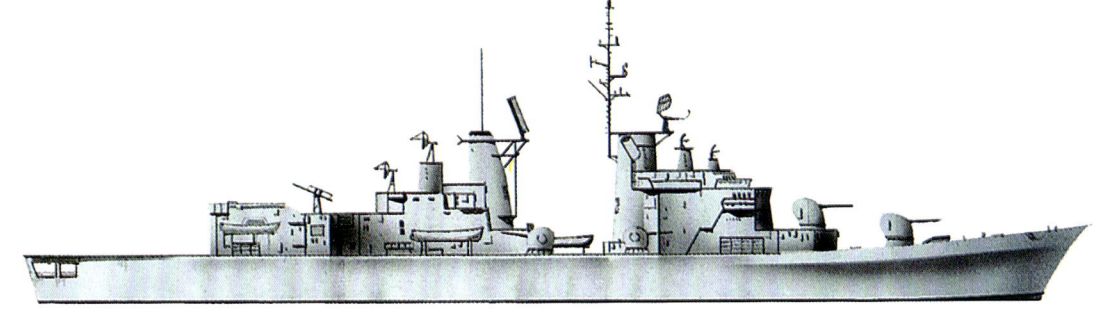

Audace was a multi-function fleet escort, its prime task being anti-submarine action. Two helicopters with weapons kit and sensors were carried. Some of its weapons had a poor arc of fire due to the height of the superstructure. Serving off Lebanon in 1982 and in the Gulf in 1990–91, it was decommissioned in 2006.

SPECIFICATIONS

Type:	Italian destroyer
Displacement:	4470 tonnes (4400 tons)
Dimensions:	135.9 x 14.6m x 4.5m (446ft x 48ft x 15ft)
Machinery:	Twin screws, geared turbines
Top speed:	33 knots
Main armament:	Two 127mm (5in) guns, one SAM launcher
Launched:	1971

Haruna

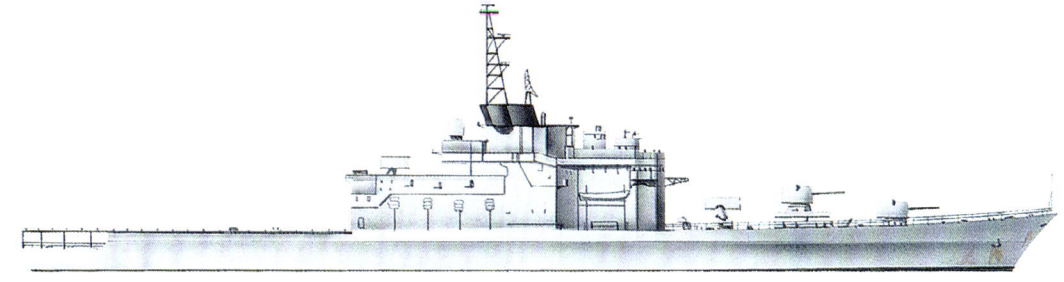

Haruna is a command ship for anti-submarine escort groups, its entire aft part devoted to facilities for three Sea King helicopters. The hangar occupies the full beam of the vessel. Automatic guns fire up to 40 rounds per minute. In 1986–87, Haruna underwent a major refit to improve its anti-aircraft defences.

SPECIFICATIONS

Type:	Japanese destroyer
Displacement:	5029 tonnes (4950 tons)
Dimensions:	153m x 17.5m x 5.2m (502ft x 57ft 5in x 17ft)
Machinery:	Twin screws, turbines
Main armament:	Two 127mm (5in) guns, Sea Sparrow missile launcher, six 324mm (12.75in) torpedo tubes
Launched:	February 1972

1971  1972

Destroyers & Frigates (anti-submarine)

Around 200 crew are needed for a destroyer – not a large number in relation to the size of the ship and its firepower, and made possible by the use of technology. Post-2010 destroyers will include a 3-D phased array radar system in their equipment.

Gurkha

Gurkha was the third of seven all-purpose frigates of the 'Tribal' class, among the first British warships to be air-conditioned in all crew areas and most working spaces. Decommissioned in 1979, reactivated at the time of the Falklands conflict in 1982, *Gurkha* was sold to Indonesia in 1985, and laid up in 1999.

SPECIFICATIONS	
Type:	British frigate
Displacement:	2743 tonnes (2700 tons)
Dimensions:	109m x 12.9m x 5.3m (360ft x 42ft 4in x 17ft 6in)
Machinery:	Single screw, turbine and gas turbine
Top speed:	28 knots
Main armament:	Two 114mm (4.5in) guns, one Limbo three-barrelled anti-submarine mortar
Complement:	253
Launched:	July 1960

Georges Leygues

Georges Leygues and its seven sister destroyers are France's prime anti-submarine force. Gas turbine engines develop 52,000hp, while diesels develop 10,400hp; the cruising range at 18 knots on diesels is 18,050km (9500 miles). *Georges Leygues* carries two Lynx helicopters and has full hangar facilities.

SPECIFICATIONS	
Type:	French destroyer
Displacement:	4236 tonnes (4170 tons)
Dimensions:	139m x 14m x 5.7m (456ft x 46ft x 18ft 8in)
Machinery:	Twin screws, gas turbines and diesel engines
Top speed:	30 knots
Main armament:	One 100mm (3.9in) gun, Exocet missiles
Launched:	December 1976

TIMELINE

1960 1976

DESTROYERS & FRIGATES (ANTI-SUBMARINE)

Glasgow

Glasgow was active in the Falklands War, and served at East Timor and in the South Atlantic patrol. It had air-search radar, and a fire control system, and carried one helicopter. Two triple 324mm (12.75in) anti-submarine torpedo tube sets were fitted. Decommissioned in 2005, it was broken up in Turkey in 2009.

SPECIFICATIONS

Type:	British destroyer
Displacement:	4165 tonnes (4100 tons)
Dimensions:	125m x 14.3m x 5.8m (410ft x 47ft x 19ft)
Machinery:	Twin screws, gas turbines
Top speed:	30 knots
Main armament:	One 114mm (4.5in) gun, one twin Sea Dart mount
Launched:	April 1976

Cushing

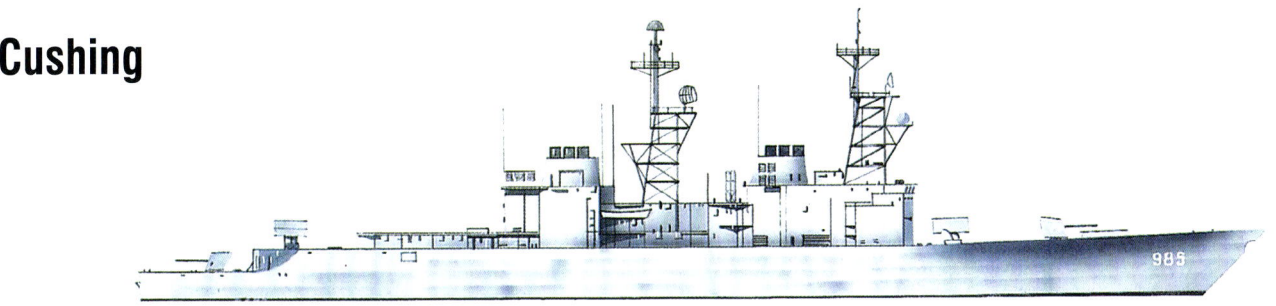

Cushing was the last survivor of the Spruance class, sunk as a target in 2005. Anti-submarine ships, they carried two helicopters, plus the Phalanx CIWS air defence system, and Harpoon and Sparrow missiles. The first US Navy surface vessels fitted with gas turbines, they could run on a single engine at 19 knots.

SPECIFICATIONS

Type:	US destroyer
Displacement:	7924 tonnes (7800 tons)
Dimensions:	161m x 17m x 9m (529ft 2in x 55ft 1in x 28ft 10in)
Machinery:	Twin screws, gas turbines
Top speed:	30 knots
Main armament:	Two 127mm (5in) guns, six 322mm (12.75in) torpedo tubes
Launched:	June 1978

Hatsuyuki

A radical departure from previous Japanese anti-submarine destroyer designs, Hatsuyuki resembles the French Georges Leygues class in its layout, although its weapons systems are of US origin. The propulsion machinery is British: two groups of gas turbines, one set developing 56,780hp, and the other 10,680hp.

SPECIFICATIONS

Type:	Japanese destroyer
Displacement:	3760 tonnes (3700 tons)
Dimensions:	131.7m x 13.7m x 4.3m (432ft x 45ft x 14ft)
Machinery:	Twin screws, gas turbines
Top speed:	30 knots
Main armament:	One 76mm (3in) gun, one eight-cell Sea Sparrow launcher, two 20mm (0.79in) Phalanx CIWS
Complement:	190
Launched:	November 1980

1978    1980

Tromp

Tromp and sister-ship *De Ruyter* acted as flagships to two long-range NATO task groups, for operating in the Eastern Atlantic. An octuple Sea Sparrow launcher with 60 reloads provided short-range anti-aircraft and anti-missile defence, and a later refit added a Goalkeeper point defence gun system. A single Lynx helicopter is carried.

Tromp

Replacing two cruisers in service with the Royal Netherlands navy, HNLMS *Tromp* and *De Ruyter* were among the largest and most capable of frigates afloat. Weapons fitted included Harpoon, Standard and Sea Sparrow missiles.

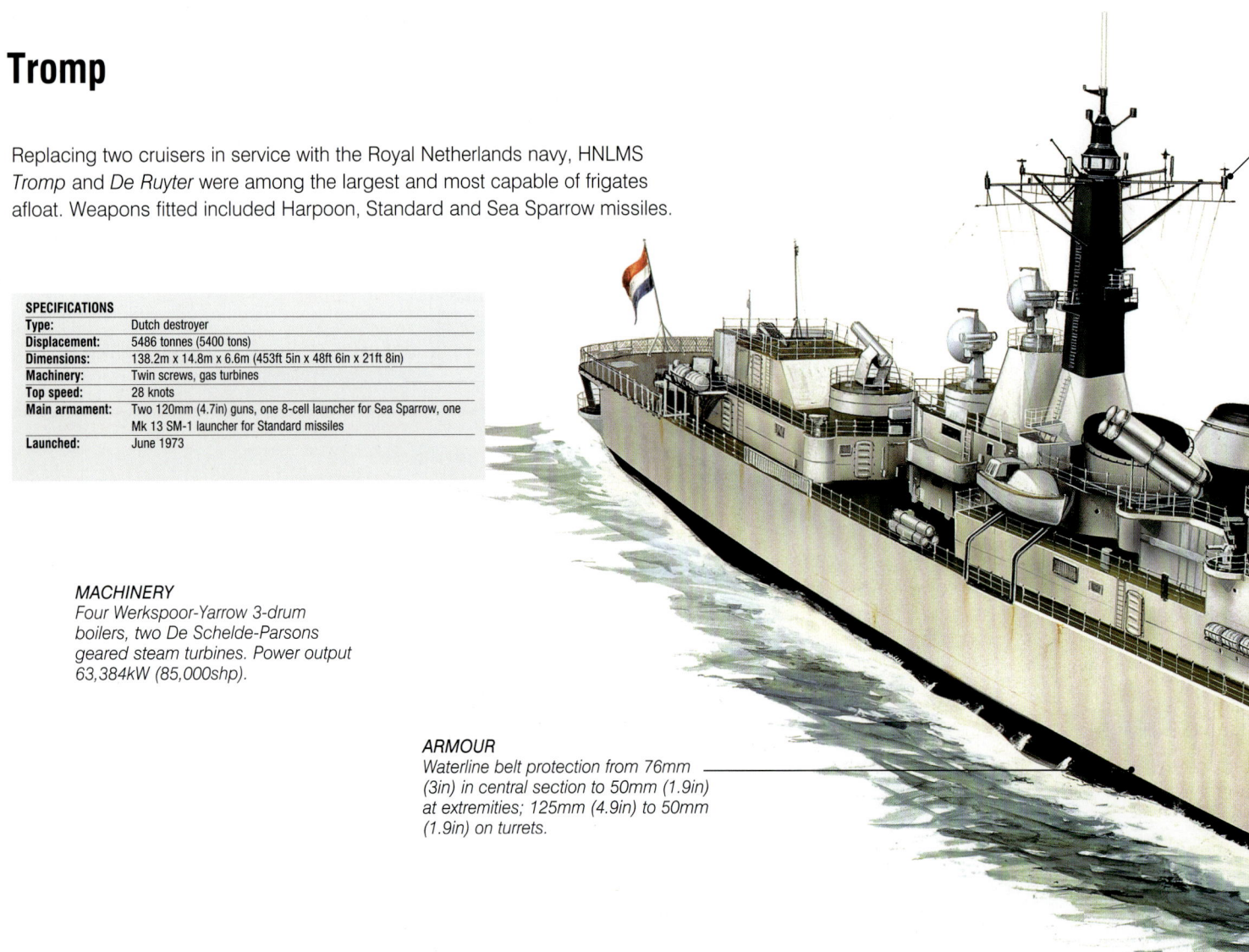

SPECIFICATIONS	
Type:	Dutch destroyer
Displacement:	5486 tonnes (5400 tons)
Dimensions:	138.2m x 14.8m x 6.6m (453ft 5in x 48ft 6in x 21ft 8in)
Machinery:	Twin screws, gas turbines
Top speed:	28 knots
Main armament:	Two 120mm (4.7in) guns, one 8-cell launcher for Sea Sparrow, one Mk 13 SM-1 launcher for Standard missiles
Launched:	June 1973

MACHINERY
Four Werkspoor-Yarrow 3-drum boilers, two De Schelde-Parsons geared steam turbines. Power output 63,384kW (85,000shp).

ARMOUR
Waterline belt protection from 76mm (3in) in central section to 50mm (1.9in) at extremities; 125mm (4.9in) to 50mm (1.9in) on turrets.

TROMP

MASTS
In original form the ship was unusual in having no separate masts, with extensions to the control tower and funnel serving instead.

UPGRADE
A major upgrade in 1985-88 brought new search and fire-control radar systems, decoy launchers and other countermeasures systems, and data links.

MISSILES
In 1993 eight Otomat Mk2 SSMs were installed and in 1996 the Bofors guns were replaced by Oto Melara twin 40L70 DARDO compact gun mountings.

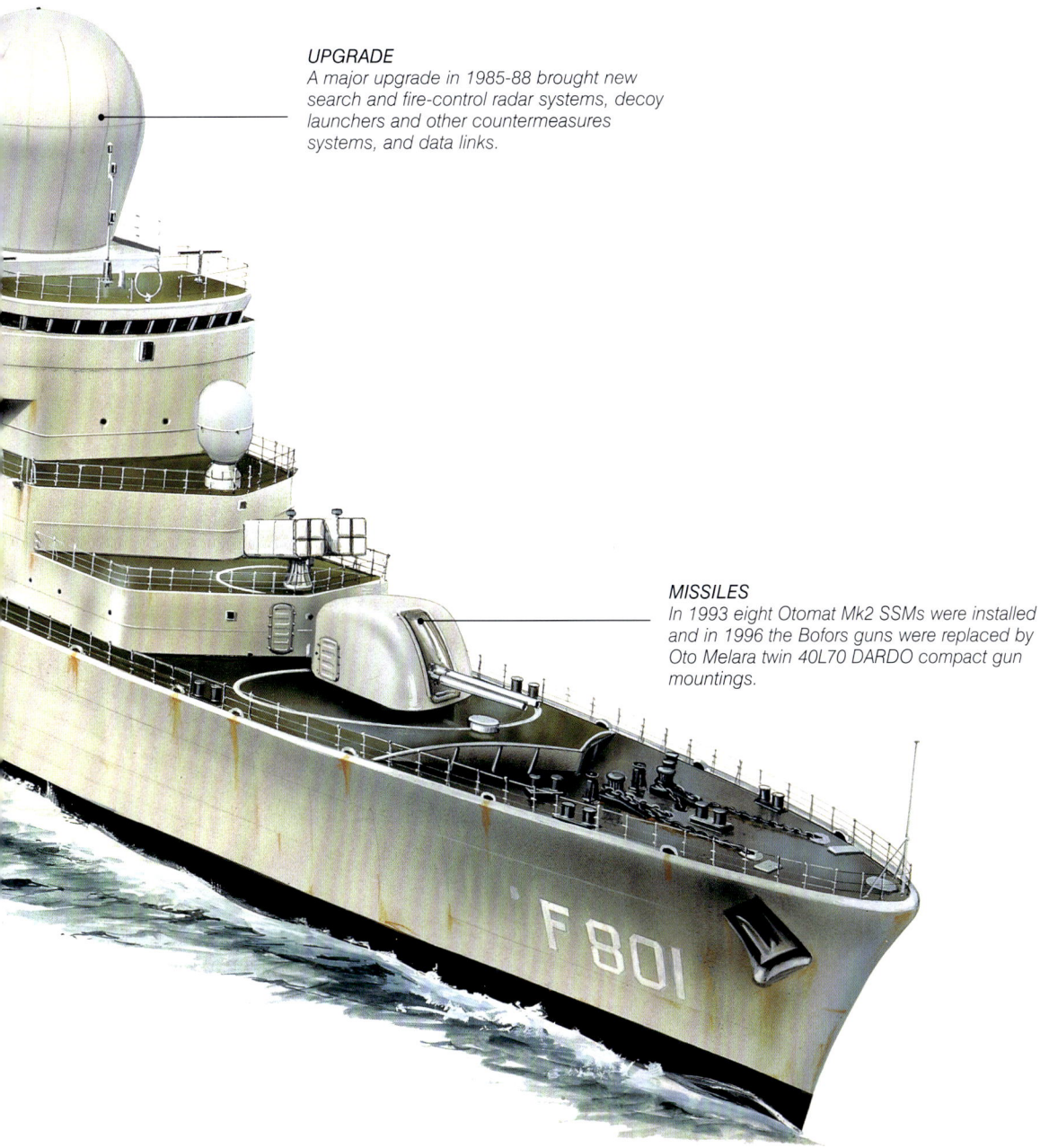

Frigates of the 1950s

It is not always easy to attach type-names to modern warships. 'Frigate' is a case in point. Earlier in the twentieth century, this described an escort ship, especially for merchant convoys. It was generally smaller and slower than a destroyer, and not armed with torpedoes. In the 1950s, this description was already ceasing to fit.

Grafton

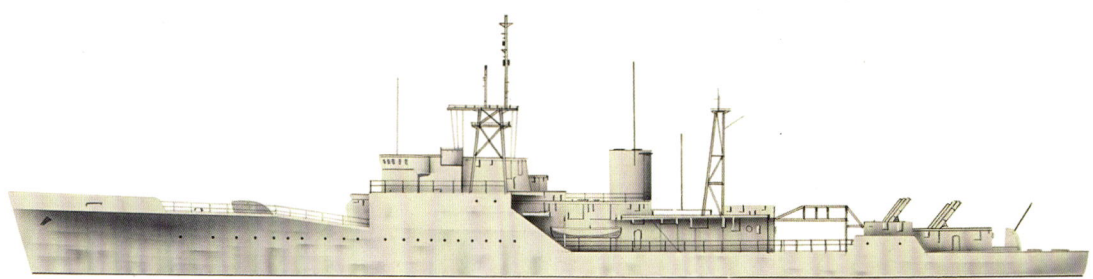

Largely prefabricated, the 12 frigates of *Grafton*'s class were too lightly gunned to be effective as escort ships. Their anti-submarine weapons consisted of two Limbo three-barrelled depth-charge launchers firing a pattern of large depth charges with great accuracy over a wide area. *Grafton* was broken up in 1971.

SPECIFICATIONS	
Type:	British frigate
Displacement:	1480 tonnes (1456 tons)
Dimensions:	94.5m x 10m x 4.7m (310ft x 33ft x 15ft 6in)
Machinery:	Single screw, turbines
Top speed:	27.8 knots
Main armament:	Two 40mm (1.6in) guns
Launched:	September 1954

Centauro

Centauro was one of a class of four vessels built with US funds and equipped with automatic anti-submarine and medium anti-aircraft armament. The guns were mounted one above the other in the twin turrets, but this arrangement was later changed to conventional placing. *Centauro* was stricken in 1984.

SPECIFICATIONS	
Type:	Italian frigate
Displacement:	2255 tonnes (2220 tons)
Dimensions:	104m x 11m x 4m (339ft x 38ft x 11ft 6in)
Machinery:	Twin screws, geared turbines
Main armament:	Four 76mm (3in) guns
Launched:	April 1954

TIMELINE
1954   1955

FRIGATES OF THE 1950S

Cigno

Of the same class as *Centauro* (see oppsite) but of greater displacement, *Cigno* shared the same features. The Italian-made 76mm (3in) guns were mounted in twin turrets and could fire 60 rounds per minute. In the 1960s, the turrets were replaced by three single 76mm (3in) mounts. *Cigno* was broken up in 1983.

SPECIFICATIONS

Type:	Italian frigate
Displacement:	2455 tonnes (2220 tons)
Dimensions:	103m x 12m x 4m (339ft 3in x 38ft x 11ft 6in)
Machinery:	Twin screws, geared turbines
Main armament:	Four 76mm (3in) guns
Launched:	March 1955

Gatineau

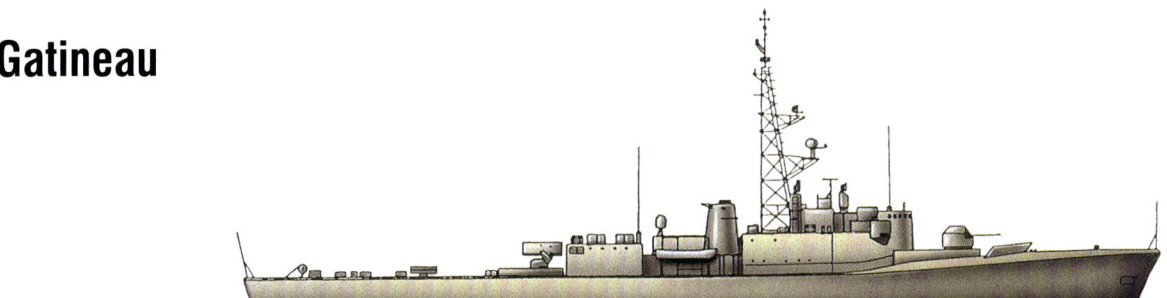

Developed from the *St Laurent* class, *Gatineau* was among four of its class to be modernized between 1966 and 1973, with variable-depth sonar and an ASROC launcher replacing an anti-submarine mortar and one 76mm (3in) gun. It served during the Gulf War. Decommissioned in 1996, *Gatineau* was broken up in 2009.

SPECIFICATIONS

Type:	Canadian frigate
Displacement:	2641.6 tonnes (2600 tons)
Dimensions:	111.6m x 12.8m x 4.2m (366ft x 42ft x 13ft 2in)
Machinery:	Twin screws, geared turbines
Top speed:	28 knots
Main armament:	One Harpoon octuple SSM, two 76mm (3in) guns, one Mk15 Phalanx 20mm (0.79in) CIWS, six 324mm (12.75in) torpedo tubes
Complement:	290
Launched:	1957

Gemlik

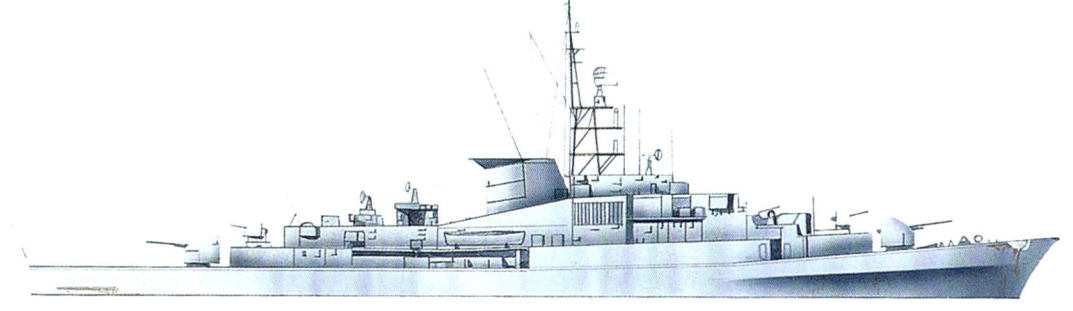

Emden, a *Köln*-class West German frigate, was transferred to the Turkish Navy in September 1983. It carried anti-submarine weapons, sensors and electronic counter-measures. Four launch tubes fired acoustic homing torpedoes. It could also lay up to 80 mines. After a fire, *Emden* was scrapped in 1994.

SPECIFICATIONS

Type:	Turkish frigate
Displacement:	2743 tonnes (2700 tons)
Dimensions:	109.9m x 11m x 5.1m (360ft 7in x 36ft x 16ft 9in)
Machinery:	Twin screws, gas turbines/diesel engines
Top speed:	28 knots
Main armament:	Two 100mm (3.9in) guns, four 533mm (21in) torpedo tubes
Launched:	March 1959

Frigates of the 1960s & '70s: Pt1

Frigates, by the 1960s, were multi-task ships that fulfilled the role of the former light cruiser, able to provide a sufficiently powerful naval presence, whether helpful or punitive, in the event of some maritime incident or localized trouble-spot. Their missile armament equipped them for anti-aircraft or anti-submarine action, or both.

Carlo Bergamini

As well as fully automatic 76mm (3in) guns, this small but effective frigate carried a new type of single-barrelled mortar automatic depth charge discharger capable of firing 15 rounds per minute to a range of 920m (1000yd); two types of 304mm (12in) torpedo tube; and a helicopter. It was broken up in 1981.

SPECIFICATIONS	
Type:	Italian frigate
Displacement:	1676 tonnes (1650 tons)
Dimensions:	94m x 11m x 3m (308ft 3in x 37ft 3in x 10ft 6in)
Machinery:	Twin screws, diesel motors
Main armament:	Three 76mm (3in) guns
Launched:	June 1960

Dido

Third of the *Leander* class frigates, *Dido* was one of eight to receive the GWS 40 Ikara ASW missile system, from 1978. It also carried a Wasp (later replaced by a Lynx) light helicopter. In 1983, *Dido* was sold to New Zealand to become HMNZS *Southland*. It was stricken in 1995 and broken up in India.

SPECIFICATIONS	
Type:	British frigate
Displacement:	2844 tonnes (2800 tons)
Dimensions:	113m x 12m x 5.4m (372ft x 41ft x 18ft)
Machinery:	Twin screws, turbines
Top speed:	30 knots
Main armament:	Two 114mm (4.5in) guns, one quad launcher for Seacat missiles
Complement:	263
Launched:	December 1961

TIMELINE
1960 1961

Doudart de Lagrée

Doudart de Lagrée, intended for escort work and colonial patrol, could carry a force of 80 commandos. In the late 1970s, one gun turret was replaced by four Exocet missile launchers. Developments in building techniques and equipment design meant that the class was superseded, and the ship was stricken in 1991.

SPECIFICATIONS	
Type:	French frigate
Displacement:	2235 tonnes (2200 tons)
Dimensions:	102m x 11.5m x 3.8m (334ft x 37ft 6in x 12ft 6in)
Machinery:	Twin screws, diesels
Top speed:	25 knots
Main armament:	Three 100mm (3.9in) guns, twin anti-aircraft weapons
Complement:	210
Launched:	April 1961

Galatea

Galatea was an improvement of the Type 12 Rothesay class frigate. The missile system was housed on the extended superstructure forward of the bridge. The class was updated in the 1970s and 1980s. After serving in the Far East and Persian Gulf, Galatea was decommissioned in 1987 and sunk as a target in 1988.

SPECIFICATIONS	
Type:	British frigate
Displacement:	2906 tonnes (2860 tons)
Dimensions:	113.4m x 12.5m x 4.5m (372ft x 41ft x 14ft 9in)
Machinery:	Twin screws, turbines
Top speed:	28 knots
Main armament:	One anti-submarine Ikara missile launcher
Launched:	May 1963

Yubari

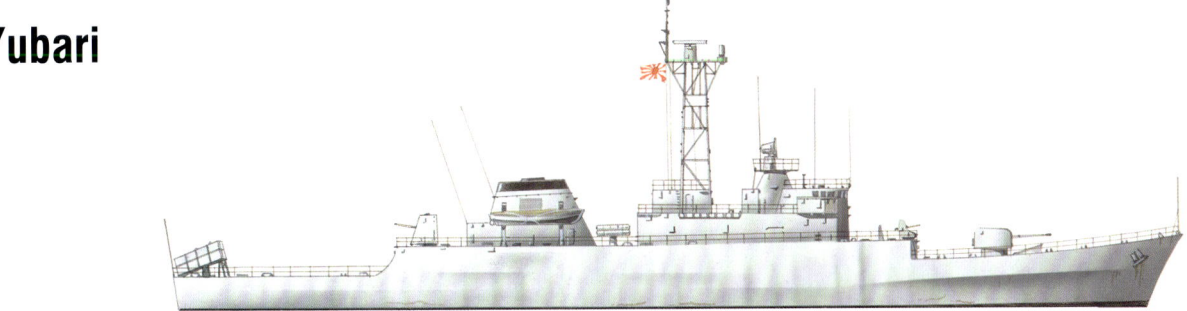

The Japanese Navy's Ishikari class was too small to carry the equipment, weapon stocks and electronic equipment vital for a late twentieth-century escort vessel, and the Yubari class is an enlarged version, with more fuel capacity and greater cruising range. A Phalanx CIWS was intended but not fitted.

SPECIFICATIONS	
Type:	Japanese frigate
Displacement:	1777 tonnes (1690 tons)
Dimensions:	91m x 10.8m x 3.5m (298ft 6in x 35ft 5in x 11ft 6in)
Machinery:	Twin screws, gas turbines and diesels
Top speed:	25 knots
Main armament:	Two quad Harpoon SSM launchers, one 76mm (3in) gun, one 375mm (14.75in) mortar, six 324mm (12.75in) torpedo tubes
Complement:	98
Launched:	1982

1963    1982

Frigates of the 1960s & '70s: Part 2

Distinctions between destroyers and frigates were increasingly blurred. Both types now routinely carried a helicopter, to aid in reconnoitring, anti-submarine attacks, and search-and-rescue missions. Many frigates were now equipped for anti-submarine warfare, with sonar detection and ASROC missile launchers.

Davidson

A *Garcia*-class destroyer escort, re-rated as a frigate from 1975, *Davidson* had gyro-driven stabilizers, enabling it to operate in heavy seas. A large box launcher held eight Asroc anti-submarine missiles. Twin torpedo tubes were later removed. Sold to Brazil in 1989 as *Paraibo*, it was decommissioned in 2002.

SPECIFICATIONS
Type:	US frigate
Displacement:	3454 tonnes (3400 tons)
Dimensions:	126m x 13.5m x 7m (414ft 8in x 44ft 3in x 24ft)
Machinery:	Single screw, turbines
Top speed:	27 knots
Main armament:	Two 127mm (5in) dual-purpose guns
Complement:	270
Launched:	1964

Carabiniere

Supplementary gas turbines gave *Carabiniere* extra speed when needed. Mast and funnel were an integrated structure. Anti-submarine weapons were a single semi-automatic depth charge mortar and six torpedo tubes, and two helicopters. Anti-missile defence was provided by SCLAR rockets. It was withdrawn in 2008.

SPECIFICATIONS
Type:	Italian frigate
Displacement:	2743 tonnes (2700 tons)
Dimensions:	113m x 13m x 4m (371ft x 43ft 6in x 12ft 7in)
Machinery:	Twin screws, diesels, gas turbines
Top speed:	20 knots (diesel), 28 knots (diesel and turbines)
Main armament:	Six 76mm (3in) guns
Launched:	September 1967

TIMELINE

1964   1967   1968

Alvand

Known as the *Saam* class until the lead ship's name changed to *Alvand* in 1985, these craft were a British Vosper Thornycroft design, in effect a scaled-down version of the Type 21 frigate. The Seacat launcher was later removed. One ship of this class was sunk by US aircraft in 1988, but *Alvand* remains in service.

SPECIFICATIONS	
Type:	Iranian frigate
Displacement:	1564.7 tonnes (1540 tons)
Dimensions:	94.5m x 10.5m x 3.5m (310ft x 34ft 5in x 11ft 6in)
Machinery:	Twin screws, gas turbines and diesels
Top speed:	40 knots
Main armament:	One SSM launcher, one Seacat SAM system, one 114mm (4.5in) gun, one Limbo Mk10 anti-submarine mortar
Complement:	135
Launched:	1968

Downes

A large class (46 in total), these frigates were criticized for limited manoeuvrability and low anti-submarine capability. The midships tower structure was intended to carry an advanced electrical array, but this was not developed; *Downes* carried standard sea and air search radars instead. It was sunk as as a target in 2003.

SPECIFICATIONS	
Type:	US frigate
Displacement:	4165 tonnes (4100 tons)
Dimensions:	126.6m x 14m x 7.5m (415ft 4in x 46ft 9in x 24ft 7in)
Machinery:	Single screw, turbines
Top speed:	28 + knots
Main armament:	One 127mm (5in) gun, one eight-tube Sea Sparrow missile launcher plus 20mm (0.79in) Phalanx CIWS
Launched:	December 1969

Chikugo

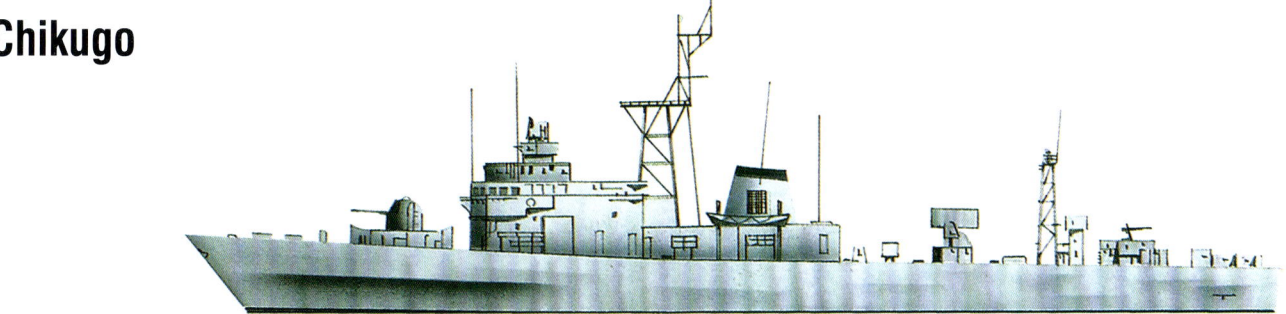

Chikugo was one of 11 units laid down in 1968. These frigates were the smallest ships to carry the anti-submarine weapon ASROC. Light anti-aircraft armament was installed, as the class were intended for inshore patrolling, protected by land-based fighter aircraft and missiles. *Chikugo* was decommissioned in 1996.

SPECIFICATIONS	
Type:	Japanese frigate
Displacement:	1493 tonnes (1470 tons)
Dimensions:	93m x 11m x 4m (305ft 5in x 35ft 5in x 11ft 6in)
Machinery:	Twin screws, diesels
Main armament:	Two 76mm (3in) guns
Launched:	1970

1969  1970

Frigates of the 1970s

While the functions of frigates and destroyers were tending to merge in deep-sea operations, a clear role for smaller versions of the frigate remained in coastal patrol work. Here there was a case for continued employment of the forward gun, in work that requires stop-and-search techniques and the interception of fast-moving craft.

Athabaskan

Athabaskan and three sister-ships were designed for anti-submarine warfare. Two hangars housed Sea King helicopters, giving the ships more flexibility than other anti-submarine vessels of the period. The armament now includes a SAM launcher, a Mk15 20mm (0.79in) Phalanx CIWS, and six 324mm (12.75in) torpedo tubes.

SPECIFICATIONS	
Type:	Canadian frigate
Displacement:	4267 tonnes (4200 tons)
Dimensions:	129.8m x 15.5m x 4.5m (426ft x 51ft x 15ft)
Machinery:	Twin screws, gas turbines
Top speed:	30 knots
Main armament:	One 127mm (5in) gun, one triple mortar
Complement:	285
Launched:	1970

Izumrud

Used by the KGB Border Guard on inshore patrol, *Izumrud* carried twin 533mm (21in) torpedo tubes, SAM SA-N-4 missiles and rocket launchers. Turbines developed 24,000hp; the diesels produced 16,000hp. Range was 1805km (950 miles) at 27 knots, 8550km (4500 miles) at 10 knots. Disposal details are unknown.

SPECIFICATIONS	
Type:	Soviet frigate
Displacement:	1219 tonnes (1200 tons)
Dimensions:	72m x 10m x 3.7m (236ft 3in x 32ft 10in x 12ft 2in)
Machinery:	Triple screws, one gas turbine, two diesel engines
Main armament:	Two 57mm (2.24in) guns, SAM missiles
Complement:	310
Launched:	1970

TIMELINE

1970   1972

FRIGATES OF THE 1970S

Najin

Equipped with Russian guns of World War II vintage and SS-N-2A missiles removed from redundant Soviet vessels, this was one of two similar ships built in North Korea in the early 1970s. Their later history is unclear. The names may have been changed and it is likely that both have been withdrawn from service.

SPECIFICATIONS	
Type:	North Korean frigate
Displacement:	1524 tonnes (1500 tons)
Dimensions:	100m x 9.9m x 2.7m (328ft x 32ft 6in x 8ft 10in)
Machinery:	Twin screws, diesels
Top speed:	33 knots
Main armament:	Two 100mm (3in) guns, three 533mm (21in) torpedo tubes
Complement:	180
Launched:	1972

Baptista de Andrade

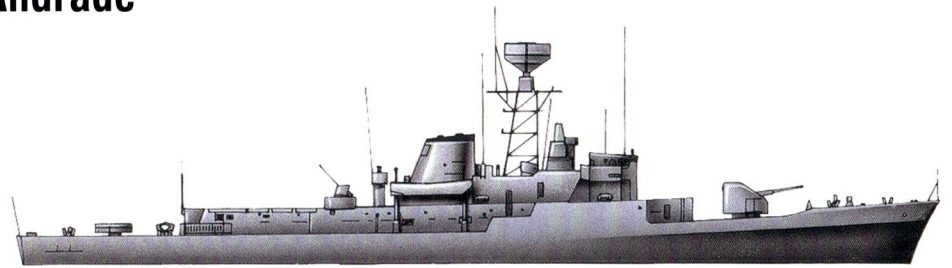

The four frigates of this class were ill-armed by contemporary standards, being deficient in anti-aircraft and anti-submarine defences. Portugal hoped to sell the quartet to Colombia in 1977, but the deal did not materialize. The *Andrades* are used only as coastal patrol vessels and are not deployed with NATO ships.

SPECIFICATIONS	
Type:	Portuguese frigate
Displacement:	1423.4 tonnes (1401 tons)
Dimensions:	84.6m x 10.3m x 3.3m (277ft 8in x 33ft 10in x 10ft 10in)
Machinery:	Twin screws, diesels
Top speed:	24.4 knots
Main armament:	One 100mm (3.9in) gun, six 324mm (12.75in) torpedo tubes
Complement:	113
Launched:	1973

Broadsword

Broadsword was the first general-purpose frigate designed to follow the *Leander* class. It was planned to build 26 units armed with missiles only, the main anti-submarine weapon being the Lynx helicopter. Later groups were fitted with extra weapons and sensors. *Broadsword* was sold to Brazil as *Greenhalgh* in 1995.

SPECIFICATIONS	
Type:	British frigate
Displacement:	4470 tonnes (4400 tons)
Dimensions:	131m x 15m x 4m (430ft 5in x 48ft 8in x 14ft)
Machinery:	Twin screws, gas turbines
Main armament:	Four M38 Exocet launchers, two 40mm (1.6in) guns
Complement:	407
Launched:	1975

Frigates of the 1970s & '80s

Most navies saw the frigate as the core-vessel of the surface fleet, capable of most tasks. Britain, for example, had 55 destroyers and 84 frigates in 1960. Twenty years later, it had 13 destroyers but still retained 53 frigates. The trend was very much towards fewer, but more versatile and comprehensively armed, warships.

Lupo

The *Lupo* class is an effective design, used by the Italian and other navies. Integral to the operating of *Lupo* was the SADOC automated combat control system, which enabled it to work with similarly fitted ships in an integrated group. Two helicopters are carried. *Lupo* was sold to Peru as *Palacios* in 2005.

SPECIFICATIONS	
Type:	Italian frigate
Displacement:	2540 tonnes (2500 tons)
Dimensions:	112.8m x 12m x 3.6m (370ft 2in x 39ft 4in x 12ft)
Machinery:	Twin screws, gas turbines plus diesels
Top speed:	35 knots
Main armament:	Eight Otomat SSM, one Sea Sparrow SAM launcher, one 127mm (5in) gun, six 324mm (12.75in) torpedo tubes
Complement:	185
Launched:	1976

Mourad Rais

Mourad Rais was Soviet-built and equipped, of the 'Koni' class of light frigate intended for export to nations allied with or friendly to Soviet Russia. Three were supplied to Algeria between 1978 and 1984, intended mainly for anti-submarine use. A Russian-managed modernization programme was under way in 2009–10.

SPECIFICATIONS	
Type:	Algerian frigate
Displacement:	1930.4 tonnes (1900 tons)
Dimensions:	95m x 12.8m x 4.2m (311ft 8in x 42ft x 13ft 9in)
Machinery:	Triple screws, diesels plus gas turbine
Top speed:	27 knots
Main armament:	One twin SA-N-4 SAM launcher, four 76mm (3in) guns, two RBU-6000 anti-submarine rocket launchers, two depth-charge racks
Complement:	110
Launched:	1978

TIMELINE

1976   1978   1980

FRIGATES OF THE 1970s & 80s

Godavari

SPECIFICATIONS	
Type:	Indian frigate
Displacement:	4064 tonnes (4000 tons)
Dimensions:	126.5m x 14.5m x 9m (415ft x 47ft 7in x 29ft 6in)
Machinery:	Twin screws, turbines
Top speed:	27 knots
Main armament:	Two 57mm (2.24in) guns, four SS-N-2C Styx missiles, SA-N-4 Gecko missiles
Complement:	313
Launched:	May 1980

Godavari is a modified British *Leander*-class frigate, but with Russian and Indian weapon systems as well. Two Sea King or Chetak helicopters can be housed in the hangar. The ship is an early example of the trend towards a smooth profile to minimize visibility on radar. Most of the armament is mounted on the foredeck.

Admiral Petre Barbuneanu

SPECIFICATIONS	
Type:	Romanian frigate
Displacement:	1463 tonnes (1440 tons)
Dimensions:	95.4m x 11.7m x 3m (303ft 1in x 38ft 4in x 9ft 8in)
Machinery:	Twin screws, diesels
Top speed:	24 knots
Main armament:	Four 76mm (3in) guns, two rocket launchers, four 533mm (21in) torpedo tubes
Complement:	95
Launched:	1981

Romania's navy operates in the Black Sea, and this vessel was intended for anti-submarine work and fitted with 16-tube RBU-2500 anti-submarine mortars. With three similar ships, it forms the 'Tetal' class. Sensor equipment included hull-mounted sonar, air/surface search radar and fire control radar.

Doyle

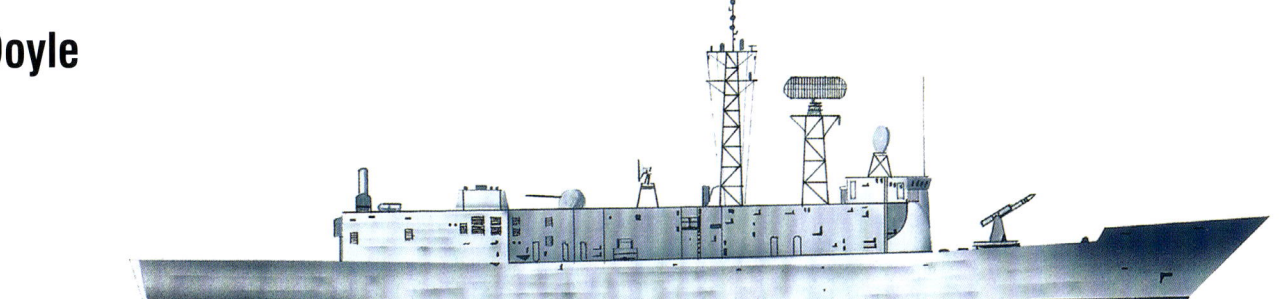

SPECIFICATIONS	
Type:	US frigate
Displacement:	3708 tonnes (3650 tons)
Dimensions:	135.6m x 14m x 7.5m (444ft 10in x 45ft x 24ft 7in)
Machinery:	Single screw, gas turbines
Top speed:	28 knots
Main armament:	One 76mm (3in) gun, Harpoon missile launcher
Launched:	May 1982

Doyle's profile is unlike that of previous frigates, an almost complete departure from earlier post-World War II designs. It reveals a warship reliant on missiles (not guns), built to have a minimal radar image, and with sophisticated radar and sonar detection systems. Its two helicopters carry anti-submarine weapons.

1981  1982

405

Frigates of the 1980s

Advances in missile technology made ships of frigate type less dependent on the traditional 127mm (5in) guns, and in the 1980s they often carried only a single main gun. Some even dispensed with that, though guns are useful for firing salutes. CIWS systems, based on 20mm (0.79in) or 30mm (1.18in) gun combinations, were introduced.

Jacob van Heemskerck

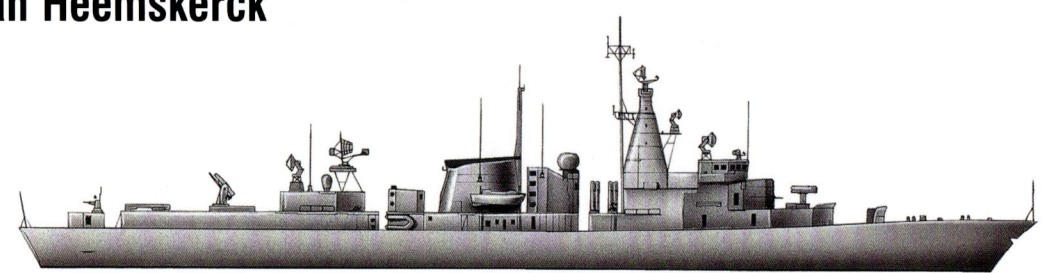

The Chilean *Admiral Latorre* since 2005, this Dutch L-class missile frigate was completed in 1986 for air defence. Dispensing with the usual forward gun, it has only two 20mm guns. Surface-search, air/surface search and fire control radars are fitted, along with hull-mounted sonar. Its sister-ship was also sold to Chile.

SPECIFICATIONS	
Type:	Dutch frigate
Displacement:	3810 tonnes (3750 tons)
Dimensions:	130.2m x 14.4m x 6m (427ft x 47ft x 20ft)
Machinery:	Twin screws, gas turbines
Top speed:	30 knots
Main armament:	Eight Harpoon SSM, Standard SM-1MR SAM, Sea Sparrrow octuple launcher, Goalkeeper 30mm (1.18in) CIWS, four 324mm (12.75in) torpedo tubes
Complement:	197
Launched:	1983

Al Madina

Designed and built in France, *Al Madina* is one of four frigates supplied to Saudi Arabia in the mid-1980s and intended for general-purpose use, but chiefly ship-to-ship fighting. They can operate a SA 365F Dauphin helicopter, though it is not regularly carried. A full suite of radar equipment is fitted, and hull-mounted sonar.

SPECIFICATIONS	
Type:	Saudi Arabian frigate
Displacement:	2651.8 tonnes (2610 tons)
Dimensions:	115m x 12.5m x 4.9m (377ft 3in x 41ft x 16ft)
Machinery:	Twin screws, diesels
Top speed:	30 knots
Main armament:	Eight Otomat Mk 2 SSM, one Crotale SAM launcher, one 100mm (3.9in) gun, four 440mm (17.33in) torpedo tubes
Complement:	179
Launched:	1983

TIMELINE

1983   1984

Kotor

Built on the general plan of the Soviet 'Koni'-class frigates, *Kotor* and *Pula* were larger, with various structural variations. Armament and mechanical equipment were Russian and Western, reflecting the non-aligned staus of Yugoslavia (as it then was). *Kotor* is now part of Montenegro's navy, but not operational.

SPECIFICATIONS

Type:	Yugoslav frigate
Displacement:	1930.4 tonnes (1900 tons)
Dimensions:	96.7m x 12.8m x 4.2m (317ft 3in x 42ft x 13ft 9in)
Machinery:	Triple screws, diesels plus gas turbine
Top speed:	27 knots
Main armament:	Four SS-N-2C SSM, one twin SA-N-4 SAM launcher, two 76mm (3in) guns, six 324mm (12.75in) torpedo tubes, two RBU-6000 anti-submarine rocket launchers
Complement:	110
Launched:	1984

Jianghu III

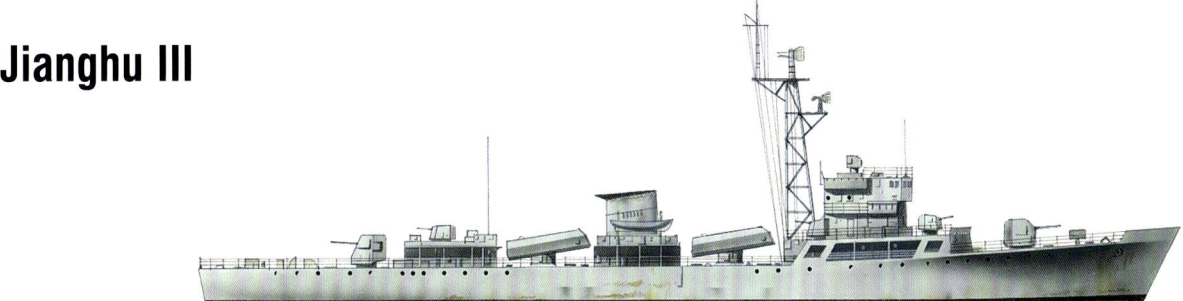

Following on from 'Jianghu I and II', this class incorporates more up-to-date anti-ship weaponry and a roomier superstructure. Low-powered, they are intended for coastal anti-submarine patrol work rather than deep-sea duty. Many boats have been sold to other navies, including Pakistan, Egypt and Thailand.

SPECIFICATIONS

Type:	Chinese frigate
Displacement:	1895 tonnes (1865 tons)
Dimensions:	103.2m x 10.83m x 3.1m (338ft 7in x 35ft 6in x 10ft 2in)
Machinery:	Twin screws, diesels
Top speed:	25.5 knots
Main armament:	Eight UJ-1 Eagle Strike SSM, four 100mm (3.9in) guns, two anti-submarine mortars, two depth-charge racks
Complement:	180
Launched:	1986

Inhaúma

Inhaúma was intended as the first of 16 light patrol ships, forming a major element in the Brazilian Navy. Designed in Germany, they carry a range of equipment, including a Swedish fire-control system, British combat data system, American engines, and French missile systems. The flight deck takes a Lynx helicopter.

SPECIFICATIONS

Type:	Brazilian frigate
Displacement:	2001.5 tonnes (1970 tons)
Dimensions:	95.8m x 11.4m x 5.5m (314ft 3in x 37ft 5in x 18ft)
Machinery:	Twin screws, diesels and gas turbine
Top speed:	27 knots
Main armament:	Four MM40 Exocet SSM, one 114mm (4.5in) gun, six 324mm (12.75in) torpedo tubes
Complement:	162
Launched:	1986

Frigates of the 1980s & '90s

Stealth technology is standard in modern frigates. Superstructures and hulls are designed to offer a minimal radar cross section, with low profiles and large areas of smooth walling. This reduces air resistance, improving speed and manoeuvrability. At the same time, air and surface search radar has widened its detection capacity.

Halifax

Canada's 'City' class are big frigates with a large funnel offset to port. *Halifax* was completed in June 1992. The armament carried is chiefly for surface and aerial defence, but the helicopter that each ship carries is normally equipped for anti-submarine action. A comprehensive ship-by-ship class refit began in 2007.

SPECIFICATIONS	
Type:	Canadian frigate
Displacement:	4826 tonnes (4750 tons)
Dimensions:	134.1m x 16.4m x 4.9m (440ft x 53ft 9in x 16ft 2in)
Machinery:	Twin screws, gas turbine and diesel
Top speed:	28 knots
Main armament:	Eight Harpoon SSM, two VLS for Sea Sparrow SAM, one 57mm (2.24in) gun, one Mk 15 Phalanx 20mm (0.79in) CIWS, four 324mm (12.75in) torpedo tubes
Complement:	225
Launched:	1988

Neustrashimyy

Introduced to improve the Soviet Navy's anti-submarine capacities, this class has four ships. The flat-flared hull, divided superstructure and funnel shape reduce and disperse the ship's radar returns. Its own sensors include towed sonar. In 2008–09, *Neustrashimyy* was deployed to the Somali coast to help combat piracy.

SPECIFICATIONS	
Type:	Soviet frigate
Displacement:	3556 tonnes (3500 tons)
Dimensions:	130m x 15.5m x 5.6m (426ft 6in x 50ft 11in x 18ft 5in)
Machinery:	Twin screws, gas turbines
Top speed:	32 knots
Main armament:	One SS-N-25 SSM launcher, one SA-N-9 SAM launcher, two CADS-N-1 gun/missile CIWS, one RBU-12000 anti-submarine rocket launcher
Complement:	210
Launched:	1988

TIMELINE   1988   1989

Thetis

SPECIFICATIONS	
Type:	Danish patrol ship
Displacement:	3556 tonnes (3500 tons)
Dimensions:	112.5m x 14.4m x 6m (369ft 1in x 47ft 3in x 19ft 8in)
Machinery:	Single screw, diesels
Top speed:	21.5 knots
Main armament:	One 76mm (3in) gun, depth-charge racks
Complement:	61
Launched:	1989

The *Thetis* class of four ships was intended to strengthen Denmark's fleet in the late twentieth and early twenty-first centuries. It was intended to mount Harpoon and Sea Sparrow missiles, but this plan was abandoned at the end of the Cold War and the class serve as lightly armed, outsize fishery protection ships.

Floréal

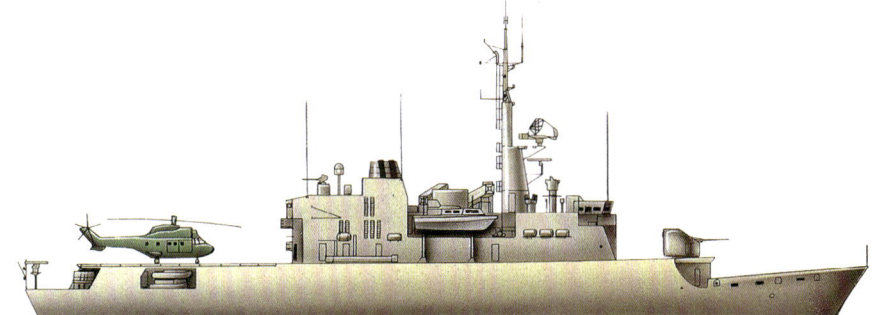

SPECIFICATIONS	
Type:	French frigate
Displacement:	2997 tonnes (2950 tons)
Dimensions:	93.5m x 14m x 4.3m (307ft x 46ft x 14ft)
Machinery:	Twin screws, diesels
Top speed:	20 knots
Main armament:	Two MM38 Exocet SSM, one 100mm (3.9in) gun
Complement:	80 plus 24 armed troops
Launched:	1990

Described as 'surveillance frigates' to protect France's 'exclusive economic zone', this class has been built using mercantile ship techniques and modular assembly, rather than the construction typical of warships. It can embark a helicopter up to Super Puma size. Two ships of the class were built for Morocco.

Nareusan

SPECIFICATIONS	
Type:	Thai frigate
Displacement:	3027.7 tonnes (2980 tons)
Dimensions:	120m x 13m x 3.81m (393ft 8in x 42ft 8in x 12ft 6in)
Machinery:	Twin screws, gas turbines and diesels
Top speed:	32 knots
Main armament:	Eight Harpoon SSM, one Mk 41 VLS for Sea Sparrow SAM, one 127mm (5in) gun, six 324mm (12.75in) torpedo tubes
Complement:	150
Launched:	1993

Built in China, to the design of the 'Jianghu' class, *Nareusan* was fitted out in Thailand, using Western machinery, weapons and electronic systems that make it more effective than the 'Jianghus'. Air/surface radar, navigation radar, fire-control radar and hull-mounted sonar are fitted, and it can embark a Lynx-type helicopter.

Minehunters & Sweepers

In the second half of the twentieth century, mines became more varied in type and in the methods of laying. New generation 'smart' mines could be programmed in various ways to sense and counter the activity of minehunters. Several American ships were damaged by mines during operations in the Persian Gulf region.

Edera

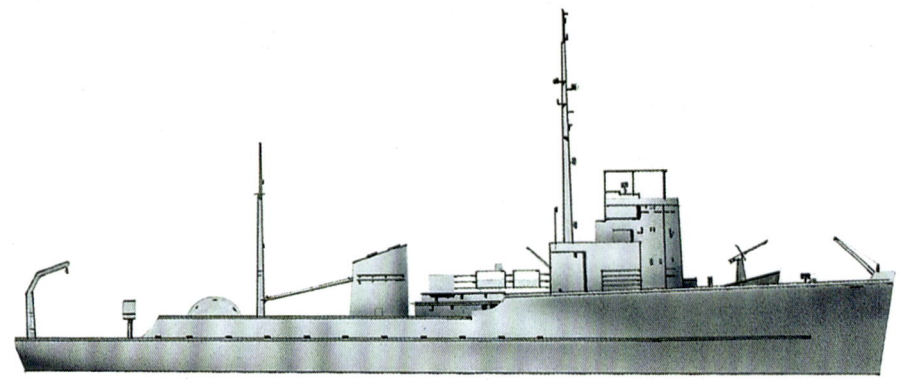

Edera was one of the 19-strong *Agave* class of minesweepers, of non-magnetic wood and alloy composite construction, and designed for inshore minesweeping duties. During the 1960s, the class was part of Italy's countermining force. Fuel carried was 25 tonnes (25 tons), enough for 4750km (2500 miles) at 10 knots.

SPECIFICATIONS	
Type:	Italian minesweeper
Displacement:	411 tonnes (405 tons)
Dimensions:	44m x 8m x 2.6m (144ft x 26ft 6in x 8ft 6in)
Machinery:	Twin screws, diesel engine
Top speed:	14 knots
Main armament:	Two 20mm (0.8in) anti-aircraft guns
Complement:	38
Launched:	1955

Bambú

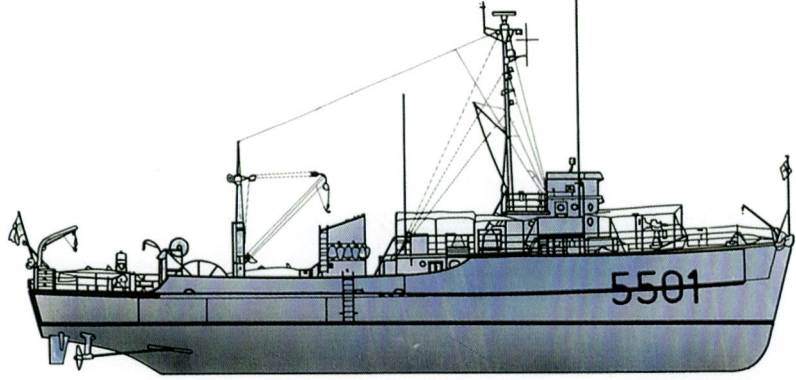

One of four converted American *Adjutant*-class minesweepers, Bambú entered service in 1956. Equipped with radar and sonar, it was wooden-hulled to defeat magnetic mines. Ships of this class were often assigned to United Nations' coastal patrol work in troubled regions. They were phased out in the 1990s.

SPECIFICATIONS	
Type:	Italian coastal minesweeper
Displacement:	375 tonnes (370 tons)
Dimensions:	44.1m x 8.5m x 2.6m (144ft 5in x 28ft x 8ft 6in)
Machinery:	Twin screws, diesel engine
Top speed:	13 knots
Main armament:	Two 20mm (0.79in) anti-aircraft guns
Complement:	31
Launched:	1956

TIMELINE

1955  1956  1957

Dromia

Dromia was one of 20 inshore minesweepers of the British 'Ham' class built in Italy between 1955 and 1957. They were designed to operate in shallow waters, rivers and estuaries, and when first built they were a new type of minesweeper, embodying many of the lessons learned during World War II and later hostilities.

SPECIFICATIONS	
Type:	Italian minesweeper
Displacement:	132 tonnes (130 tons)
Dimensions:	32m x 6.4m x 1.8m (106ft x 21ft x 6ft)
Machinery:	Twin screws, diesel engine
Top speed:	14 knots
Main armament:	One 20mm (0.79in) gun
Launched:	1957

Eridan

In the late 1970s, France, Belgium and the Netherlands combined to build 35 minehunters to a design that could be adapted by each nation. *Eridan* could be used for minehunting, minelaying, extended patrols, training, directing unmanned minesweeping craft, and as an HQ ship for diving operations.

SPECIFICATIONS	
Type:	French minehunter
Displacement:	552 tonnes (544 tons)
Dimensions:	49m x 8.9m x 2.5m (161ft x 29ft 2in x 8ft 2in)
Machinery:	Single screw, diesel engine
Top speed:	15 knots
Main armament:	One 20mm (0.79in) gun
Launched:	February 1979

Aster

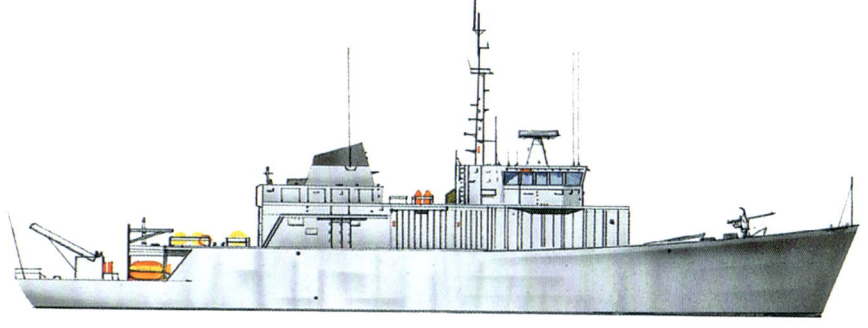

Aster is a Belgian example of the Tripartite minehunter design designed for NATO service. France, Belgium and the Netherlands each built its own hulls, which were fitted out in Belgium with French electronics and Dutch machinery. All vessels carry full nuclear, biological and chemical (NBC) protection and minesweeping equipment, and can be used as patrol and surveillance craft.

SPECIFICATIONS	
Type:	Belgian minehunter
Displacement:	605 tonnes (595 tons)
Dimensions:	51.5m x 8.9m x 2.5m (169ft x 29ft x 8ft)
Machinery:	Single screw, diesel engine, two manoeuvring propellers and one bowthruster
Top speed:	15 knots
Main armament:	One 20mm (0.79in) anti-aircraft gun
Launched:	1981

1979    1981

Specialized Naval Ships

As in previous decades, fleet support required a variety of dedicated craft. The main difference was that the complexity of modern operating systems made it imperative to build ships for purpose rather than to adapt existing ones, though there were exceptions. Other specialisms included assault and landing ships.

Filicudi

Filicudi and its sister *Alicudi* were based on a standard NATO design, and could lay nets of various depths across harbour entrances. A large, open deck for handling the nets is situated in the low bow section. A boom attached to the foremast controls the lifting and lowering of nets.

SPECIFICATIONS	
Type:	Italian net layer
Displacement:	847 tonnes (834 tons)
Dimensions:	50m x 10m x 3.2m (165ft 4in x 33ft 6in x 10ft 6in)
Machinery:	Twin screws, diesel-electric motors
Top speed:	12 knots
Main armament:	One 40mm (1.57in) gun
Launched:	September 1954

Caorle

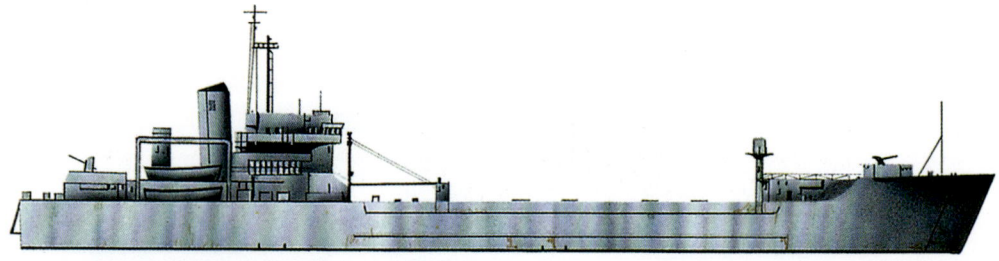

In 1972, USS *York County* was sold to Italy, and renamed *Caorle*. It could carry up to 575 fully equipped assault troops, or a mixture of troops, tanks and other vehicles. A flat bottom and shallow draught allowed the bow to be grounded on a shallow beach for unloading. It was scrapped at Naples in 1999.

SPECIFICATIONS	
Type:	Italian landing ship
Displacement:	8128 tonnes (8000 tons)
Dimensions:	135m x 19m x 5m (444ft x 62ft x 16ft 6in)
Machinery:	Twin screws, diesel engines
Top speed:	17.5 knots
Main armament:	Six 76mm (3in) guns
Launched:	March 1957

TIMELINE

Chazhma

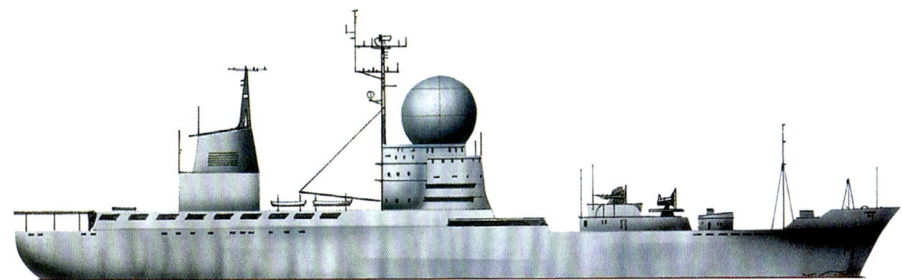

A 7381-tonne (7265-ton) bulk ore carrier of the *Dshankoy* class, *Chazhma* was converted into a missile range ship in 1963 to serve in the Pacific. A 'Ship Globe' radar was mounted in the dome above the bridge. A helicopter platform and hangar, built into the aft superstructure, let it operate one 'Hormone' helicopter.

SPECIFICATIONS

Type:	Soviet missile range ship
Displacement:	13,716 tonnes (13,500 tons)
Dimensions:	140m x 18m x 8m (458ft x 59ft x 26ft)
Machinery:	Twin screws, diesel engines
Top speed:	15 knots
Launched:	1959

Deutschland

The first West German ship to exceed the post-war limit of 3048 tonnes (3000 tons), *Deutschland* carried a range of armaments for training purposes, including 100mm (3.9in) and 40mm (1.57in) guns, depth-charge launchers, mines and torpedoes. It was towed to India for scrapping in 1994.

SPECIFICATIONS

Type:	German training ship
Displacement:	5588 tonnes (5500 tons)
Dimensions:	145m x 18m x 4.5m (475ft 9in x 59ft x 14ft 9in)
Machinery:	Triple screws, diesel motors, turbines
Top speed:	22 knots
Main armament:	Four 100mm (3.9in) guns
Complement:	500 including 267 cadets
Launched:	1960

Alligator class

The Project 1171 Nosorog large landing craft were designated 'Alligator' by NATO. Sixteen were built, with bow and stern ramps. All carried some weapons and at least one crane. Up to 30 armoured personnel carriers and their troops could be transported. Some were used in the South Ossetia War of 2008.

SPECIFICATIONS

Type:	Soviet landing ship
Displacement:	4775 tonnes (4700 tons) full load
Dimensions:	112.8m x 15.3m x 4.4m (370ft 6in x 50ft 2in x 14ft 5in)
Machinery:	Twin screws, diesels
Top speed:	18 knots
Main armament:	Two/three SA-N-5 SAM launchers, one 122mm (4.8in) rocket launcher
Complement:	75 plus 300 combat troops
Launched:	1964

1960    1964

Post-War Conventional Submarines: Part 1

The German and Japanese submarine fleets had been eliminated by the end of World War II, but the Russians, Americans and British still had substantial numbers. Post-war building programmes began only in the 1950s, stimulated by the Cold War.

Whiskey

About 240 of these attack submarines were built between 1951 and 1958. Four units were converted to early warning boats between 1959 and 1963, but from 1963 the long-range Bear aircraft reduced their strategic importance in some areas. By the 1980s, these submarines had disappeared from the effective list.

SPECIFICATIONS

Type:	Soviet submarine
Displacement:	1066 tonnes (1050 tons) [surface], 1371 tonnes (1350 tons) [submerged]
Dimensions:	76m x 6.5m x 5m (249ft 4in x 21ft 4in x 16ft)
Machinery:	Twin screws, diesel engines [surface], electric motors [submerged]
Top speed:	18 knots [surface], 14 knots [submerged]
Main armament:	Two 406mm (16in), four 533mm (21in) torpedo tubes
Launched:	1956

Golf I

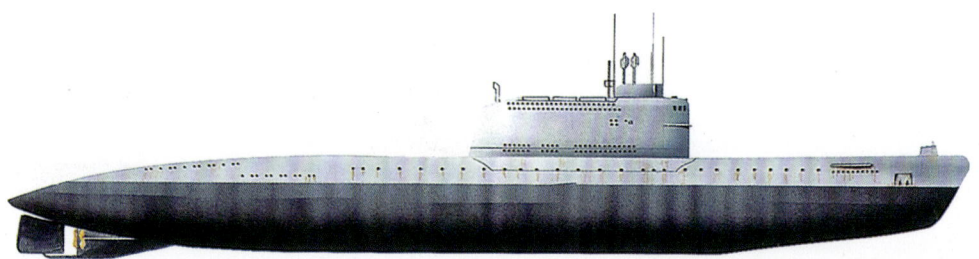

Twenty-three Golf I-class submarines were completed between 1958 and 1962, entering service at a rate of six to seven per year. The ballistic missiles were housed vertically in the rear section of the extended fin. Many boats in the class were modified after commissioning. All were withdrawn by 1990.

SPECIFICATIONS

Type:	Russian missile submarine
Displacement:	2336 tonnes (2300 tons) [surface], 2743 tonnes [2700 tons] [submerged]
Dimensions:	100m x 8.5m x 6.6m (328ft x 27ft 11 in x 21ft 8in)
Machinery:	Triple screws, diesel engine (surface), electric motors [submerged]
Top speed:	17 knots (surface), 12 knots (submerged)
Main armament:	Three SS-N-4 ballistic missiles, ten 533mm (21in) torpedo tubes
Launched:	1957

TIMELINE

1956 1957

Grayback

During construction, *Grayback* was altered to carry the first naval cruise missile, Regulus, and did so until 1964. Recommissioned in 1968 as an amphibious transport submarine for undercover missions, it carried 67 marines and their SEAL swimmer delivery vehicles. *Grayback* was sunk as a target in April 1986.

SPECIFICATIONS	
Type:	US submarine
Displacement:	2712 tonnes (2670 tons) [surface], 3708 tonnes (3650 tons) [submerged]
Dimensions:	102m x 9m (335ft x 30ft)
Machinery:	Twin screws, diesel engines [surface], electric motors [submerged]
Main armament:	Four Regulus missiles, eight 533mm (21in) torpedo tubes
Launched:	1957

Daphné

Eleven submarines of this class were launched between 1964 and 1970. The double hull had a deep keel to improve stability. They had good manoeuvrability, low noise, a small crew and were easy to maintain; several navies bought them. *Daphné* was decommissioned in 1989; the rest of the class followed by 1996.

SPECIFICATIONS	
Type:	French submarine
Displacement:	884 tonnes (870 tons) [surface], 1062 tonnes (1045 tons) [submerged]
Dimensions:	58m x 7m x 4.6m (189ft 8in x 22ft 4in x 15ft)
Machinery:	Twin screws, diesel [surface], electric drive [submerged]
Top speed:	13.5 knots [surface], 16 knots [submerged]
Main armament:	Twelve 552mm (21.7in) torpedo tubes
Launched:	June 1959

Dolfijn

Dolfijn and three sister boats were built to a unique triple-hulled design – three cylinders arranged in a triangular shape. The upper cylinder housed the crew, navigational equipment and armament; the lower cylinders, the powerplant. Maximum diving depth was almost 304m (1000ft). *Dolfijn* was broken up in 1985.

SPECIFICATIONS	
Type:	Dutch submarine
Displacement:	1518 tonnes (1494 tons) [surface], 1855 tonnes (1820 tons) [submerged]
Dimensions:	80m x 8m x 4.8m (260ft 10in x 25ft 9in x15ft 9in)
Machinery:	Twin screws, diesel [surface], electric motors [submerged]
Main armament:	Eight 533mm (21in) torpedo tubes
Launched:	May 1959

Post-War Conventional Submarines: Part 2

In the 1960s, Japan, Italy and West Germany resumed the building of submarines, placing their forces within the NATO and US-Japanese defence agreements. Several non-aligned nations, notably Sweden, also produced effective submarine types.

Enrico Toti

The four vessels of this class were the first submarines to be built in Italy since World War II. The design was revised several times before the hunter/killer model, for use in shallow and confined waters, was finally approved. Withdrawn in 1992, *Enrico Toti* is now a museum vessel in Milan.

SPECIFICATIONS	
Type:	Italian submarine
Displacement:	532 tonnes (524 tons) [surface], 591 tonnes (582 tons) [submerged]
Dimensions:	46.2m x 4.7m x 4m (151ft 7in x 15ft 5in x 13ft)
Machinery:	Single screw, diesel engines [surface], electric motors [submerged]
Top speed:	14 knots surfaced, 15 knots submerged
Main armament:	Four 533mm (21in) torpedo tubes
Launched:	March 1967

Harushio

This was Japan's first post-World War II fleet submarine class, named for Oshio, though it had a different bow shape to its successors. The primary role, in Japan's Maritime Defence Force, was to act as targets in anti-submarine training exercises. *Harushio* was withdrawn and scrapped in 1984.

SPECIFICATIONS	
Type:	Japanese submarine
Displacement:	1676.4 tonnes (1650 tons) surfaced, 2184.4 tonnes (2150 tons) [submerged]
Dimensions:	88m x 8.2m x 4.9m (288ft 8in x 26ft 11in x 16ft)
Machinery:	Twin screws, diesel [surface], electric motors [submerged]
Top speed:	18 knots surfaced, 14 knots submerged
Main armament:	Eight 533mm (21in) torpedo tubes
Complement:	80
Launched:	1967

TIMELINE

1967  1968

POST-WAR CONVENTIONAL SUBMARINES: PART 2

U-12

U-12 was one of the first class of German submarines built after World War II. They were a successful type, with over 40 boats serving in foreign navies. The hull was of a non-magnetic steel alloy. Diesel engines developed 2300hp, and the single electric motor developed 1500hp. U-12 was decommissioned in 2005.

SPECIFICATIONS
Type:	German submarine
Displacement:	425 tonnes (419 tons) [surface], 457 tonnes (450 tons) [submerged]
Dimensions:	43.9m x 4.6m x 4.3m (144ft x 15ft x 14ft)
Machinery:	Single screw, diesel engine [surface], electric motors [submerged]
Top speed:	10 knots [surface], 17 knots [submerged]
Main armament:	Eight 533mm (21in) torpedo tubes
Launched:	1968

Näcken

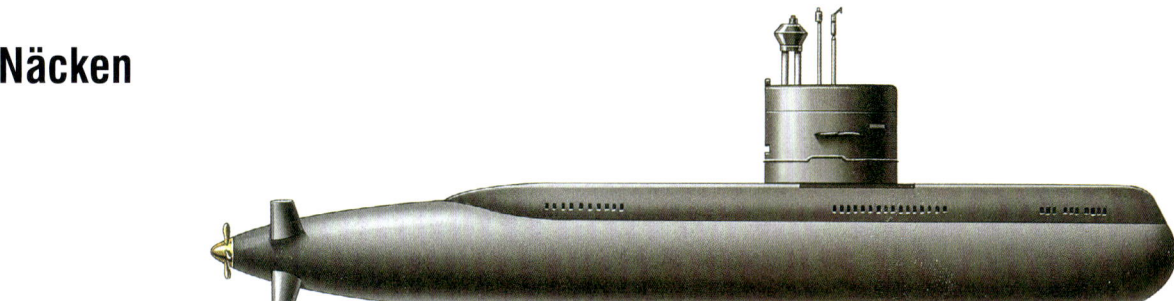

Näcken and its two sister craft were re-engined in 1987–88 with a closed-air independent propulsion system (AIP), which let them operate underwater for up to 14 days without surfacing. Withdrawn in the 1990s, they have been replaced by the Swedish-built Vastergötland class, which uses the same technology.

SPECIFICATIONS
Type:	Swedish submarine
Displacement:	995.7 tonnes (980 tons) [surface], 1169 tonnes (1150 tons) [submerged]
Dimensions:	49.5m x 5.7m x 5.5m (162ft 5in x 18ft 8in x 18ft)
Machinery:	Single screw, diesel engine [surface], electric motors [submerged]
Top speed:	20 knots [surface], 25 knots [submerged]
Main armament:	Six 533mm (21in), two 400mm (15.75in) torpedo tubes
Complement:	19
Launched:	1978

Kilo

The 'Kilo' class were the first Soviet boats to use a modern teardrop hull form, which gives a good underwater speed-power ratio. Double-hulled, they are fast and highly manoeuvrable, well suited to operations in restricted waters. About 17 continue in service with Russia, and 33 in other fleets.

SPECIFICATIONS
Type:	Russian submarine
Displacement:	2336 tonnes (2300 tons) [surface], 2946 tonnes (2900 tons) [submerged]
Dimensions:	73m x 10m x 6.5m (239ft 6in x 32ft 10in x 21ft 4in)
Machinery:	Single screw, diesel engine [surface], electric motor [submerged]
Top speed:	12 knots [surface], 18 knots [submerged]
Main armament:	Six 533mm (21in) torpedo tubes
Launched:	1981

Post-War Conventional Submarines: Part 3

While by the 1980s the nuclear submarine had taken up the major strategic role, its deployment was confined to five navies. A significant tactical part was still played by conventionally powered boats, particularly for coastal patrol.

Galerna

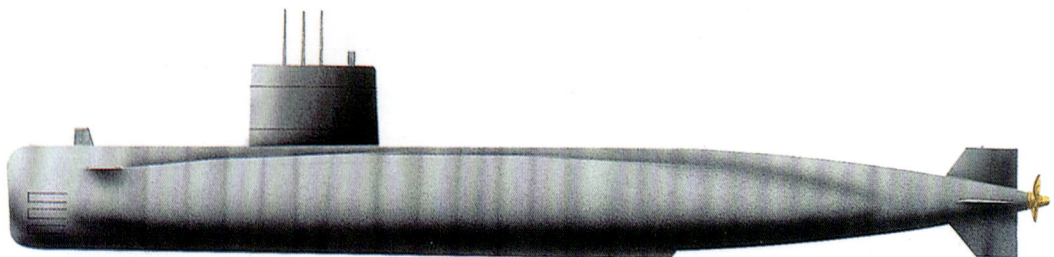

One of four medium-range submarines built in Spain to the design of the French *Agosta* class, *Galerna* marked a major step forward in Spanish submarine technology. The original weapon-stock was 16 reload torpedoes or nine torpedoes and 19 mines. State-of-the-art sonar kit is carried.

SPECIFICATIONS	
Type:	Spanish submarine
Displacement:	1473 tonnes (1450 tons) [surface], 1753 tonnes (1725 tons) [submerged]
Dimensions:	67.6m x 6.8m x 5.4m (221ft 9in x 22ft 4in x 17ft 9in)
Machinery:	Single screw, diesel engine [surface], electric motor [submerged]
Top speed:	12 knots surfaced, 20 knots submerged
Main armament:	Four 551mm (21.7in) torpedo tubes
Launched:	December 1981

Walrus

Walrus was not completed until 1991, due to a fire, and *Zeeleeuw*, commissioned in 1989, became class leader. The use of high-tensile steel gave a diving depth of 300m (985ft). New Gipsy fire control and electronic command systems reduced crew numbers to 49. A class refit in 2007 extended the service life of these boats.

SPECIFICATIONS	
Type:	Dutch submarine
Displacement:	2490 tonnes (2450 tons) [surface], 2845 tonnes (2800 tons) [submerged]
Dimensions:	67.5m x 8.4m x 6.6m (222ft x 27ft 7in x 21ft 8in)
Machinery:	Single screw, diesel engines [surface], electric motors [submerged]
Top speed:	13 knots [surface], 20 knots [submerged]
Main armament:	Four 533mm (21in) torpedo tubes
Launched:	October 1985

TIMELINE

1981   1985   1986

Hai Lung

SPECIFICATIONS	
Type:	Taiwanese submarine
Displacement:	2414 tonnes (2376 tons) [surface], 2702 tonnes (2660 tons) [submerged]
Dimensions:	66.9m x 8.4m x 6.7m (219ft 5in x 27ft 6in x 22ft)
Machinery:	Single screw, diesel engines [surface], electric motors [submerged]
Top speed:	11 knots [surface], 20 knots [submerged]
Main armament:	Six 533mm (21in) torpedo tubes
Complement:	67
Launched:	October 1986

Hai Lung is a modified Dutch *Zwaardvis*-class, probably the most efficient design of the 1970s. They are quiet boats, with all machinery mounted on anti-vibration mountings. They carried up to 28 Tigerfish acoustic homing torpedoes, and in 2005 were upgraded to carry UGM-84 Harpoon anti-ship missiles.

Upholder

SPECIFICATIONS	
Type:	British submarine
Displacement:	2220 tonnes (2185 tons) [surface], 2494 tonnes (2455 tons) [submerged]
Dimensions:	70.3m x 7.6m x 5.5m (230ft 8in x 25ft x 18ft)
Machinery:	Single screw, diesel engine [surface], electric motors [submerged]
Top speed:	12 knots [surface], 20 knots [submerged]
Main armament:	Six 533mm (21in) torpedo tubes
Launched:	December 1986

Designed as a new Royal Navy conventionally-powered patrol submarine, all four in the Upholder class were transferred to Canada in 1998. The teardrop-shaped hull is of high tensile steel, enabling dives to 200m (656ft). The class has had technical problems, and three were undergoing refits in 2009–10.

Collins

SPECIFICATIONS	
Type:	Australian submarine
Displacement:	2220 tonnes (3051 tons) [surface], 2494 tonnes (3353 tons) [submerged]
Dimensions:	77.5m x 7.8m x 7m (254ft x 25ft 7in x 23ft)
Machinery:	Single screw, diesel-electric engines [surface], electric motor [submerged]
Top speed:	10 knots [surface], 20 knots [submerged]
Main armament:	Six 533mm (21in) torpedo tubes
Launched:	1993

Swedish-designed and Australian-built, this attack submarine was marred by mechanical and electronic problems when introduced in 1995 and the combat data management system was replaced. Three of the six boats in the *Collins* class are active, the others in reserve. Plans for replacement were put in hand in 2007.

 1993

Nuclear Submarines: Part 1

Until 1954, the submarine was essentially a surface ship that could submerge. Experiments had been going on with power sources that would allow for indefinite submergence. A hydrogen peroxide motor was tried in the 1940s, but nuclear reaction offered the answer: in 1954 USS *Nautilus* became the first 'true' submarine.

Nautilus

The world's first nuclear-powered submarine, *Nautilus* was of conventional design. Early trials established many records, including nearly 2250km (1400 miles) submerged in 90 hours at 20 knots, and a passage beneath the ice over the North Pole. Stricken in 1980, *Nautilus* is preserved at Groton, Connecticut.

SPECIFICATIONS	
Type:	US submarine
Displacement:	4157 tonnes (4091 tons) [surface], 4104 tonnes (4040) [submerged]
Dimensions:	98.7m x 8.4m x 6.6m (323ft 9in x 27ft 8in x 21ft 9in)
Machinery:	Twin screws, nuclear reactor, turbines
Top speed:	23 knots [submerged]
Main armament:	Six 533mm (21in) torpedo tubes
Complement:	105
Launched:	January 1954

Skipjack

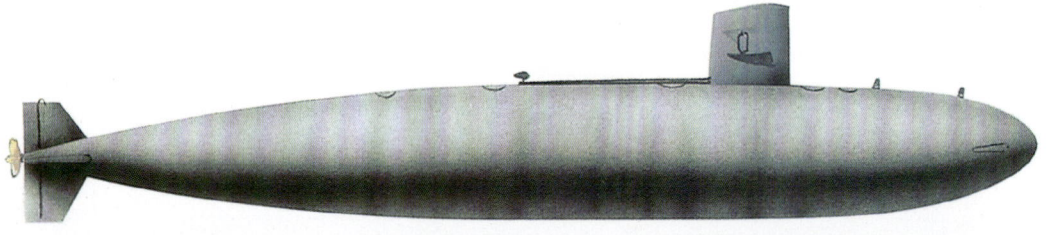

With a teardrop-form hull, and diving planes on the fin, *Skipjack* was fast and manoeuvrable. No stern tubes were fitted: the aft hull shape tapered sharply. It introduced the S5W fast-attack propulsion plant used in all subsequent attack and ballistic submarines until the *Los Angeles*. The class was withdrawn by the 1990s.

SPECIFICATIONS	
Type:	US submarine
Displacement:	3124 tonnes (3075 tons) [surface], 3570 tonnes (3513 tons) [submerged]
Dimensions:	76.7m x 9.6m x 8.9m (251ft 8in x 31ft 6in x 29ft 2in)
Machinery:	Single screw, nuclear reactor, turbines
Top speed:	18 knots surfaced, 30 knots submerged
Main armament:	Six 533mm (21in) torpedo tubes
Complement:	93
Launched:	May 1958

TIMELINE

1954   1958   1959

NUCLEAR SUBMARINES: PART 1

USS Halibut

First deployed with cruise missiles, *Halibut* was used for secret intelligence work, often involving the retrieval of objects of military interest from the sea-bed. Midget submarines, carried in the former Regulus missile space, were used for this work. *Halibut* was decommissioned in 1976, and broken up in 1994.

SPECIFICATIONS

Type:	US submarine
Displacement:	(3846 tons) surfaced, (4895 tons) submerged
Dimensions:	106.7m x 9m x 6.3m (350ft x 29ft 6in x 20ft 9in)
Machinery:	Twin screws, one nuclear reactor
Top speed:	15 knots [surfaced], 15.5 knots [submerged]
Main armament:	Five SSM-N-8 Regulus 1 or two SSM-N-9 Regulus II, six 533mm (21in) torpedo tubes
Complement:	111
Launched:	1959

George Washington

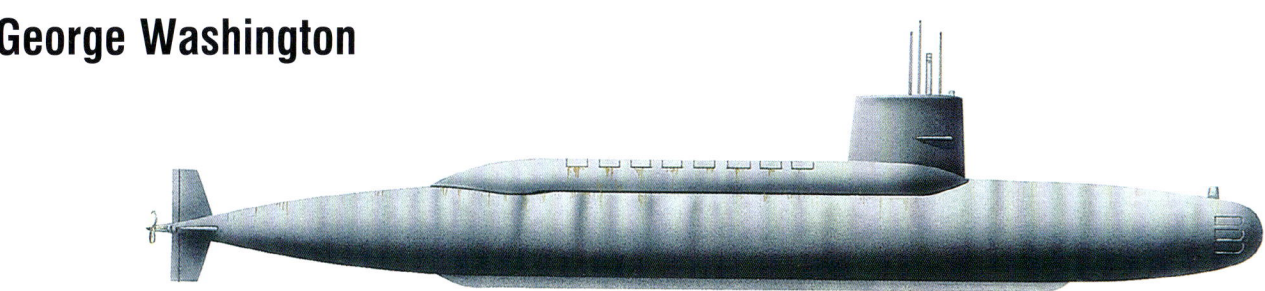

In 1955, the Soviet Union began modifying submarines to carry nuclear-tipped ballistic missiles. At the time, the United States was developing the Polaris A1 missile, and the submarine *Scorpion* was adapted to carry it. Renamed *George Washington*, it was 'de-missiled' in the 1980s, and decommissioned in 1986.

SPECIFICATIONS

Type:	US ballistic missile submarine
Displacement:	6115 tonnes (6019 tons) [surface], 6998 tonnes (6888 tons) [submerged]
Dimensions:	116.3m x 10m x 8.8m (381ft 7in x 33ft x 28ft 10in)
Machinery:	Single screw, one pressurized water-cooled reactor, turbines
Top speed:	20 knots [surface], 30.5 knots [submerged]
Main armament:	Sixteen Polaris missiles, six 533mm (21in) torpedo tubes
Launched:	June 1959

Dreadnought

Britain's first nuclear-powered submarine, *Dreadnought* was a detect-and-destroy vessel. The form of the hull was based on the shape of a whale. Its power-plant was an American S5W reactor as fitted to the US *Skipjack* submarines. Laid up in 1982, *Dreadnought* was towed to Rosyth for disposal in the following year.

SPECIFICATIONS

Type:	British submarine
Displacement:	3556 tonnes (3500 tons) [surface], 4064 tonnes (4000 tons) [submerged]
Dimensions:	81m x 9.8m x 8m (265ft 9in x 32ft 3in x 26ft)
Machinery:	Single screw, nuclear reactor, steam turbines
Top speed:	20 knots [surface], 30 knots [submerged]
Main armament:	Six 533mm (21in) torpedo tubes
Complement:	88
Launched:	October 1960

 1960

Resolution

Lead boat of a class of four, designed to carry Britain's nuclear deterrent armament of US Polaris missiles, *Resolution* went on its first patrol in 1968. It was intended that one of the four submarines would always be on active service. With the deployment of the *Vanguard* class with Trident missiles, *Resolution* was retired in 1994.

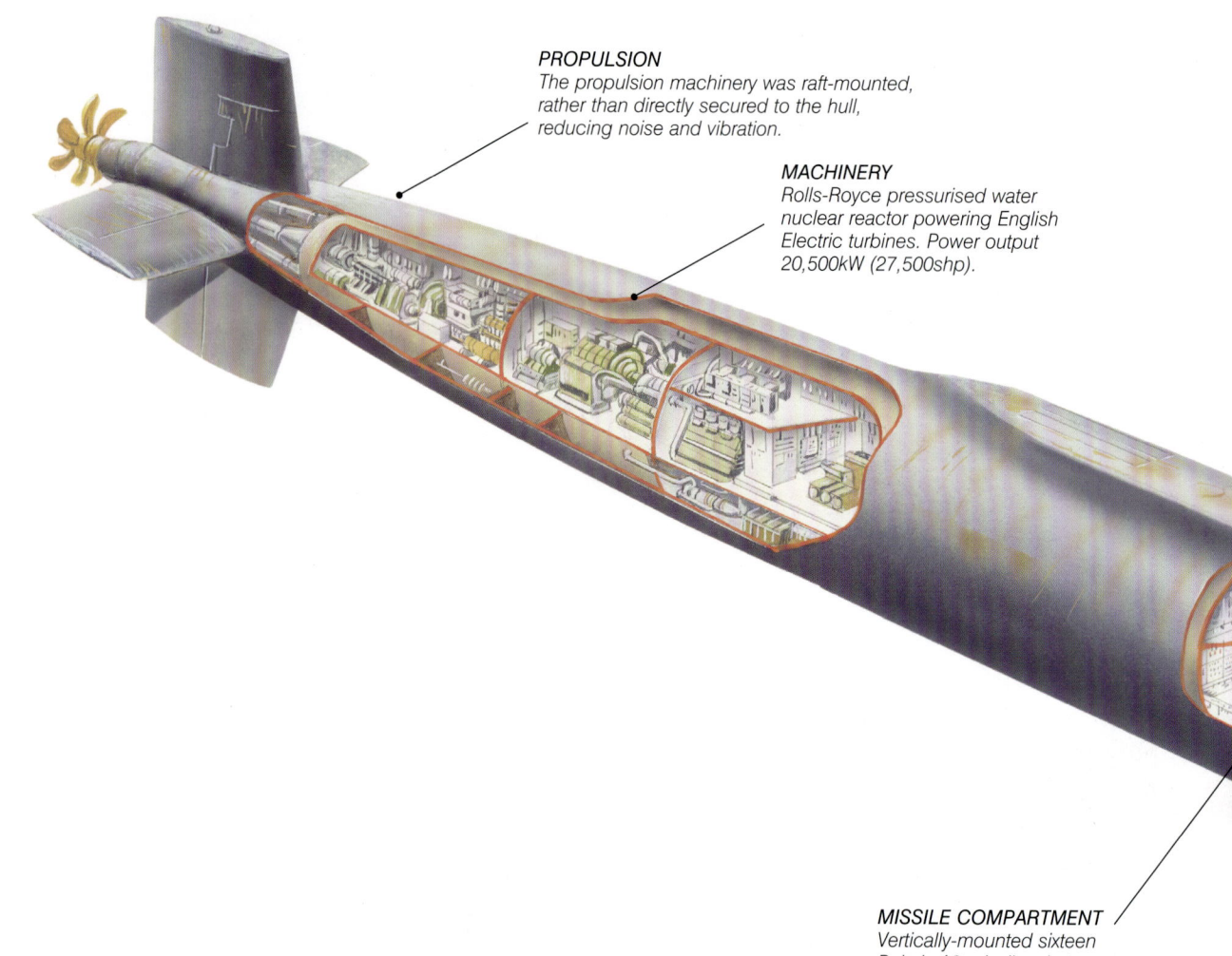

PROPULSION
The propulsion machinery was raft-mounted, rather than directly secured to the hull, reducing noise and vibration.

MACHINERY
Rolls-Royce pressurised water nuclear reactor powering English Electric turbines. Power output 20,500kW (27,500shp).

MISSILE COMPARTMENT
Vertically-mounted sixteen Polaris A3 missiles, in two rows of eight. These had a range of 4,631km (2,500 nautical miles) and multiple nuclear warheads.

Resolution

The *Resolution* class submarines were armed with the American Polaris SLBM and they took over the British deterrent role from the RAF in 1968. Their characteristics were very similar to the American *Lafayettes*.

SPECIFICATIONS

Type:	British submarine
Displacement:	7620 tonnes (7500 tons) [surface], 8636 tonnes (8500 tons) [submerged]
Dimensions:	129.5m x 10m x 9.1m (425ft x 33ft x 30ft)
Machinery:	Single screw, pressurized water reactor, geared steam turbines
Top speed:	20 knots [surfaced], 25 knots [submerged]
Main armament:	Sixteen UGM-27C Polaris A-3 SLBM, six 533mm (21in) torpedo tubes
Complement:	143
Launched:	September 1966

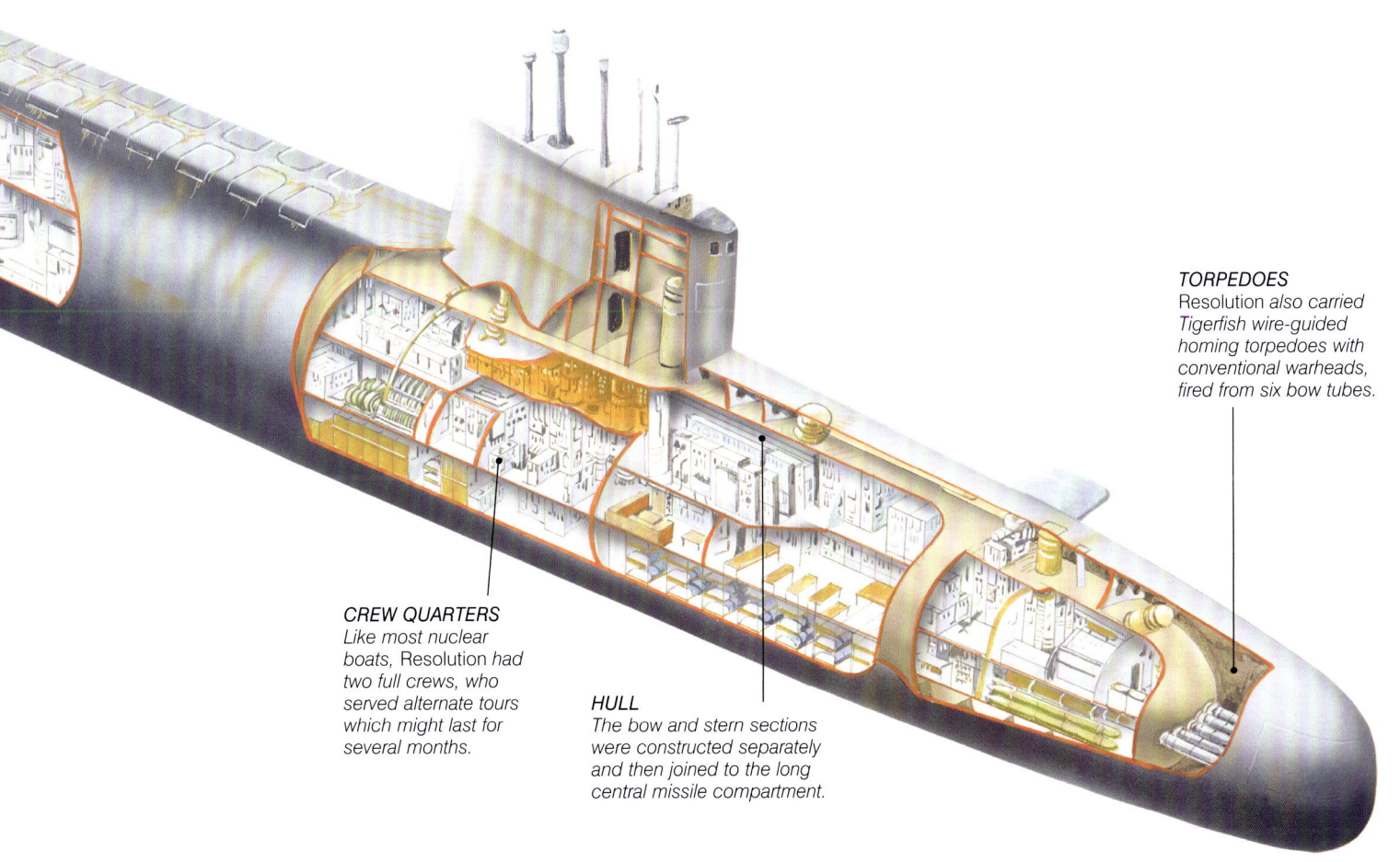

CONVERSION
The four Resolution class boats were adapted in the mid-1980s to carry new Polaris AT-K missiles with British Chevaline multiple re-entry warheads.

TORPEDOES
Resolution also carried Tigerfish wire-guided homing torpedoes with conventional warheads, fired from six bow tubes.

CREW QUARTERS
Like most nuclear boats, Resolution had two full crews, who served alternate tours which might last for several months.

HULL
The bow and stern sections were constructed separately and then joined to the long central missile compartment.

Nuclear Submarines: Part 2

Two US nuclear submarines, *Thresher* (1963) and *Scorpion* (1968), were lost in the 1960s. The loss of *Thresher* prompted the introduction of the Deep Sea Rescue Vessel (DSRV). Nuclear boats first carried conventional torpedoes but developments in rocket technology led to the introduction of the missile-firing submarine.

Daniel Boone

Daniel Boone was one of a sub-class of the *Lafayette* strategic missile nuclear submarines. Completed in April 1964, and fitted to carry UGM 73A Poseidon missiles, it was one of 12 boats adapted for the more reliable Trident type in 1980. With the advent of the *Ohio* class, *Daniel Boone* was retired in 1994.

SPECIFICATIONS

Type:	US submarine
Displacement:	7366 tonnes (7250 tons) [surface], 8382 tonnes (8250 tons) [submerged]
Dimensions:	130m x 10m x 10m (425ft x 33ft x 33ft)
Machinery:	Single screw, single water-cooled nuclear reactor, turbines
Top speed:	20 knots [surface], 35 knots [submerged]
Main armament:	Sixteen Polaris missiles, four 533mm (21in) torpedo tubes
Launched:	June 1962

Warspite

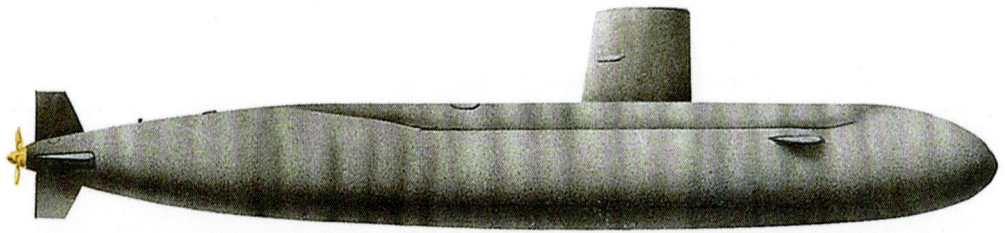

Using old battleship names confirmed nuclear submarines as the new capital ships. One of five boats in Britain's first class of nuclear submarines, *Warspite* also had an emergency battery, diesel generator and electric motor. It was withdrawn in 1991, when hairline cracks were found in its primary coolant circuit.

SPECIFICATIONS

Type:	British submarine
Displacement:	4368 tonnes (4300 tons) [surface], 4876 tonnes (4800 tons) [submerged]
Dimensions:	87m x 10m x 8.4m (285ft x 33ft 2in x 27ft 7in)
Machinery:	Single screw, pressurized water-cooled nuclear reactor, turbines
Top speed:	28 knots [submerged]
Main armament:	Six 533mm (21in) torpedo tubes
Launched:	1965

TIMELINE

1962　　1965

George Washington Carver

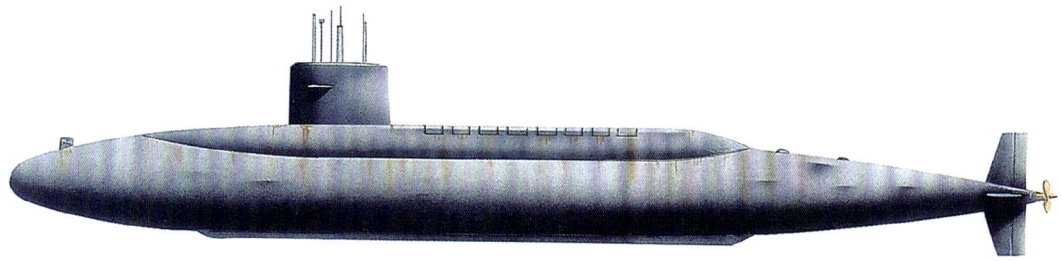

One of 29 vessels in the *Lafayette* class, *George Washington Carver* could dive to depths of 300m (985ft), and the nuclear core provided enough energy to propel the boat for 760,000km (400,000 miles). Its missile tubes were deactivated in 1991 and it became an attack boat. It was stricken and sent for recycling in 1993.

SPECIFICATIONS	
Type:	US submarine
Displacement:	7366 tonnes (7250 tons) [surface], 8382 tonnes (8250 tons) [submerged]
Dimensions:	129.5m x 10m x 9.6m (424ft 10in x 33ft 2in x 31ft 6in)
Machinery:	Single screw, one pressurized water-cooled nuclear reactor
Speed:	20 knots [surface], 30 knots [submerged]
Main armament:	Sixteen Trident C4 missiles, four 533mm (21in) torpedo tubes
Launched:	August 1965

Narwhal

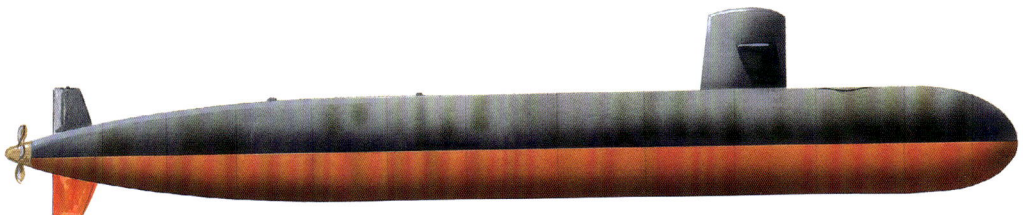

The United States' 100th nuclear submarine, *Narwhal* was an attack boat, of the *Sturgeon* class. These were larger than the *Thresher* and *Permit* class, powered by an improved reactor. *Narwhal* was notably quiet and much of its activity was in reconnaissance, or eavesdropping. It was stricken in 1999.

SPECIFICATIONS	
Type:	US submarine
Displacement:	4374 tonnes (4246 tons) [surface], 4853.4 tonnes (4777 tons) [submerged]
Dimensions:	89.1m x 9.6m x 7.8m (292ft 3in x 31ft 0in x 25ft 6in)
Machinery:	Single screw, one pressurized water-cooled nuclear reactor
Speed:	26 knots [submerged]
Main armament:	Four 533mm (21in) torpedo tubes
Launched:	August 1965

Yankee

Project 667A, known to NATO as the 'Yankee' class, were more powerful than previous Russian nuclear submarines. Thirty-four boats were built, capable of firing missiles from underwater, and patrolled the US eastern seaboard. By the SALT arms limitation agreement, all strategic 'Yankees' were withdrawn by 1994.

SPECIFICATIONS	
Type:	Soviet submarine
Displacement:	7823 tonnes (7700 tons) [surfaced], 9450 tonnes [9300 tons] [submerged]
Dimensions:	132m x 11.6m x 8m (433ft 10in x 38ft 1in x 26ft 4in)
Machinery:	Twin screws, nuclear reactors, turbines
Top speed:	13 knots [surface], 27 knots [submerged]
Main armament:	Sixteen SS-N-6 missile tubes, six 533mm (21in) torpedo tubes
Complement:	120
Launched:	1967

1967

Nuclear Submarines: Part 3

France launched its first class of nuclear submarines armed with ballistic missiles with *Le Redoutable* in 1967, and China produced the 'Han' class in 1972. The invisibility of the nuclear submarine and the secrecy surrounding its cruising missions, together with its destructive power, added to the tensions of the Cold War.

Charlie II

Following from the 12 smaller 'Charlie I' (NATO code) as the Soviet Navy's Project 670M, the 'Charlie II' class carried SS-N-9 Siren anti-ship missiles, which could be fitted with nuclear warheads, and two sizes of torpedo. The class shadowed US carrier battle groups. Six were built, serving until the mid-1990s.

SPECIFICATIONS	
Type:	Soviet missile submarine
Displacement:	4368.8 tonnes (4300 tons) [surface], 5181.6 tonnes (5100 tons) [submerged]
Dimensions:	103.6m x 10m x 8m (340 x 32ft 10in x 28ft)
Machinery:	Single screw, nuclear reactor
Top speed:	24 knots surfaced
Main armament:	Eight SS-N-9 cruise missiles, four 533mm (21in) and four 406mm (16in) torpedo tubes
Complement:	98
Launched:	1967

Delta I

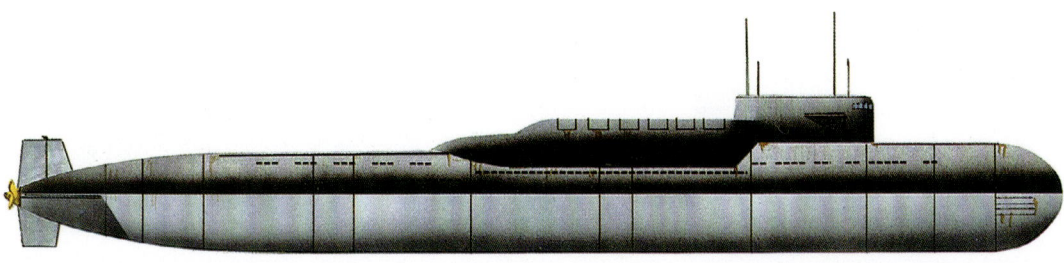

Between 1972 and 1977, Russia moved ahead in the Cold War with Project 667B – 18 large 'Delta'-class vessels, armed with new missiles that could out-range the US Poseidons. Initial tests showed the SS-N-48 missiles had a range of over 7600km (4000 miles). The class was scrapped between 1995 and 2004.

SPECIFICATIONS	
Type:	Russian submarine
Displacement:	11,176 tonnes (11,000 tons) submerged
Dimensions:	150m x 12m x 10.2m (492ft x 39ft 4in x 33ft 6in)
Machinery:	Twin screws, two nuclear reactors, turbines
Top speed:	19 knots [surface], 25 knots [submerged]
Main armament:	Twelve SS-N-48 missile tubes, six 457mm (18in) torpedo tubes
Launched:	1971

TIMELINE

1967  1971  1972

NUCLEAR SUBMARINES: PART 3

Han

The Chinese Navy went nuclear in the early 1970s with the 'Han'-class attack submarines. The highly streamlined hull shape, based upon the US vessel *Albacore*, is a departure from previous Chinese submarine designs. Five boats were completed, of which two or three were considered still operational in 2010.

SPECIFICATIONS

Type:	Chinese submarine
Displacement:	5080 tonnes (5000 tons) [submerged], surface displacement tonnage unknown
Dimensions:	90m x 8m x 8.2m (295ft 3in x 26ft 3in x 27ft)
Machinery:	Single screw, one pressurized-water nuclear reactor with turbine drive
Top speed:	25 knots [submerged]
Main armament:	Six 533mm (21in) torpedo tubes
Launched:	1972

Los Angeles

Los Angeles was lead boat in the world's most numerous class of nuclear submarines, with 45 still serving in 2010. Later members of the class (built up to 1996) have been modified. Though intended as hunter-killer boats, they are capable of land attack with Tomahawk missiles, shown in Iraq and Afghanistan.

SPECIFICATIONS

Type:	US submarine
Displacement:	6096 tonnes (6000 tons) [surface], 7010.4 tonnes (6900 tons) [submerged]
Dimensions:	109.7m x 10m x 9.8m (360ft x 33ft x 32ft 4in)
Machinery:	Single screw, pressurized-water nuclear reactor, turbines
Top speed:	31 knots [submerged]
Main armament:	Four 533mm (21in) torpedo tubes, up to eight Tomahawk cruise missiles
Complement:	127
Launched:	April 1976

Ohio

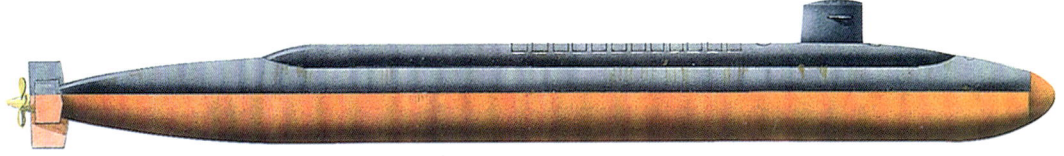

The *Ohio* class were the largest submarines built in the West, surpassed only by the Soviet 'Typhoon' class. The size was determined by the size of the reactor plant. *Ohio* was commissioned in 1981. Originally the class carried the Trident C-4 missile, but from the ninth boat they were built to carry the D-5 version.

SPECIFICATIONS

Type:	US submarine
Displacement:	16,360 tonnes (16,764 tons) [surface], 19,050 tonnes (18,750 tons) [submerged]
Dimensions:	170.7m x 12.8m x 11m (560ft x 42ft x 36ft 5in)
Machinery:	Single screw, pressurized-water nuclear reactor, turbines
Top speed:	28 knots [surface], 30+ knots [submerged]
Main armament:	24 Trident missiles, four 533mm (21in) torpedo tubes
Launched:	April 1979

1976

1979

Nuclear Submarines: Part 4

Improved reactors encouraged construction of large nuclear submarines. Boats of the United State's *Ohio* class were giants, but dwarfed by the Russian 'Typhoon' boats. Later types were smaller, on grounds of efficiency as well as cost (it was estimated that the United States had spent $700,000,000 on development by 1998).

Victor III

Russia's nuclear-powered hunter-killer submarines were codenamed 'Victor' by NATO. 'Victor III' could fire the SS-N-16 missile, which delivers a conventional homing torpedo to a greater range than otherwise possible. At least 43 'Victors' were launched, 26 of them being 'Victor IIIs', with some still operational in 2010.

SPECIFICATIONS	
Type:	Soviet submarine
Displacement:	6400 tonnes (6300 tons) [submerged]
Dimensions:	104m x 10m x 7m (347ft 8in x 32ft 10in x 23ft)
Machinery:	Single screw, pressurized water-cooled nuclear reactor, turbines
Top speed:	30 knots
Main armament:	Six 533mm (21in) torpedo tubes
Launched:	1978

Typhoon

'Typhoon' is the largest submarine yet built, nearly half as big again as the US *Ohio* class. The missile tubes are situated in two rows in front of the fin. It can force a way up through ice up to 3m (9ft 10in) thick. *Dmitry Donskoi,* of this class, was test-firing new Bulava-M missiles in 2008–09.

SPECIFICATIONS	
Type:	Russian submarine
Displacement:	25,400 tonnes (24,994 tons) [surface], 26,924 tonnes (26,500 tons) [submerged]
Dimensions:	170m x 24m x 12.5m (562ft 6in x 78ft 8in x 41ft)
Machinery:	Twin screws, pressurized water-cooled nuclear reactors, turbines
Top speed:	27 knots [submerged]
Main armament:	Twenty SS-N-20 nuclear ballistic missiles, two 533mm (21in) and four 650mm (25.6in) torpedo tubes
Launched:	1979

TIMELINE 1978 1979

NUCLEAR SUBMARINES: PART 4

San Francisco

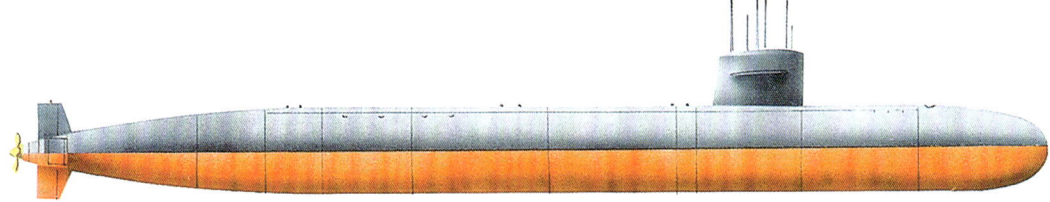

A *Los Angeles* class large attack submarine, *San Francisco* ran at full speed into an uncharted undersea mountain beneath the Pacific Ocean in January 2005, damaging the bows, but it managed to surface safely. Repaired with parts of withdrawn members of the class, it was restored to the fleet in 2008.

SPECIFICATIONS	
Type:	US submarine
Displacement:	6300 tonnes (6200 tons) [surface], 7010 tonnes (6900 tons) [submerged]
Dimensions:	110m x 10m x 9.8m (360ft x 33ft x 32ft 4in)
Machinery:	Single screw, nuclear powered pressurized-water reactor, turbines
Top speed:	30+ knots [submerged]
Main armament:	Four 533mm (21in) torpedo tubes. Harpoon and Tomahawk missiles
Launched:	October 1979

Oscar I

Project 949 was for a class of submarines combining the 'Typhoon' class's Arktika reactor with the 'Victor III' sonar systems, able to launch cruise missiles while submerged and to spend up to 50 days under water. The missiles were placed between the inner and outer pressure hulls, at an angle of 40° from the vertical.

SPECIFICATIONS	
Type:	Soviet submarine
Displacement:	tonnes (12,500 tons) [surface], 14,122 tonnes (13,900 tons) [submerged]
Dimensions:	143m x 18.21m x 8.99m (469ft 2in x 59ft 9in x 29ft 6in)
Machinery:	Twin screws, two pressurized water-cooled nuclear reactors
Top speed:	23 knots [submerged]
Main armament:	Four 533mm (21in) and four 650mm (25.6in) torpedo tubes launching SS-N-19, SS-N-15 and SS-N-16 Stallion missiles
Launched:	April 1981

Xia

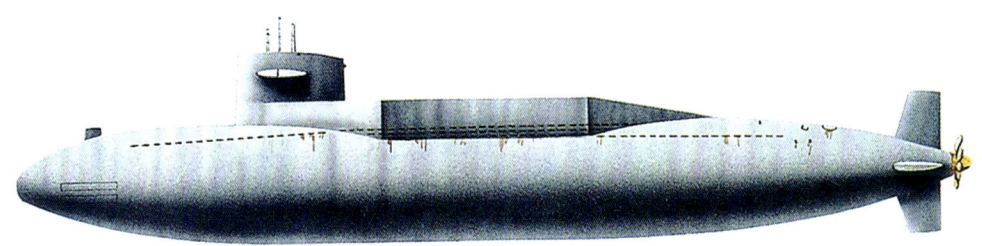

Type 092, 'Xia' was laid down in 1978. China's first ballistic missile submarine, it was an experimental craft. Two were built, of which one remains in service. The missiles were two-stage solid fuel rockets with inertial guidance for ballistic flight to 8000km (5000 miles), fitted with a nuclear warhead of two megatons.

SPECIFICATIONS	
Type:	Chinese submarine
Displacement:	8128 tonnes (8000 tons), submerged
Dimensions:	120m x 10m x 8m (393ft 8in x 32ft 10in x 26ft 3in)
Machinery:	Single screw, pressurized water-cooled nuclear reactor
Top speed:	22 knots [submerged]
Main armament:	Twelve tubes for CSS-N-3 missiles, six 533mm (21in) torpedo tubes
Launched:	April 1981

Nuclear Submarines: Part 5

In the Cold War, nuclear submarines shadowed enemy fleet groups, eavesdropping on naval exercises. Doubtless the practice continues, but operational boats fall into two categories: strategic missile submarines armed with long-range nuclear weapons, and 'hunter-killer' submarines to intercept and destroy enemy vessels.

Georgia

An *Ohio*-class boat, *Georgia* was redesignated SSGN (guided missile) when modified to carry cruise missiles in 2004. A major refit and overhaul followed in 2008. The fin is set far forward, ahead of the missile tubes. Its nuclear reactor is shielded from the engine, control centre and living quarters.

SPECIFICATIONS	
Type:	US submarine
Displacement:	16,865 tonnes (16,600 tons) [surface], 19,000 tonnes (18,700 tons) [submerged]
Dimensions:	170.7m x 12.8m x 10.8m (560ft x 42ft x 35ft 5in)
Machinery:	Single screw, pressurized water-cooled nuclear reactor, turbines
Top speed:	28 knots [surface], 30+ knots [submerged]
Main armament:	Twenty-four Trident missiles (C4), four 533mm (21in) torpedo tubes
Launched:	November 1982

Sierra

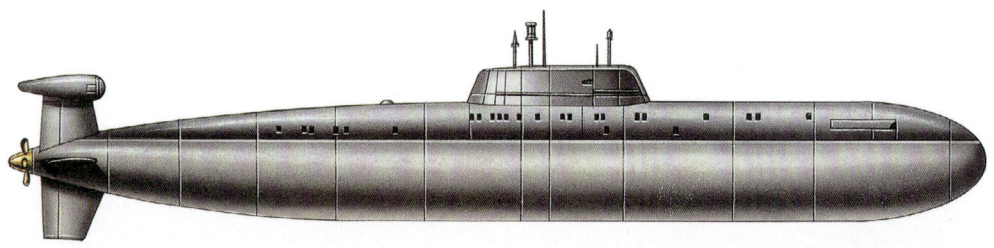

This was Project 945 and four boats were built before production switched to the 'Akula' boats of Project 971. Named 'Sierra' by NATO, the class used the Arktika reactor. Titanium-hulled, they had better safety provisions than previous Soviet nuclear submarines, including a crew escape pod. Three have been withdrawn.

SPECIFICATIONS	
Type:	Russian submarine
Displacement:	7315 tonnes (7200 tons) [surface], 10,262 tonnes (10,100 tons) [submerged]
Dimensions:	107m x 12m x 8.8m (351ft x 39ft 5in x 28ft 11in)
Machinery:	Single screw, pressurized water-cooled nuclear reactor, turbines
Top speed:	8 knots [surfaced], 36 knots [submerged]
Main armament:	Four 533mm (21in) and four 650mm (25.6in) torpedo tubes with provision for SS-N-22 and SS-N-16 missiles
Complement:	61
Launched:	1983

TIMELINE

1982  1983  1985

Torbay

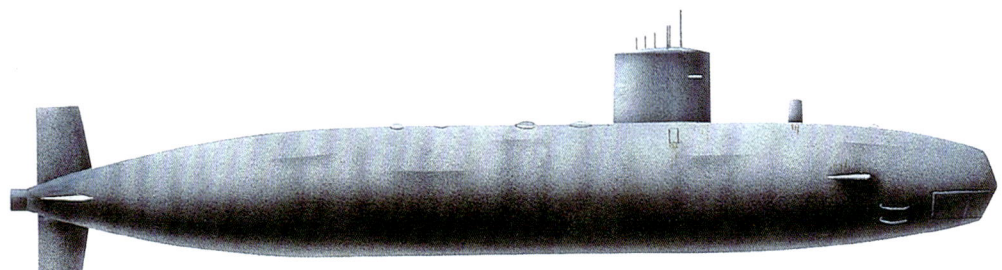

Torbay was one of the *Trafalgar* class of fleet submarine ordered in 1977, with a longer-life nuclear reactor. The main propulsion and auxiliary machinery raft are suspended from transverse bulkheads to maximize sound insulation. Anechoic tiles also reduce the acoustic signature. Modernizing of the class is taking place.

SPECIFICATIONS

Type:	British submarine
Displacement:	4877 tonnes (4800 tons) [surface], 5384 tonnes (5300 tons) [submerged]
Dimensions:	85.4m x 10m x 8.2m (280ft 2in x 33ft 2in x 27ft)
Machinery:	Pump jet, pressurized water-cooled reactor, turbines
Main armament:	Five 533mm (21in) tubes for Tigerfish torpedoes
Launched:	March 1985

Vanguard

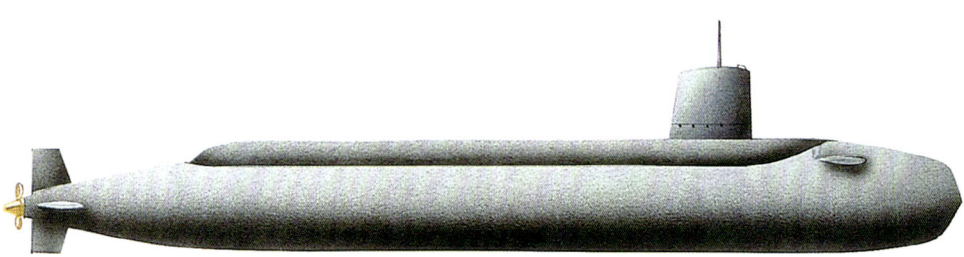

Vanguard carries 16 missiles in vertical launch tubes aft of the sail. Each can bear up to 14 warheads to targets more than 12,350km (6500 miles) distant. Like all submarines of this type, it operates independently, remaining submerged for months. The nuclear reactor is refitted and re-cored every eight years.

SPECIFICATIONS

Type:	British submarine
Displacement:	15,240 tonnes (15,000 tons) [submerged]
Dimensions:	148m x 12.8m x 12m (486ft 6in x 42ft x 39ft 4in)
Machinery:	Single screw, pressurized water-cooled nuclear reactor
Top speed:	25+ knots [submerged]
Main armament:	Sixteen Trident D5 missiles, four 533mm (21in) torpedo tubes
Complement:	135
Launched:	1990

Le Triomphant

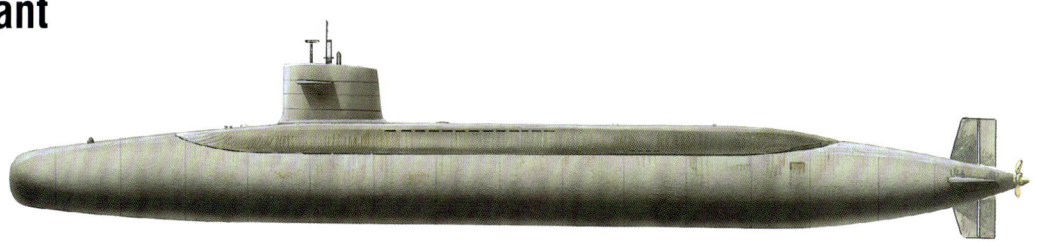

Le Triomphant is the first of France's 'new generation' missile submarines, commissioned in 1997 to replace the *Le Redoutable* class. It uses a new form of propeller. In February 2009, it and HMS *Vanguard* 'scraped' each other while on independent secret patrol beneath the Atlantic, but without consequences.

SPECIFICATIONS

Type:	French submarine
Displacement:	12,842 tonnes (12,640 tons) [surface], 14,564.4 tonnes (14,335 tons) [submerged]
Dimensions:	138m x 17m x 12.5m (453ft x 55ft 8in x 41ft)
Machinery:	Single screw, nuclear reactor with pump jet propulsor
Top speed:	20 knots [surface] 25 knots [submerged]
Main armament:	M51 nuclear missiles from 2010
Launched:	1993

1990    1993

Kursk

Kursk was an 'Oscar II' class nuclear submarine, an attack boat with nuclear missiles. In 1999, it was deployed in the Mediterranean. It sank after internal explosions while on exercises with the Northern Fleet in the Barents Sea, on 12 August 2000. All the crew perished. A complex salvage operation retrieved the wreck in October 2001.

Kursk

Kursk sank down to 354ft (108m). Russia, Britain and Norway launched a rescue operation, but ten days after the explosions, the remaining crew was declared dead. Twenty-three out of the crew of 118 survived the explosions, but they were trapped in a compartment and died when their air ran out.

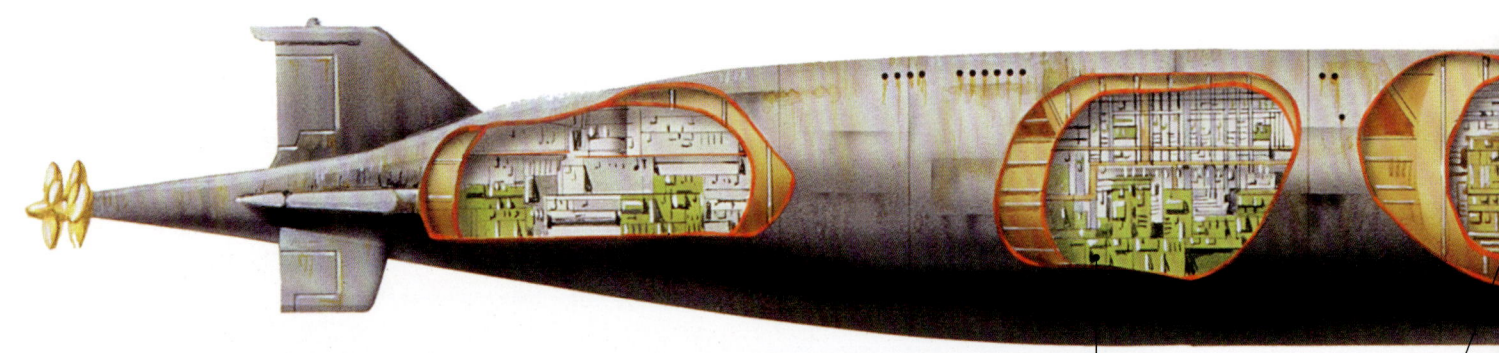

MACHINERY
Two pressurised water-cooled nuclear reactors powering two steam turbines. Power output 73,070kW (98,000shp).

FIFTH COMPARTMENT
This housed the nuclear reactors, and was protected by armoured steel walls 130mm (5.1in) thick, which withstood the blast.

KURSK

SPECIFICATIONS

Type:	Russian submarine
Displacement:	14,834 tonnes (14,600 tons) [surface], 16,256 tonnes (16,000 tons) [submerged]
Dimensions:	154m x 18.21m x 8.99m (505ft 2in x 59ft 9in x 29ft 6in)
Machinery:	Twin screws, two pressurized water-cooled reactors powering steam turbines
Top speed:	16 knots [surface], 32 knots [submerged]
Main armament:	24 Granit cruise missiles, four 533mm (21in) and two 650mm (25.5in) torpedo tubes
Complement:	118
Launched:	1994

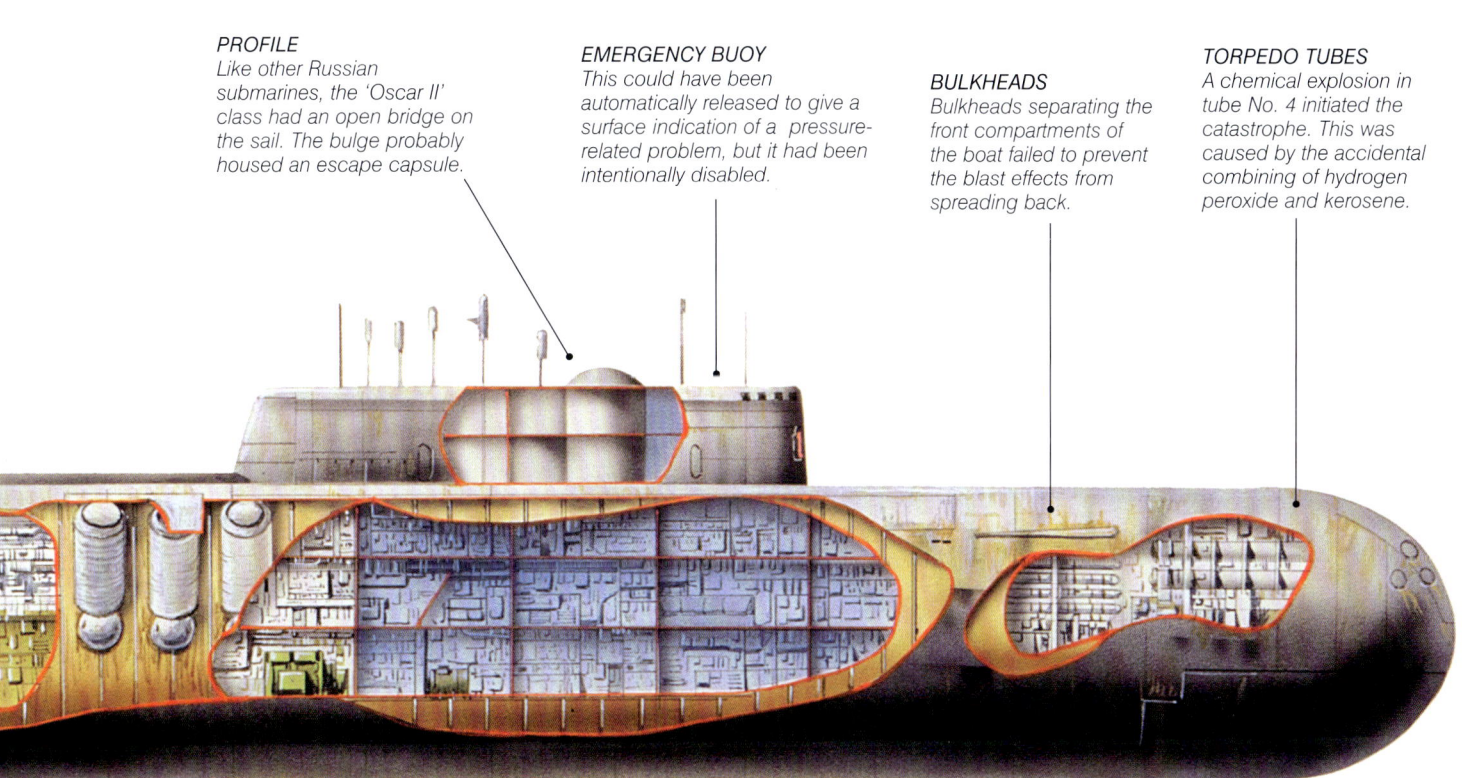

PROFILE
Like other Russian submarines, the 'Oscar II' class had an open bridge on the sail. The bulge probably housed an escape capsule.

EMERGENCY BUOY
This could have been automatically released to give a surface indication of a pressure-related problem, but it had been intentionally disabled.

BULKHEADS
Bulkheads separating the front compartments of the boat failed to prevent the blast effects from spreading back.

TORPEDO TUBES
A chemical explosion in tube No. 4 initiated the catastrophe. This was caused by the accidental combining of hydrogen peroxide and kerosene.

SECOND EXPLOSION
135 seconds after the first, a second larger explosion ripped open the third and fourth compartments.

Support and Repair Ships

The concept of the rapid-deployment task force – able to move at short notice to remote destinations, for military reasons or to provide post-disaster aid – depends on support and repair ships, which can keep up with other vessels in the force and provide the essentials of an operating base, often with command functions as well.

Hunley

Hunley and its sister Holland were designed to provide repair and supply services to fleet ballistic missile submarines. With 52 workshops, Hunley could deal with the requirements of several submarines at once. It carried a helicopter for at-sea delivery. Decommissioned in 1994, it was broken up in 2007.

SPECIFICATIONS	
Type:	US submarine tender
Displacement:	19,304 tonnes (19,000 tons)
Dimensions:	182.6m x 25.3m x 8.2m (599ft x 83ft x 27ft)
Machinery:	Single screw, diesel-electric engines
Main armament:	Four 20mm (0.79in) guns
Complement:	2490
Launched:	September 1961

Engadine

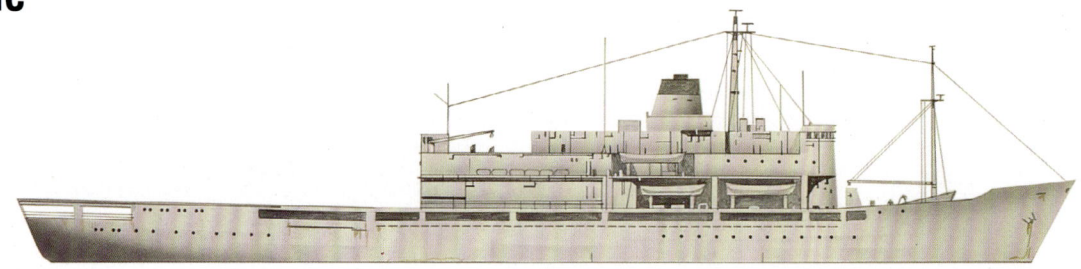

Laid down in August 1965, the Fleet Auxiliary Engadine was designed to train helicopter crews in deep-water operations. The large hangar aft of the funnel held four Wessex and two WASP helicopters, or two of the larger Sea Kings. Engadine could also operate pilotless target aircraft. It was scrapped in 1996.

SPECIFICATIONS	
Type:	British helicopter support ship
Displacement:	9144 tonnes (9000 tons)
Dimensions:	129.3m x 17.8m x 6.7m (424ft 3in x 58ft 5in x 22ft)
Machinery:	Single screw, diesel engine
Top speed:	16 knots
Complement:	81 plus 113 training crew
Launched:	1966

TIMELINE

1961    1966    1970

SUPPORT AND REPAIR SHIPS

Basento

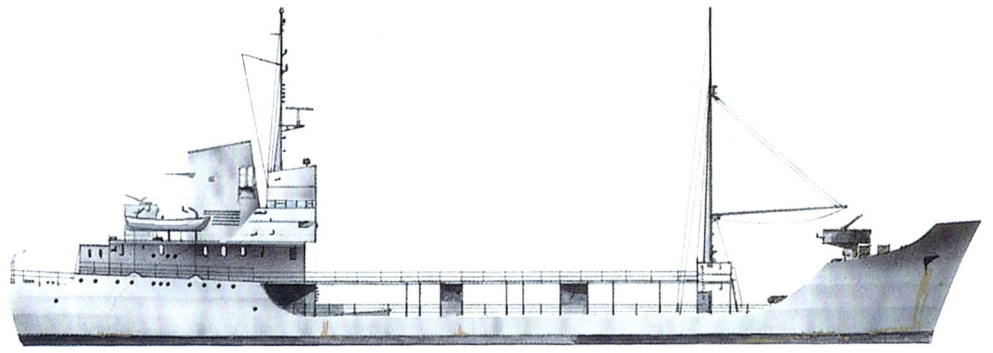

Basento was an auxiliary vessel supplying fresh water to the Italian fleet. Tank capacity is 1016 tonnes (1000 tons) and the ship's range is nearly 5700km (3000 miles) at 7 knots. The machinery space is placed aft. In 2009, Basento was given to Ecuador, to become a water supply ship for the arid Galapagos Islands.

SPECIFICATIONS

Type:	Italian naval water tanker
Displacement:	1944 tonnes (1914 tons)
Dimensions:	66m x 10m x 4m (216ft 6in x 33ft x 13ft)
Machinery:	Twin screws, diesels
Top speed:	12.5 knots
Armament:	Two light anti-aircraft guns
Launched:	1970

Fort Grange

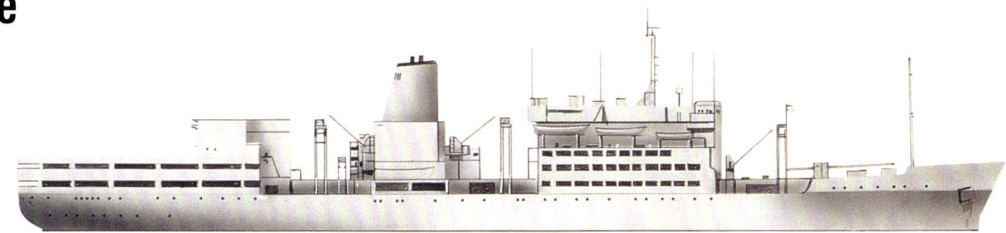

Up to 3500 tonnes of stores can be carried on board, to refuel and restock ships at sea in operational conditions. In addition. it can fly up to four Sea King helicopters on combat missions. *Fort Grange* spent 1994–2000 in the Adriatic Sea. Renamed *Fort Rosalie* in 2000, it was refitted in 2008–09.

SPECIFICATIONS

Type:	British fleet replenishment ship
Displacement:	23,165 tonnes (22,800 tons)
Dimensions:	183.9m x 24.1m x 8.6m (603ft x 79ft x 28ft 2in)
Machinery:	Single screw, diesels
Top speed:	22 knots
Main armament:	Two 20mm (0.79in) cannon
Complement:	140 plus 36 aircrew
Launched:	1976

Frank Cable

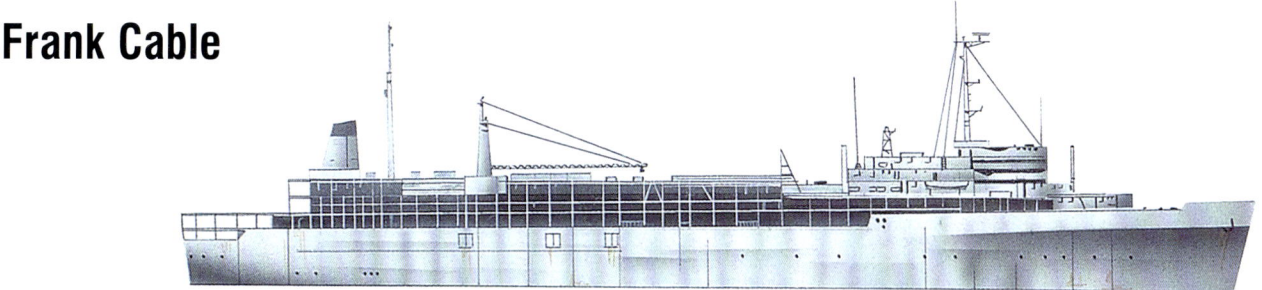

One of three improved versions of the *Spear* class, *Frank Cable* was equipped to support the *Los Angeles* class of submarine; up to four of this type can be handled simultaneously. These vessels are a far cry from World War II tenders, which were often old warships on their last role before scrapping.

SPECIFICATIONS

Type:	US submarine tender
Displacement:	23,368 tonnes (23,000 tons)
Dimensions:	196.9m x 25.9m x 7.6m (646ft x 85ft x 25ft)
Machinery:	Single screw, turbines
Top speed:	18 knots
Main armament:	Two 40mm (1.57in) guns
Launched:	1978

WARSHIPS 1945–PRESENT

Twenty-First Century Warships: 1

The US Navy has focused its attention on two key types of surface ship. First is the large aircraft carrier, capable of providing a powerful platform for air strikes anywhere in the world's seas. Second is the amphibious assault ship, equipped to transport and land combat troops and their mechanized and armoured support vehicles.

Charles de Gaulle

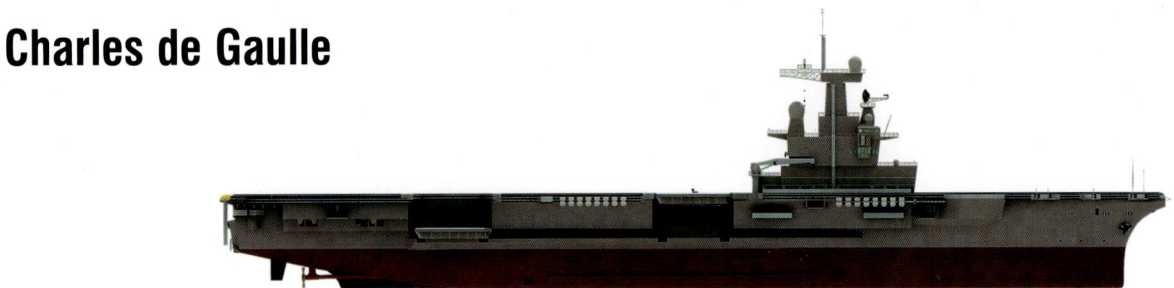

The only non-US nuclear-powered carrier, flying Super Etendard, Rafale, and E-2C Hawkeye aircraft. Propeller problems delayed commissioning until 2001 and reduced its operating speed. It saw active service in Operation Enduring Freedom in 2001-2, flying missions against al-Qaeda targets. It had a 15-month refit in 2007-8 including new propellers, but required further repairs in 2009.

SPECIFICATIONS	
Type:	French aircraft carrier
Displacement:	42,672 tonnes (42,000 tons)
Dimensions:	261.5m x 64.36m x 9.43m (858ft x 211ft 2in x 30ft 10in)
Machinery:	Twin screws, two pressurised water nuclear reactors, four diesel-electric motors
Top speed:	27 knots
Main armament:	Four SYLVER launchers, MBDA Aster SAM, Mistral short-range missiles
Aircraft:	40
Complement:	1350 plus 600 air wing
Launched:	1994

Ronald Reagan

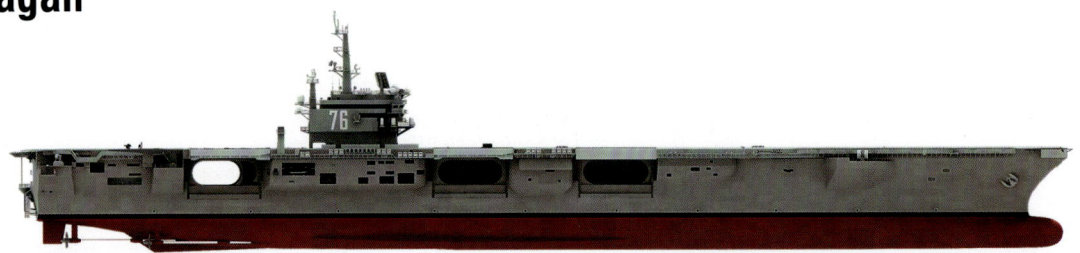

This ninth and penultimate ship in the *Nimitz* class has many new design features. Its island is placed further aft to give more open flight deck. Air traffic control and instrument landing guidance are included in a comprehensive range of systems. In 2008 *Reagan* launched over 1,140 sorties into Afghanistan. Its home port is San Diego, California.

SPECIFICATIONS	
Type:	US aircraft carrier
Displacement:	103,000 tonnes (101,000 tons)
Dimensions:	332.8m x 76.8m x 11.3m (1092ft x 252ft x 37ft)
Machinery:	Quadruple screws, two nuclear reactors, four turbines
Top speed:	30+ knots
Main armament:	Two Mk 29 ESSM launchers, two RIM-116 Rolling Airframe Missile launchers
Armour:	Classified
Aircraft:	90 aircraft
Complement:	3200 plus 2480 air wing
Launched:	March 2001

TIMELINE 1994 2001 2004

Mistral

Built partly in St Nazaire and partly in Brest, where it was completed, *Mistral* was commissioned in February 2006, and deployed off Lebanon later that year. For short-range missions, troop capacity can double to 900. Two landing barges and 70 vehicles are also carried. In 2010 it was announced that Russia was to buy a *Mistral*-type ship.

SPECIFICATIONS	
Type:	French amphibious assault ship
Displacement:	21,300 tonnes (20,959 tons) full load
Dimensions:	199m x 32m x 6.33m (652ft 9in x 105ft x 20ft 8in)
Machinery:	Twin screws, diesel-electric motors
Top speed:	18.8 knots
Main armament:	Two Simbad systems
Aircraft:	16 heavy or 35 light helicopters
Complement:	310 plus 450 troops
Launched:	Ooctober 2004

New York

Steel salvaged from the World Trade Centre was incorporated in the construction of this fifth ship in the *San Antonio* class, though it is actually named for NY state. It carries two air-cushion attack craft (LCAC) or one LCU (Landing Craft Utility). Delivered in August 2009, it has been troubled by main bearing failures in its engines.

SPECIFICATIONS	
Type:	US amphibious transport dock ship
Displacement:	25,298.4 tonnes (24,900 tons) full load
Dimensions:	208.5m x 31.9m x 7m (684ft x 105ft x 23ft)
Machinery:	Twin screws, turbo diesels
Top speed:	22 knots
Main armament:	Two Bushmaster II 30mm (0.18in) Close In guns, two RAM missile launchers
Aircraft:	Two CH-53E Super Stallion, two MV-22B Osprey tiltrotor aircraft, four CH-46 Sea Knight, four AH-1 Sea Cobra or UH-1 Iroquois helicopters.
Complement:	360 plus 700 marines
Launched:	December 2007

Zumwalt

New and still developing technologies are incorporated in this $1.1 billion-plus ship. Its hull, recalling the old ironclads, is claimed to have the radar print of a fishing boat. Automated systems in every department reduce crew numbers dramatically. Land attack as much as sea-based targets is envisaged as a potential mission. Three are currently under construction.

SPECIFICATIONS	
Type:	US multimission destroyer
Displacement:	14,797 tonnes (14,564 tons)
Dimensions:	182.9m x 24.6m x 8.4m (600ft x 80ft 8in x 27ft 7in)
Machinery:	Gas turbines, emergency diesels, powering advanced induction motors (AIM)
Top speed:	30.3 knots
Main armament:	20 Mk 57 VLS modules, Evolved Sea Sparrow and Tomahawk missiles, two 155mm (2.24in) advanced gun systems, two 57mm (6.1in) guns (CIWS)
Sensors:	AN/SPY multi-function radar, volume search radar, dual band sonar
Aircraft:	1 helicopter
Complement:	140
Launch date:	2015

Twenty-First Century Warships - 2

It takes great resources and a substantial industrial and naval infrastructure to sustain the design, building and operation of modern warships. Increasingly, navies and constructors of allied powers are joining forces to plan and fund the development of new warships: a trend particularly noticeable among the countries of the European Union.

Sachsen

The *Sachsen* class air-defence frigates are an enhanced version of the *Brandenburg* class and share features with the Spanish F100 class. Extensive countermeasures equipment include chaff and flare launchers, and detection systems include long-range air and surface surveillance and target indication radar. An STN Atlas Elektronik DSQS-24B bow sonar is fitted.

SPECIFICATIONS	
Type:	German frigate
Displacement:	5690 tonnes (5599 tons)
Dimensions:	143m x 17.4m x 5m (469ft x 57ft x 16ft 4in)
Machinery:	Twin controllable pitch screws, gas turbines and diesels (CODAG)
Top speed:	29 knots
Main armament:	Two Harpoon anti-ship missile systems, Sea Sparrow SAM system, two Mk32 double torpedo launchers, one 76mm (3in) Oto Melara gun
Aircraft:	Two NH90 helicopters
Complement:	230 plus 13 aircrew
Launched:	October 2000

Ma'anshan

This 'stealth' frigate carries formidable missile armament and a full range of radar systems, with many features developed from French originals. The sonar is however believed to be a Russian MGK-335 fixed sonar suite. Only two were built before the even more sophisticated *Type 054A* was introduced: an indication of the speed of Chinese naval development.

SPECIFICATIONS	
Type:	Chinese Type 054 frigate
Displacement:	4118 tonnes (4053 tons)
Dimensions:	134m x 16m (439ft 7in x 52ft)
Machinery:	Twin screw CODAD drive from four SEMT Pielstick diesel engines, most likely Type 16 PA6 STC
Top speed:	27-30 knots
Main armament:	Two quadruple launchers for YJ83 anti-ship cruise missiles, eight-cell Hong Qi7 short-range SAM system, four 6-barrel 30mm (1.18in) AK-630 CIWS, two Type 87 6-tube ASROC launchers, one 100mm (3.94in) gun
Aircraft:	One Kamov Ka 28 'Helix' or Harbin Z-9C helicopter
Launched:	September 2003

TIMELINE

2000  2003  2004

Houbei class

Navies are adopting missile-armed fast attack craft, more heavily armed than earlier patrol boats. This class, also known as Type 002, catamaran-hulled with stealth features, is in serial production in China. Detection equipment includes surface search and navigational radars and HEOS300 electro-optics. The two diesel engines generate 5,119 kW (6,865hp). Duties are coastal and inshore patrols.

SPECIFICATIONS	
Type:	Chinese missile boat
Displacement:	223.5 tonnes (220 tons) full load
Dimensions:	42.6m x 12.2m x 1.5m (139ft 7in x 40ft x 4ft 10in)
Machinery:	Twin water jet propulsors, diesels
Top speed:	36 knots
Main armament:	Eight C-801/802/803 anti-ship, or eight Hongniao long-range cruise missiles, twelve QW MANPAD surface-to-air missiles, one 30mm (1.18in) gun
Complement:	12
Launched:	April 2004

Daring

First of eight planned Type 45 destroyers, replacing Type 42, *Daring* was constructed in six 'blocks' in three different yards before final assembly. Its primary purpose is air defence and it has SAMPSON multi-function air tracking and S1850M three-dimensional air surveillance radars. But it is also fitted with MFS-7000 sonar and SSTDS underwater decoy systems.

SPECIFICATIONS	
Type:	British destroyer
Displacement:	8092 tonnes (7962.5 tons) full load
Dimensions:	152.4m x 21.2m x 7.4m (500ft x 69ft 6in x 24ft 4in)
Machinery:	Twin screws, integrated full electric propulsion, gas turbines
Top speed:	29+ knots
Main armament:	SYLVER missile launcher, MBDA Aster 15 and 30 missiles, 2 Phalanx 20mm (0.79in) CIWS
Aircraft:	One Lynx HMA 8 or Merlin HM 1 helicopter
Complement:	190
Launched:	February 2006

Aquitaine

Product of French-Italian co-operation, the FREMM multi-purpose frigate can be adapted to anti-air, anti-ship, and anti-submarine operation. *Aquitaine* and eight sister ships are due for launching from 2012; a further two will be deployed as anti-aircraft ships. Italy plans to build ten; six as general-purpose, four primarily for ASW action. Other NATO navies are likely to acquire them.

SPECIFICATIONS	
Type:	French FREMM-type frigate
Displacement:	6000 tonnes (5544 tons)
Dimensions:	142m x 20m x 5m (465ft 9in x 65ft 6in x 16ft 4in)
Machinery:	Twin screws, integrated full electric propulsion, gas turbines
Top speed:	27+ knots
Main armament:	MM-40 Exocet block 3, MU 90 torpedoes
Aircraft:	One NH90 helicopter
Complement:	108
Launched:	February 2012

2006

2012

Aircraft Index

Aeritalia
   G.222 184
   Lockheed F-104 ASA Starfighter 73
Airtech CN-235 185
Antonov An-72 'Coaler' 185
Armstrong Whitworth, Argosy C.1 59
Atlas Cheetah D 69
Avro
   Arrow 121
   Shackleton AEW.2 144
   Shackleton MR.2 104–5
   Vulcan B.1 27
   Vulcan B.2 27, 111
Avro Canada CF-100 Mk 4b 29

BAC TSR.2 121
Beech RC-12D 113
Bell
   X-1A 34
   X-2 34
   X-4 35
   X-5 35
   XV-15 201
Bell/Boeing
   MV-22 Osprey 201
   V-22 Osprey 201
Beriev
   A-50 'Mainstay' 145
   Be-10 'Mallow' 94
   Be-12 'Mail' 95
Blackburn
   Beverley C.1 59
   Buccaneer S.2 101
   Firebrand IV 47
Boeing
   B-17G Fortress 16
   B-17H Fortress 30
   B-29B Superfortress 18
   B-47E Stratojet 62
   B-52D Stratofortress 63, 85
   B-52G 188
   B-52H 188

E-3A Sentry 145
E-3D Sentry AEW.1 173
KC-135A Stratotanker 63
RB-29A Superfortress 52
RB-47H Stratojet 53
RC-135V 112
British Aerospace
   EAP 193
   Harrier GR.5 149
   Harrier GR.7 149
   Hawk T.51 187
   Hawk T.52 187
   Hawk T.Mk 1 186
   Hawk T.Mk 1A 186
   Nimrod MR.2P 111
   Nimrod R.1 113
   Sea Harrier FA.2 172
   Sea Harrier FRS Mk 2 149
   see also McDonnell Douglas-BAe
British Aerospace/McDonnell Douglas, Harrier GR.7 172

Canadair CL-84-1 200
CASA 212 Aviocar 184
Cessna O-2 Skymaster 80
Chengdu
   F-7 57
   F-7P Airguard 191
Convair
   B-36J Peacemaker 62
   B-58 Hustler 63
   F-102 Delta Dagger 96
   F-106A Delta Dart 97
   F-106B Delta Dart 97
   TF-102 Delta Dagger 97

Dassault
   MD.450 Ouragon 51
   Mirage 5M 67
   Mirage 5SDE 67
   Mirage 50C (Pantera) 69
   Mirage 2000-01 178
   Mirage 2000B 179
   Mirage 2000C 157, 178
   Mirage 2000H Vajra 197

Mirage 2000N 179
Mirage 2000P 179
Mirage F1 CK 154
Mirage IIIC 66
Mirage IIICJ 75
Mirage IIIEA 108
Mirage IIIO 66
Mirage IIIR 67
Mystère IVA 51, 74
Rafale A 199
Rafale M 198–9
Super Etendard 107
Super Mystère B.2 76
De Havilland
   Mosquito FB.VI 16
   Sea Vixen FAW.2 100
   Spider Crab (Vampire) 10
   Vampire FB.5 48
   Vampire FB.9 48
   Vampire (SNCASE Mistral) 50
   Vampire T.11 49
   Venom FB.1 49
   Venom FB.4 44
   Venom FB.50 49
Dornier
   Do 24T-2 31
   Do 29 200
Douglas
   A-1H Skyraider 85
   A-4C Skyhawk 116
   A-4F Skyhawk 82
   A-4G Skyhawk 116
   A-4M Skyhawk 117
   A-4N Skyhawk 76
   A-4PTM Skyhawk 117
   D-558-1 Skystreak 36
   Dakota 43
   Dakota III 42
   Dakota IV 43
   EKA-3 Skywarrior 83
   F3D-2 Skynight 25
   LC-47 Skiplane 43
   SC-54D Skymaster 31
   Skyraider AEW.1 46
   TA-4J Skyhawk 117

X-3A Stiletto 35
see also McDonnell Douglas

English Electric
   Canberra B.2 14, 44, 109
   Canberra B.66 71
   Canberra B(I).8 15
   Canberra PR.3 14
   Canberra PR.9 15
   Canberra TT.18 15
Eurofighter
   Typhoon 132–3
   Typhoon DA-2 192
   Typhoon T.1 192–3

Fairchild
   AC-119K Stinger 84
   C-1119G Packet 58
   R4Q-1 Boxcar 58
Fairchild Republic
   A-10A 158
   A-10A N/AW Thunderbolt II 181
   A-10A Thunderbolt II 180, 181
   OA-10A 181
   YA-10A Thunderbolt II 180
Fairey
   Firefly AS.5 46
   Gannet AS.1 101
FMA IA-58 Pucara 109
Fouga CM.170 Magister 126–7

General Dynamics
   EF-111A 177
   F-16/79 168
   F-16A 163, 168
   F-16A Block 15 ADF 169
   F-16A Fighter 152
   F-16C Block 50D 169
   F-16N 169
   F-111 85
   F-111A/TACT 182
   F-111C 183

INDEX

F-111E 183
F-111F 142
F-111G 183
FB-111A 182
Gloster
  Meteor F.8 19
  Meteor NF.11 45
Grumman
  A-6 Intruder 140
  A-6E Intruder 83, 141
  C-2A Greyhound 61
  E-2 Hawkeye 61
  E-2C Hawkeye 144
  EA-6A Intruder 140
  EA-6B Prowler 141, 176
  F-14A Tomcat 142, 160–1
  F-14B Tomcat 161
  F-14D Tomcat 161
  F7F-2N Tigercat 12
  F7F-3N Tigercat 25
  F8F-1 Bearcat 12
  F9F-2 Panther 21, 40
  HU-16 Albatross 31
  KA-6D Prowler 141
  S-2E Tracker 107
  S-2F Tracker 60
  S-2T Tracker 61
  US-2A Tracker 60
  X-29 37
  see also Northrop Grumman
Grumman (General Dynamics)
  EF-111A 177

HAL 748M 196
Handley Page
  Victor B.2 26
  Victor K.2 27
Harbin H-5 190
Hawker
  Hunter F.1 32
  Hunter F.5 32
  Hunter F.56 71
  Hunter F.58 33
  Hunter F.70 153
  Hunter T.8M 33
  Hunter T.75A 33
  Sea Fury FB.2 47
  Sea Fury FB.11 20
  Sea Hawk FB.Mk.3 45, 71
  Sea Hawk FGA.4 47
Hawker Siddeley
  AV-8A Harrier 92
  AV-8S Matador 93
  Buccaneer S.2B 157
  Harrier GR.3 92
  Harrier T.4 93
  Kestrel FGA.1 91
  Nimrod AEW.3 145
  P.1127 91
  Sea Harrier FRS.1 110
  Sea Harrier FRS.Mk 1 93
Hawker Siddeley/HAL 748M 196

IAI
  Kfir 68
  Kfir C1 (F-21A) 68
  Kfir TC2 69
Ilyushin
  Il-28 'Beagle' 21, 98
  Il-38 'May' 105
  Il-76 'Candid' 115
  Il-76MD 'Candid' 143
  Il-76TD Gajaraj 'Candid' 196

Kawasaki
  C-1A 115
  EC-1 205
  P-2J Neptune 105
  T-4 205

Lisunov Li-2 42
Lockheed
  AC-130 Spectre 84
  C-5 Galaxy 114
  C-130B Hercules 118
  C-130E Hercules 109
  C-130K Hercules C.1 111
  C-141B StarLifter 114
  CF-104G Starfighter 73
  EC-130E Hercules 118
  EC-130Q Hercules 119
  EP-3E Aries II 89
  F-80C 19, 38
  F-94 Starfire 39
  F-94B Starfire 24
  F-104C Starfighter 72
  F-117A Nighthawk 164–5, 165
  KC-130F Hercules 119
  L-188PF Electra 106
  P-3 Orion AEW&C 89
  P-3A Orion 88
  P-3B Orion 88
  P-80 Shooting Star 10
  QF-80 Shooting Star 38
  RF-80A Shooting Star 39
  RP-3A Orion 89
  SP-2H Neptune 106
  SR-71A Blackbird 112
  T-33A 39
  TF-104G Starfighter 72, 73
  TR-1A 113
  TriStar K Mk 1 115
  U-2A 64
  U-2C 64
  U-2CT 65
  U-2D 65
  U-2R 65
  WC-130H Hercules 119
  XST 'Have Blue' 165
Lockheed Martin
  F-22A 210–11
  YF-22 Raptor 211

Macchi MC.205 Veltro 17
Martin
  AM-1 Mauler 13
  P4M-1 Mercator 52
  P5M Marlin 94
  PBM-5 Mariner 30
  see also Lockheed Martin
McDonnell
  F-101B Voodoo 29
  F2H-2 Banshee 13, 41
  F4H-1 Phantom II 40
McDonnell Douglas
  A-4KU Skyhawk 155
  A-4P Skyhawk 108
  A-4Q Skyhawk 107
  AV-8B Harrier II 148
  CF-18A Hornet 173
  F-4 Phantom II 8–9
  F-4C Phantom II 78
  F-4E Phantom II 77
  F-4EJ Kai Phantom 204
  F-4G Phantom II 82, 177
  F-4S Phantom II 79
  F-15 Eagle 152
  F-15A Eagle 146, 147
  F-15B Eagle 146
  F-15B Strike Eagle 147
  F-15C Eagle 157
  F-15C MSIP 158
  F-15DJ Eagle 204
  F-15J Eagle 147
  F/A-18A Hornet 202
  F/A-18B Hornet 203
  F/A-18C Hornet 203
  F/A-18D 159
  F/A-18E Super Hornet 203
  F4H-1 Phantom II 78
  Phantom FG.1 101
  Phantom FGR.2 79
  RF-4EJ Phantom II 79
  T-45A Goshawk 187
  TAV-8B Harrier 148
McDonnell Douglas-BAe AV-8B Harrier 159
Mikoyan-Gurevich
  MiG-9 'Fargo' 11
  MiG-15 'Fagot' 21
  MiG-15bis 'Fagot' (S-103) 128
  MiG-15UTI 'Midget' 128
  MiG-17 'Fresco' 75, 87, 129
  MiG-17 'Fresco-C' 162
  MiG-17F 'Fresco' 77
  MiG-17PF 'Fresco-D' 129
  MiG-19 'Farmer' 129
  MiG-19S 'Farmer' 54, 87
  MiG-21 'Fishbed' 56
  MiG-21 MF 'Fishbed' 75, 87
  MiG-21bis 'Fishbed' 171
  MiG-21FL 'Fishbed E' 197
  MiG-21MF 'Fishbed-J' 166
  MiG-21PF 'Fishbed' 154
  MiG-21PF 'Fishbed-D' 56
  MiG-21PFMA 'Fishbed' 57
  MiG-21U 'Mongol' 57
  MiG-23 'Flogger-E' 143

441

INDEX

MiG-23BN 155
MiG-23BN 'Flogger-F' 151, 153
MiG-23BN 'Flogger-H' 167
MiG-23M 'Flogger-B' 150
MiG-23MF 'Flogger-B' 150
MiG-23MLD 'Flogger-K' 163
MiG-23MS 'Flogger-E' 153
MiG-23UB 'Flogger-C' 151
MiG-25 'Foxbat' 206–7
MiG-25BM 'Foxbat-F' 176
MiG-25P 'Foxbat-A' 55, 143
MiG-25R 'Foxbat' 207
MiG-27 'Flogger-J' 151
MiG-29 155
MiG-29 'Fulcrum A' 167, 194, 195
MiG-29B 'Fulcrum' 171
MiG-29K 195
MiG-29M 'Fulcrum D' 195
MiG-29UB 'Fulcrum-B' 194
MiG-31 'Foxhound-A' 55, 207
Mitsubishi XF-2B 205
Myasishchev
   M-4 'Bison C' 98
   M-50 'Bounder' 121

Nord 2501D Noratlas 59
North American
   AJ-2 Savage 41
   B-45A Tornado 53
   F-51D Mustang 18
   F-82G Twin Mustang 24
   F-86A-5 Sabre 22
   F-86D Sabre Dog 28–9
   F-86E-10 Sabre 23
   F-86F-30 Sabre 23
   F-86F Sabre 70
   F-100F Super Sabre 81
   FJ-1 Fury 13
   FJ-3M Fury 41
   OV-10A Bronco 80

RA-5C Vigilante 53
Sabreliner TP 86 209
X-15 37
Northrop
   B-2 189
   F-5A Freedom Fighter 81, 102
   F-5B Freedom Fighter 103
   F-5E Tiger II 103
   RF-5E Tigereye 103
   T-38 Talon 102
   X-24 37
   YF-17 202
   YF-23A 211
Northrop Grumman
   B-2A 173
   E-8A J-STARS 159

On Mark B-26K Counter Invader 81

Panavia
   Tornado ECR 177
   Tornado F.Mk 3 136
   Tornado GR.1 136–7, 156
   Tornado IDS 137

Republic
   F-84E Thunderjet 19
   F-84F Thunderstreak 45
   F-105B Thunderchief 86
   F-105F Thunderchief 86
   XF-91 Thunderceptor 120
Rockwell
   B-1A 189
   B-1B 189
Ryan
   FR-1 Fireball 11
   X-14B 36
   XV-5A 90

Saab
   A 32A Lansen N 208
   91 Safir 122
   105/Sk 60 125
   B 17 122
   J 21 123
   J 21R 11
   J 29 Tunnan 123
   J 32B Lansen 123

J 35F Draken 124
JA 37 Viggen 208
JAS 39 Gripen 125
JAS 39A Gripen 209
MFI-15 124
SF 37 Viggen 125
SK 37 Viggen 209
Saunders Roe SR.53 120
SEPECAT
   Jaguar 197
   Jaguar A 138
   Jaguar GR.1 138
   Jaguar GR.1A 156
   Jaguar IM (Shamsher) 139
   Jaguar International 139
   Jaguar T.2 139
Shenyang
   F-6 191
   F-6 (MiG-19 'Farmer') 70
   FT-6 190
   J-6 191
Shin Meiwa PS-1 95
Short Sunderland 95
SOKO
   G-2A Galeb 170
   J-21 Jastreb 170
   UTVA 75 171
Sud Aviation, Sud-Ouest
   Aquilon 203 50
   SO.4050 Vautour 74
   SO.4050 Vautour IIB 51
Sukhoi
   Su-7BM 'Fitter-A' 77, 130
   Su-7BMK 'Fitter-A' 130
   Su-7UM 'Moujik-A' 131
   SU-15 'Flagon' 55
   Su-17M-4 'Fitter-K' 131, 166
   Su-20 'Fitter-C' 131
   Su-24 'Fencer-B' 175
   Su-24M 'Fencer-D' 175
   Su-24MR 'Fencer-E' 167
   Su-25 'Frogfoot' 162
   Su-25 'Frogfoot-A' 174
   Su-25K 'Frogfoot-A' 174
   Su-25UTG 'Frogfoot B' 175
   Su-27 T-10-1 'Flanker-A' 212

Su-27IB/Su-34 'Fullback' 213
Su-27UB 'Flanker-C' 212
Su-33/Su-27K 'Flanker-D' 213
Su-35 (Su-27M) 213
Supermarine
   Scimitar F.1 100
   Spitfire 17

Transall C.160 185
Tupolev
   Tu-16R 'Badger-K' 99
   Tu-22PD 'Blinder-E' 163
   Tu-22R 'Blinder-C' 99
   Tu-95 'Bear' 99

VFW-Fokker VAK-191B 90
Vickers Valiant B(K).1 26
Vought
   A-7B Corsair II 83
   A-7D Corsair II 135
   A-7H Corsair II 135
   F-8D Crusader 134
   F-8E(N) Crusader 135
   F4U-4B Corsair 20
   F4U-5N Corsair 25
   RF-8G Crusader 134

Yakovlev
   Yak-28P 'Firebar' 54
   Yak-38 'Forger-A' 91

Tank Index

2S7 307
76mm Otomatic Air Defence Tank 272
155/45 Norinco SP Gun 338

AAAV 347
AAAVR7A1 327
AAV7 324–5
Abbot 264
ABRA/RATAC 349
ADATS 341
AIFV 285
Al Khalid Main Battle Tank 332
AMX-10 PAC 90 302
AMX 10P 285
AMX-13 228–9
AMX-13 DCA 277
AMX-30 250–1
AMX-30 Tractor 291
AMX-40 300
AMX VCI 280
Arisgator 314
Arjun 334
Armoured Starstreak 341
AS-90 338
ASU-57 230
ASU-85 Tank Destroyer 268

Bandkanon 265
BAT-2 326
BAV-485 242
Bionix 25 346
BM-21 235
BM-24 234
BMD 284
BMP-1 283
BMP-2 319
BMP-3 Infantry Combat Vehicle 347
BMS-1 Alacran 318
Bradley M2 342–3
BRDM-1 SA-9 Gaskin 241
BTR-50 281
Bv 206 323

C1 Ariete 300
CAMANF 322
Centurion A41 216–17
Centurion AVRE 290
Centurion Mk 5 225
Challenger 1 296–7
Challenger 2 335
Chieftain AVRE 291
Chieftain Mk 5 254–5
Close Combat Vehicle - Light 303
Cobra 321
Combat Engineer Tractor 326
Conqueror Heavy Tank 219
Crotale 279

DANA 306

EE-3 Jararaca 287
ENGESA EE-T1 Osorio 301

Flakpanzer 1 Gepard 278
FROG-7 270
FUG 287
FV102 Striker 269
FV432 282
FV433 Stormer 319

G6 Rhino 307
GAZ-46 MAV 242
GCT 155mm 306
GDF-CO3 277
Gillois PA 243
Green Archer 315
GT-S 243

Honest John 270
Hornet Malkara 268

Jagdpanzer Jaguar 309
Jagdpanzer Kanone (JPK) 231

K-61 243
Khalid 257

KIFV K-200 320

L33 304
LARS II 235
Leclerc 336–7
Leopard 1 A3 253
Leopard 1 Armoured Engineer Vehicle 252, 291
Leopard 2 330–1
Leopard 2A6 335
LVTP7 322
Lynx CR 286

M1 Abrams 328–9
M1 Tunguska 313
M1A1 Abrams 298–9
M1A2 Abrams 299
M1A2 with TUSK 334
M2 Bradley 342–3
M4 C2V 349
M9 ACE 327
M26 Pershing 214–15
M41 Walker Bulldog 262
M42 240
M47 Patton I Medium Tank 219
M48 222–3
M48 Chaparral 276
M48A1 224
M48A2 224
M48A5 225
M50 Ontos 230
M53/59 240
M59 232
M-60P 281
M60 247
M60A1 249
M60A2 246–7
M75 232
M-80 MICV 320
M103 Heavy Tank 218
M107 264
M109 266–7
M109A6 Paladin 339
M110A2 305
M113 233, 236–7

M113A1 233
M114A1E1 314
M163 Vulcan 273
M548 Carrier with Lance Missile 271
M551 Sheridan 262
M577 348
M727 HAWK 241
M728 290
M901 315
M980 346
M992 FAASV 349
M1973 304
M1974 305
Magach (Israeli M48) 225
Marder 284, 316–17
Merkava 292–3
MIM-104 (GE) Patriot 341
MK 61 231
Mk F3 155mm 265
MLRS 310–11
MT-LB 318

Norinco Type 63 Amphibious Light Tank 263

OF 40 257
Olifant Mk 1A 257
OT-62 282
OT-810 280

Palmaria 307
Panzerhaubitzer 2000 339
Panzerjäger 90 309
Pbv 283
Pegaso VAP 3550/1 323
Pershing 271
Pionierpanzer Dachs 2 327
PRAM-S 348
PT-76 226–7
PT-91 Main Battle Tank 333
Pz 61 and Pz 68 Main Battle Tanks 248

Raketenjadgpanzer 1 269
Raketenjadgpanzer 2 308

INDEX

Ramses II 332
Rascal 339
Roland 273

SA-4 Ganef 273
SA-6 Gainful 276
SA-8 Gecko 277
SA-10 Grumble 312
SA-11 Gadfly 312
SA-13 Gopher 279
Saladin 238–9
Saurer 4K 4FA-G1 281
Schützenpanzer Spz 11-2 kurz 286
Scorpion 288–9
Scorpion 90 302
Sergeant York 313
Shahine 279
Sho't (Israeli Centurion) 256
Sidam 25 340
SM-240 305
Spartan 278
SPz lang HS30 231
SS-1c Scud B Launcher 271
Steyr SK 105 Light Tank 263
Stingray 303
Stridsvagn 103 (S-tank) 263
SU 60 233

T-10 Heavy Tank 218
T54/55 Main Battle Tank 220–1
T-62 244–5, 248
T-64 Main Battle Tank 249
T-72 258
T-72 'Ural' 258–9
T-72CZ M4 261
T-72G 260
T-72M 260
T-72S 261
T-80 294
T-80BV 295
T-80U 295
T-90 Main Battle Tank 333
TAM 303
Tracked Rapier 313
Type 54-1 265
Type 59 219
Type 60 269

Type 63 272
Type 69 294
Type 70 234
Type 73 (command variant) 285
Type 74 256
Type 77 319
Type 80 301
Type 85 301
Type 87 AWSP 340
Type 88 295
Type 89 Infantry Combat Vehicle 321
Type 90-II 335
Type 90 Main Battle Tank 333
Type 6640A 323

VBL 309
VCC-1 Armoured Personnel Carrier 321
VCC-80/Dardo IFV 347
Vijayanta (Vickers Mark 1 and Mark 3) 249

Walid 235
Warrior 344–5
Wiesel 1 TOW 308

YP-104 287
YW 531 283

Zelda 315
ZSU-23-4 274–5
ZSU-57-2 241
ZTS T-72 261

Warships Index

Warships Name Index

Admiral Petre Barbuneanu 405
Admiral Senyavin 376
Al Madina 406
Alligator class 413
Almirante Riveros 389
Alvand 401
America 353
Aquitaine 439
Aragua 390
Arleigh Burke 386
Asagumo 391
Aster 411
Athabaskan 402
Audace 391

Babur 378
Badr 369
Bambú 410
Baptista de Andrade 403
Basento 435
Beskytteren 368
Boykiy 381
Brandenburg 387
Broadsword 403
Bunker Hill 373

Caio Duilio 379
Caorle 412
Carabiniere 400
Carl Vinson 7
Carlo Bergamini 398
Centauro 396
Chakri Naruebet 361
Charles de Gaulle 436
Charlie II 426
Chazhma 413
Chikugo 401
Cigno 397
Clemenceau 356
Collins 419
Coronel Bolognesi 378, 390
Cushing 393

Daniel Boone 424
Daphné 415
Dardo 366
Daring 439
Davidson 400
Delta I 426
Denver 364
D'Estienne d'Orves 367
Deutschland 413
Devonshire 381
Didi 398
Dmitri Donskoi 377
Dmitry Pozharsky 376
Dolfijn 415
Doudart de Lagrée 399
Downes 401
Doyle 405
Dreadnought 421
Dromia 411
Duguay-Trouin 383
Dupleix 384
Duquesne 382

Edera 410
Edinburgh 386
Eilat 369
Engadine 434
Enrico Toti 416
Enterprise 352
Eridan 411
Euro 385

Farragut 380
Filicudi 412
Floréal 409
Forrestal 352
Fort Grange 435
Frank Cable 435
Fremantle 368

Galatea 399
Galerna 418
Galveston 372
Gatineau 397
Gemlik 397
George Washington 353, 421

444

INDEX

George Washington Carver 425
Georges Leygues 392
Georgia 430
Giuseppe Garibaldi 358–9
Glasgow 393
Godavari 405
Golf I 414
Grafton 396
Grayback 415
Gremyaschiy 380
Grom 389
Groningen 389
Gurkha 392

Hai Lung 419
Halibut 421
Halifax 408
Hamayuki 385
Han 427
Haruna 391
Harushio 416
Hatsuyuki 393
Hermes 356
Houbei class 439
Hunley 434

Impavido 381
Inhaúma 407
Intrepid 364
Invincible 354–5
Ivan Rogov 365
Iwo Jima 360
Izumrud 402

Jacob van Heemskerck 406
Jeanne D'Arc 360
Jianghu III 407
John F. Kennedy 353

Kaszub 369
Kerch 377
Kiev 357
Kilo 417
Kirov 374–5
Kongo 387
Kotor 407
Kursk 432–3

Le Triomphant 431
Long Beach, USS 370–1
Los Angeles 427
Lupo 404

Ma'anshan 438
Mississippi 373
Mistral 437
Moskva 361
Mount Whitney 365
Mourad Rais 404
Murasame 387
Mutenia 385

Näcken 417
Najin 403
Nanuchka I 367
Nareusan 409
Narwhal 425
Nautilus 350–1, 420
Neustrashimyy 388, 408
New York 437

Ognevoy 382
Ohio 427
Oscar I 429

Prat 379

Resolution 422–3
Ronald Reagan 436

Sachsen 438
San Francisco 429
Santisima Trinidad 384
Shanghai 366
Sierra 430
Skipjack 420
Slava 377
Spica 367
Spruance 383
St Laurent 388

Tachikaze 383
Tarawa 362–3
Thetis 409
Ticonderoga 373
Tiger 379
Torbay 431

Tromp 394–5
Typhoon 428

U-12 417
Upholder 419

Vanguard 431
Veinticinco de Mayo 357
Victor III 428
Vikrant 357
Vittorio Venento 361

Walrus 418
Warspite 424
Whidbey Island 365
Whiskey 414
Worden 372

Xia 429

Yankee 425
Yubari 399

Zumwalt 437

General Index

aerobatic teams 33, 51, 119, 126, 127, 186
AEW (airborne early warning) aircraft 46, 89, 144–5
Afghanistan
   aircraft 162, 174
   wars in 162–3, 436
air sea rescue 30–1
aircraft carriers 352–63, 436
Algeria, frigates 404
ammunition support vehicle 349
amphibious assault ships 362–3, 437
amphibious vehicles
   anti-aircraft vehicles 241, 277
   armoured personnel carriers (APCs) 282, 285, 314, 318, 319, 321, 347
   combat engineer vehicles 327
   post-war vehicles 242–3, 263
   reconnaissance vehicles 242, 288–9
   self-propelled guns 266–7
   tanks/AFVs 226–7, 236–7, 322–5
anti-aircraft vehicles 240–1, 272–9, 312–13, 340–1
anti-submarine aircraft 60, 61, 88–9, 101, 107, 111, 145
anti-submarine ships 388–93, 396–7, 400–2, 404–5, 407, 408
anti-tank vehicles 268–9, 302, 308–9
Argentina
   aircraft 43, 106–9
   ships 357, 384
   tanks 303
   see also Falklands War (1982)
armoured cars 238–9
armoured personnel carriers (APCs) 284–5, 316–17, 344–5
   tracked 232–3, 280–3, 314, 318–21, 346–7
assault ships 362–5, 437

Australia
   aircraft 116, 183
   patrol ships 368
   submarines 419
Austria, tanks/AFV 263, 281

Balkans, air wars 170–3
Belgium
   aircraft 43, 127
   minehunters/sweepers 411
   tracked APCs 321
Bordelon, Guy 25
Brazil
   frigates 407
   tanks/AFV 287, 301, 322

Canada
   aircraft 121, 173, 200
   ships 388, 397, 402, 408
   North American Air Defense Command (NORAD) 28–9
Carmichael, Peter 19
Chile
   aircraft 69
   ships 379, 389
   tracked APCs 318
China
   aircraft 57, 73, 190–1
   anti-aircraft vehicles 272
   rocket systems 234
   self-propelled guns 265, 338
   ships 366, 407, 438, 439
   submarines 427, 429
   tanks 219, 263, 294, 301, 332, 335
   tracked APCs 283
combat engineer vehicles 290–1, 326–7
Command and Control Vehicle (C2V) 349
command ships 364–5
corvettes 366–9
cruisers 370–9
Czechoslovakia/Czech Republic
   aircraft 56, 128, 150, 174
   anti-aircraft vehicles 240
   self-propelled guns 306

   tanks 261
   tracked APCs 280, 282

delta-wings 96–7, 124
Denmark
   aircraft 73
   ships 368, 409
Desert Storm 140, 154–9, 165, 373
destroyers 380–95, 437, 439
   anti-submarine 388–93
   guided missiles 380–7

Egypt
   aircraft 17, 67, 75, 128
   anti-aircraft vehicles 275
   patrol ships 369
   rocket systems 235
   Six-Day War (1967) 74–5
   Suez crisis 44–5
   tanks 332
   Yom Kippur War (1973) 76–7
engineering vehicles 326–7
Eurofighter 192–3

FAC (Forward Air Control) 80, 81
Falklands War (1982)
   aircraft 106–11
   reconnaissance vehicles 288–9
   ships 354–5, 356, 357, 384, 393
Fields, Joseph E. 22
Finland, aircraft 187
Fleet Air Arm 46–7, 100–1
flying boats 30–1, 94–5
flying boxcars 58–9
flying broadcast station 118
Forward Air Control (FAC) 80, 81
Foster, Cecil 23
France
   aircraft 45, 50–1, 66–7, 126–7, 138–9, 157, 178–9, 185, 197–9
   aircraft carriers 356, 360, 436
   amphibious assault ships 437
   anti-aircraft vehicles 273, 277, 279

   anti-tank vehicles 309
   armoured personnel carriers (APCs) 280, 285
   combat engineer vehicles 291
   destroyers 382–3, 384, 392
   frigates 399, 409, 439
   minehunters/sweepers 411
   patrol ships 367
   self-propelled guns 231, 265, 306
   submarines 415, 431
   tanks/AFV 250–1, 228–9, 231, 243, 300, 302, 336–7
frigates 367, 392, 394–409, 438, 439

Georgia, aircraft 174
Germany
   aircraft 31, 54, 127, 167, 185, 200
   anti-tank vehicles 308–9
   combat engineer vehicles 327
   main battle tanks 248, 335
   observation vehicle 349
   reconnaissance vehicles 286
   rocket systems 235
   self-propelled guns 231, 339
   ships 387, 413, 438
   submarines 417
   see also West Germany
Glenn, John 23
Greece, aircraft 96, 135
guided missile destroyers 380–7
Gulf War *see* Iraq, wars in

heavy lifters 114–15
Heinemann, Ed 117
helicopter carriers 360–1
Hungary, reconnaissance vehicles 287

India
   aircraft 58, 139, 196–7, 207

INDEX

aircraft carriers 357
tanks 249, 334
Indo-Pakistan War 56, 70–1, 191, 223
Indonesia, aircraft 187
intelligence 112–13, 159
see also reconnaissance
Iran
aircraft 161, 195
frigates 401
Iraq
aircraft 75, 154–5
wars in 140, 154–9, 298–9, 342–3, 356, 386, 310–11
Israel
aircraft 16, 17, 68–9, 74–7, 127, 144
and Lebanon 152–3
patrol ships 369
self-propelled guns 304, 339
Six-Day War (1967) 68, 74–5, 225
tanks 225, 245, 256, 292–3, 315
War of Independence 16–17
Yom Kippur War (1973) 68, 76–7, 245, 275
ISTAR 112–13
Italy
aircraft 73, 136–7, 184, 192–3
aircraft carriers 358–9, 361
amphibious tanks 323
anti-aircraft vehicles 272, 340
armoured personnel carriers (APCs) 314, 321, 347
cruisers 379
destroyers 381, 385, 391
frigates 396–8, 400, 404–5
main battle tanks 257, 300
minehunters/sweepers 410–11
net layer 412
patrol ships 366
self-propelled guns 307
submarines 416, 435

jamming 176–7
Japan
aircraft 95, 105, 115, 204–5
anti-aircraft vehicles 340
anti-tank vehicles 269
armoured personnel carriers (APCs) 233, 321
destroyers 383, 385, 387, 391, 393
frigates 399, 401
submarines 416
tanks 256, 333
Jordan, aircraft 48, 184

Korean War (1950–3) 18–25
Kosovo war (1999) 171
Kuwait, aircraft 154–5, 203

landing ships 365, 412, 413
Lebanon, aircraft 126, 152–3
Lebanon War (1982) 152–3
Libya, aircraft 142–3
light tanks 262–3, 302–3
Long Range Aviation Force (USSR) 98–9

Macedonia, aircraft 174
Mahurin, Walker 'Bud' 23
main battle tanks
1970s 256–61
1980s 292–301, 330–7
Cold War 246–55
Malaysia, aircraft 117
maritime patrol aircraft 88–9, 104–5
minehunters/sweepers 410–11
MLRS 310–11
Morgan, Dave 110
Mullins, Arnold 'Moon' 18

net layer 412
Netherlands
aircraft 90
reconnaissance vehicles 287
ships 389, 394–5, 406
submarines 415, 418
New Zealand, aircraft 88, 95
night fighters (aircraft) 24–5, 45, 50
NORAD (North American Air Defense Command) 28–9
North Korea
frigates 403
Korean War 18–25
North Vietnamese Air Force (NVAF) 87
Norway, aircraft 31
nuclear submarines 420–33

one-offs (aircraft) 120–1
Operation Desert Storm 140, 154–9, 165, 373
Operation El Dorado Canyon 142

Pakistan
aircraft 163, 191
cruisers 378
Indo-Pakistan War 56, 70–1, 191, 223
tanks 332
Panama, invasion of 164
patrol ships 366–9
Peru
aircraft 179
ships 378, 390
Poland
aircraft 57
ships 369, 389
tanks 260, 333
polar exploration 43
Portugal, frigates 403

reconnaissance
aircraft 14, 39, 52–3, 64 5, 67, 79, 99, 103, 112–13, 119, 125, 134, 167, 207
tanks/AFVs 286–9
rocket systems
MLRS 310–11
post-war 234–5
Romania, ships 385, 405
Russia see USSR/Russia

SAC (Strategic Air Command) 62–3
Saudi Arabia 67
aircraft 185
frigates 406
Operation Desert Storm 157
SEAD (Suppression of Enemy Air Defences) 176–7
self-propelled guns 230–1, 264–7, 304–7, 338–9, 348
Serbia, aircraft 171
Singapore
aircraft 33
tracked APCs 346

Six-Day War (1967) 68, 74–5, 225
Slovakia
aircraft 167
self-propelled guns 348
South Africa
aircraft 43, 69
main battle tanks 257
self-propelled guns 307
South Korea, tanks/AFV 295, 320
Soviet Union see USSR/Russia
Spain
aircraft 93, 192–3
amphibious tanks 323
submarines 418
specialized naval ships 412–13
spyplanes 64–5, 112–13
stealth attack aircraft 164–5, 173, 210–11
stealth ships 438–9
Strategic Air Command (SAC) 62–3
strategic bombers 188–9
submarine support ships 434, 435
submarines 414–33
conventional 414–19
nuclear-powered 350–1, 420–33
Suez crisis 44–5, 46
support and repair ships 434–5
Suppression of Enemy Air Defences (SEAD) 176–7
Sweden
aircraft 11, 122–5, 208–9
patrol ships 367
self-propelled guns 265
submarines 417
tanks/AFV 263, 323, 283
Switzerland
aircraft 33, 49
anti-aircraft vehicles 277, 341
anti-tank vehicles 309
Syria, aircraft 75, 76–7, 153

tactical liaison vehicle 348
tactical missile launchers 270–1
tactical transports 184–5
Taiwan, submarines 419
Thailand, ships 361, 409
tiltrotors 200–1

447

INDEX

tracked APCs 232–2, 280–3, 314, 318–21, 346–7
tracked armoured vehicles 348–9
Turkey
　aircraft 96
　frigates 397

Ukraine
　combat engineer vehicles 326
　tanks 332
United Kingdom
　AEW (airborne early warning) aircraft 144–5
　aircraft, cold war 10, 14–15, 20, 26–7, 32–3, 44–9, 59, 91–3, 100–1, 104–5, 110–11, 113, 115, 120
　aircraft, modern era 136–9, 144–5, 148–9, 156–7, 172, 186–7, 192–3, 196–7
　aircraft carriers 354–5, 356
　anti-aircraft vehicles 278, 313, 341
　anti-tank vehicles 268, 269
　armoured cars 238–9
　armoured personnel carriers (APCs) 282, 319, 344–5
　assault ships 364
　combat engineer vehicles 290–1, 326
　cruisers 379
　destroyers 381, 386, 393, 439
　frigates 392, 396, 398, 399, 403
　heavy tanks 219
　light tanks 302
　main battle tanks 216–17, 249, 254–5, 257, 296–7, 335
　maritime patrol aircraft 104–5
　medium tanks 225
　reconnaissance vehicles 288–9
　self-propelled guns 264, 338
　submarines 419, 421–4, 431

support and repair ships 434–5
vertical take-off 91–3
United States of America
　air sea rescue 30–1
　aircraft, Cold War 10–13, 18–25, 28–31, 34–43, 52–3, 58, 60–5, 72–3, 78–86, 88–90, 94, 96–7, 102–3, 112–14, 116–21
　aircraft, modern era 134–5, 140–3, 146–9, 158–61, 168–9, 173, 176–7, 180–3, 188–9, 210–11
　aircraft carriers 352–3, 360, 436
　ammunition support vehicle 349
　amphibious tanks 322, 324–5
　amphibious transport ship 437
　anti-aircraft vehicles 240, 241, 273, 276, 313, 341
　anti-tank 315
　armoured personnel carriers (APCs) 232, 233, 236–7, 285, 347
　assault ships 362–3, 364–5
　Bradley M2 342–3
　combat engineer vehicles 290, 327
　Command and Control Vehicle (C2V) 349
　cruisers 370–3
　destroyers 380, 383, 386, 393, 437
　flying boats 94
　flying boxcars 58
　frigates 400, 401, 405
　heavy tanks 214–15, 218
　light tanks 262, 303
　main battle tanks 246–7, 249, 298–9, 334
　maritime patrol aircraft 88–9
　medium tanks 219, 222–5
　North American
　　Air Defense Command (NORAD) 28–9
　reconnaissance 52–3,

64–5, 286, 314
rocket systems 310–11
SEAD (Suppression of Enemy Air Defences) 176–7
self-propelled guns 230, 264, 266–7, 305, 339
spyplanes 64–5
stealth attack aircraft 210–11
Strategic Air Command (SAC) 62–3
submarines 415, 420–1, 424–5, 427, 429–30
support and repair ships 434–5
tactical liaison vehicle 348
tactical missile launchers 270–1
tiltrotors 201
vertical take-off 90
X-Planes 34–7
USSR/Russia
　AEW (airborne early warning) aircraft 145
　Afghan wars 162–3
　aircraft, Cold War 11, 21, 54–7, 91, 94–5, 98–9, 105, 115, 121, 128–31
　aircraft, modern 145, 150–1, 162–3, 174–6, 194–7, 206–7, 212–13
　aircraft carriers 357, 361
　amphibious tanks 226–7, 242–3
　anti-aircraft vehicles 241, 273, 274–5, 276, 277, 279, 312, 313
　anti-tank vehicles 268
　armoured personnel carriers (APCs) 233, 281, 283, 284, 318, 319, 347
　assault ships 365
　corvettes 367
　cruisers 374–7
　destroyers 380, 381, 382, 388
　flying boats 94–5
　frigates 402, 408
　heavy tanks 218
　main battle tanks 220–1, 244–5, 248, 249, 258–61, 294–5, 333
　maritime patrol aircraft 105

rocket systems 234–5
SEAD (Suppression of Enemy Air Defences) 176
self-propelled guns 230, 304–5, 307
specialized naval ships 413
submarines 414, 417, 425–6, 428–9, 430, 432–2
tactical missile launchers 270–1
vertical take-off 91

V-Bombers 26–7
Venezuela, destroyers 390
vertical take-off 90–3
Vietnam War (1955–75)
　aircraft 72, 80–7, 94, 97, 116, 134, 135, 140
　anti-aircraft vehicles 275
　tanks 223, 233, 262

Warsaw Pact, aircraft 166–7
weather reconnaissance 119
West Germany
　aircraft 72, 90, 136–7, 192–3
　anti-aircraft vehicles 273, 278
　anti-tank vehicles 269, 308
　armoured personnel carriers (APCs) 284
　combat engineer vehicles 291
　main battle tanks 252–3, 330–1
　self-propelled guns 231
　ships 413
　see also Germany

X-Planes 34–7

Yeager, Charles 'Chuck' 34
Yom Kippur War (1973) 68, 76–7, 245, 275
Yugoslavia
　aircraft 170–1
　frigates 407
　tracked APCs 281, 320, 346
　war in 170–3, 354–5